AF 35654 6

EEG Signal Processing for
Biomedical Applications

EEG Signal Processing for Biomedical Applications

Editor

Yvonne Tran

MDPI • Basel • Beijing • Wuhan • Barcelona • Belgrade • Manchester • Tokyo • Cluj • Tianjin

Editor
Yvonne Tran
Macquarie University
Australia

Editorial Office
MDPI
St. Alban-Anlage 66
4052 Basel, Switzerland

This is a reprint of articles from the Special Issue published online in the open access journal *Sensors* (ISSN 1424-8220) (available at: https://www.mdpi.com/journal/sensors/special_issues/ EEG_Biomedical).

For citation purposes, cite each article independently as indicated on the article page online and as indicated below:

LastName, A.A.; LastName, B.B.; LastName, C.C. Article Title. *Journal Name* **Year**, *Volume Number*, Page Range.

ISBN 978-3-0365-6535-4 (Hbk)
ISBN 978-3-0365-6536-1 (PDF)

Contents

Editorial

EEG Signal Processing for Biomedical Applications

Yvonne Tran

Department of Linguistics, Macquarie University Hearing, Macquarie University, Sydney, NSW 2109, Australia; yvonne.tran@mq.edu.au

1. Introduction

Electroencephalography (EEG) signals are used widely in clinical and research settings. Electrical activity generated from large populations of neurons in the brain is measured using scalp-mounted EEG sensors. As a result, we can obtain information regarding brain activity in various cognitive and emotional states. Due to their ability to provide this type of information, EEG signals are used in applications such as monitoring levels of alertness and mental engagement, investigating chronic conditions, and as signals for biofeedback or assistive devices. Innovations in this field have led to advancements in signal processing methods and the development of novel applications ranging from brain–computer interfaces (BCIs) to neuromarketing. EEG signals can be processed in time, frequency, or spatial domains, providing multi-dimensional means to interpret brain activities. Aside from providing invaluable information, EEG signals also have the advantage of capturing complex neural patterns at a high rate of speed. As a reliable, portable, and non-invasive way to measure the electrical activity in the brain, EEG is a central methodology for affordable and practical research and a promising clinical healthcare tool. This Special Issue focuses on EEG signal processing for biomedical engineering applications with original research, communication, and review papers demonstrating broad methodologies and applications. Fifteen papers address various informative themes. These range from examining physical innovations for the development of EEG sensors to studies in clinical populations such as individuals with epilepsy, spinal cord injury, and Amyotrophic Lateral Sclerosis (ALS). In this Special Issue, many novel EEG signal-processing strategies and analysis techniques are explored.

2. Overview of Contribution

Two communication papers are included in the Special Issue, with the first highlighting a new concept for EEG sensor development [1]. As EEG signals are acquired from the scalp, this paper presented an anatomically realistic textile-based head phantom for the assessment of EEG sensors. A gelatin-based head phantom is long-lasting and can accurately mimic body electrode frequencies, allowing for stable and accurate measurements of EEG signals. The outcomes from this paper will add to this field by allowing newly developed EEG electrodes to be validated. The second communication paper [2] presented a novel network analysis approach using a multi-layer model. Traditionally, in graph analysis, models are based on single layers. However, with the brain being a multi-layer network, analysis will be constrained when conveying brain topologies through single-layer models. Multi-layer networks produce more reliable approximations of the topology and dynamics of motor functions from the brain.

Within the theme of graph analysis, papers by Hag et al., Perez-Ortiz et al., and Šverko et al. all examined functional connectivity from EEG signals [3–5]. Friston (1994) defines functional connectivity as the temporal coincidence of spatially distant neurophysiological events [6]. It is said to have a measurable statistical relationship that captures two things occurring together which are related to each other [7]. Hag et al. used hybrid multi-domain EEG-based machine learning feature sets to assess mental stress. The functional connectivity network showed a statistically significant decrease during mental stress. Results from

Citation: Tran, Y. EEG Signal Processing for Biomedical Applications. *Sensors* **2022**, *22*, 9754. https://doi.org/10.3390/s22249754

Received: 8 December 2022
Accepted: 12 December 2022
Published: 13 December 2022

the time, frequency, and functional connectivity domains showed that the accuracy in detecting mental stress from EEG signals was highest with functional connectivity. However, combining the features from all three domains improved the overall accuracy, demonstrating greater nuance when using multiple EEG processing methods. Perez-Ortiz et al. examined functional connectivity and frequency power alterations in evoked potentials, specifically P300, in patients with ALS. P300 signals were utilized in a BCI device to control a robotic arm. People with ALS had overactivated beta bands and under-activated alpha bands in connectivity measures compared to the control participants. The results indicated that connectivity in EEG signals may be a valuable tool for monitoring disease progress and measuring cognitive atrophy. In their study, Šverko et al. presented a method for analyzing EEG connectivity. In this paper, they proposed the complex Pearson correlation coefficient (CPCC) as a unique single measure to provide information on phase locking and weighted phase lag. This proposed connectivity measure could accelerate the computation of brain connectivity and enhance our understanding of brain processes. A review paper in this issue also showed the importance of connectivity measures in mental stress assessment. In their review [8], Katmah et al. found that the selection of the most appropriate features is crucial to successful mental stress detection. Features with additional connectivity network measures and deep learning approaches could improve detection accuracy in terms of mental stress.

The examination of EEG signals in clinical populations can contribute to a better understanding of brain processes in people with neurological disorders. Tran et al. explored the effects of virtual reality (VR) intervention on the brain activity of people with neuropathic pain and spinal cord injury [9]. A significant reduction in pain intensity was reported after VR intervention, corresponding to statistically significant changes in EEG signals, specifically in the alpha and low gamma bands. Guo and Wang [10] examined brain activity associated with acupuncture. As the scientific explanation for the effects of acupuncture is still unknown, in this research, they studied the power spectrum changes during acupuncture manipulation. They found acupuncture manipulations were associated with delta and alpha rhythms. The neural responses from this study may have implications for the use of acupuncture as a complementary treatment for improving symptoms in neurological disorders. EEG signals in epilepsy were examined in two other studies [11,12], in which novel analysis techniques were assessed. Sánchez-Hernández et al. evaluated dimensionality reduction for feature selection methods with classification methods for epileptic seizures from EEG signals. They found that reducing selected features increased the classifier's performance. Obukhov et al. used wavelet ridges as a diagnostic EEG feature for the detection of epileptic seizures. It was shown that the application of this methodology will reduce the total duration and number of fragments needed for analysis. Additionally, Hossain et al. examined wavelet decomposition for the correction of movement artifacts in single-channel EEG with fNIRS signals [13]. This method combined wavelet packet decomposition with canonical correlation analysis. This proposed method outperformed comparative methods in removing motion artifacts from a single EEG channel.

Novel EEG signal-processing methods for various applications were also examined in four additional papers which focused on other topics. Zhang et al., in their review paper, discussed the application of transfer learning for EEG signals and BCIs [14]. In machine learning, transfer learning refers to using a model developed for one task as a starting point for constructing another model. The decoding performance in classification and regression tasks was found to be effective with this method. Kamrud et al. [15] investigated the detection of vigilance decrement in both cross-participant and cross-task modes, that is, robust models which can perform in unseen conditions. The research from this paper demonstrated that models could be built for EEG as a marker of vigilance levels even from unseen tasks. Charuthamrong et al. [16] used both auditory- and visual-based event-related potential to assess speech discrimination. Both the visual and auditory methods achieved reasonable accuracy rates and were shown to be potentially suitable for use in an automatic speech discrimination assessment system. Additionally, Zhou et al. used evoked potentials to investigate repetitive transcranial magnetic stimulation (rTMS) [17]. The goal was to develop rTMS EEG-evoked potentials as biomarkers for cortical excitability from rTMS. The changes found in the evoked potentials may have reflected GABAergic-mediated inhibition in specific brain regions.

3. Conclusions

A primary focus of this Special Issue was the demonstration of new methods for the analysis of EEG signals for biomedical engineering applications. The examination of various analysis methods led to the presentation of a diverse range of novel strategies. Through their results, the authors of these papers have provided a better understanding of cognitive states and brain activity based on different EEG signal processing methodologies and machine learning strategies.

Conflicts of Interest: The author declares no conflict of interest.

References

1. Tseghai, G.B.; Malengier, B.; Fante, K.A.; Van Langenhove, L. A Long-Lasting Textile-Based Anatomically Realistic Head Phantom for Validation of EEG Electrodes. *Sensors* **2021**, *21*, 4658. [CrossRef] [PubMed]
2. Covantes-Osuna, C.; López, J.B.; Paredes, O.; Vélez-Pérez, H.; Romo-Vázquez, R. Multilayer Network Approach in EEG Motor Imagery with an Adaptive Threshold. *Sensors* **2021**, *21*, 8305. [CrossRef] [PubMed]
3. Hag, A.; Handayani, D.; Pillai, T.; Mantoro, T.; Kit, M.H.; Al-Shargie, F. EEG Mental Stress Assessment Using Hybrid Multi-Domain Feature Sets of Functional Connectivity Network and Time-Frequency Features. *Sensors* **2021**, *21*, 6300. [CrossRef] [PubMed]
4. Perez-Ortiz, C.X.; Gordillo, J.L.; Mendoza-Montoya, O.; Antelis, J.M.; Caraza, R.; Martinez, H.R. Functional Connectivity and Frequency Power Alterations during P300 Task as a Result of Amyotrophic Lateral Sclerosis. *Sensors* **2021**, *21*, 6801. [CrossRef] [PubMed]
5. Šverko, Z.; Vrankić, M.; Vlahinić, S.; Rogelj, P. Complex Pearson Correlation Coefficient for EEG Connectivity Analysis. *Sensors* **2022**, *22*, 1477. [CrossRef] [PubMed]
6. Friston, K.J. Functional and effective connectivity in neuroimaging: A synthesis. *Hum. Brain Mapp.* **1994**, *2*, 56–78. [CrossRef]
7. Eickhoff, S.B.; Müller, V.I. Functional Connectivity. In *Brain Mapping*; Toga, A.W., Ed.; Academic Press: Waltham, MA, USA, 2015; pp. 187–201.
8. Katmah, R.; Al-Shargie, F.; Tariq, U.; Babiloni, F.; Al-Mughairbi, F.; Al-Nashash, H. A Review on Mental Stress Assessment Methods Using EEG Signals. *Sensors* **2021**, *21*, 5043. [CrossRef]
9. Tran, Y.; Austin, P.; Lo, C.; Craig, A.; Middleton, J.W.; Wrigley, P.J.; Siddall, P. An Exploratory EEG Analysis on the Effects of Virtual Reality in People with Neuropathic Pain Following Spinal Cord Injury. *Sensors* **2022**, *22*, 2629. [CrossRef] [PubMed]
10. Guo, X.; Wang, J. Low-Dimensional Dynamics of Brain Activity Associated with Manual Acupuncture in Healthy Subjects. *Sensors* **2021**, *21*, 7432. [CrossRef] [PubMed]
11. Sánchez-Hernández, S.E.; Salido-Ruiz, R.A.; Torres-Ramos, S.; Román-Godínez, I. Evaluation of Feature Selection Methods for Classification of Epileptic Seizure EEG Signals. *Sensors* **2022**, *22*, 3066. [CrossRef] [PubMed]
12. Obukhov, Y.V.; Kershner, I.A.; Tolmacheva, R.A.; Sinkin, M.V.; Zhavoronkova, L.A. Wavelet Ridges in EEG Diagnostic Features Extraction: Epilepsy Long-Time Monitoring and Rehabilitation after Traumatic Brain Injury. *Sensors* **2021**, *21*, 5989. [CrossRef] [PubMed]
13. Hossain, M.S.; Chowdhury, M.E.H.; Reaz, M.B.I.; Ali, S.H.M.; Bakar, A.A.A.; Kiranyaz, S.; Khandakar, A.; Alhatou, M.; Habib, R.; Hossain, M.M. Motion Artifacts Correction from Single-Channel EEG and fNIRS Signals Using Novel Wavelet Packet Decomposition in Combination with Canonical Correlation Analysis. *Sensors* **2022**, *22*, 3169. [CrossRef] [PubMed]
14. Zhang, K.; Xu, G.; Zheng, X.; Li, H.; Zhang, S.; Yu, Y.; Liang, R. Application of Transfer Learning in EEG Decoding Based on Brain-Computer Interfaces: A Review. *Sensors* **2020**, *20*, 6321. [CrossRef] [PubMed]
15. Kamrud, A.; Borghetti, B.; Schubert Kabban, C.; Miller, M. Generalized Deep Learning EEG Models for Cross-Participant and Cross-Task Detection of the Vigilance Decrement in Sustained Attention Tasks. *Sensors* **2021**, *21*, 5617. [CrossRef] [PubMed]
16. Charuthamrong, P.; Israsena, P.; Hemrungrojn, S.; Pan-ngum, S. Automatic Speech Discrimination Assessment Methods Based on Event-Related Potentials (ERP). *Sensors* **2022**, *22*, 2702. [CrossRef] [PubMed]
17. Zhou, J.; Fogarty, A.; Pfeifer, K.; Seliger, J.; Fisher, R.S. EEG Evoked Potentials to Repetitive Transcranial Magnetic Stimulation in Normal Volunteers: Inhibitory TMS EEG Evoked Potentials. *Sensors* **2022**, *22*, 1762. [CrossRef] [PubMed]

Communication

A Long-Lasting Textile-Based Anatomically Realistic Head Phantom for Validation of EEG Electrodes

Granch Berhe Tseghai [1,2,*], Benny Malengier [1], Kinde Anlay Fante [2] and Lieva Van Langenhove [1]

[1] Department of Materials, Textiles and Chemical Engineering, Ghent University, 9000 Gent, Belgium; Benny.Malengier@UGent.be (B.M.); Lieva.VanLangenhove@UGent.be (L.V.L.)

[2] Jimma Institute of Technology, Jimma University, Jimma, Ethiopia; kinde.anlay@ju.edu.et

* Correspondence: GranchBerhe.Tseghai@UGent.be; Tel.: +32-465-570-635

Abstract: During the development of new electroencephalography electrodes, it is important to surpass the validation process. However, maintaining the human mind in a constant state is impossible which in turn makes the validation process very difficult. Besides, it is also extremely difficult to identify noise and signals as the input signals are not known. For that reason, many researchers have developed head phantoms predominantly from ballistic gelatin. Gelatin-based material can be used in phantom applications, but unfortunately, this type of phantom has a short lifespan and is relatively heavyweight. Therefore, this article explores a long-lasting and lightweight (-91.17%) textile-based anatomically realistic head phantom that provides comparable functional performance to a gelatin-based head phantom. The result proved that the textile-based head phantom can accurately mimic body-electrode frequency responses which make it suitable for the controlled validation of new electrodes. The signal-to-noise ratio (SNR) of the textile-based head phantom was found to be significantly better than the ballistic gelatin-based head providing a 15.95 dB $\pm$ 1.666 ($\pm 10.45\%$) SNR at a 95% confidence interval.

Keywords: e-textile; head phantom; electroencephalography; conductive material

Citation: Tseghai, G.B.; Malengier, B.; Fante, K.A.; Van Langenhove, L. A Long-Lasting Textile-Based Anatomically Realistic Head Phantom for Validation of EEG Electrodes. *Sensors* **2021**, *21*, 4658. https://doi.org/10.3390/s21144658

Academic Editor: Yvonne Tran

Received: 10 May 2021
Accepted: 6 July 2021
Published: 7 July 2021

Publisher's Note: MDPI stays neutral with regard to jurisdictional claims in published maps and institutional affiliations.

1. Introduction

Measuring the electrical activity in the brain, heart, muscles, etc., using electrodes to know the health condition of humans and/or animals is a common clinical practice. However, such electrodes have to be validated prior to being employed in clinical practices. For instance, PEDOT/PSS-based and silver-based electrocardiography (ECG) electrodes have been developed [1] to measure heart activity but a scientific validation was not performed as part of that research as ECG signals were different from person to person and even for the same person over time. Electroencephalography (EEG) measurements to monitor brain activity are much more variable with changes over seconds.

For the validation of EEG electrodes, it is, therefore, required to develop head phantoms as maintaining a constant brain activity is hardly possible. Hence, it is required to conduct a test in an environment as realistic as possible with a known ground truth of source location and brain activity. This can be performed via digital phantoms by modeling the propagation of the signal originating within the brain to the electrodes [2]. However, the studies via digital head phantom are hardly suited to mimic motion artifacts of a realistic EEG, electromagnetic interference noise generated by the power lines, and high power electronic equipment [3]. For that reason, many researchers have developed head phantoms predominantly from ballistic gelatin [4–8]. Gelatin-based materials are a good material to be used in phantom applications, but unfortunately, this type of phantom has a short life span [9] and is too heavyweight. Examples of gelatin-based head phantoms are shown in Figure 1.

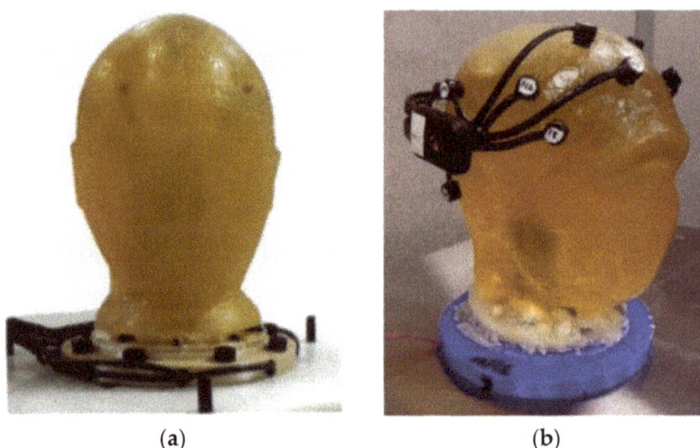

(a)   (b)

Figure 1. Examples of gelatin-based phantoms: (**a**) from [7]; (**b**) from [8].

Recently, Tsizin et al. developed a realistic head phantom mimicking the electromagnetic properties of the head where the internal volume of a human skull was filled with a conductive gel [10]. However, the lifetime of the phantom was only about a month. Other EEG head phantoms [11,12] prepared by casting were also introduced but still, the casting process is complicated, the phantoms are heavy and expensive. Therefore, developing a simple lightweight and long-lasting textile-based head phantom would be an important improvement.

The emergence of electrically conductive textiles led textile materials to a versatile application in the electronic and medical industries [13]. Electrically conductive textiles can be developed by different techniques and in different forms [14]. Moreover, the electrical and physical properties of the textile substrate can be easily controlled, and the required extent of stretchability, flexibility, and conductivity can be imparted by regulating the substrate, textile construction, and application of the conductive component. Therefore, this work explores the use of e-textiles for a head phantom.

2. Materials and Methods

2.1. Head Phantoms Construction

A textile-based head phantom was constructed by placing a bi-directional stretchy nylon/spandex (18:7) EeonTex conductive stretchable fabric (obtained from MANDU, Finland) over an anatomically realistic 3D-print polylactic acid (PLA) skull. The conductive fabric has a surface resistivity that can be custom-tuned for specific requirements in the range of 10^4 to 10^7 Ω/square. To mimic the neurons, twenty (20) 3.5 mm stereo male–male dipole wires were installed underneath the conductive fabric per the 10–20 EEG placement system as shown in Figure 2a. Side to side, a gelatin-based head phantom was also constructed from 900 g gelatin, 40.5 g table salt, and 4.5 L demineralized water according to [15], for comparison. Thirty-seven (37) dipole wires were installed inside the ballistic gelatin as shown in Figure 2b. The skull, base-ring, inner-post, and guiding wires have been constructed from PLA using an FDM 3D printer at Ingegno Maker Space (Drongen, Belgium). The photographic images of the constructed textile and gelatin-based head phantoms and their components are shown in Figure 2a,b, respectively.

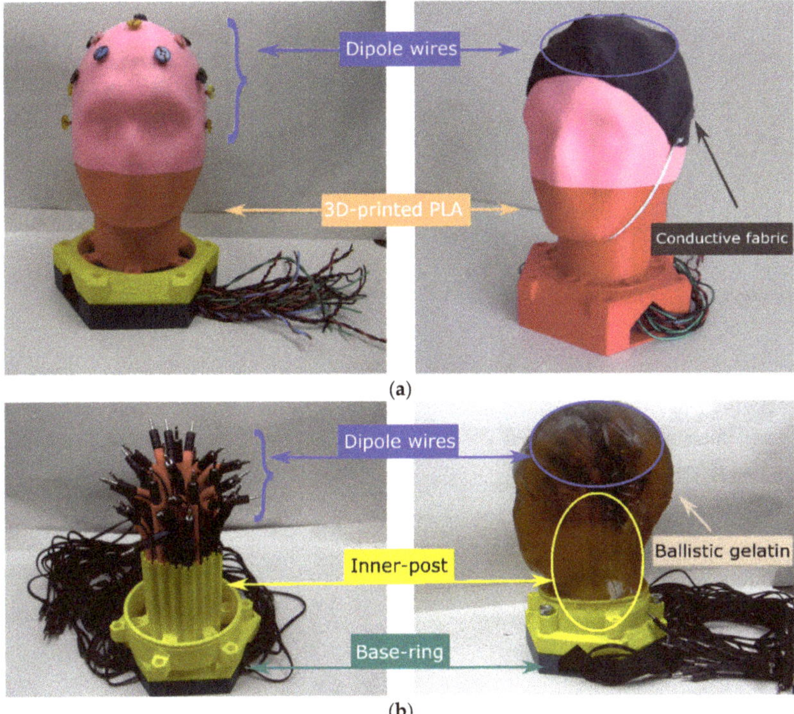

Figure 2. Head Phantom: (**a**) textile-based; (**b**) ballistic gelatin-based.

2.2. Head Phantom Validation

To validate the head phantoms, a synthetic sine wave (360 mV peak to peak voltage, 168 mV maximum voltage, −192 mV minimum voltage, 9.925 Hz frequency) was generated using a function generator DDS Function Signal Generator and recorded with a handheld tablet digital oscilloscope (Micsig TO1104). This was then injected into the head phantoms as shown in Figure 3. To impersonate events, the electroencephalography (EEG) phantom signal parameters were set in the alpha wave range and the amplitude was varied with the function generator to mimic a neurological event.

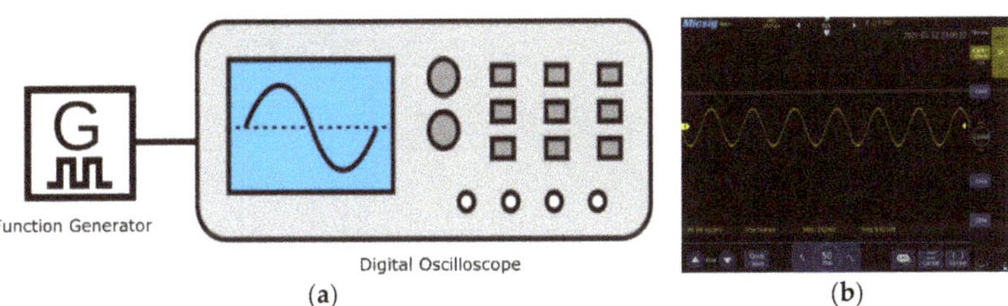

Figure 3. Synthetic sine wave generation: (**a**) wave generation setup using a function generator and digital oscilloscope; (**b**) the photographic image of the generated synthetic sine wave.

The head phantom replaces a real human head, and EEG electrodes can be attached as one would do on a human. In this test, the generated EEG wave was measured on both types of head phantoms using an active reusable snap Ag/AgCl dry electrode connected to a Cyton biosensing Board (8-channels) of OpenBCI according to the setup in Figure 4.

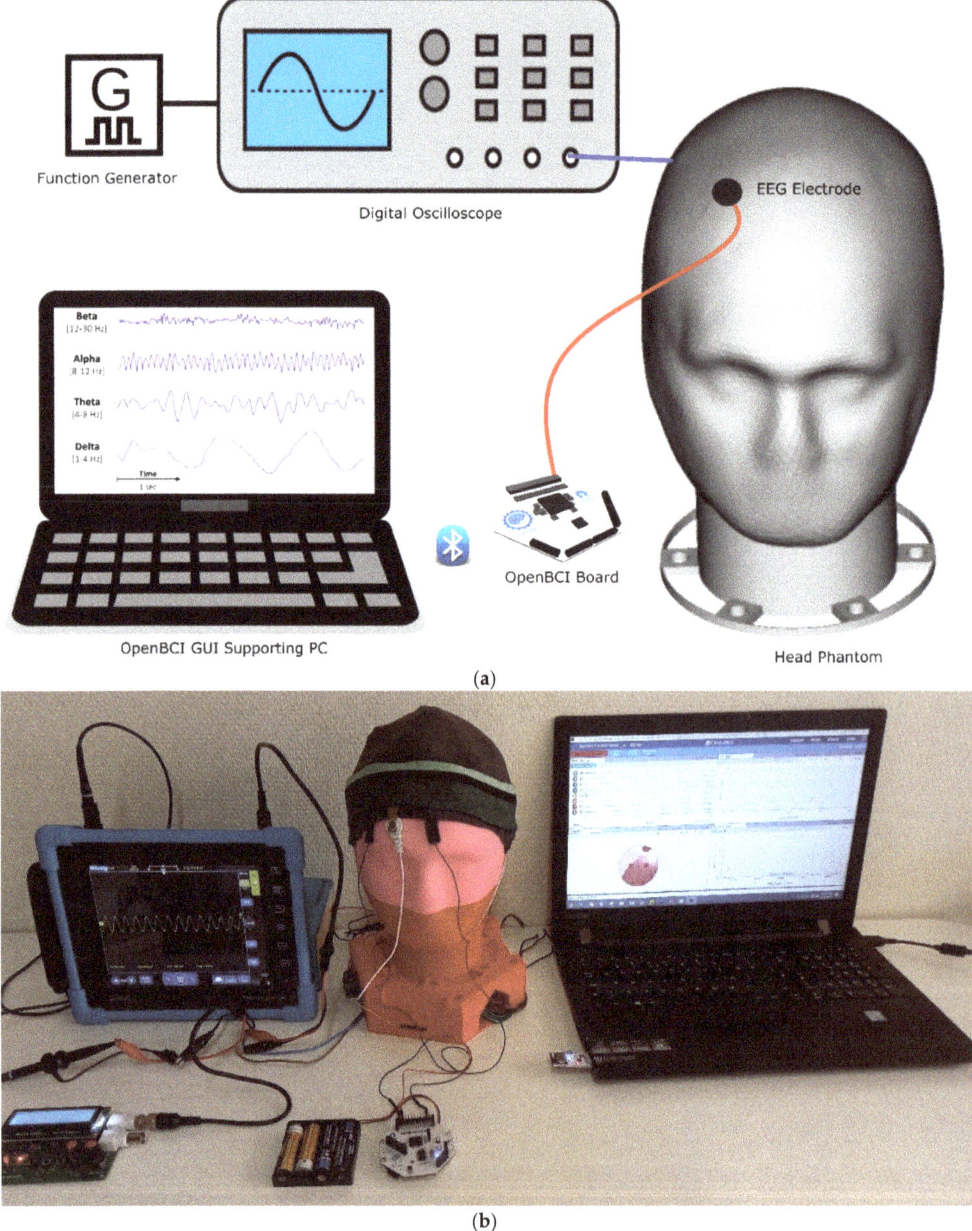

Figure 4. Measurement-setup: (**a**) schematic illustration; (**b**) actual.

2.3. Phantom-to-Electrode Impedance

The head phantom-to-electrode impedance was measured using a three-electrode configuration (reference, counter, and active electrodes), also with the Cyton Biosensing (OpenBCI) board and reusable snap Ag/AgCl dry EEG electrodes to study the difference between the ballistic gelatin and textile-based head phantoms. The system was adopted from OpenBCI and was suggested to measure skin-to-electrode impedance as the OpenBCI Cython board has an installed ADS1299 to measure impedance. A 5 kΩ resistor is built into the OpenBCI board in series to each electrode and has to be taken into account. The ADS1299 has a feature called "Lead Off Detection" that can do the impedance measurement by injecting a known current into each electrode. A 6 nA current is forced into the electrode line by a current source built into the ADS1299 [16], regardless of how much resistance or impedance there is between the current source and the ground (within reason). Hence, a 6 nA current will be present through the electrode to the ground during this test. For this work, only the head phantoms were used, no humans. Therefore, the impedance was calculated using Equation (1), where the current is 6×10^{-9} A. Then, the phantom-to-electrode impedance was analytically calculated.

$$Average\ Impedance(\Omega) = \frac{Average\ Voltage(V)}{Current(I)} \tag{1}$$

However, the average voltages collected during the test are in root mean square voltages (*Vrms*). Thus, the *average voltage* was calculated using Equation (2).

$$Average\ Voltage = \frac{Vrms \times 2\sqrt{2}}{\pi} = \frac{Vrms}{1.1107} \tag{2}$$

Finally, the average impedance here is the series resistance of the head phantom-to-electrode interface and the 5 kΩ resistor built into the OpenBCI board. So, to obtain the actual impedance of just the phantom-to-electrode interface, one needs to subtract 5 kΩ from the average impedance as in Equation (3).

$$Actual\ Average\ Impedance(\Omega) = Average\ Impedance(\Omega) - 5000 \tag{3}$$

2.4. Signal Analysis

The quality of signals collected was mathematically analyzed in terms of Signal-to-Noise Ratio (SNR) using Equation (4). The peak-to-peak voltage signal is the synthetic peak-to-peak voltage injected from the digital oscilloscope to the head phantom and the peak-to-peak voltage signal is the difference between the injected and collected back peak-to-peak voltage signal.

$$SNR(dB) = 10log\left(\frac{Peak\ to\ Peak\ Voltage\ Signal}{Peak\ to\ Peak\ Voltage\ Noise}\right) \tag{4}$$

The event-related spectral perturbation (ERSP) and inter-trial coherence (ITC) time-frequency measurements were then processed and analyzed via EEGLAB software that is treated as in Equation (5) according to spectral and coherence estimates on EEG recordings [17]. ITC is computed from single-trial EEG to reflect the temporal and spectral synchronization within EEG, explaining the extent to which underlying phase-locking occurs [17].

$$ITC(f,t) = \frac{1}{n}\sum_{k=1}^{n}\frac{F_k(f,t)}{F_k(f,t)\vee} \tag{5}$$

where F, t and n denote frequency, time and amount of data, respectively.

3. Results and Discussion

The new textile-based head phantom has a much lighter weight than the gelatin-based i.e., 0.5 and 6 kg, respectively. Therefore, the weight reduction is 91.67% which makes it more suitable for handling and moving from place to place. In addition, it is not delicate like the ballistic gelatin-based, where the shape of ballistic gelatin could be distorted and decays fast even when kept in a refrigerator. In our case, the gelatin-based head phantom begun decaying after a week of its construction which may also depend on the weather where it is placed during testing. In contrast, the textile-based head phantom does not decay at all.

3.1. Phantom-to-Electrode Impedance

The results in Table 1 indicate that the impedance of the textile-based head phantom is significantly lower with an f-ratio value of 2123.35 and a *p*-value of <0.001 at a 95% confidence interval according to one-way ANOVA. It is 1863 Ω for the textile-based head phantom and 2297 Ω, so they are in the same operating range. For comparison, a skin-to-electrode impedance measurement was performed on a human with the OpenBCI board and was found to be in the range of 3239.55 Ω to 1991.09 Ω, which is in the same range as the textile-based head phantom. The lower impedance means the long-lasting and lightweight textile-based head phantom can collect somewhat better-quality signals than the gelatin-based head phantom which would make it preferable for validating EEG electrodes in particular and other bio-potential electrodes in general. The head phantom can also potentially be used during modeling and simulation work related to brain neurological activities.

Table 1. Head phantom to electrode impedance.

Test	Time Counter (s)	Textile-Based Head Phantom			Gelatin-Based Head Phantom		
		V_{raw}	Z_{raw}	Z_{act}	V_{raw}	Z_{raw}	Z_{act}
1	30	41.18	6863	1863	43.93	7321	2321
2	60	43.81	7301	2301	43.98	7330	2330
3	90	42.66	7110	2110	44.54	7423	2423
4	120	42.92	7153	2153	43.87	7311	2311
5	150	43.13	7188	2188	42.49	7081	2081
6	180	42.44	7073	2073	44.12	7353	2353
7	210	41.33	6888	1888	43.72	7286	2286
8	240	42.72	7120	2120	43.63	7271	2271
Mean		**41.18**	**6863**	**1863**	**43.79**	**7297**	**2297**

V_{raw} = Raw Average Voltage (μV), Z_{avg} = Raw Average Impedance (Ω), Z_{act} = Actual Average Impedance (Ω).

3.2. Electroencephalogram (EEG) Signal

EEG is a term for the electrical signals of the brain [18] and was introduced by Hans Berger in 1929 [19]. Electrodes located outside (noninvasive brain-computer interface) of our brain, i.e., on the human scalp, are used to measure EEG. The frequency is the most common method for classifying EEG waveforms, to the point that EEG waves are denoted using Greek numerals based on their frequency spectrum. Delta (0.5 to 4 Hz), theta (4 to 7 Hz), alpha (8 to 12 Hz), sigma (12 to 16 Hz), and beta are the most widely studied waveforms (13 to 30 Hz).

The textile-based head phantom allowed for the injection of well-defined synthetic waves using a digital oscilloscope, and collection of the EEG waveform using an OpenBCI board, strongly similar and matching to the gelatin-based. The EEG wave collected from the textile-based head phantom predominantly lays in the alpha band, the same as the injected sine wave. Whereas, from the ballistic gelatin, a very small theta band was observed where an injected band power was generated. From the EEG band powers in Figure 5, the noise in the textile-based head phantom was less, however, statistically, the root-mean-square voltages (Vrms) from the time series in Figure 5 in both phantoms were not significantly different at 95% of confidence interval according to one-way ANOVA. The frequency vs. FFT (Fast Fourier Transform) plot showed that the amplitude and frequencies were strongly similar and in the same range, in addition, the head plot was also quite similar. Therefore, this textile-based head phantom can potentially replace the gelatin-based head for validating EEG electrodes.

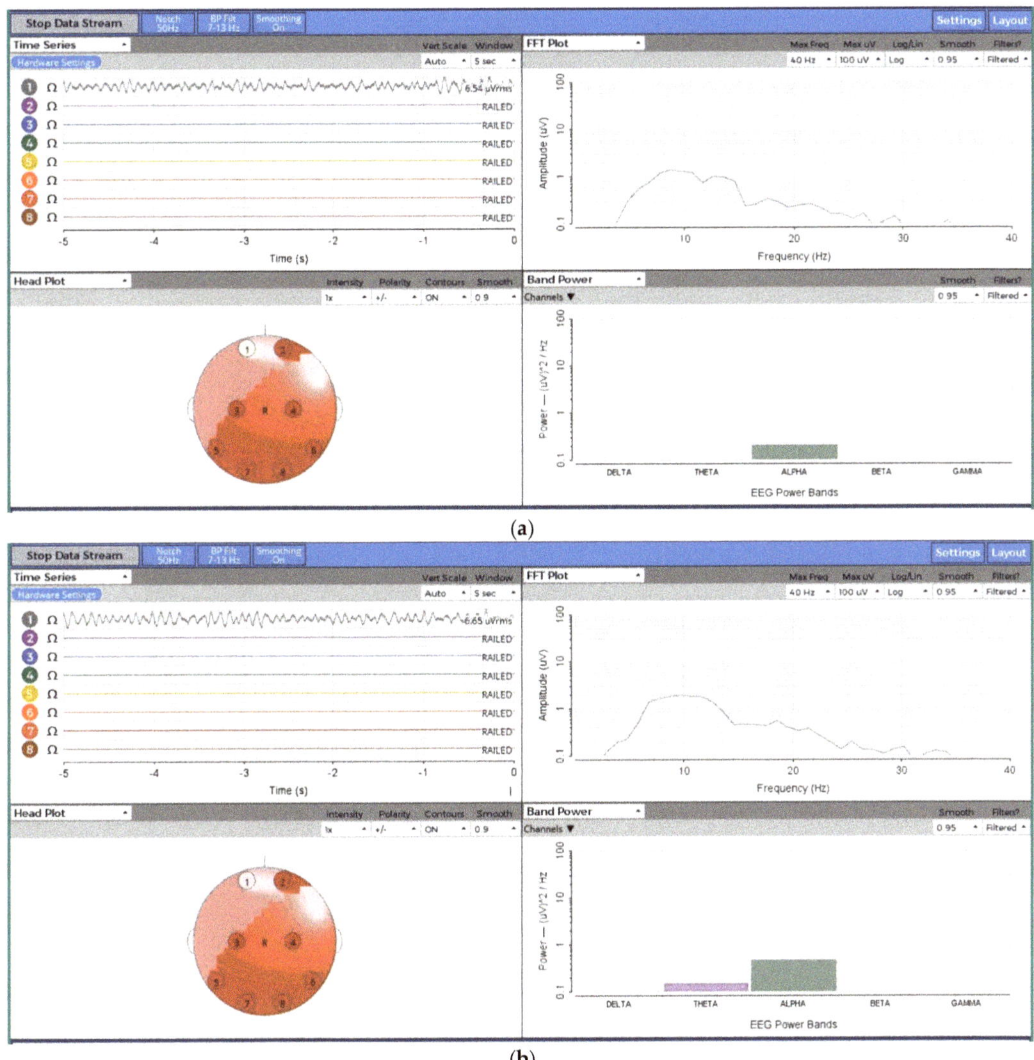

Figure 5. EEG signal from OpenBCI board: (**a**) textile-based head phantom; (**b**) gelatin-based head phantom.

3.3. SNR Analysis

From Table 2, the SNR of the textile-based head phantom was found to be significantly better than the gelatin-based one. The marginal error was 15.95 dB $\pm$ 1.666 ($\pm$10.45%) with a 95% confidence interval. Therefore, textile-based head phantoms are preferable.

Table 2. Injected wave, acquired signal, and SNR of the head phantoms.

	Wave	V Max (mV)	V Min (mV)	V Pk-Pk (mV)	SNR (dB)
Function generator	Synthetic Signal	168.00	−192.00	360.00	
	Signal	164.92	−184.30	349.22	
Gelatin-based head phantom	Noise	3.08	−7.701	10.78	15.1
	Signal/Noise	0.054	0.024	0.032	
	Signal	166.83	−185.80	352.63	
Textile-based head phantom	Noise	1.17	−6.20	7.37	16.8
	Signal/Noise	0.142	0.03	47.84	

3.4. Inter-Trial Coherence (ITC) and Event-Related Spectral Perturbation (ERSP)

The frequency and time ranges are plotted on the y-axis and x-axis, respectively, and a color scale is used, with green representing non-significant ITC and red representing significant ITC at a 99% confidence interval. The averaged ERP response for that person (in blue) is plotted beneath each ITC plot. The ERP response amplitude scale for both phantoms is somewhat close in this study. From EEGLAB software analysis, the log power spectral density for both the CDE and TE was ~90 dB. However, the distribution of spectral powers was more uniform in the textile-based main phantom. The ITC and ERP plots of the textile-based and gelatin-based head phantoms are shown in Figure 6.

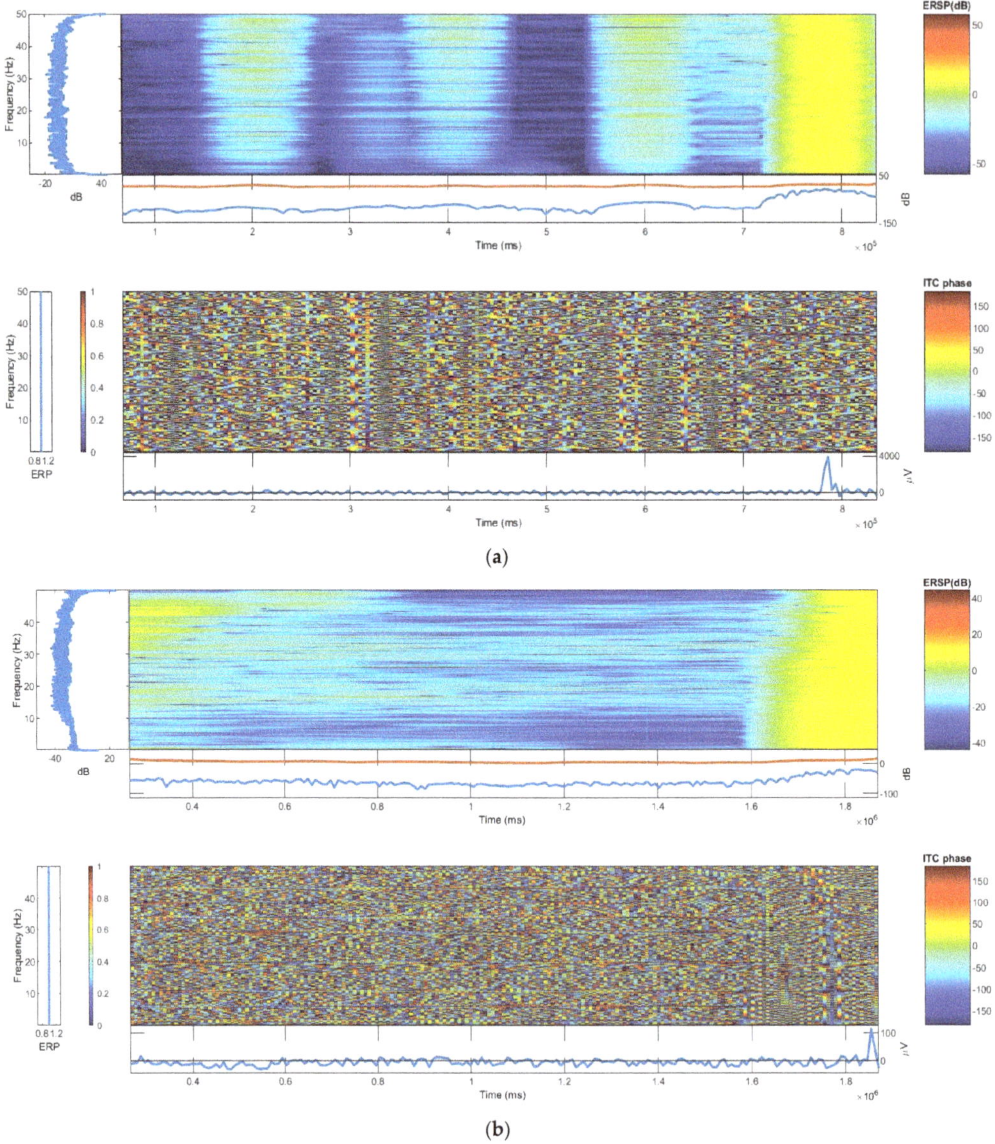

Figure 6. ITC and ERSP: (**a**) gelatin-based head phantom; (**b**) textile-based head phantom.

4. Conclusions

Keeping the human brain constant is hardly possible. Therefore, anatomically realistic head phantoms should be used to validate bio-potential electrodes such as for an electroencephalogram (EEG). In this work, we explored a long-lasting and lightweight head phantom that allows synthetic wave injection and measuring at a performance similar to the commonly used ballistic gelatin-based head phantoms. It was found to perform similarly, and for some users even better than the gelatin-based one. While the textile-based phantom was designed for EEG, it can also be adapted to electrocardiogram, electromyogram, electrooculogram, and other related studies as well.

Author Contributions: Conceptualization, G.B.T.; methodology, G.B.T. and B.M.; validation, G.B.T.; formal analysis, G.B.T.; investigation, G.B.T.; resources, G.B.T. and B.M.; data curation, G.B.T.; writing—original draft preparation, G.B.T.; writing—review and editing, B.M.; visualization, G.B.T., B.M., K.A.F., and L.V.L.; project administration, K.A.F. and L.V.L.; funding acquisition, L.V.L. All authors have read and agreed to the published version of the manuscript.

Funding: This research and APC were funded by The research was funded by NASCERE and IUPEPPE projects and European Commission (Smartex Project), grant number 610465-EPP-1-2019-1-EL-EPPKA2-CBHE-JP.

Institutional Review Board Statement: Not applicable.

Acknowledgments: The authors would like to express appreciation for the support of the NASCERE and IUPEPPE Projects sponsored with funds from the Ethiopian government and Smartex project funded with support from the European Commission. This publication reflects the views only of the authors, and the Ethiopian government and the European Commission cannot be held responsible for any use which may be made of the information contained therein. The authors would also like to thank the Ingegno Maker Space, Drongen, Belgium for the use of their makerlab and equipment in the creation of the 3D prints.

Conflicts of Interest: The authors declare no conflict of interest. The funders had no role in the design of the study; in the collection, analyses, or interpretation of data; in the writing of the manuscript, or in the decision to publish the results.

References

1. Tseghai, G.B.; Malengier, B.; Fante, K.A.; Nigusse, A.B.; Etana, B.B.; Van Langenhove, L. PEDOT:PSS/PDMS-Coated Cotton Fabric for ECG Electrode. In Proceedings of the IEEE International Conference on Flexible and Printable Sensors and Systems (FLEPS), Manchester, UK, 16 August 2020. [CrossRef]
2. Hallez, H.; Vanrumste, B.; Grech, R.; Muscat, J.; De Clercq, W.; Vergult, A.; D'Asseler, Y.; Camilleri, K.P.; Fabri, S.G.; Van Huffel, S.; et al. Review on Solving the Forward Problem in EEG Source Analysis. *J. Neuroeng. Rehabil.* **2007**, *4*, 46. [CrossRef] [PubMed]
3. Oliveira, A.S.; Schlink, B.R.; Hairston, W.D.; König, P.; Ferris, D.P. Induction and Separation of Motion Artifacts in EEG Data Using a Mobile Phantom Head Device. *J. Neural Eng.* **2016**, *13*, 036014. [CrossRef] [PubMed]
4. Kohli, S.; Casson, A.J. Removal of Gross Artifacts of Transcranial Alternating Current Stimulation in Simultaneous EEG Monitoring. *Sensors* **2019**, *19*, 190. [CrossRef] [PubMed]
5. Richardson, C.; Bernard, S.; Dinh, V.A. A Cost-Effective, Gelatin-Based Phantom Model for Learning Ultrasound-Guided Fine-Needle Aspiration Procedures of the Head and Neck. *J. Ultrasound Med.* **2015**, *34*, 1479–1484. [CrossRef] [PubMed]
6. Md Said, M.S.; Seman, N.; Sulaiman, N.R.; Abd Rahman, T. Modeling of Gelatin-Based Head Phantom Based on Its Electrical Properties for Wideband Microwave Imaging Application. *Appl. Mech. Mater.* **2015**, *781*, 608–611. [CrossRef]
7. Symeonidou, E.-R.; Nordin, A.; Hairston, W.; Ferris, D. Effects of Cable Sway, Electrode Surface Area, and Electrode Mass on Electroencephalography Signal Quality during Motion. *Sensors* **2018**, *18*, 1073. [CrossRef]
8. Owda, A.Y.; Casson, A.J. Electrical Properties, Accuracy, and Multi-Day Performance of Gelatine Phantoms for Electrophysiology. *BioRxiv* **2020**. [CrossRef]
9. Said, M.S.M.; Seman, N. Preservation of Gelatin-Based Phantom Material Using Vinegar and Its Life-Span Study for Application in Microwave Imaging. *IEEE Trans. Dielectr. Electr. Insul.* **2017**, *24*, 528–534. [CrossRef]
10. Tsizin, E.; Mund, T.; Bronstein, A. Printable Anisotropic Phantom for EEG with Distributed Current Sources. In Proceedings of the IEEE International Symposium on Biomedical Imaging (ISBI), Washington, DC, USA, 4–7 April 2018.
11. Collier, T.J.; Kynor, D.B.; Bieszczad, J.; Audette, W.E.; Kobylarz, E.J.; Diamond, S.G. Creation of a Human Head Phantom for Testing of Electroencephalography Equipment and Techniques. *IEEE Trans. Biomed. Eng.* **2012**, *59*, 2628–2634. [CrossRef]
12. Audette, W.E.; Bieszczad, J.; Allen, L.V.; Diamond, S.G.; Kynor, D.B. *Design and Demonstration of a Head Phantom for Testing of Electroencephalography (EEG) Equipment*; Report number: TN-1113; Creare, Inc.: Hanover, NH, USA, 2020. [CrossRef]

13. Tseghai, G.B.; Malengier, B.; Fante, K.A.; Van Langenhove, L. The Status of Textile-Based Dry EEG Electrodes. *Autex Res. J.* **2021**, *21*, 63–70. [CrossRef]
14. Tseghai, G.B.; Malengier, B.; Fante, K.A.; Nigusse, A.B.; Langenhove, L.V. Integration of Conductive Materials with Textile Structures, an Overview. *Sensors* **2020**, *20*, 6910. [CrossRef] [PubMed]
15. Yu, A.; Hairston, W.D. Open EEG Phantom. Available online: https://osf.io/qrka2/ (accessed on 8 October 2020).
16. Texas Instruments. *ADS1299-x Low-Noise, 4-, 6-, 8-Channel, 24-Bit, Analog-to-Digital Converter for EEG and Biopotential Measurements*; Texas Instruments: Dallas, TX, USA, 2017.
17. Nash-Kille, A.; Sharma, A. Inter-Trial Coherence as a Marker of Cortical Phase Synchrony in Children with Sensorineural Hearing Loss and Auditory Neuropathy Spectrum Disorder Fitted with Hearing Aids and Cochlear Implants. *Clin. Neurophysiol.* **2014**, *125*, 1459–1470. [CrossRef]
18. Chatterjee, R.; Datta, A.; Sanyal, D.K. Ensemble Learning Approach to Motor Imagery EEG Signal Classification. In *Machine Learning in Bio-Signal Analysis and Diagnostic Imaging*; Elsevier: Amsterdam, The Netherlands, 2019; pp. 183–208. ISBN 978-0-12-816086-2.
19. Berger, H. Uber Das Elektrenkephalogramm Des Menschen. *Eur. Arch. Psychiatr.* **1929**, *87*, 527–570. [CrossRef]

Communication

Multilayer Network Approach in EEG Motor Imagery with an Adaptive Threshold

César Covantes-Osuna, Jhonatan B. López, Omar Paredes, Hugo Vélez-Pérez and Rebeca Romo-Vázquez *

Departamento de Bioingeniería Traslacional, CUCEI, Universidad de Guadalajara, Guadalajara 44430, Mexico; cesar.covantes@alumnos.udg.mx (C.C.-O.); jhonatan.lopez@alumnos.udg.mx (J.B.L.); omar.paredes@academicos.udg.mx (O.P.); hugo.velez@academicos.udg.mx (H.V.-P.)
* Correspondence: rebeca.romo@academicos.udg.mx

Abstract: The brain has been understood as an interconnected neural network generally modeled as a graph to outline the functional topology and dynamics of brain processes. Classic graph modeling is based on single-layer models that constrain the traits conveyed to trace brain topologies. Multilayer modeling, in contrast, makes it possible to build whole-brain models by integrating features of various kinds. The aim of this work was to analyze EEG dynamics studies while gathering motor imagery data through single-layer and multilayer network modeling. The motor imagery database used consists of 18 EEG recordings of four motor imagery tasks: left hand, right hand, feet, and tongue. Brain connectivity was estimated by calculating the coherence adjacency matrices from each electrophysiological band (δ, θ, α and β) from brain areas and then embedding them by considering each band as a single-layer graph and a layer of the multilayer brain models. Constructing a reliable multilayer network topology requires a threshold that distinguishes effective connections from spurious ones. For this reason, two thresholds were implemented, the classic fixed (average) one and Otsu's version. The latter is a new proposal for an adaptive threshold that offers reliable insight into brain topology and dynamics. Findings from the brain network models suggest that frontal and parietal brain regions are involved in motor imagery tasks.

Keywords: adaptive threshold; coherence; functional connectivity; multilayer network; otsu

Citation: Covantes-Osuna, C.; López, J.B.; Paredes, O.; Vélez-Pérez, H.; Romo-Vázquez, R. Multilayer Network Approach in EEG Motor Imagery with an Adaptive Threshold. *Sensors* **2021**, *21*, 8305. https://doi.org/10.3390/s21248305

Academic Editor: Yvonne Tran

Received: 2 October 2021
Accepted: 6 December 2021
Published: 12 December 2021

Publisher's Note: MDPI stays neutral with regard to jurisdictional claims in published maps and institutional affiliations.

1. Introduction

The brain is a complex system with spatio-temporal dynamics that can be mapped by techniques that measure brain activity: electroencephalography (EEG), magnetoencephalography (MEG), and functional magnetic resonance imaging (fMRI) [1]. These techniques have been widely used to model brain networks that represent the structural and functional connectivity of the brain. Among all those techniques, EEG is an accessible, widespread method that measures the electrical activity of the brain on the scalp with a time resolution in milliseconds [2]. EEG analyses have divided brain waves into five major frequency bands: delta, δ (0.5–4 Hz); theta, θ (4–8 Hz); alpha, α (8–13 Hz); beta, β (13–30 Hz); and gamma, γ (30–128 Hz) [3]. Network models based on these frequency bands have revealed distinctive patterns and brain dynamics that have been used to study both normal and pathological mental states [4–6]. These network models can be analyzed using graphs built from an adjacency matrix that results from a brain connectivity analysis.

Brain connectivity analyses estimate the interaction strength among local information processing areas of the brain. Current state-of-the-art reports three types of connectivity: structural, based on the anatomical structure of the brain; functional, that measures the statistical dependence of different brain areas; and effective, which estimates causal relations among brain regions [7]. Concerning functional connectivity, literature describes various methods of estimation; including correlation (time domain dependence), and coherence (frequency domain dependence) [7]. Coherence measures the statistical relationship between two signals in the frequency domain [8] and it has been widely used in cerebral activity analyses involving memory [9], mathematical [5,6], and reasoning [10] task studies. It has also been applied to analyze differences at specific frequencies in patients with brain disorders [11], such as Parkinson's [12] and Alzheimer's diseases [13] and epilepsy [14]. In this work, adjacency matrices calculated from coherence between brain areas in electrophysiological bands were used to estimate functional connectivity.

Motor imagery is a cognitive-motor process widely studied by coherence analysis that has the potential to trigger and control actuators in brain-machine interface systems without any external motor action. Such systems aim to control a device through the brain activity of a user. Recent studies have focused on characterizing EEG through graph analysis to pinpoint not only brain areas but also interactions between them [15].

A graph is a mathematical tool used to describe the brain as a set of nodes (brain regions) and edges (connections) [16]. In Graph theory, there are different kinds of graphs, among which we can mention single-layer and multilayer ones. In single-layer networks, the edges represent the same type of connections between nodes. The associations between zones depend on a single character, which may be directed or undirected [17]. Some studies of brain connectivity have examined the brain as a single-layer graph linked by a single temporal or frequency property [18,19]. In cases where nodes can be linked based on multiple characters, associations are treated independently to build multiple single-layer networks that ignore the synergy between characters. Multilayer networks are suitable for these scenarios because they have the flexibility required to integrate multiple types of interactions in a single model.

The brain is currently considered a multilayer network [20]. As it was pointed out by [21], brain networks are intrinsically multilayers. There is not a single neuronal connectivity pattern able to fully represent brain functioning. Then, a multilayer framework is suitable for analyzing brain connectivity without either throwing away or combining different information. This focus improves understanding of brain complexity and interaction spectra with no need to discard electrophysiological data. This approach has proved to be a powerful tool in describing the complex organization and evolution of the human brain and its relationship to cognition [22]. Multilayer networks have been applied in brain analyses [23] using fMRI [24,25], MEG [4,26], gene expression [27] and EEG [20,21,28,29] techniques. The range of topological properties to be explored is, therefore, wider than in classic single-layer modeling [30]. Here, the efficiency of information flow results from multilayer interdependence within the network, rather than being an effect of each layer individually [31].

In the workflow of graph analysis, a common practice consists of thresholding networks to eliminate spurious connections [32]. That is because functional connectivity analysis, through measuring the statistical dependence among brain areas, yields a continuous weight range for interaction strength. Since some of these interactions should be labeled as spurious by the randomness of the signal, it is critical to exclude them from the brain connectivity analysis.

In this study, two thresholds were tested: the fixed (average) threshold, which is widely used in the literature, and a recently proposed threshold called Otsu. The fixed threshold method establishes a single, absolute threshold value over the entire network, typically fixed by averaging the adjacency matrices [33]. Values above this average are considered connections and are assigned a value of 1, while values below the average are discarded and receive a value of 0 that results in a binarized adjacency matrix. The main disadvantage of this approach is that a fixed threshold based on averages is conditioned by the weight distribution in the adjacency matrices, but this means that it will behave unreliably in the presence of outliers and non-normal distributions.

In contrast to the fixed threshold, Otsu's approach involves optimizing the threshold value by evaluating how well the binarization process identifies two types of data (i.e., pixels, voxels, etc.) [34]. Some applications of Otsu's methodology include structural segmentation in fMRI [35–38], and noise removal in EEG recordings using wavelet decomposition [39]. In our case, Otsu's methodology was implemented for image segmentation and binarization [40]. To the best of our knowledge, and after an exhaustive literature search, Otsu's method has not been applied to estimate the threshold of adjacency matrices in brain connectivity analyses. In this context, and considering the adjacency matrices as images that contain information about brain connectivity gathered from EEG recordings, this work proposes to apply Otsu's threshold to these matrices to estimate an optimal threshold for brain connectivity analyses.

In light of the foregoing, this study aimed to analyze EEG dynamics by classical single-layer and multilayer network models for a motor imagery dataset. This was conducted to feature the movement and its dynamics, and thus pinpoint patterns capable of feeding a BCI system. The coherence adjacency matrices for each electrophysiological band (δ, θ, α and β) of the brain areas were analyzed individually on a single-layer approach, and then integrated, considering each band as a layer, to build a brain network model following the multilayer approach. Both approaches were built with fixed and Otsu's thresholds.

Our results show that multigraph models cluster the four studied movements and lead to pinpointing the key electrodes for the motor imagery task that are located mainly on the frontal and parietal cortex. These brain zones coincide with the results presented in [15,41–44]. These works model brain connectivity with single-layer approach and a known threshold. However, our work explores a proof-of-concept EEG multilayer brain connectivity with an adaptative threshold. For this purpose, the paper is organized as follows: Section 2 addresses the material and methods, including the database description, the EEG signal preprocessing, and the connectivity estimation; in Section 3 the threshold, and single-layer and multilayer networks approaches are introduced, concluding with the results and discussion of the single-layer and the multilayer brain models with both thresholds in Section 4. The paper ends with the conclusions.

2. Materials and Methods

2.1. Database

In this study, the open access BNCI Horizon 2020 dataset (2a of BCI Competition IV) [45] was retrieved to pinpoint patterns of motor imagery. This dataset consists of 18 EEG recordings (Figure 1a) taken from 9 subjects (recorded in two sessions on different days) for four different motor imagery tasks (Figure 1b): left hand (class 1), right hand (class 2), feet (class 3), and tongue (class 4). The signals were recorded at a 250 Hz sampling rate and then band-pass filtered between 0.5–100 Hz. Electrodes were placed according to the 10–10 International System at Fz, Fc3, Fc1, Fcz, Fc2, Fc4, C5, C3, C1, Cz, C2, C4, C6, Cp3, Cp1, Cpz Cp2, Cp4, P1, Pz, P2, and POz.

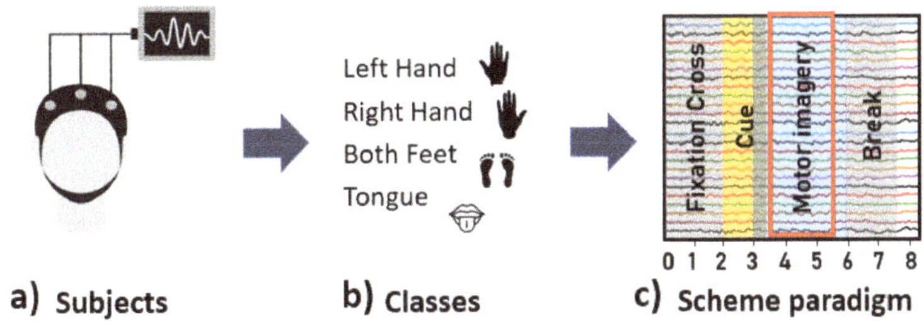

Figure 1. Time and scheme paradigm.

The experimental paradigm for each trial is illustrated in Figure 1c [46]. On the trials, subjects began by focusing their eyes on a black screen ($t = 0$ s). After two seconds $t = 2$ s, an arrow image pointing left, right, down, or up (representing one of the four classes) appeared and remained on the screen for 1.25 s. Subjects then carried out the corresponding motor imagery task until the arrow image on the screen disappeared at $t = 6$ s, indicating a brief pause before the beginning the next trial. The time window corresponding to motor imagery (MI) onset) $t = 3.5$–5.5 s of the experimental paradigm was analyzed.

2.2. Preprocessing

Each EEG recording was composed of 6 runs (Figure 2a) separated by a short break. Each run consisted of 48 trials (12 for each class), resulting in 288 total trials of 2 s each (72 for each class).

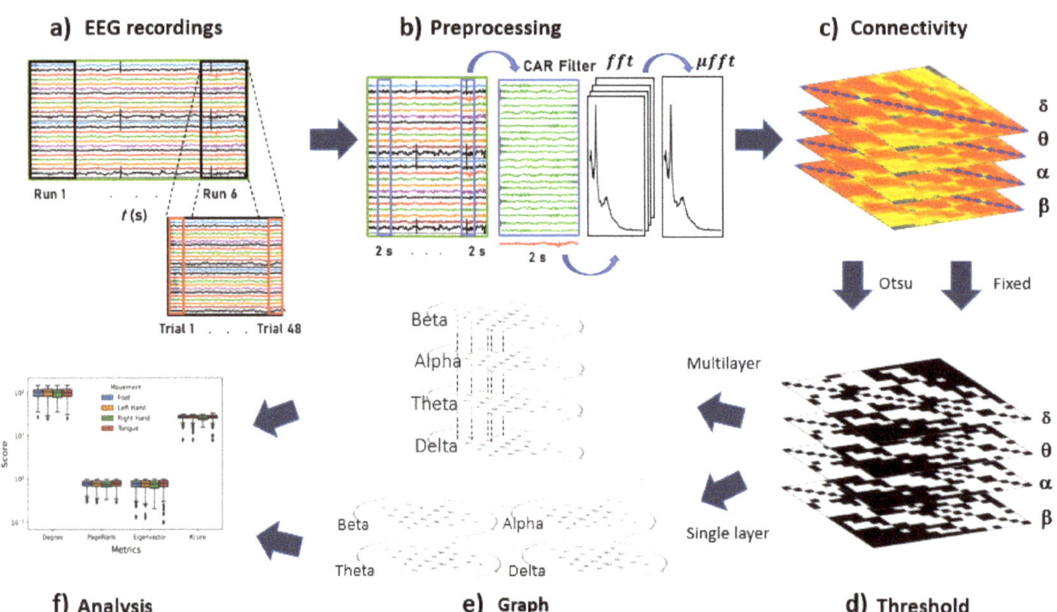

Figure 2. Schematic flowchart of the study methodology. Here, (**a**) correspond to the acquisition paradigm of all four classes of motor imagery, (**b**) data prepossessing, while (**c**–**f**) outline the stages of the connectivity graph analysis.

To reduce the EEG spatial interference, a Common Average Reference (CAR) filter (Equation (1)) was applied for each of the 288 two-second EEG windows.

$$V_i^{CAR} = V_i^{CR} - \frac{1}{N} \sum_{j=1} V_j^{CR} \tag{1}$$

where V_i^{CR} represents the potential between electrode i and the reference electrode, and N is the total number of electrodes.

Once the 288 windows filtered, each two-second window was transformed into the frequency domain. The power spectral for the 72 windows of each motion class was averaged to obtain 4 two-second frequency-averaged EEG windows. This process was carried out on each of the 18 recordings (Figure 2b).

2.3. Connectivity Estimation

The coherence index values between two signals range from 0 to 1. A value close to 1 indicates a strong relationship, while a value close to 0 represents weak interactions between signals. Coherence index is defined as (Equation (2)):

$$C_{xy}(f) = \frac{|S_{xy}(f)|^2}{S_{xx}(f)S_{yy}(f)}, \tag{2}$$

where x and y are two signals or channels, $\mathbf{C}_{xy}(f)$ is the coherence spectrum matrix as a function of a given frequency f, $\mathbf{S}_{xy}(f)$ is the cross-power spectrum, and $\mathbf{S}_{xx}(f)$ and $\mathbf{S}_{yy}(f)$ are the auto-power spectra of x and y, respectively [47].

3. EEG Processing

3.1. Layers Construction

To generate the single- and multilayer network models for the used motor imagery dataset, four layers were estimated, each corresponding to the main electrophysiological bands (δ, θ, α and β). Each layer was built by estimating the coherence among the EEG electrodes, then averaging the magnitude of the frequencies that comprised each band. This approach generated an adjacency matrix for each band (Figure 2c).

As mentioned above, four two-second averaged windows were obtained from the 18 EEG recordings for each MI class. After that, functional brain connectivity was estimated in each window by calculating the pairwise coherence indices among the 22 electrodes. This allowed us to obtain 22×22 weighted adjacency matrices for each class as a layer. Figure 3 shows an example of a β band-coherence adjacency matrix for the left-hand IM. Red indicates a high coherence value, while blue represents weakly connected areas. These layers were evaluated by the approaches of a single layer, where the layers of electrophysiological bands were analyzed separately; and the multilayer, where each class layer was integrated to build a multiple network model.

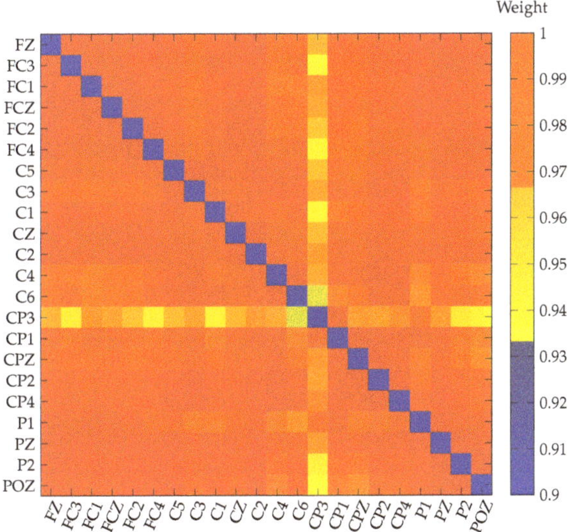

Figure 3. Example of a coherence adjacency matrix in the β electrophysiological band (13–30 Hz) for the IM of the left-hand.

3.2. Threshold Estimation

The threshold stage (Figure 2d) is a key step in graph analysis that provides reliable estimates of the network topology [48] and preserves the local topological features of the network measures [16,49]. In this study, the adjacency matrices were thresholded to build the connectivity networks using two methods: the widely used fixed threshold approach (i.e., average degree across the network) [33], and a proposal for a novel method based on image segmentation the Otsu's method [40].

Otsu's Threshold

This threshold uses the adjacency matrix data to calculate data distribution represented as a histogram Figure 4). In brain networks, histograms such as this one correspond to the scores of the weighted adjacency matrix. In our case, the matrices consisted of 22×22 values from 0 to 1 (coherence range values).

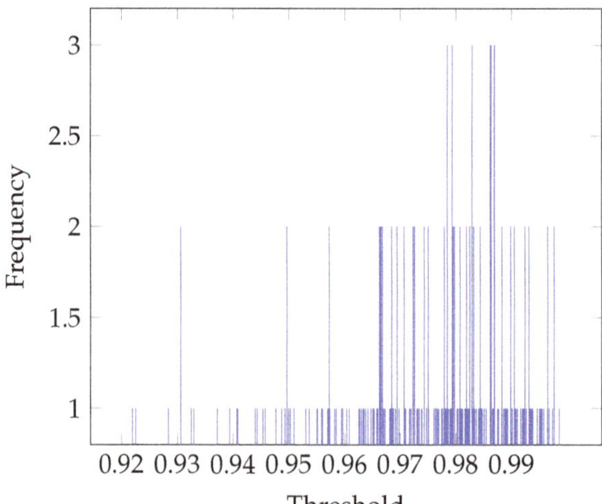

Figure 4. An adjacency matrix weight histogram for use in Otsu's method. The data correspond to the β band (13–30 Hz) for the IM of the left-hand.

For example, if we fix the threshold at $T = 0.01$, then adjacency values below T can be classified as class C_1 and correspond to spurious connections. Values above T are classified as class C_2 and correspond to effective connections. Thus, connections in C_1 are counted and divided by the total number of connections, N (22×22), to obtain the intensity w_1, and, likewise, for C_2 to estimate the intensity, w_2. The means, μ_1 and μ_2, and variances, σ_1^2 and σ_2^2, of these intensity values are also estimated, and the procedure is repeated for each increment of T until the range of values is completed. Obviously, all connections for C_2 are 1.

Next, the "Within-Class Variance (WCV)" (Equation (3)) and the "Between-Class Variance (BCV)" (Equation (4)) were computed in this threshold.

$$WCV = w_1\sigma_1^2 + w_2\sigma_2^2 \tag{3}$$

$$BCV = w_1 w_2 (\mu_1 - \mu_2)^2 \tag{4}$$

The optimal threshold is the value that minimizes WCV while maximizing BCV. Figure 5 shows an example of the distributions for an adjacency matrix with a maximum BCV and a minimum WCV. As can be seen, the optimal threshold is $T = 0.9698$. Once calculated, the weighted adjacency matrix is binarized. An example of this procedure is shown in Figure 5.

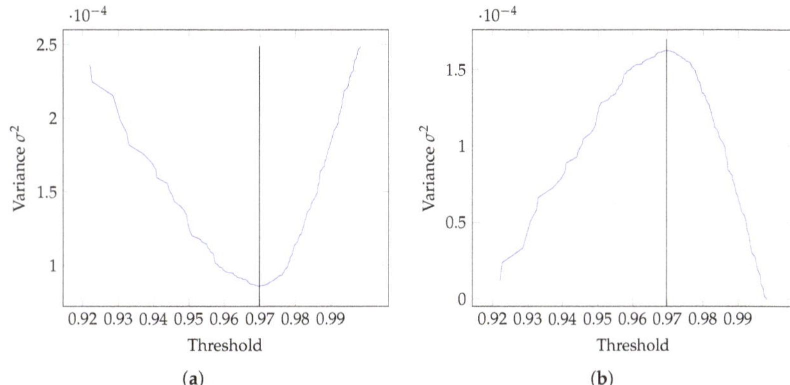

Figure 5. WCV and BCV histograms for the β band (13–30 Hz) for the IM right-hand. The optimum threshold value is $T = 0.9698$, indicating the minimum WCV value and the maximum BCV value. (**a**) Within-Class Variance (WCV) with the minimum value of 8.61×10^{-05} at threshold position $T = 0.9698$. (**b**) Between-Class Variance (BCV) with the maximum value of 1.62×10^{-04} at threshold position $T = 0.9698$.

Comparing the values of the thresholds obtained by the fixed ($T = 0.9747$) and Otsu's methods ($T = 0.9698$) we find that they tend to be similar. Therefore, the binarized matrices obtained from these thresholds (Figure 6) are close related. This suggests that both methods could generate similar topologies. However, as mentioned above, Otsu's threshold has the advantage of estimating an optimized threshold based on the distribution of the weights in the adjacency matrix, while the fixed threshold average is sensitive to outliers and non-normal distributions.

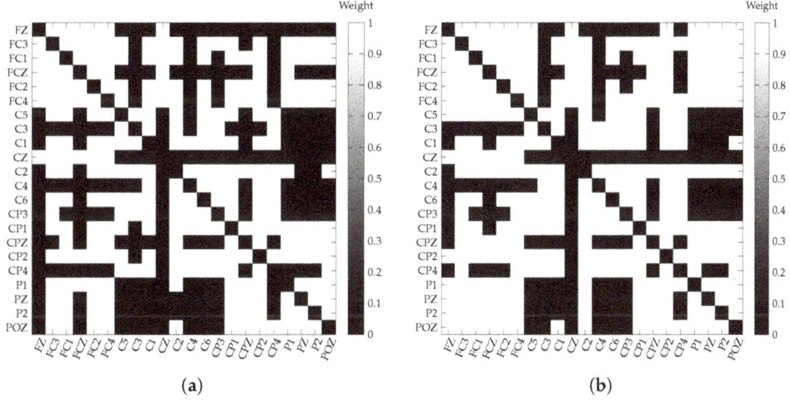

Figure 6. Example of the binarized adjacency matrix from Figure 3; (**a**) fixed threshold (0.9809) and (**b**) Otsu's threshold (0.9750) for the β band (13–30 Hz) for the IM right-hand. (**a**) Example of the binarized adjacency matrix by a fixed threshold (0.9747), in the β band (13–30 Hz) for the IM of the left-hand. (**b**) Example of the binarized adjacency matrix by the Otsu's threshold (0.9698), in the β band (13–30 Hz) for the IM of the left-hand.

3.3. Single-Layer Network Estimation

To model brain dynamics in motor imaginary tasks by the single-layer approach, multiple single-layer graphs were built for each class for all 18 EEG recordings. Each graph corresponds to a band network representation (δ, θ, α, and β) of the 22 EEG electrodes as the graph nodes, and the brain wiring or graph edges corresponding to the effective band-coherence score between electrodes. Notice that such graphs are independent one of another, despite in nature, the brain oscillome is not compartmentalized but modulates electrophysiological bands as a whole. Then, 72 graphs were obtained for each MI class that corresponds to the four frequency bands of the 18 EEG MI recordings.

Then, four graph metrics were estimated: degree (Equation (5)), that measures the electrode neighborhood by adding all j-column a_{ij} adjacency matrix coefficients for the i-node v; eigenvector centrality x_v (Equation (6)), that evaluates the neighborhood ($M(v)$) integration by estimating the eigenvalues λ and their eigenvector x_t; k-core number (Equation (7)), that represents the electrode coreness level where each node's score k is the subgraph $G(C)$ to which it belongs with degree nodes $d_{G(C)}(v)$ greater than k; and PageRank (Equation (8)), that ranks the node importance by averaging the ratio of its neighbors' pagerank $P_R(v)$ and their degree $d(v)$.

$$d(v) = \sum_{i,j \in V} a_{ij} \tag{5}$$

$$x_v = \frac{1}{\lambda} \sum_{t \in M(v)} x_t \tag{6}$$

$$\forall v \in C : d_{G(C)}(v) \geq k \tag{7}$$

$$P_R(v) = \sum_{u \in B_v} \frac{P_R(u)}{d(u)} \tag{8}$$

3.4. Multilayer Network Estimation

For the multilayer approach, the layers that correspond to each electrophysiological band were retrieve, and then integrated into multi-level graph models for each class of all 18 EEG recordings. For these graphs, the intra-layer edges were considered to be present between the nodes themselves, since all electrophysiological bands operate simultaneously. In the next step, multilayer metrics were estimated (Figure 2e,f) using the MuxViz framework in R language [50].

The metrics considered were degree, PageRank, eigenvector centrality, and k-core. The degree (Equation (9)) is the number of links through the layers, ignoring the interlayer link nodes themselves. PageRank (Equation (10)) is the probability of a node reaching any other node $\frac{(1-r)}{NL}$, so it ranks the nodes based on the latter probability [51]. As in a single-layer model, those probabilities are uniform, $u_{j\beta}^{i\alpha}$, through all nodes, and are interactively updated. However, in the multilayer case, the probabilities. $u_{j\beta}^{i\alpha}$, are considered to be the initial values of the next layer. For eigenvector centrality (Equation (11)), the suprajacency matrix is encoded into an aggregate matrix, $M_{j\beta}^{i\alpha}$ via an eigentensor $\Theta_{j\alpha}$. The eigenvector centrality is the dot product of the leading eigenvector, λ_1^{-1} and the neighborhood of each node [52]. Finally, k-core (Equation (12)) represents the ratio of the coreness n_{k-core} for the probability of specific degree-node $n_k(q)$ through all the layers [53].

$$k^i = M_{j\beta}^{i\alpha} U_{\alpha}^{\beta} u^j \tag{9}$$

$$R_{j\beta}^{i\alpha} = r T_{j\beta}^{i\alpha} + \frac{(1-r)}{NL} u_{j\beta}^{i\alpha} \tag{10}$$

$$\Theta_{j\beta} = \lambda_1^{-1} M_{j\beta}^{i\alpha} \Theta_{i\alpha} \tag{11}$$

$$P_k(q) = \frac{n_k(q)}{n_{k-core}} \tag{12}$$

4. Results and Discussion

4.1. Single-Layer Network

Statistical analysis for each band was performed to evaluate which electrode metric differs among the MI classes. Thus, the electrode metric distributions for each MI class were considered to be dependent variables of such class. Thereafter, a MANOVA was performed to determine the significative electrodes and followed by a post hoc test on each electrode.

MANOVA post hoc test consists of applying a one-way ANOVA on the significative electrodes and a posterior Games-Howell post hoc test, to locate the motions that have a significant difference at these electrodes.

The single-layer network results presented in Table 1 and Figure 7 show that the significant electrodes correspond to the frontal and parietal cortex in β band (Figure 7c). Post hoc analysis points that the significant electrodes on these brain areas corresponding to each graph metrics are: degree—C3, FC4, POZ, CP2 and CP3 (Figure 7d); eigenvector—C3, POZ, CP2 and CP4 (Figure 7e); k-core and PageRank metrics were not significative for the MANOVA. From these results, and considering those electrodes that were significative in at least two metrics, we first labeled as key electrodes: C3, POZ and CP2. Later, from these electrodes, we identified which were higher than the fixed and Otsu's thresholds. Thus, the fixed threshold retrieved the C3, POZ and CP2 electrodes (Figure 7b), while Otsu's threshold only retained C3 and POZ electrodes (Figure 7a).

Table 1. p-values < 0.05 for the post hoc Games-Howell test in single layer of the fixed and Otsu's threshold.

Metric	Band	Electrode	Movement 1	Movement 2	Threshold Fixed	Threshold Otsu
Degree	Beta	C3	Foot	Right Hand	*	*
		POz	Right Hand	Tongue	*	*
		CP2	Left Hand	Right Hand	*	−
		CP3	Right Hand	Tongue	*	−
		FC4	Right Hand	Tongue	*	−
Eigenvector	Beta	C3	Foot	Right Hand	*	*
		POz	Right Hand	Tongue	*	*
		CP2	Left Hand	Right Hand	*	−
		CP4	Foot	Left Hand	*	−

Entries '−' indicate no data according to threshold (average or Otsu); '*' indicates significant data found in either the same threshold individually, or in both with the same threshold (fixed or Otsu).

Our results support that the frontal and parietal brain areas drive MI, as reported by Shenoy and Vinod [54]. In the latter study, the authors analyzed the same database for the four MI movements as in the present work. The common electrodes in both studies are C3, FC4, CP3 and CP4. These areas have been reported as the main MI electrodes in several connectivity analyses [15,41–44]. Most of these works are subject-wise analyzed, and their findings slightly deviate. However, all coincided with the brain zones (frontal and parietal) and the electrophysiological bands and sub-bands (mainly α and β) involved in MI.

The aforementioned picture suggests that an integrative analysis for all electrophysiological bands can retrieve the driver nodes on the MI brain dynamics. Multilayer network analysis is a model that meets the above-mentioned constraints.

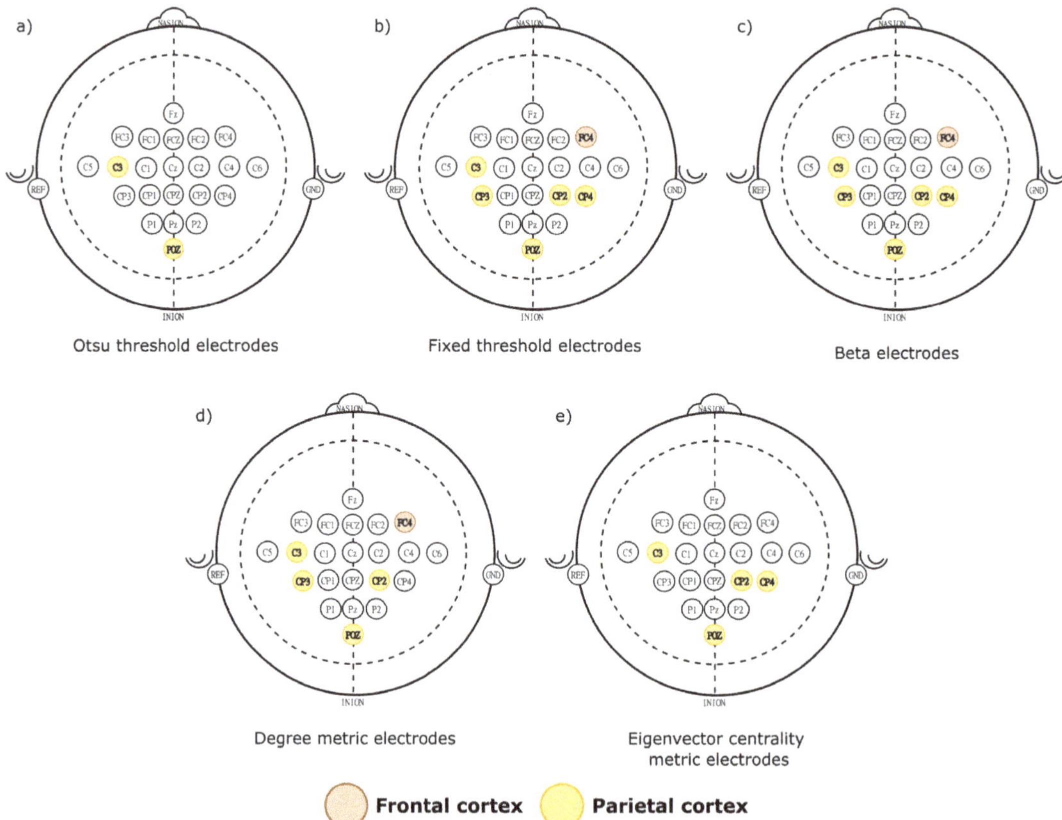

Figure 7. The electrodes with significative differences according to the post hoc MANOVA. In (**a**) all electrodes obtained with the post hoc analysis; (**b**,**c**) display the significative electrodes for the single-layer metrics with significative differences, while (**d**,**e**) identify the electrodes that were significative for both thresholds.

4.2. Multilayer Network

To analyze the dynamics of MI in EEG recordings through a multilayer network model, a one-way ANOVA was performed to evaluate the multilayer metrics estimated for both thresholds. The metric distributions for each movement were obtained independently of its associated electrode; that is, all the metrics per electrode were concatenated. In this analysis, PageRank, Eigenvector and k-core were significatives ($p < 0.05$) for the Otsu's threshold, while the fixed threshold in k-core was only significative (Table 2). This points to a difference between movements in the topology of the brain nucleus. Figure 8 shows the metric distribution for each movement.

To eliminate familywise errors, a post hoc paired t-test was performed using Benjamin-Hochberg FDR correction. This resulted in significative differences between the left-hand movement when k-core distributions were compared to the other movements for both thresholds (Figure 9).

Next, two analyses were conducted, a clustering to verify whether the movements are distinguishable based on all multilayer metrics, and statistical analysis to unveil the significant electrodes between imaginary movements.

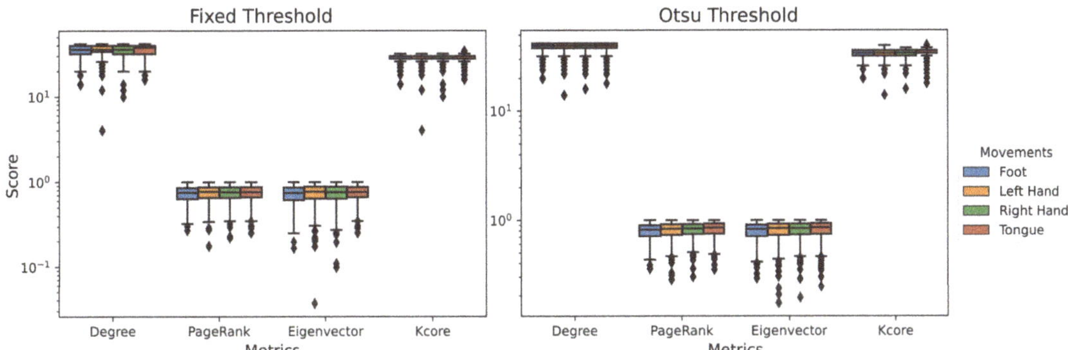

Figure 8. Multilayer graph metric distributions for all four MI classes. After applying a one-way ANOVA test, Otsu's threshold showed significant values ($p < 0.05$) for k-core and degree, while fixed threshold only k-core was significantly different ($p < 0.05$).

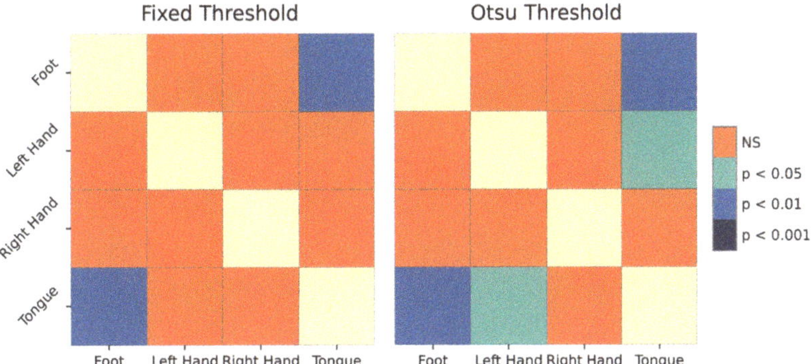

Figure 9. Post hoc test scores in k-core for Otsu's threshold on window 3.5–5.5 s.

4.2.1. Clustering

To assess whether the imaginary movements diverged between them by multilayer metrics, an unsupervised approach was performed for the evaluated window data, i.e., the time from 3.5–5.5 s. Thus, all electrode metrics were concatenated and linear discriminant analysis (LDA) was carried out to lower the data dimensionality into a 3D mapping (Figure 10).

Afterward, k-means clustering was developed with four clusters, assuming that each cluster will depict each of the four movements. To evaluate the intersection between the estimated k-means cluster and the real targets (the imagery movements), the completeness score was calculated yielding a score equal to one of both thresholds.

For each threshold, the clusters mapped differently for movements. For the fixed threshold, cluster 1 (red) represents the left hand, cluster 2 (green) the right hand, cluster 3 (blue) the foot, and cluster 4 (black) the tongue. Meanwhile, for the Otsu's threshold, cluster 1 (red) maps to the foot movement, cluster 2 (green) to the left hand, cluster 3 (blue) to the tongue, and cluster 4 (black) to the right hand. This finding points out that the multilayer graph metrics despite the threshold do illustrate the topological connectivity dynamics all during imaginary movements.

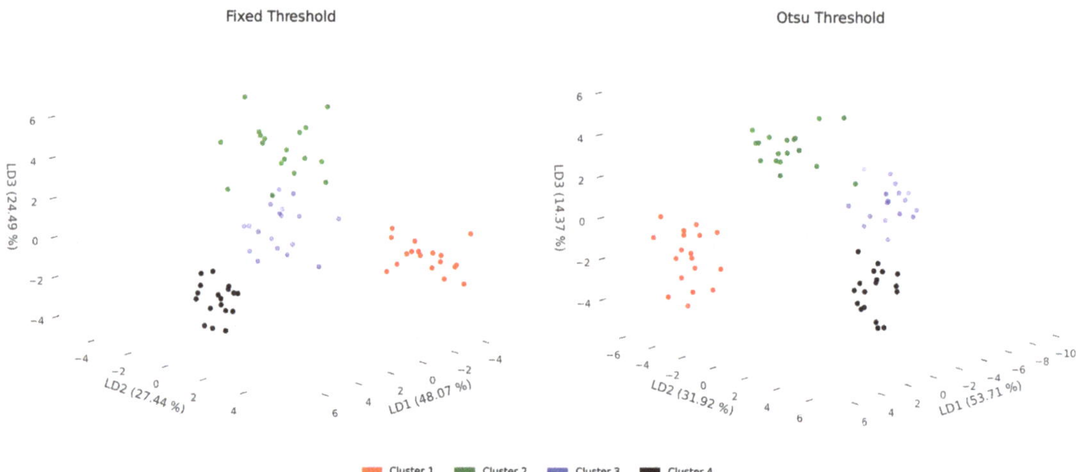

Figure 10. 4-cluster k-means for a low dimensionality representation (LDA) of multilayer graph metrics, on the left for the fixed threshold and right for the Otsu's threshold. The four estimated clusters mapped to the imaginary movements in the window 3.5–5.5 s studied in the present work.

4.2.2. Key Electrodes

To elucidate the electrodes most likely associated with the movements, an electrode-wise statistical analysis was conducted for all multilayer metrics among the four movements. This was evaluated by considering that the electrodes in brain dynamics are the dependent variable among the movements. In this case, a multivariate analysis of variance (MANOVA) was performed for both thresholds. The k-core metric was discarded in both cases since it did not comply with the MANOVA assumption that the data must be normally distributed between groups. The three remaining metrics showed significant differences (Table 3). Table 3 shows significative p for the eigenvector, PageRank, and degree metrics for both thresholds. This can illustrate that brain topology during imaginary movement is driven by key brain electrodes that switch to control distinct movements.

Table 2. p-values from the ANOVA test on window 3.5–5.5 s.

	Fixed	**Otsu**
Degree	9.42×10^{-01}	9.83×10^{-01}
PageRank	2.29×10^{-02}	3.3×10^{-03}
Eigenvector	8.72×10^{-02}	1.77×10^{-02}
Kcore	2.7×10^{-03}	3.7×10^{-03}

Table 3. p-values from the MANOVA test of the average and Otsu's thresholds on window 3.5–5.5 s.

	Fixed	**Otsu**
Degree	3.72×10^{-01}	9.56×10^{-01}
PageRank	5.94×10^{-02}	6.2×10^{-02}
Eigenvector	3.48×10^{-02}	1.71×10^{-02}
Kcore	Error	Error

After the MANOVA analysis, a one-way ANOVA for each of the 22 electrodes was performed to identify the key electrodes that contributed to the significant differences found in the MANOVA.

These 22 one-way ANOVAs were applied for the metrics with significant p-value of Table 3; that is, degree, eigenvector and PageRank for both thresholds. After the one-way ANOVAs, a post hoc Games-Howell test was conducted to determine the electrodes involved in the changes in brain dynamics for the four movements. Results of this analysis are presented in Table 4.

Table 4. *p*-values < 0.05 for the post hoc Games-Howell test in multilayer of the fixed and Otsu's threshold on window 3.5–5.5 s.

Electrode	Movement 1	Movement 2	Metric			Threshold	
			Degree	Eigenvector	PageRank	Fixed	Otsu
CP2	Left Hand	Tongue	—	*	—	*	—
P2	Foot	Left Hand	—	*	—	—	*

Entries '−' indicate no data according to the metric (degree, eigenvector or PageRank) or threshold (average or Otsu); '*' indicates significative data found in either the same graph metric individually, or in both with the same threshold (fixed or Otsu).

Table 4 shows that 2 electrodes most likely drive the brain dynamics of the MI dataset analyzed. These electrodes are P2 and CP2 (Figure 11a). Among these electrodes, multilayer eigenvector metric point that the significant ones are: P2 and CP2 (Figure 11d). Degree, *k*-core and PageRank metrics were not significative for the MANOVA.

The key nodes that the fixed threshold gather the P2 (Figure 11c) electrode, while the Otsu threshold identified the electrode CP2 (Figure 11b).

These results suggests that both Fixed and Otsu's thresholds are selective to yield the pivotal electrodes from the multilayer network. Despite that the electrodes imaged by both thresholds differ, these electrodes are neighbors and localized over the same brain area. Thus, it suggests that the Otsu threshold can recognize the underlying dynamics which widely tested thresholds as Fixed have also distinguished. Figure 12 shows an example of a multilayer graph.

The multilayer approach outlined in this study allowed us to cluster the dynamics linked to all the studied imaginary movements. Based on this finding, we pinpointed the key electrodes for such dynamics. Our results are congruent with the state-of-the-art analyses [15,41–44] that reported the frontal and parietal areas as the main brain areas in MI. In more detail, Babiloni et al. [55] indicated that sensorimotor events are correlated via the coherence with a functional coupling between parietal and central areas. All these works applied a single-layer approach for different frequency bands. To the best of our knowledge in the literature, there was not reported MI analysis based on multilayer graph models. For future work, our multilayer workflow will be tested in practical BCI applications. Our proposal, which couples an adaptive threshold with a multilayer network model, shall be cross-validated on new databases to validate its advantages over the widespread single-layer analysis.

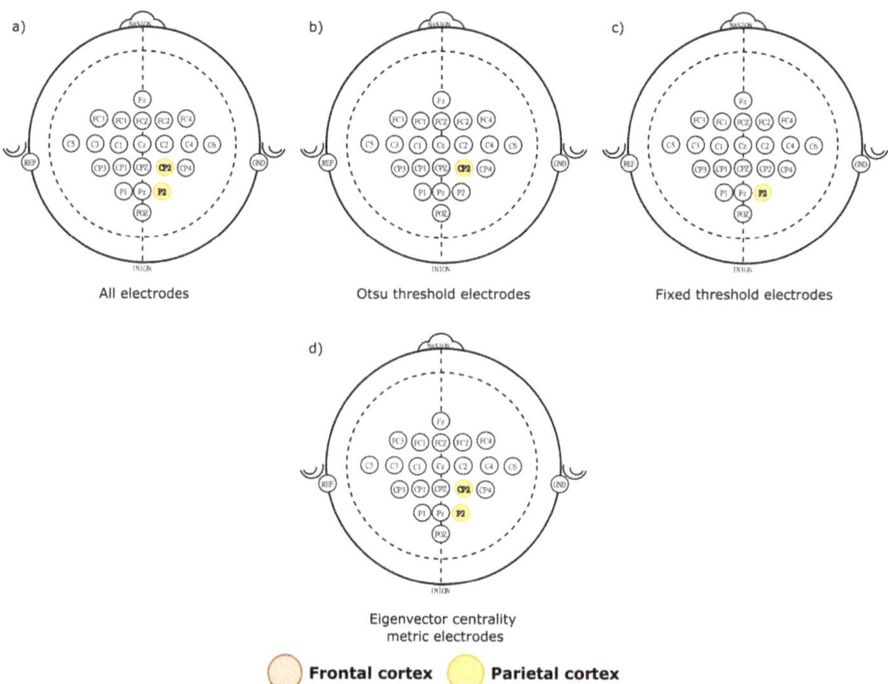

Figure 11. The electrodes with significative differences according to the post hoc MANOVA. In (**a**) all electrodes obtained with the post hoc analysis; (**b**,**c**) display the significative electrodes for the multilayer metrics with significative differences, while (**d**) identify the electrodes that were significatives for both thresholds.

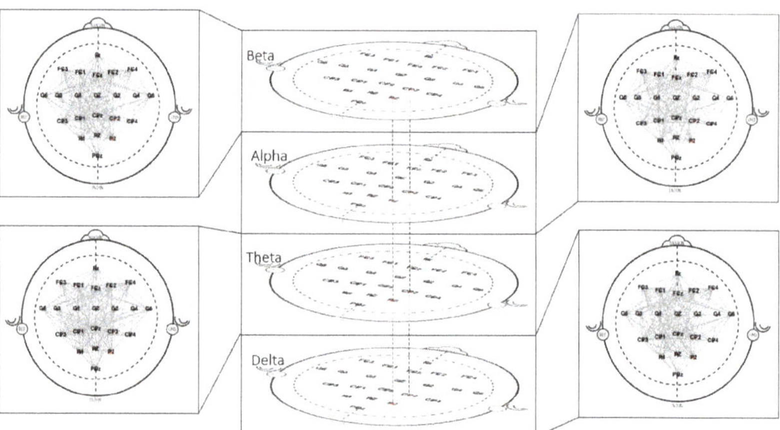

Figure 12. Multilayer graph for right-hand movement intention. Each electrophysiological band is a layer of the graph. Those electrodes with significative difference identified (CP2 and P2) are highlighted in red in the present study. Cross-layer edges are found for all nodes, yet only those corresponding to the significant electrodes are plotted. Each individual layer is shown separately to provide a more detailed picture of the intra-layer connectivity.

5. Conclusions

In this study, we modeled single-layer and multilayer network models to analyze MI in EEG recordings. Our analysis shows that the regions activated in MI tasks are located mainly in the frontal and parietal cortex for the single-layer approach and in the parietal cortex for the multilayer approach. To pinpoint the effective connections in MI graphs a proof-of-concept threshold approach known as Otsu was proposed. The present work illustrates that combining an adaptive threshold, such as Otsu, together with integrative graph models, such as multilayer networks, produces a more reliable approximation of both the topology and dynamics associated with cognitive and motor brain functions.

Finally, future work should aim to implement this methodology to study brain connectivity in other kinds of EEG databases.

Author Contributions: Conceptualization, R.R.V., C.C.-O. and H.V.-P.; methodology, C.C.-O., J.B.L., O.P., R.R.-V. and H.V.-P.; software, C.C.-O., J.B.L. and O.P.; validation, C.C.-O. and O.P. formal analysis, C.C.O. and O.P.; investigation, C.C.-O., O.P., R.R.-V. and H.V.-P.; data curation, C.C.-O., J.B.L. and O.P.; writing—original draft preparation, C.C.-O. and O.P.; writing—review and editing, R.R.-V. and H.V.-P.; supervision R.R.-V. and H.V.-P.; project administration, R.R.-V. and H.V.P. All authors have read and agreed to the published version of the manuscript.

Funding: This research received no external funding. This work was supported by the Consejo Nacional de Ciencia y Tecnología—CONACyT [Scholarship to C.C.-O. scholarship 480527, O.P. with CVU 713526 and J.B.L. with CVU 745514].

Institutional Review Board Statement: Not applicable.

Informed Consent Statement: Not applicable.

Data Availability Statement: http://www.bbci.de/competition/iv/.

Conflicts of Interest: The authors declare no conflict of interest.

References

1. Park, H.J.; Friston, K. Structural and functional brain networks: From connections to cognition. *Science* **2013**, *342*, 6158. [CrossRef] [PubMed]
2. Subasi, A.; Ercelebi, E. Classification of EEG signals using neural network and logistic regression. *Comput. Methods Programs Biomed.* **2005**, *78*, 87–99. [CrossRef]
3. Sanei, S.; Chambers, J. Brain–Computer Interfacing. *EEG Signal Process.* **2007**, 239–265. [CrossRef]
4. Mandke, K.; Meier, J.; Brookes, M.J.; O'Dea, R.D.; Van Mieghem, P.; Stam, C.J.; Hillebrand, A.; Tewarie, P. Comparing multilayer brain networks between groups: Introducing graph metrics and recommendations. *NeuroImage* **2018**, *166*, 371–384. [CrossRef]
5. Klados, M.A.; Pandria, N.; Micheloyannis, S.; Margulies, D.; Bamidis, P.D. Math anxiety: Brain cortical network changes in anticipation of doing mathematics. *Int. J. Psychophysiol.* **2017**, *122*, 24–31. [CrossRef]
6. González-Garrido, A.A.; Gómez-Velázquez, F.R.; Salido-Ruiz, R.A.; Espinoza-Valdez, A.; Vélez-Pérez, H.; Romo-Vazquez, R.; Gallardo-Moreno, G.B.; Ruiz-Stovel, V.D.; Martínez-Ramos, A.; Berumen, G. The analysis of EEG coherence reflects middle childhood differences in mathematical achievement. *Brain Cogn.* **2018**, *124*, 57–63. [CrossRef] [PubMed]
7. Sakkalis, V. Review of advanced techniques for the estimation of brain connectivity measured with EEG/MEG. *Comput. Biol. Med.* **2011**, *41*, 1110–1117. [CrossRef] [PubMed]
8. Bowyer, S.M. Coherence a measure of the brain networks: Past and present. *Neuropsychiatr. Electrophysiol.* **2016**, *2*, 1. [CrossRef]
9. Jiang, Z. Study on EEG power and coherence in patients with mild cognitive impairment during working memory task. *J. Zhejiang Univ. Sci. B* **2005**, *6*, 1213–1219. [CrossRef]
10. Vecchiato, G.; Susac, A.; Margeti, S.; Fallani, F.D.V.; Maglione, A.G.; Supek, S.; Planinic, M.; Babiloni, F. High-resolution EEG analysis of power spectral density maps and coherence networks in a proportional reasoning task. *Brain Topogr.* **2013**, *26*, 303–314. [CrossRef] [PubMed]
11. Özerdem, A.; Güntekin, B.; İlhan, A.; Turp, B.; Başar, E. Reduced long distance gamma (28–48 Hz) coherence in euthymic patients with bipolar disorder. *J. Affect. Disord.* **2011**, *132*, 325–332. [CrossRef]
12. Moazami-Goudarzi, M.; Sarnthein, J.; Michels, L.; Moukhtieva, R.; Jeanmonod, D. Enhanced frontal low and high frequency power and synchronization in the resting EEG of parkinsonian patients. *NeuroImage* **2008**, *41*, 985–997. [CrossRef]
13. Hidasi, Z.; Czigler, B.; Salacz, P.; Csibri, É.; Molnár, M. Changes of EEG spectra and coherence following performance in a cognitive task in Alzheimer's disease. *Int. J. Psychophysiol.* **2007**, *65*, 252–260. [CrossRef] [PubMed]
14. Song, J.; Tucker, D.; Gilbert, T.; Hou, J.; Mattson, C.; Luu, P.; Holmes, M. Methods for Examining Electrophysiological Coherence in Epileptic Networks. *Front. Neurol.* **2013**, *4*, 55. [CrossRef]

15. Gaxiola-Tirado, J.A.; Salazar-Varas, R.; Gutiérrez, D. Using the partial directed coherence to assess functional connectivity in electroencephalography data for brain–computer interfaces. *IEEE Trans. Cogn. Dev. Syst.* **2017**, *10*, 776–783. [CrossRef]
16. Dennis, E.L.; Jahanshad, N.; Toga, A.W.; McMahon, K.L.; De Zubicaray, G.I.; Martin, N.G.; Wright, M.J.; Thompson, P.M. Test-retest reliability of graph theory measures of structural brain connectivity. In *International Conference on Medical Image Computing and Computer-Assisted Intervention*; Springer: Berlin/Heidelberg, Germany, 2012; pp. 305–312.
17. Kinsley, A.C.; Rossi, G.; Silk, M.J.; VanderWaal, K. Multilayer and Multiplex Networks: An Introduction to Their Use in Veterinary Epidemiology. *Front. Vet. Sci.* **2020**, *7*, 596. [CrossRef]
18. Ahmadlou, M.; Adeli, H. Functional community analysis of brain: A new approach for EEG-based investigation of the brain pathology. *Neuroimage* **2011**, *58*, 401–408. [CrossRef]
19. Zippo, A.G.; Della Rosa, P.A.; Castiglioni, I.; Biella, G.E. Alternating dynamics of segregation and integration in human EEG functional networks during working-memory task. *Neuroscience* **2018**, *371*, 191–206. [CrossRef] [PubMed]
20. Puxeddu, M.; Petti, M.; Mattia, D.; Astolfi, L. The optimal setting for multilayer modularity optimization in multilayer brain networks. In Proceedings of the 2019 41st Annual International Conference of the IEEE Engineering in Medicine and Biology Society (EMBC), Berlin, Germany, 23–27 July 2019; pp. 624–627.
21. Puxeddu, M.G.; Petti, M.; Astolfi, L. A Comprehensive Analysis of Multilayer Community Detection Algorithms for Application to EEG-Based Brain Networks. *Front. Syst. Neurosci.* **2021**, *15*, 624183.. [CrossRef] [PubMed]
22. Muldoon, S.F.; Bassett, D.S. Network and multilayer network approaches to understanding human brain dynamics. *Philos. Sci.* **2016**, *83*, 710–720. [CrossRef]
23. Vaiana, M.; Muldoon, S.F. Multilayer brain networks. *J. Nonlinear Sci.* **2018**, *30*, 2147–2169. [CrossRef]
24. Braun, U.; Schäfer, A.; Walter, H.; Erk, S.; Romanczuk-Seiferth, N.; Haddad, L.; Schweiger, J.I.; Grimm, O.; Heinz, A.; Tost, H.; et al. Dynamic reconfiguration of frontal brain networks during executive cognition in humans. *Proc. Natl. Acad. Sci. USA* **2015**, *112*, 11678–11683. [CrossRef] [PubMed]
25. De Domenico, M.; Sasai, S.; Arenas, A. Mapping multiplex hubs in human functional brain networks. *Front. Neurosci.* **2016**, *10*, 326. [CrossRef]
26. Brookes, M.J.; Tewarie, P.K.; Hunt, B.A.; Robson, S.E.; Gascoyne, L.E.; Liddle, E.B.; Liddle, P.F.; Morris, P.G. A multi-layer network approach to MEG connectivity analysis. *Neuroimage* **2016**, *132*, 425–438. [CrossRef] [PubMed]
27. Paredes, O.; López, J.B.; Covantes-Osuna, C.; Ocegueda-Hernández, V.; Romo-Vázquez, R.; Morales, J.A. A Transcriptome Community-and-Module Approach of the Human Mesoconnectome. *Entropy* **2021**, *23*, 1031. [CrossRef]
28. Frolov, N.; Maksimenko, V.; Hramov, A. Revealing a multiplex brain network through the analysis of recurrences. *Chaos Interdiscip. J. Nonlinear Sci.* **2020**, *30*, 121108. [CrossRef]
29. Naro, A.; Maggio, M.G.; Leo, A.; Calabrò, R.S. Multiplex and Multilayer Network EEG Analyses: A Novel Strategy in the Differential Diagnosis of Patients with Chronic Disorders of Consciousness. *Int. J. Neural Syst.* **2020**, *31*, 2050052. [CrossRef] [PubMed]
30. Havlin, S.; Stanley, H.E.; Bashan, A.; Gao, J.; Kenett, D.Y. Percolation of interdependent network of networks. *Chaos Solitons Fractals* **2015**, *72*, 4–19. [CrossRef]
31. Yu, M.; Engels, M.M.; Hillebrand, A.; Van Straaten, E.C.; Gouw, A.A.; Teunissen, C.; Van Der Flier, W.M.; Scheltens, P.; Stam, C.J. Selective impairment of hippocampus and posterior hub areas in Alzheimer's disease: An MEG-based multiplex network study. *Brain* **2017**, *140*, 1466–1485. [CrossRef] [PubMed]
32. Van Den Heuvel, M.P.; Fornito, A. Brain networks in schizophrenia. *Neuropsychol. Rev.* **2014**, *24*, 32–48. [CrossRef]
33. Van Wijk, B.C.; Stam, C.J.; Daffertshofer, A. Comparing brain networks of different size and connectivity density using graph theory. *PLoS ONE* **2010**, *5*, e13701. [CrossRef]
34. Sezgin, M.; Sankur, B. Survey over image thresholding techniques and quantitative performance evaluation. *J. Electron. Imaging* **2004**, *13*, 146–166.
35. Roy, S.; Nag, S.; Maitra, I.K.; Bandyopadhyay, S.K. A review on automated brain tumor detection and segmentation from MRI of brain. *arXiv* **2013**, arXiv:1312.6150.
36. Sandhya, G.; Babu Kande, G.; Savithri, T.S. Multilevel thresholding method based on electromagnetism for accurate brain MRI segmentation to detect white matter, gray matter, and CSF. *BioMed Res. Int.* **2017**, *2017*, 6783209. [CrossRef]
37. Fortin, M.; Omidyeganeh, M.; Battié, M.C.; Ahmad, O.; Rivaz, H. Evaluation of an automated thresholding algorithm for the quantification of paraspinal muscle composition from MRI images. *Biomed. Eng. Online* **2017**, *16*, 61. [CrossRef]
38. Sharma, A.; Kumar, S.; Singh, S.N. Brain tumor segmentation using DE embedded OTSU method and neural network. *Multidimens. Syst. Signal Process.* **2019**, *30*, 1263–1291. [CrossRef]
39. Geetha, G.; Geethalakshmi, S. Artifact removal from eeg using spatially constrained independent component analysis and wavelet denoising with otsu's thresholding technique. *Procedia Eng.* **2012**, *30*, 1064–1071. [CrossRef]
40. Otsu, N. A threshold selection method from gray-level histograms. *IEEE Trans. Syst. Man Cybern.* **1979**, *9*, 62–66. [CrossRef]
41. Astolfi, L.; Cincotti, F.; Mattia, D.; Marciani, M.G.; Baccala, L.A.; de Vico Fallani, F.; Salinari, S.; Ursino, M.; Zavaglia, M.; Ding, L.; et al. Comparison of different cortical connectivity estimators for high-resolution EEG recordings. *Hum. Brain Mapp.* **2007**, *28*, 143–157. [CrossRef]

42. Rathee, D.; Cecotti, H.; Prasad, G. Estimation of effective fronto-parietal connectivity during motor imagery using partial granger causality analysis. In Proceedings of the 2016 International Joint Conference on Neural Networks (IJCNN), Vancouver, BC, Canada, 24–29 July 2016; pp. 2055–2062.
43. Cardoso, V.F.; Delisle-Rodriguez, D.; Romero-Laiseca, M.A.; Loterio, F.A.; Gurve, D.; Floriano, A.; Valadão, C.; Silva, L.; Krishnan, S.; Frizera-Neto, A.; et al. Effect of a Brain–Computer Interface Based on Pedaling Motor Imagery on Cortical Excitability and Connectivity. *Sensors* **2021**, *21*, 2020. [CrossRef]
44. Alanis-Espinosa, M.; Gutiérrez, D. On the assessment of functional connectivity in an immersive brain-computer interface during motor imagery. *Front. Psychol.* **2020**, *11*, 1301. [CrossRef]
45. Tangermann, M.; Müller, K.R.; Aertsen, A.; Birbaumer, N.; Braun, C.; Brunner, C.; Leeb, R.; Mehring, C.; Miller, K.J.; Mueller-Putz, G.; et al. Review of the BCI competition IV. *Front. Neurosci.* **2012**, *6*, 55.
46. Brunner, C.; Leeb, R.; Müller-Putz, G.; Schlögl, A.; Pfurtscheller, G. *BCI Competition 2008–Graz Data Set A*; Institute for Knowledge Discovery, Graz University of Technology: Graz, Austria, 2008; Volume 16, pp. 1–6.
47. De Vico Fallani, F.; Astolfi, L.; Cincotti, F.; Mattia, D.; la Rocca, D.; Maksuti, E.; Salinari, S.; Babiloni, F.; Vegso, B.; Kozmann, G.; et al. Evaluation of the Brain Network Organization From EEG Signals: A Preliminary Evidence in Stroke Patient. *Anat. Rec.* **2009**, *292*, 2023–2031. [CrossRef]
48. Tsai, S.Y. Reproducibility of structural brain connectivity and network metrics using probabilistic diffusion tractography. *Sci. Rep.* **2018**, *8*, 11562. [CrossRef]
49. Stefan, S.; Schorr, B.; Lopez-Rolon, A.; Kolassa, I.T.; Shock, J.P.; Rosenfelder, M.; Heck, S.; Bender, A. Consciousness indexing and outcome prediction with resting-state EEG in severe disorders of consciousness. *Brain Topogr.* **2018**, *31*, 848–862. [CrossRef] [PubMed]
50. Domenico, M.D.; Porter, M.A.; Arenas, A. MuxViz: A tool for multilayer analysis and visualization of networks. *J. Complex Networks* **2014**, *3*, 159–176. [CrossRef]
51. Brin, S.; Page, L. The anatomy of a large-scale hypertextual Web search engine. *Comput. Netw. ISDN Syst.* **1998**, *30*, 107–117. [CrossRef]
52. Domenico, M.D.; Solé-Ribalta, A.; Omodei, E.; Gómez, S.; Arenas, A. Ranking in interconnected multilayer networks reveals versatile nodes. *Nat. Commun.* **2015**, *6*, 6868. [CrossRef]
53. Azimi-Tafreshi, N.; Gómez-Gardeñes, J.; Dorogovtsev, S.N. k-core percolation on multiplex networks. *Phys. Rev. E* **2014**, *90*, 032816. [CrossRef]
54. Shenoy, H.V.; Vinod, A.P. An iterative optimization technique for robust channel selection in motor imagery based brain computer interface. In Proceedings of the 2014 IEEE International Conference on Systems, Man, and Cybernetics (SMC), San Diego, CA, USA, 5–8 October 2014; pp. 1858–1863.
55. Babiloni, C.; Brancucci, A.; Vecchio, F.; Arendt-Nielsen, L.; Chen, A.C.; Rossini, P.M. Anticipation of somatosensory and motor events increases centro-parietal functional coupling: An EEG coherence study. *Clin. Neurophysiol.* **2006**, *117*, 1000–1008. [CrossRef]

Review

Application of Transfer Learning in EEG Decoding Based on Brain-Computer Interfaces: A Review

Kai Zhang [1,2], Guanghua Xu [1,2,*], Xiaowei Zheng [1,2], Huanzhong Li [1], Sicong Zhang [1], Yunhui Yu [1] and Renghao Liang [1]

[1] School of Mechanical Engineering, Xi'an Jiaotong University, Xi'an 710049, China; zhangkai0912@stu.xjtu.edu.cn (K.Z.); hlydx1314@stu.xjtu.edu.cn (X.Z.); lihuanzhong@stu.xjtu.edu.cn (H.L.); zhsicong@mail.xjtu.edu.cn (S.Z.); yuyunhui@stu.xjtu.edu.cn (Y.Y.); lrh8131@stu.xjtu.edu.cn (R.L.)

[2] State Key Laboratory for Manufacturing Systems Engineering, Xi'an Jiaotong University, Xi'an 710049, China

* Correspondence: ghxu@xjtu.edu.cn

Received: 18 October 2020; Accepted: 4 November 2020; Published: 5 November 2020

Abstract: The algorithms of electroencephalography (EEG) decoding are mainly based on machine learning in current research. One of the main assumptions of machine learning is that training and test data belong to the same feature space and are subject to the same probability distribution. However, this may be violated in EEG processing. Variations across sessions/subjects result in a deviation of the feature distribution of EEG signals in the same task, which reduces the accuracy of the decoding model for mental tasks. Recently, transfer learning (TL) has shown great potential in processing EEG signals across sessions/subjects. In this work, we reviewed 80 related published studies from 2010 to 2020 about TL application for EEG decoding. Herein, we report what kind of TL methods have been used (e.g., instance knowledge, feature representation knowledge, and model parameter knowledge), describe which types of EEG paradigms have been analyzed, and summarize the datasets that have been used to evaluate performance. Moreover, we discuss the state-of-the-art and future development of TL for EEG decoding. The results show that TL can significantly improve the performance of decoding models across subjects/sessions and can reduce the calibration time of brain–computer interface (BCI) systems. This review summarizes the current practical suggestions and performance outcomes in the hope that it will provide guidance and help for EEG research in the future.

Keywords: EEG; transfer learning; review; decoding; classification

---

1. Introduction

A brain–computer interface (BCI) is a communication method between a user and a computer that does not rely on the normal neural pathways of the brain and muscles [1]. According to the methods of electroencephalography (EEG) signal collection, BCIs can be divided into three types, namely, non-invasive, invasive, and partially-invasive BCIs. Among them, non-invasive BCIs realize the control of external equipment via EEG and by transforming EEG recordings into a command, which have been widely used due to their convenient operation. Figure 1 shows a typical non-invasive BCI system framework based on EEG, which usually consists of three parts: EEG signal acquisition, signal decoding, and external device control. During this process, signal decoding is the key step to ensure the operation of the whole system.

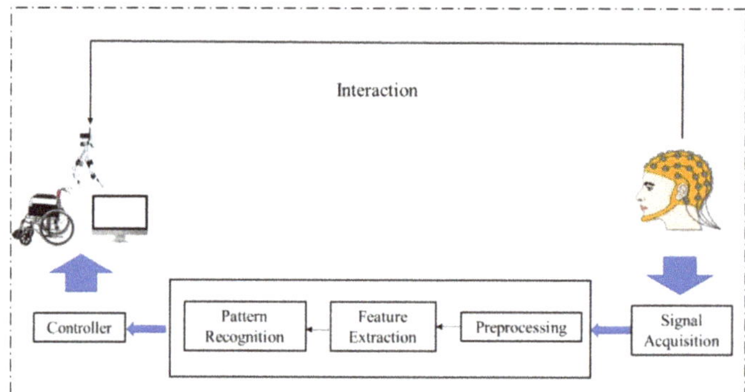

Figure 1. Framework of an electroencephalography (EEG)-based brain–computer interface (BCI) system.

The representation of EEG typically takes the form of a high-dimensional matrix, which includes the information of sampling points, channels, trials, and subjects [2]. Meanwhile, the most common features of EEG-based BCIs include spatial filtering, band power, time points, and so on. Recently, machine learning (ML) has shown its powerful ability for feature extraction in EEG-based BCI tasks [3,4].

BCI technology based on EEG has made great progress, but the challenges of weak robustness and low accuracy greatly hinder the application of BCIs in practice [5]. From the perspective of signal decoding, the reasons are as follows: First, one of the main assumptions of ML is that training and test data belong to the same feature space and are subject to the same probability distribution. However, this assumption is often violated in the field of bioelectric signal processing, because differences in physiological structure and psychological states may cause obvious variation in EEG. Therefore, signals from different sessions/subjects on the same task show different features and distribution.

Second, EEG signals are extremely weak and are always accompanied by unrelated artifacts from other areas of the brain, which potentially mislead discriminant results and decrease the classification accuracy. Third, the strict requirements for the experimental conditions of BCI systems make it difficult to obtain large and high-quality datasets in practice. It is difficult for a classification model based on small-scale samples to obtain strong robustness and high classification accuracy. However, large-scale and high-quality datasets are the basis for guaranteeing the decoding accuracy of models.

One promising approach to solve these problems is transfer learning (TL). The principle of TL is realizing the knowledge transfer from different but related tasks, i.e., using existing knowledge learned from accomplished tasks to help with new tasks. The definition of TL is as follows: A given domain D consists of a feature space X and a marginal probability distribution $P(X)$. A task T consists of a label space y and a prediction function f. A source domain D_s and a target domain D_T may have different feature spaces or different marginal probability distributions, i.e., $X_s \neq X_T$ or $P_s(X) \neq P_T(X)$. Meanwhile, tasks T_s and T_T are subject to different label spaces. The aim of TL is to help improve the learning ability of the target predictive function $f_T(\cdot)$ in D_T using the knowledge in D_s and T_s [6].

There are two main scenarios in EEG-based BCIs, namely, cross-subject transfer and cross-session transfer. The goal of TL is to find the similarity between new and original tasks and then to realize the discriminative and stationary information transfer across domains [7]. In this study, we attempted to summarize the transferred knowledge for EEG based on following three types: Knowledge of instance, knowledge of feature representation, and knowledge of model parameters.

This review of TL applications for EEG classification attempted to address the following critical questions: What problems does TL solve for EEG decoding? (Section 3.1); which paradigms of EEG are used for TL analysis? (Section 3.2); what kind of datasets can we refer to in order to verify the performance of these methods? (Section 3.3); what types of TL frameworks are available? (Section 3.4).

First, the search methods for the identification of studies are introduced in Section 2. Then, the principle and classification criteria of TL are analyzed in Section 3. Next, the TL algorithms for EEG from 2010 to 2020 are described in Section 4. Finally, the current challenges of TL in EEG decoding are discussed in Section 5.

2. Methodology

A wide literature search from 2010 to 2020 was conducted, resorting to the main databases, such as Web of Science, PubMed, and IEEE Xplore. The keywords used for the electronic search were TL, electroencephalogram, brain–computer interface, inter-subject, and covariate shift. Table 1 lists the collection criteria for inclusion or exclusion.

Table 1. Inclusion and exclusion criteria.

Inclusion Criteria	Exclusion Criteria
• Published within the last 10 years (as transfer learning (TL) for EEG has been proposed and developed in recent years).	• A focus on the processing of invasive EEG, electrocorticography (ECoG), magnetoencephalography (MEG), source imaging, fMRI, and so on, or joint studies with EEG.
• A focus on non-invasive EEG signals (as the object for discussion in this review). • A specific explanation of how to apply TL to EEG signal processing.	• No specific description of TL for EEG processing.

The search method of this review is shown in Figure 2, which was used to identify and to narrow down the collection of TL-based studies, resulting in a total of 246 papers. Duplicates between all datasets and studies without full-text links were excluded. Finally, 80 papers that meet the inclusion criteria were included.

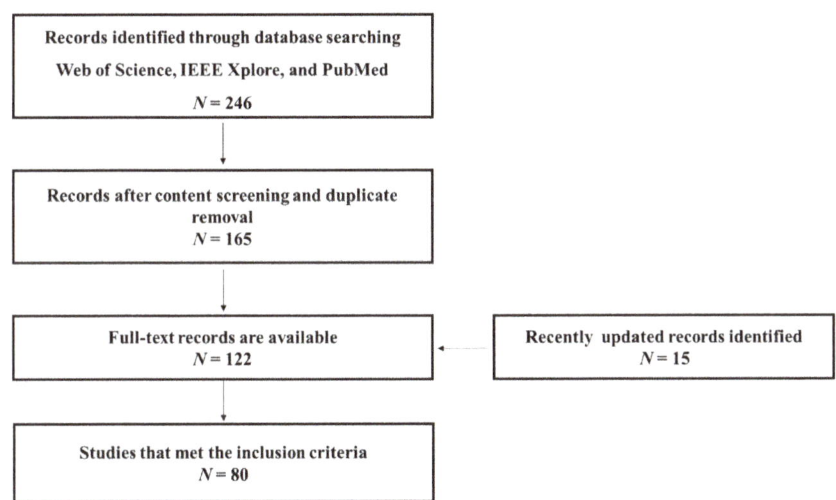

Figure 2. The search method for identifying relevant studies.

3. Results

3.1. What Problems Does Transfer Learning Solve?

This review of the literature on TL applications for EEG attempted to address the following critical questions:

3.1.1. The Problem of Differences across Subjects/Sessions

Although advanced methods such as machine learning have been proven to be a critical tool in EEG processing or analysis, they still suffer from some limitations that hinder their wide application in practice. Consistency of the feature space and probability distribution of training and test data is an important prior condition of machine learning. However, in the field of biomedical engineering, such as EEG based on BCIs, this hypothesis is often violated. Obvious variation in feature distribution typically occurs in representations of EEG across sessions/subjects. This phenomenon results in a scattered distribution of EEG signal features, an increase in the difficulty of feature extraction, and a reduction in the performance of the classifier.

3.1.2. The Problem of Small Sample Size

In recent years, machine learning and deep neural networks have provided good results for the classification of linguistic features, images, sounds, and natural texts. A main reason for its success is that their massive amount of data guarantees the performance of the classifier. However, in practical applications of BCI, it is difficult to collect high-quality and large EEG datasets due to the limitations of strict requirements for the experimental environment and available subjects. The performance of these methods is highly sensitive to the number of samples; a small sample size tends to lead to overfitting during model training, which adversely affects the classification accuracy [8].

3.1.3. The Problem of Time-Consuming Calibration

A large amount of data are required to calibrate a BCI system when a subject performs a specific EEG task. This requirement commonly takes a long calibration session, which is inevitable for a new user. For example, when a subject performs a steady-state visually evoked potential (SSVEP) speller task, the various commands cause a long calibration time. However, collecting calibration data is time-consuming and laborious, which reduces the efficiency of the BCI system.

3.2. EEG Paradigms for Transfer Learning

There are four paradigms of EEG-BCIs discussed in this paper and the percentage of these paradigms across collected studies are shown in Figure 3.

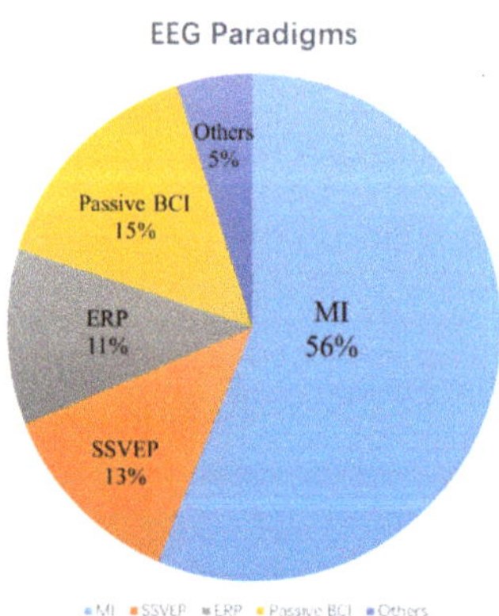

Figure 3. The percentage of different EEG pattern strategies across collected studies.

3.2.1. Motor Imagery

Motor imagery (MI) is a mental process that imitates motor intention without real motion output [9], which activates the neural potential in primary sensorimotor areas. Different imagery tasks will induce potential activity in different regions of the brain. Thus, this response can be converted into a classification task. The feature of MI signals is often expressed in the form of frequency or band energy [10]. Due to task objectives and various feature representations, a variety of machine learning algorithms (e.g., deep learning and Riemannian geometry) can be applied to the decoding of MI [11,12].

3.2.2. Steady-State Visually Evoked Potentials

When a human receives a fixed frequency of flashing visual stimuli, the potential activity of the cerebral cortex is modulated to produce a continuous response related to the frequency (same or multiples) of these stimuli. This physiological phenomenon is referred to a SSVEP [13]. Due to their stable and obvious representation of signals, BCI systems based on SSVEP are widely used to control equipment such as mobile devices, wheelchairs, and spellers.

3.2.3. Event-Related Potentials

Event-related potentials are responses for multiple or diverse stimuli corresponding to specific meanings [14]. P300 is the most representative type of ERP, which occurs about 300 ms after a visual or auditory stimulus. A feature classification model can be used for decoding P300.

3.2.4. Passive BCIs

A passive BCI is a form of interaction that does not rely on external stimuli. It achieves a brain control task by encoding the mental activity from different states of the brain [15]. Common types of passive BCI tasks include driver drowsiness, emotion recognition, mental workload assessment, and epileptic detection [16], which can be decoded by regression and classification models [17,18].

3.3. Case Studies on a Shared Dataset

Analysis between different datasets is not valid because they use different equipment or communication protocols. In addition, different mental tasks and collecting procedures also bring great differences to EEG. Therefore, the reviewed studies mainly concentrate on the TL across subjects/sessions in the same dataset. In Table 2, we briefly summarize the publicly available EEG dataset in this review.

Table 2. Dataset.

Datasets	Task	Subject	Channel	Amount of Data (Per Subject)	Sampling Rate	Reference
BCIC-II-IV	2 MI classes	1	28	3 sessions/416 trials	1000 Hz	[19]
BCIC-III-II	P300	2	64	5 sessions	240 Hz	[20]
BCIC-III-IVa	2 MI classes	5	118	4 sessions/280 trials	1000 Hz	[21]
BCIC-IV-2a	MI	9	25	2 sessions/288 trials	250 Hz	[22]
BCIC-IV-2b	2 MI classes	9	6	720 trials	250 Hz	[23]
P300 speller	P300	8	8	5 sessions/20 trials	256 Hz	[24]
DEAP	ER	32	40	125 trials	128 Hz	[25]
BCIC -III-IVC	MI	1	118	630 trials	200 Hz	[26]
SEED	ER	15	64	3 sessions/15 trials	200 Hz	[27]
OpenMIIR	Music Imagery	10	64	5 sessions/12 trials in four tasks	512 Hz	[28]
CHB-MIT	ED	22	23	844 hours' collection	256 Hz	[29]

3.4. Transfer Learning Architecture

In this review, we summarize previous studies according to "what knowledge should be transferred in EEG processing." Multi-step processing for EEG across subjects/sessions results in discriminative information in different steps. Therefore, determining what should be transferred is the key problem according to different EEG tasks. Pan et al. [6] proposed authoritative classification approaches based on "what to transfer." All papers collected in this review were classified according to this method (Figure 4). In the following sections, we have selected several representative methods for analysis.

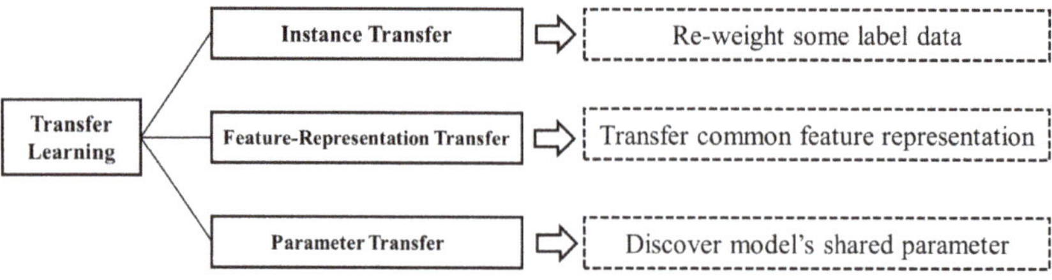

Figure 4. Different approaches to transfer learning.

3.4.1. Transfer Learning Based on Instance Knowledge

It is often assumed that we can easily obtain large amounts of markup data from a source domain, but this data cannot be directly reused. Instance transfer approaches re-weight some source domain data as a supplement for the target domain. Based on instance transfer, the majority of the literature utilized the measurement method to evaluate the similarity between data from the source and target domains. The similarity metric was then converted into the transfer weight coefficient, which was directly used to instance transfer by re-weighting the source domain data [30–32]. Herein, we have listed a few typical methods based on instance transfer.

Reference [33] proposed an instance TL method based on K–L divergence measurements. They measured the similarity of the normal distribution between two domains and transformed this similarity into a transfer weight coefficient for the target subject.

Suppose that the normal distribution from the two datasets N_0 and N_1 can be expressed as:

$$N_0 \sim N(\mu_0, \Sigma_0), N_1 \sim N(\mu_1, \Sigma_1) \tag{1}$$

where μ_i and Σ_i are the mean value and variance ($i = 1/0$), respectively. The K–L divergence of the two distributions can be expressed as:

$$KL[N_0][N_1] = 0.5[(\mu_1 - \mu_0)]^T \Sigma_1^{-1}(\mu_1 - \mu_0) + trace\left(\Sigma_1^{-1}\Sigma_0\right) - ln\left(\frac{det\Sigma_0}{det\Sigma_1}\right) - K] \tag{2}$$

where K denotes the dimension of the data, μ represents the mean value, and Σ is the variance, *det* represents calculation of the determinant.

The similarity weight δ_s can be calculated by:

$$\delta_s = \frac{1/\overline{(KL[N_0, N_1] + \partial)^4}}{\Sigma_{i=1}^{m}\left(1/\left(\overline{KL[N_0, N_1] + \partial}\right)^4\right)} \tag{3}$$

where ∂ is the balancing coefficient and KL is the summed divergence of the distribution characteristics of the target subjects. The results show that instance transfer can effectively reduce the calibration time and can significantly improve the average classification accuracy of MI tasks.

Li et al. [34] proposed importance-weighted linear discriminant analysis (IWLDA) with bootstrap aggregation. They defined the ratio $r(x)$ of test and training input densities as transfer weight:

$$r(x) = \frac{P_{te}(x)}{P_{tr}(x)} \tag{4}$$

where P_{tr} and P_{te} represent the marginal probability distribution of the training set and the test set, respectively.

Then, they optimized the parameters of the LDA model by adding a regularization coefficient and transfer weights:

$$\min \sum_{i=1}^{N} r(x_i)(y_i - \hat{f}(x_i; \theta)^2) + \lambda\|\theta\| \tag{5}$$

where y_i refers to the target labels corresponding to the feature vectors x_i for i-th trials. Parameter θ is learned by least-squares.

$$\min \sum_{i=1}^{N} (y_i - \hat{f}(x_i; \theta))^2 \tag{6}$$

where

$$X = \begin{pmatrix} 1, x_1 \\ 2, x_2 \\ . \\ . \\ . \\ . \\ n, x_n \end{pmatrix} \tag{7}$$

The least-squares solution can be obtained by:

$$\hat{\theta}_{IWLDA} = \left(X^T DX + \lambda I\right)^{-1} X^T Dy \tag{8}$$

where λ (≥ 0) is the regularization parameter, D is the diagonal matrix with the i-th diagonal element, I is the identity matrix and $\hat{\theta}_{IWLDA}$ is the least-squares solution. They also combined the bagging method that independently constructs accurate and diverse base learners to improve the classification accuracy and to reduce the variance. The weighted parameters of the LDA model in the target domain can thus be optimized.

Covariate shift [35] is a common phenomenon in EEG processing across subjects/sessions. It is defined as follows: Given an input space X and an output space Y, the marginal distribution of D_s is inconsistent with D_T, i.e., $P_S(x) \neq P_T(x)$. However, the conditional distribution of the two domains is the same, $P_S(y/x) = P_T(y/x)$. Covariate shift obviously affects the unbiasedness of a model in standard model selection, which reduces the generalization ability of the machine model during EEG decoding [30].

To address this issue, research has proposed covariate shift adaptation. For example, Raza et al. [36] proposed a transductive learning model based on the k-nearest neighbor principle. They initialized the classifier using data from the calibration stage and trained the optimal classification boundary. Then, adaptation was executed to update the classifier. The updated rules are as follows:

First, the Euclidean distance is used to measure unlabeled and labeled data:

$$dist_{(p,q)} = \sqrt{\sum_{j=1}^{m}\left(q_j - p_j\right)^2} \tag{9}$$

where p and q refer to the unlabeled and labeled data points, respectively, and $dist$ is the Euclidean distance. Then, the k-nearest neighbors are selected based on the Euclidean distance. Next, this distance is converted to inverse form $dist_{inv(i)}$, which represents the corresponding pattern in the training database that is closer to the current unlabeled feature set.

$$dist_{inv(i)} = \frac{1}{d^i_{(q,p)} + \epsilon} \tag{10}$$

where i is the label and $\epsilon = 0.001$ is the bias. To decide if the current trial's features and estimated label should be added to the existing knowledge base, a confidence ratio CR is calculated:

$$CR_j = \frac{\sum_1^k dist_{inv(i)} \left(l(i) == j\right)}{\sum_1^k dist_{inv(i)}} \tag{11}$$

The CR index is calculated to predict the label for the unlabeled test data. The predicted test data are then added into the knowledge database, following which the decision boundary is recalculated to realize the update.

3.4.2. Transfer Learning Based on Feature Representation

TL based on feature representation can be achieved by reducing the difference between two domains by feature transformation or projecting the feature from two domains into the uniform feature space [37–39]. Unlike instance transfer, feature representation TL aims to encode the shared information across subjects/sessions into a feature representation. For example, spatial filtering and time–frequency transformation are used to transform the raw data into feature representations.

Nakanish et al. proposed a spatial filtering approach called the task-related component analysis (TRCA) method to enhance the reproducibility during SSVEP tasks and to improve the performance of an SSVEP-based BCI [40].

Suppose that two domain signals consist of two parts: A task-related signal $s(t)$ and a task-unrelated signal $z(t)$. A multichannel signal from $x(t)$ can be calculated as:

$$x_i(t) = a_{1,i}\,s(t) + a_{2,i}z(t),\ i = 1,\ 2,\ 3\ldots n \tag{12}$$

where i represents the number of channels and a refers to the project coefficients; 1 and 2 represent labels.

$$y(t) = x_i(t) \sum_{i=1}^{n} x(t) = \sum_{i=1}^{n} (a_{1,i}\,s(t) + a_{2,i}z(t)) \tag{13}$$

where $y(t)$ refers to the target data, and the optimization goal is to solve $a_{1,i} = 1$ and $a_{2,i} = 0$. The covariance between the $j_1 - th$ and the $j_2 - th$ trials is described as:

$$c_{j_1,j_2} = Cov\left(y^{(j_1)}(t), y^{(j_2)}(t)\right) = \sum_{i_1 i_2 = 1}^{n} w_{i_1} w_{i_2} Cov(x^{(j_1)}(t), x^{(j_2)}(t)) \tag{14}$$

All combinations of the trials are summed as:

$$\sum_{j_1,j_2 = 1, j_1 \neq j_2}^{N_t} c_{j_1,j_2} = \omega^T S \omega \tag{15}$$

where j represents the number of trials and ω refers to the spatial filters. Matrix s is defined as:

$$S_{i_1,i_2} = \sum_{i_1,i_2 = 1, i_1 \neq i_2}^{N_t} Cov\left(x^{j_1}_{i_1}(t), x^{j_2}_{i_2}(t)\right) \tag{16}$$

The variance of $y(t)$ is constrained to obtain a finite solution:

$$\mathrm{Var}\left(\sum(t)\right) = \omega^T Q \omega = 1 \tag{17}$$

The optimization is calculated as:

$$\widetilde{\omega} = \mathrm{argmax}\, \frac{\omega^T S \omega}{\omega^T Q \omega} \tag{18}$$

where $\widetilde{\omega}$ is the optimal spatial filter. Finally, the correlation coefficient is calculated by Pearson's correlation analysis between the data from the two domains. In their study, spatial filters as a feature representation were transferred to the target domain. The results showed that this method significantly improves the information transfer rates and classification accuracy. Based on this research, Tanaka [41] improved the TRCA method by maximizing the similarity across group of subjects, and they named this novel method group TRCA. The results showed that the group representation calculated by the group TRCA method achieve high consistency between two domains and offer effective data supplementation during brain–computer interaction.

CSP is a popular method for feature extraction, which is often used for MI classification. During calculation, a spatial filter is adopted to maximize the separation between the class variances of EEG. However, heterogeneous data across subjects/sessions causes poor classification performance of the model in the training stage. One feasible approach to solve the limitation is regularization. Lotte [42] presented regularized CSP to improve the classification accuracy across subjects. In their study, they discussed two strategies. One of them was regularizing the covariance matrix estimated. They can be, respectively, expressed as:

$$\widetilde{S}_i = (1 - \gamma)\widetilde{S}_i + \gamma I \tag{19}$$

$$\hat{S}_i = (1 - \beta)c_i S_i + \beta D_i \tag{20}$$

where S_i represents the initial spatial covariance matrix for class i, $\widetilde{S}_i$ is the regularized estimate, I is the identity matrix, c_i is a constant scaling parameter, and D_i represents the generic covariance matrix. The regularization parameters can be defined as γ and β. This strategy aims to optimize the covariance matrix by transforming other subjects' data into covariance combined with the regularization parameters and by transferring this feature to the target subject.

Another approach is regularizing the CSP objective function. CSP uses spatial filters ω to extremize the function:

$$J(\omega) = \frac{\omega^T C_1 \omega}{\omega^T C_2 \omega} \tag{21}$$

where C_i is spatial covariance matrix from class i. This approach optimizes CSP algorithms by regularizing the CSP objective function itself:

$$G_{P_1}(\omega) = \frac{\omega^T S_1 \omega}{\omega^T S_2 \omega + \partial P(\omega)} \tag{22}$$

where $P(\omega)$ represents a penalty function for the measurement distance between the spatial filter and the prior information. The goal of the objective function is to maximize $G_{P_1}(\omega)$ and to minimize $P(\omega)$. ∂ is a user-defined regularization parameter. The prior information from the source domain provides a good solution to guide the optimization direction of the estimation of spatial filters.

In addition, adaptation regularization is a typical feature TL method based on the structural risk minimization principle and the regularization theory. Cross-domain feature transfer is mainly operated by three methods: (1) Utilize the structural risk minimization principle and minimize the structural risk functional; (2) minimize the distribution difference between the joint probability distributions; (3) maximize the manifold consistency underlying the marginal distributions [43]. In recent research, Chen et al. [44] developed an efficient cross-subject TL framework for driving status detection. They used adaptation regularization to measure and reduce the difference of the features from the two domains and to extract the features by filtering algorithms. The results showed that this framework can achieve high recognition accuracy and good transfer ability.

3.4.3. Transfer Learning Based on Model Parameters

The assumption of model parameter TL is that individual models across subjects/sessions should share some parameters. The key step of this approach is to find shared parameter information and to realize knowledge transfer. The domain adaption (DA) of a classifier is the common method of model parameter transfer. The knowledge of the parameter information from D_s is reused and adjusted according to the prior distribution of D_T [45]. A DA method, named adaptive extreme learning machine (ELM), was proposed by Bamdadian et al. [46]. ELM is a single-hidden layer feedforward neural network, which determines the output weights by operating the inverse operation of the hidden layer weight matrices [47]. This method has two steps: First, the classifier is initialized by data from the calibration session. Then, the update rule for the output weight based on least-square minimization is calculated. The update rule is calculated as follows:

The initial output weight α can be defined as:

$$\alpha = H^+T = \varphi^{-1}H^TT \tag{23}$$

where H is the output matrix of hidden layer, $\varphi = H^TH$ and H^+ refer to the Moore–Penrose pseudo-inverse of H, and T represents the label category. The updated weight α^{m+1} is calculated as:

$$\alpha^{m+1} = \alpha^m + \varphi_{k+1}^{-1}H_{k+1}^T(T_{k+1} - H_{k+1}\alpha^m) \tag{24}$$

$$\varphi_{k+1} = \varphi_k + H_{k+1}^TH_{k+1} \tag{25}$$

$$\varphi_{k+1}^{-1} = \varphi_k^{-1} - \varphi_k^{-1}H_{k+1}^T\left[I + H_{k+1}\varphi_k^{-1}H_{k+1}^T\right]H_{k+1}\varphi_k^{-1} \tag{26}$$

where k is k-th hidden node, φ is orthogonal matrix calculated by H. The experiential results showed that adaptive ELM can significantly improve the classification accuracy in MI classification across subjects.

Another strategy is ensemble learning, which combines multiple weak classifiers from the source domain into a strong classifier. Dalhoumi et al. [48] proposed a novel ensemble strategy based on Bayesian model averaging. They calculated the probability of having a class label y_{q+1} given a feature vector h_{q+1}:

$$P(y_{q+1}/x_{q+1}) = \sum_{n=1}^{N} P\left(\frac{y}{x_{q+1}^n, j_n}\right)P\left(\frac{j_n}{T}\right) \tag{27}$$

where x_{q+1}^n is the logarithmic variance feature vector, j_n is a set of hypotheses from the source domain, and T is the test set. The hypothesis prior $P\left(\frac{j_n}{T}\right)$ is estimated in the following method:

$$w^* = argmin \sum_{p=1}^{p} l\left(\sum_{n=1}^{N} j^n(x_p^n)y_p\right) \tag{28}$$

$$p\left(\frac{j_n}{T}\right) = w_n^* \tag{29}$$

where x_p^k is the projection of the feature vector x on the spatial filters of subject k. The learned ensemble classifier can be used to predict labels for the target user:

$$h^* = \sum_{n=1}^{N} w_n^* j_n \tag{30}$$

The results showed that this ensemble strategy can improve the classification performance in small-scale EEG data by evaluation on a real dataset.

In recent years, deep neural networks have provided good results for the processing of EEG signals [49,50]. Due to their end-to-end model structure and automatic feature extraction ability, deep neural networks minimize the interference of redundant information and improve the classification performance. Inspired by computer vision, a deep neural network learns generic feature representations by lower layers of the model. Specific feature representations with the relevant specific subjects or sessions are learned by the high layer [51]. Therefore, freezing lower layers and fine-tuning higher layers is a good way to realize model parameter transfer based on deep learning.

Zhao et al. [52] proposed an end-to-end deep convolution network for MI classification. To avoid the limitation of a small sample and overfitting, they utilized the data from D_s to pre-train the source network and to transfer the parameters of several layers to initialize the target network. First, the network was pre-trained using data from the source domain. Then, they used the M source subjects W^s to initialize the nth layer's target network by a weight average:

$$W_n^t = \sum_{m=1}^{M} \rho_m W_{mn}^s \tag{31}$$

where ρ represents the strength of the source network and W_{mn}^s refers to the connecting weights of the nth layer to the next layer. The next stage is to fine-tune the target initialized network by data from D_T. The results showed that the parameter transfer strategy can reduce the calibration time for new subjects and can help the deep convolution network to obtain better classification performance.

Raghu et al. used CNN combined with TL to recognize epileptic seizures [53]. They proposed two different transfer methods: To finetune a pre-trained network and then extract image features by said pre-trained network, and to classify the status of brain using an SVM. Popular networks such as Alexnet, VGG16net, VGG19net, and Squeezenet, were used to verify the performance of the proposed framework.

The summary of collected studies is shown in Table 3.

Table 3. Summary of transfer learning for EEG decoding.

Pattern	Reference	Type	Transfer Method	Feature Extraction	Datasets	Results
MI	[33]	ITL	Similarity measurement with KL divergence	LR+CSP	19 subjects' data, BCIC-IV-2a, and BCIC-III-4a	70.3% and 75% and 75%
MI	[54]	ITL	Informative subspace transferring and selective ITL with active learning	LDA	BCIC-IV-2b	\
MI	[55]	FTL	Ensemble learning and adaptive learning	LDA	NIPS	68.1%
MI	[56]	ITL	Similarity measurement with Jensen Shannon ratio and rule adaptation TL	CSP+LDA	BCIC-IV-2a	77%
MRP	[57]	ITL	MMD and regularized discriminative spatial pattern	Linear RR	BCIC-I-1 and BCIC-II-4	\
P300	[58]	ITL	Ensemble learning generic information	Bayesian LDA	8 participants' data	62.5%
SSVEP	[59]	ITL	Variability assessment Fisher's discriminant ratios	Cluster	8 subjects' data	\
P300	[60]	ITL	Dynamically adjusts weights of instances	Liner SVM	BCIC-III-2 and dataset of P300 speller	74.9%
MI	[61]	ITL	Selective informative with normalized entropy	LDA	BCIC-IV-2b	75.8%
MI	[62]	ITL	Selective informative expected decision boundary	LDA	BCIC-IV-2a	75.6%
MI	[63]	MTL	Domain adaptation parallel transport on the cone manifold of SPD	Linear SVM	BCIC-IV-2a	\
DD	[64]	ITL	Selective transfer learning	Linear regression	15 subjects' data	66%
MI	[65]	ITL	Composite local temporal correlation CSP Frobenius distance	Liner quadratic Mahalanobis	BCIC-III-IVa	89.21%
SSVEP	[66]	ITL	Ensemble learning and similarity measurement with mutual information	LDA	10 healthy subjects' data	\
Cognitive detection	[67]	ITL	Similarity measurement by Pearson's correlation coefficient	SVM	28 subjects' data	87.6%
ERP	[68]	FTL	Probabilistic zero framework	Unsupervised adaptation	Akimpech dataset	\

Table 3. *Cont.*

Pattern	Reference	Type	Transfer Method	Feature Extraction	Datasets	Results
MI	[69]	FTL	DA with power spectral density	CNN	BCIC-IV-2a	\
MI	[70]	FTL	Many-objective optimization	Linear SVM	BCIC-III-IVa	75.8%
VEP	[71]	FTL	Active semi-supervised TL	SVM	14 subjects' experiments	\
MI	[72]	FTL	Adaptive Selective CSP	Discriminant analysis	6 participant experiments	61%
MI	[34]	FTL	CSA	Importance-weighted LDA	BCIC-III dataset	79.1%
MI	[73]	ITL	Instance TL based p-hash	CNN	BCIC-IV-2b	\
MI	[74]	FTL	Informative TL with AL	LDA	BCIC-IV-2a	67.7%
MI	[75]	MTL	Modifications of CSP	SVM	BCIC-III-IVa	\
MI	[76]	MTL	k-nearest neighbors principle	SVM+LDA	BCIC-IV-b	\
ER	[77]	FTL	Transfer recursive feature elimination	Least-squares SVM	DEAP dataset	78%
SSVEP	[78]	MTL	Least-squares transformation	\	8 participant experiments	82.1%
MI	[79]	FTL	Domain transfer multiple kernel boosting	SVM	BCIC-III-Iva 5 subjects' data	81.6% 76%
SSVEP	[80]	FTL	Spatial filtering transfer	TRCA	10 subjects' data	\
SSVEP	[81]	FTL	Reference template transfer	MestCCA	10 subjects' data	\
SSVEP	[82]	FTL	Reference template transfer	Transfer template-CCA	12 subjects' data	85%
SSVEP	[83]	FTL	Reference template transfer	Adaptive combined-CCA	10 subjects' data	83%
MI	[84]	FTL	Fuzzy TL based on generalized hidden-mapping RR	SVM	BCIC-IV-2a	89.3%
MI	[46]	MTL	Adaptive extreme learning machine	SVM+ELM	12 subjects' data	71.8%
MI	[48]	MTL	Classifier ensemble	LDA	BCIC-IV-2a	\
MI	[13]	FTL	Regularized CSP with TL	LDA	BCIC-III-IVa	78.9%
Multitask	[85]	FTL	Geometrical transformations on Riemannian Procrustes analysis	/	8 publicly available BCI datasets	\

Table 3. *Cont.*

Pattern	Reference	Type	Transfer Method	Feature Extraction	Datasets	Results
ERP	[86]	FTL	Spectral transfer using information geometry	MDRM	15 subjects' data	62%
MI	[87]	FTL	Space adaptation	LDA	BCIC-IV-2a	77.5%
MI	[88]	FTL	Feature space transformation	LDA	PhysioNet datasets	72%
MI	[89]	FTL	Tangent space-based TL	LDA	BCIC-IV-2a	\
ER	[90]	FTL	Transfer component analysis and kernel principle component analysis	SVM	SEED	77.96%
MI	[91]	FTL	Transfer kernel CSP	SVM	BCIC-III-IVa	81.14%
MI/ ERP	[92]	FTL	Affine transform	Minimum distance mean and Bayesian classifier	BCIC-IV-2a and Brain Invaders experiment	\
SSVEP	[93]	ITL	Riemannian similarities	Bootstrapping	12 subjects' data	80.9% & 81.3%
MI	[94]	FTL	Multitask learning	RR+SVM	10 healthy subjects' data and an ALS subject's data	85%
Imagined speech	[95]	ITL	Inductive transfer learning	Naïve Bayesian classifier	27 subjects' data	68.9%
MI	[96]	FTL	Transferable discriminative dimensionality reduction	KNN+SVM	5 subjects' data	74.4%
MI	[7]	FTL	Nonstationary information transfers	LDA	5 subjects' data and BCIC-III-IVa	80.4%
MI	[97]	MTL	Fine-tuned based on VGG16	CNN	BCIC-IV-2b	74.2%
MI	[98]	MTL	Fine-tuned based on pre-trained network	CNN	BCIC-IV-2a	69.71%
ErrPs	[99]	MTL	Fine-tuned based on pre-trained network	CNN	15 epilepsy patients' data	81.50%
Music Imagination	[100]	MTL	Fine-tuned based on AlexNet	LSTM	OpenMIIR dataset	30.83%
ErrPs	[101]	MTL	Fine-tuned based on pre-trained network	CNN	31 subjects' data	84.1%
DD	[102]	ITL	Source domain selection	Weighted adaptation regularization	16 subjects 'data	\

Table 3. *Cont.*

Pattern	Reference	Type	Transfer Method	Feature Extraction	Datasets	Results
Attention detection	[103]	ITL	Subject adaptation	CNN	8 subjects 'data	84.17%
MI	[52]	ITL	Subject transfer	CNN	BCIC-IV2a and BCIC-IV-2b	0.56 and 0.65 (MK)
MI	[104]	MTL	Fine-tuned based on pre-trained network	RBM	BCIC-IV2a and 12 subjects' data	88.9%
P300	[105]	MTL	Fine-tuned based on pre-trained network	CNN	BCIC-III-2	90.5%
MI	[106]	MTL	Fine-tuned based on pre-trained network	CNN	BCIC-IV-2b	0.57 (MK)
MI	[107]	MTL	Fine-tuned based on pre-trained network	Conditional variational autoencoder	PhysioNet datasets	73%
Music Imagination	[108]	FTL	Cross-domain encoder	Attention decoder-RNN	OpenMIIR datasets	37.9%
MI	[36]	ITL	Covariate shift detection and adaptation	Linear SVM	BCIC-IV-2a BCIC-IV-2b	73.8% and 69.7%
MWA	[109]	FTL	Cross-subject statistical shift	Random forest	9 subjects' data	\
MI	[110]	MTL	Fine-tuned based on multiple network	CNN	BCIC-IV-2a and BCIC-IV-2b	\
MI	[111]	MTL	Four-strategy model transfer learning	Deep neural network	BCIC-IV-2a	\
MI	[112]	ITL	Subject–subject transfer	CNN	3 subjects' data	\
P300	[113]	ITL	Subject–subject transfer	Linear SVM	22 subjects' data	68.7%
ER	[114]	ITL	Measurement on Riemannian geometry instance transfer	SVM	MDME and SDMN datasets	\
DD	[40]	FTL	Adaptation regularization based TL	Multiple classifier	23 subjects' data and NIH physio bank	89.59%
ED	[53]	MTL	Fine-tuned based on pre-trained network	GoogLeNet and Inception v3	TUH open-source database	82.85% and 88.30%
ED	[115]	MTL	Fine-tuned based on pre-trained network	CNN and bidirectional LSTM	CHB-MIT EEG dataset	99.6%

Table 3. *Cont.*

Pattern	Reference	Type	Transfer Method	Feature Extraction	Datasets	Results
ED	[116]	MTL	Fine-tuned based on pre-trained network	CNN	CHB-MIT EEG dataset	92.7%
MI	[117]	FTL	Spatial filtering transfer and matrix decomposition	ELM	BCIC-III-IVa and BCIC-IV-1	89% 62%
SSVEP	[41]	FTL	Spatial filtering transfer	Group TRCA	Benchmark dataset	\
MI	[118]	MTL	Adversarial inference	CNN	52 subjects' data	\
ER	[119]	FTL	Power spectral density feature	Polynomial/Gaussian kernels/ naïve Bayesian SVM	DEAP, MAHNOB-HCI, and DREAMER	\
MWA	[120]	FTL	Ensemble learning	Stacked denoising autoencoder	8 subjects' data	92%
MI	[121]	FTL	Data mapping and ensemble learning	LDA	BCIC-IV-2a	0.58 (MK)
MI	[122]	FTL	Center-based discriminative feature learning	CNN	BCIC-III-IVa	76%

"\" represents that there is no specific description or else multiple descriptions for the results.

4. Discussion

Based on the numerous papers surveyed herein, we briefly summarized the development of the application of TL to EEG decoding. This will help researchers scan the status of this field and receive useful guidance in future work.

According to the various studies surveyed in this paper, it is not hard to determine the points of interest that researchers focus on. As shown in Figure 2, more studies have focused on active BCI (i.e., MI, SSVEP, and ERP) among these different EEG paradigms. One possible explanation is that the goal of these mental activity decoding studies is to categorize EEG from different classes. This would allow many machine learning methods to be applied to this paradigm. From Table 2, it can be seen that the application scenarios of TL in the existing literature have focused almost only on classification and regression tasks.

The method of model parameter transfer is not applicable to only a few subjects with initially low BCI performance. The feature of EEG from these subjects exhibits inseparability in feature space. Therefore, the parameter optimization of the classifier does not significantly improve the classification results. It is worth noting that the adaptive strategy of the classifier should be considered a supplement to achieve the goal of a calibration-free mode of operation [123]. The combination of TL and the adaptive strategy may receive increasing attention in future studies.

It is also worth noting that TL showed good results across subjects/experiments, but the detail of variability across sessions/subjects was unclear. Some studies proposed that the Bayesian model is a promising approach to capture variability. This model is built based on multitask learning, and variation in some features is often extracted, such as spectral and spatial [124,125].

Due to its end-to-end structure and competitive performance, deep learning has been successful in processing EEG data [126]. However, the computational power and small-scale data are a limitation during practical operation. A hybrid structure based on TL and deep learning is a promising way to address this issue. For example, one of the methods is fine-tuning the pre-trained network, which has proven to be effective. With the development of deep learning technology, the research for such a hybrid structure is still a hot topic for future research.

As reported in the above-cited studies, TL is instrumental in EEG decoding across subjects/sessions. However, knowledge transfer across tasks/device is still a blank field. This issue is worth exploring and will make EEG-based BCI systems much more practical.

5. Conclusions

In this paper, we reviewed the research on TL for EEG decoding that was published between 2010 and 2020. We discussed numerous approaches that can be divided into three categories: Instance transfer, feature representation transfer, and parameter of classifier transfer. Based on the summary of their results, we can conclude that TL can effectively improve the decoding performance in classification and regression tasks. In addition, TL provides adequate performance in initializing BCI systems for a new subject, which reduces the length of time of the calibration process. Although there are some limitations for using TL for EEG decoding, such as the scope of application of TL and suboptimal performance on some occasions, TL shows strong robustness. Overall, TL is instrumental in EEG decoding across subjects/sessions. In addition, achieving a calibration-free model of operation and higher accuracy of decoding are worthy of further research.

Author Contributions: K.Z. and G.X. designed the study. K.Z. wrote the manuscript, X.Z., H.L. collected the relevant papers. S.Z., Y.Y., R.L. prepared the figures. All authors have read and agreed to the published version of the manuscript.

Funding: Research supported by National Key Research & Development Plan of China (Grant No.2017YFC1308500), GDAS' Project of Science and Technology Development (Grant No. 2019GDASYL-0502002) and Key Research & Development Plan of Shaanxi Province (Grant No. 2018ZDCXL-GY-06-01).

Acknowledgments: The authors would like to thank the support by Guangdong Institute of Medical Instruments & National Engineering Research Center for Healthcare Devices.

Conflicts of Interest: The authors declare that the research was conducted in the absence of any commercial or financial relationships that could be construed as a potential conflict of interest.

Abbreviations: List of Acronyms

AL	Active learning
ALS	Amyotrophic Lateral Sclerosis
ATL	Active Transfer Learning
BCIC	Brain Computer Interface Competition
CCA	Canonical Correlation Analysis
CNN	Convolution Neural Network
CSA	Covariate Shift Adaptation
CSP	Common Spatial Pattern
DA	Domain Adaptation
DD	Drowsiness Detection
ELM	Extreme Learning Machine
ED	Epileptic Detection
ER	Emotion Recognition
ERP	Event-Related Potential
ErrPs	Electroencephalography -measured error-related potentials
FTL	Feature Transfer Learning
ITL	Instance Transfer Learning
JSR	Jensen Shannon Ratio
KL	Kullback–Leibler
KNN	K-Nearest Neighbor
LDA	Linear Discriminate Analysis
LR	Logistic Regression
LSTM	Long Short-Term Memory
MDRM	Minimum Distance to Riemannian Mean classifiers
MI	Motor Imagery
MMD	Maximum mean discrepancy
MK	Means Kappa value
MTL	Model Transfer Learning
MWA	Mental-Workload Assessment
MRP	Movement Related Potentials
RNN	Recurrent Neural Network
RR	Ridge Regression
RBM	Restricted Boltzmann Machine
SMR	Sensory Motor Rhythm
SVM	Support Vector Machine
SSVEP	Steady State Visual Evoked Potential
SELM	Sigmoid Extreme Learning Machine
SPD	Symmetric Positive Definite
TL	Transfer Learning
TRCA	Task-Related Component Analysis
VEP	Visual Evoked Potential
VGG	Visual Geometry Group

References

1. Yuan, H.; He, B. Brain–Computer Interfaces Using Sensorimotor Rhythms: Current State and Future Perspectives. *IEEE Trans. Biomed. Eng.* **2014**, *61*, 1425–1435. [CrossRef] [PubMed]
2. Subasi, A.; Gursoy, M.I. EEG signal classification using PCA, ICA, LDA and support vector machines. *Expert Syst. Appl.* **2010**, *37*, 8659–8666. [CrossRef]

3. Lotte, F.; Bougrain, L.; Cichocki, A.; Clerc, M.; Congedo, M.; Rakotomamonjy, A.; Yger, F. A review of classification algorithms for EEG-based brain-computer interfaces: A 10 year update. *J. Neural Eng.* **2018**, *15*, 031005. [CrossRef] [PubMed]

4. Jiao, Y.; Zhang, Y.; Chen, X.; Yin, E.; Jin, J.; Wang, X.; Cichocki, A. Sparse Group Representation Model for Motor Imagery EEG Classification. *IEEE J. Biomed. Health Inf.* **2018**, *23*, 631–641. [CrossRef] [PubMed]

5. Krepki, R.; Blankertz, B.; Curio, G.; Müller, K.-R. The Berlin Brain Computer Interface (BBCI)—Towards a new communication channel for online control in gaming applications. *Multimed. Tools Appl.* **2017**, *33*, 73–90. [CrossRef]

6. Pan, S.J.; Yang, Q. A Survey on Transfer Learning. *IEEE Trans. Knowl. Data Eng.* **2009**, *22*, 1345–1359. [CrossRef]

7. Samek, W.; Meinecke, F.; Muller, K.-R. Transferring Subspaces between Subjects in Brain-Computer Interfacing. *IEEE Trans. Biomed. Eng.* **2013**, *60*, 2289–2298. [CrossRef] [PubMed]

8. Lecun, Y.; Bengio, Y.; Hinton, G. Deep learning. *Nature* **2015**, *521*, 436–444. [CrossRef] [PubMed]

9. Bonassi, G.; Biggio, M.; Bisio, A.; Ruggeri, P.; Bove, M.; Avanzino, L. Provision of somatosensory inputs during motor imagery enhances learning-induced plasticity in human motor cortex. *Sci. Rep.* **2017**, *7*, 1–10. [CrossRef] [PubMed]

10. Pfurtscheller, G.; Neuper, C. Motor imagery and direct brain-computer communication. *Proc. IEEE* **2001**, *89*, 1123–1134. [CrossRef]

11. Ahn, M.; Jun, S.C. Performance variation in motor imagery brain–computer interface: A brief review. *J. Neurosci. Methods* **2015**, *243*, 103–110. [CrossRef]

12. Zhang, K.; Xu, G.; Han, Z.; Ma, K.; Zheng, X.; Chen, L.; Duan, N.; Zhang, S. Data Augmentation for Motor Imagery Signal Classification Based on a Hybrid Neural Network. *Sensors* **2020**, *20*, 4485. [CrossRef]

13. Cheng, M.; Gao, X.; Gao, S.; Xu, D. Design and Implementation of a Brain-Computer Interface With High Transfer Rates. *IEEE Trans. Biomed. Eng.* **2002**, *49*, 1181–1186. [CrossRef]

14. Handy, T.C. (Ed.) *Event-Related Potentials: A Methods Handbook*; The MIT Press: Boston, MA, USA, 2005.

15. Zander, T.O.; Kothe, C. Towards passive brain–computer interfaces: Applying brain–computer interface technology to human–machine systems in general. *J. Neural Eng.* **2011**, *8*, 025005. [CrossRef]

16. Arico, P.; Borghini, G.; Di Flumeri, G.; Sciaraffa, N.; Colosimo, A.; Babiloni, F. Passive BCI in Operational Environments: Insights, Recent Advances, and Future Trends. *IEEE Trans. Biomed. Eng.* **2017**, *64*, 1431–1436. [CrossRef]

17. Cui, Y.; Xu, Y.; Wu, D. EEG-Based Driver Drowsiness Estimation Using Feature Weighted Episodic Training. *IEEE Trans. Neural Syst. Rehabi. Eng.* **2019**, *27*, 2263–2273. [CrossRef]

18. Muehl, C.; Allison, B.; Nijholt, A.; Chanel, G. A survey of affective brain computer interfaces: Principles, state-of-the-art, and challenges. *Brain-Comput. Interfaces* **2014**, *1*, 66–84. [CrossRef]

19. Blankertz, B.; Curio, G.; Müller, K. Classifying Single Trial EEG: Towards Brain Computer Interfacing. In *Advances in Neural Information Processing Systems 14 (NIPS 01)*; Diettrich, T.G., Becker, S., Ghahramani, Z., Eds.; The MIT Press: Cambridge, MA, USA, 2002.

20. Wolpaw, J.R.; Birbaumer, N.; McFarland, D.J.; Pfurtscheller, G.; Vaughan, T.M. Brain–computer interfaces for communication and control. *Clin. Neurophysiol.* **2002**, *113*, 767–791. [CrossRef]

21. Dornhege, G.; Blankertz, B.; Curio, G.; Müller, K. Boosting bit rates in non-invasive EEG single-trial classifications by feature combination and multi-class paradigms. *IEEE Trans. Biomed. Eng.* **2004**, *51*, 993–1002. [CrossRef] [PubMed]

22. Naeem, M.; Brunner, C.; Leeb, R.; Graimann, B.; Pfurtscheller, G. Seperability of four-class motor imagery data using independent components analysis. *J. Neural Eng.* **2006**, *3*, 208–216. [CrossRef]

23. Leeb, R.; Lee, F.; Keinrath, C.; Scherer, R.; Bischof, H.; Pfurtscheller, G. Brain–Computer Communication: Motivation, Aim, and Impact of Exploring a Virtual Apartment. *IEEE Trans. Neural Syst. Rehabil. Eng.* **2007**, *15*, 473–482. [CrossRef]

24. Riccio, A.; Esimione, L.; Eschettini, F.; Epizzimenti, A.; Inghilleri, M.; Belardinelli, M.E.; Mattia, D.; Cincotti, F. Attention and P300-based BCI performance in people with amyotrophic lateral sclerosis. *Front. Hum. Neurosci.* **2013**, *7*, 732. [CrossRef] [PubMed]

25. Koelstra, S.; Muhl, C.; Soleymani, M.; Lee, J.-S.; Yazdani, A.; Ebrahimi, T.; Pun, T.; Nijholt, A.; Patras, I. DEAP: A Database for Emotion Analysis; Using Physiological Signals. *IEEE Trans. Affect. Comput.* **2011**, *3*, 18–31. [CrossRef]

26. Li, Y.; Koike, Y.; Sugiyama, M. A Framework of Adaptive Brain Computer Interfaces. In Proceedings of the International Conference on Biomedical Engineering & Informatics, Tianjin, China, 17–19 October 2009.

27. Zheng, W.-L.; Lu, B.-L. Investigating Critical Frequency Bands and Channels for EEG-Based Emotion Recognition with Deep Neural Networks. *IEEE Trans. Auton. Ment. Dev.* **2015**, *7*, 162–175. [CrossRef]

28. Stober, S.; Sternin, A.; Owen, A.M.; Grahn, J.A. Towards music imagery information retrieval: Introducing the openmiir dataset of eeg recordings from music perception and imagination. In Proceedings of the 16th ISMIR Conference, Malaga, Spain, 26–30 October 2015; pp. 763–769.

29. Shoeb, A. Application of Machine Learning to Epileptic Seizure Onset Detection and Treatment. Ph.D. Thesis, Massachusetts Institute of Technology, Cambridge, MA, USA, 2009.

30. Masashi, S.; Krauledat, M.; MÃžller, K. Covariate shift adaptation by importance weighted cross validation. *J. Mach. Learn. Res.* **2007**, *8*, 985–1005.

31. Khan, M.; Heisterkamp, D.R. Adapting instance weights for unsupervised domain adaptation using quadratic mutual information and subspace learning. In Proceedings of the 2016 23rd International Conference on Pattern Recognition (ICPR), Cancun, Mexico, 4–8 December 2016; pp. 1560–1565.

32. Zadrozny, B. Learning and evaluating classifiers under sample selection bias. In Proceedings of the Twenty-First International Conference on Machine Learning—ICML '04; Association for Computing Machinery (ACM): New York, NY, USA, 2004; p. 114.

33. Azab, A.M.; Mihaylova, L.; Ang, K.K.; Arvaneh, M. Weighted Transfer Learning for Improving Motor Imagery-Based Brain–Computer Interface. *IEEE Trans. Neural Syst. Rehabil. Eng.* **2019**, *27*, 1352–1359. [CrossRef]

34. Li, Y.; Kambara, H.; Koike, Y.; Sugiyama, M. Application of covariate shift adaptation techniques in brain–computer interfaces. *IEEE Trans. Biomed. Eng.* **2010**, *57*, 1318–1324.

35. Shimodaira, H. Improving predictive inference under covariate shift by weighting the log-likelihood function. *J. Stat. Plan. Inference* **2000**, *90*, 227–244. [CrossRef]

36. Raza, H.; Cecotti, H.; Li, Y.; Prasad, G. Adaptive learning with covariate shift-detection for motor imagery-based brain–computer interface. *Soft Comput.* **2016**, *20*, 3085–3096. [CrossRef]

37. Liu, J.; Shah, M.; Kuipers, B.; Savarese, S. Cross-view action recognition via view knowledge transfer. In Proceedings of the CVPR 2011, Providence, RI, USA, 20–25 June 2011; pp. 3209–3216.

38. Zheng, V.W.; Pan, S.J.; Yang, Q.; Pan, J.J. Transferring ulti-device localization models using latent multi-task learning. In Proceedings of the Twenty-Third AAAI Conference on Artificial Intelligence, Chicago, IL, USA, 13–17 July 2008; Volume 8, pp. 1427–1432.

39. Pan, S.J.; Tsang, I.W.; Kwok, J.T.; Yang, Q. Domain Adaptation via Transfer Component Analysis. *IEEE Trans. Neural Netw.* **2010**, *22*, 199–210. [CrossRef]

40. Nakanishi, M.; Wang, Y.; Chen, X.; Wang, Y.T.; Gao, X.; Jung, T.P. Enhancing Detection of SSVEPs for a High-Speed Brain Speller Using Task-Related Component Analysis. *IEEE Trans. Biomed. Eng.* **2017**, *65*, 104–112. [CrossRef]

41. Tanaka, H. Group task-related component analysis (gTRCA): A multivariate method for inter-trial reproducibility and inter-subject similarity maximization for EEG data analysis. *Sci. Rep.* **2020**, *10*, 1–17. [CrossRef]

42. Lotte, F.; Guan, C. Regularizing Common Spatial Patterns to Improve BCI Designs: Unified Theory and New Algorithms. *IEEE Trans. Biomed. Eng.* **2011**, *58*, 355–362. [CrossRef]

43. Long, M.; Wang, J.; Ding, G.; Pan, S.J.; Yu, P.S. Adaptation Regularization: A General Framework for Transfer Learning. *IEEE Trans. Knowl. Data Eng.* **2013**, *26*, 1076–1089. [CrossRef]

44. Chen, L.; Zhang, A.; Lou, X. Cross-subject driver status detection from physiological signals based on hybrid feature selection and transfer learning. *Expert Syst. Appl.* **2019**, *137*, 266–280. [CrossRef]

45. Zhang, B.; Lu, J.; Peitao, W.; Tang, Z. A review on transfer learning for brain-computer interface classification. In Proceedings of the 5th International Conference on Information Science and Technology (ICIST), Changsha, China, 24–26 April 2015.

46. Bamdadian, A.; Guan, C.; Ang, K.K.; Xu, J. Improving session-to-session transfer performance of motor imagery-based BCI using adaptive extreme learning machine. In Proceedings of the 35th Annual International Conference of the IEEE Engineering in Medicine and Biology Society (EMBC), Osaka, Japan, 3–7 July 2013; pp. 2188–2191.

47. Huang, G.-B.; Zhu, Q.-Y.; Siew, C.-K. Extreme learning machine: Theory and applications. *Neurocomputing* **2006**, *70*, 489–501. [CrossRef]

48. Dalhoumi, S.; Dray, G.; Montmain, J. Knowledge Transfer for Reducing Calibration Time in Brain-Computer Interfacing. In Proceedings of the IEEE International Conference on Tools with Artificial Intelligence—ICTAI, Limassol, Cyprus, 10–12 November 2014.

49. Dai, M.; Zheng, D.; Na, R.; Wang, S.; Zhang, S. EEG Classification of Motor Imagery Using a Novel Deep Learning Framework. *Sensors* **2019**, *19*, 551. [CrossRef]

50. Lu, N.; Li, T.; Ren, X. A Deep Learning Scheme for Motor Imagery Classification based on Restricted Boltzmann Machines. *IEEE Trans. Neural Syst. Rehabil. Eng.* **2016**, *25*, 566–576. [CrossRef]

51. Krizhevsky, A.; Sutskever, I.; Hinton, G.E. ImageNet Classification with Deep Convolutional Neural Networks. *Commun. ACM* 2012. [CrossRef]

52. Zhao, D.; Tang, F.; Si, B.; Feng, X. Learning joint space–time–frequency features for EEG decoding on small labeled data. *Neural Netw.* **2019**, *114*, 67–77. [CrossRef]

53. Raghu, S.; Sriraam, N.; Temel, Y.; Rao, S.V.; Kubben, P.L. EEG based multi-class seizure type classification using convolutional neural network and transfer learning. *Neural Netw.* **2020**, *124*, 202–212. [CrossRef] [PubMed]

54. Hossain, I.; Khosravi, A.; Nahavandhi, S. Active transfer learning and selective instance transfer with active learning for motor imagery based BCI. In Proceedings of the 2016 International Joint Conference on Neural Networks (IJCNN), Vancouver, BC, Canada, 24–29 July 2016.

55. Tu, W.; Sun, S. A subject transfer framework for EEG classification. *Neurocomputing* **2012**, *82*, 109–116. [CrossRef]

56. Giles, J.; Ang, K.K.; Mihaylova, L.S.; Arvaneh, M. A Subject-to-subject Transfer Learning Framework Based on Jensen-shannon Divergence for Improving Brain-computer Interface. In Proceedings of the ICASSP 2019-2019 IEEE International Conference on Acoustics, Speech and Signal Processing (ICASSP), Brighton, UK, 12–17 May 2019; pp. 3087–3091.

57. Wang, P.; Lu, J.; Lu, C.; Tang, Z. An algorithm for movement related potentials feature extraction based on transfer learning. In Proceedings of the 2015 5th International Conference on Information Science and Technology (ICIST), Changsha, China, 24–26 April 2015; pp. 309–314.

58. Adair, J.; Brownlee, A.; Daolio, F.; Ochoa, G. Evolving training sets for improved transfer learning in brain computer interfaces. In *International Workshop on Machine Learning, Optimization, and Big Data*; Springer: Cham, Switzerland, 2017; pp. 186–197.

59. Wei, C.-S.; Nakanishi, M.; Chiang, K.-J.; Jung, T.-P. Exploring Human Variability in Steady-State Visual Evoked Potentials. In Proceedings of the 2018 IEEE International Conference on Systems, Man, and Cybernetics (SMC), Miyazaki, Japan, 7–10 October 2018; pp. 474–479.

60. Hou, J.; Li, Y.; Liu, H.; Wang, S. Improving the P300-based brain-computer interface with transfer learning. In Proceedings of the2017 8th International IEEE/EMBS Conference on Neural Engineering (NER), Shanghai, China, 25–28 May 2017; pp. 485–488.

61. Hossain, I.; Khosravi, A.; Hettiarachchi, I.T.; Nahavandhi, S. Informative instance transfer learning with subject specific frequency responses for motor imagery brain computer interface. In Proceedings of the 2017 IEEE International Conference on Systems, Man, and Cybernetics (SMC), Banff, AB, Canada, 5–8 October 2017; Volume 2017, pp. 252–257.

62. Hossain, I.; Khosravi, A.; Hettiarachchi, I.; Nahavandi, S. Multiclass informative instance transfer learning framework for motor imagery-based brain-computer interface. *Comput. Intell. Neurosci.* **2018**, *2018*, 6323414. [CrossRef] [PubMed]

63. Yair, O.; Ben-Chen, M.; Talmon, R. Parallel Transport on the Cone Manifold of SPD Matrices for Domain Adaptation. *IEEE Trans. Signal Process.* **2019**, *67*, 1797–1811. [CrossRef]

64. Wei, C.-S.; Lin, Y.-P.; Wang, Y.-T.; Jung, T.-P.; Bigdely-Shamlo, N.; Lin, C.-T. Selective Transfer Learning for EEG-Based Drowsiness Detection. In Proceedings of the 2015 IEEE International Conference on Systems, Man, and Cybernetics, Kowloon, China, 9–12 October 2015; pp. 3229–3232.

65. Hatamikia, S.; Nasrabadi, A.M. Subject transfer BCI based on composite local temporal correlation common spatial pattern. *Comput. Boil. Med.* **2015**, *64*, 1–11. [CrossRef]

66. Sybeldon, M.; Schmit, L.; Akcakaya, M. Transfer Learning for SSVEP Electroencephalography Based Brain–Computer Interfaces Using Learn++. NSE and Mutual Information. *Entropy* **2017**, *19*, 41. [CrossRef]

67. Wei, C.S.; Lin, Y.P.; Wang, Y.T.; Lin, C.T.; Jung, T.P. Transfer learning with large-scale data in brain-computer interfaces. In Proceedings of the 2016 38th Annual International Conference of the IEEE Engineering in Medicine and Biology Society (EMBC), Orlando, FL, USA, 16–20 August 2016; pp. 4666–4669.

68. Kindermans, P.J.; Tangermann, M.; Müller, K.R.; Schrauwen, B. Integrating dynamic stopping, transfer learning and language models in an adaptive zero-training ERP speller. *J. Neural Eng.* **2014**, *11*, 035005. [CrossRef]

69. Jeon, E.; Ko, W.; Suk, H.I. Domain Adaptation with Source Selection for Motor-Imagery based BCI. In Proceedings of the 2019 7th International Winter Conference on Brain-Computer Interface (BCI), Gangwon, Korea, 18–20 February 2019; pp. 1–4.

70. Pal, M.; Bandyopadhyay, S.; Bhattacharyya, S. A Many Objective Optimization Approach for Transfer Learning in EEG Classification. *arXiv* **2019**, arXiv:1904.04156.

71. Wu, D.; Lance, B.; Lawhern, V. Transfer learning and active transfer learning for reducing calibration data in single-trial classification of visually-evoked potentials. In Proceedings of the 2014 IEEE International Conference on Systems, Man, and Cybernetics (SMC), San Diego, CA, USA, 5–8 October 2014; pp. 2801–2807.

72. Jin, Y.; Mousavi, M.; de Sa, V.R. Adaptive CSP with subspace alignment for subject-to-subject transfer in motor imagery brain-computer interfaces. In Proceedings of the 2018 6th International Conference on Brain-Computer Interface (BCI), GangWon, Korea, 15–17 January 2018; pp. 1–4.

73. Zhang, K.; Xu, G.; Chen, L.; Tian, P.; Han, C.; Zhang, S.; Duan, N. Instance Transfer Subject-Dependent Strategy for Motor Imagery Signal Classification Using Deep Convolutional Neural Networks. *Comput. Math. Meth. Med.* **2020**, *2020*, 1683013. [CrossRef]

74. Hossain, I.; Khosravi, A.; Hettiarachchi, I.; Nahavandi, S. Calibration Time Reduction Using Subjective Features Selection Based Transfer Learning For Multiclass BCI. In Proceedings of the 2018 IEEE International Conference on Systems, Man, and Cybernetics (SMC), Miyazaki, Japan, 7–10 October 2018; pp. 491–498.

75. Kang, H.; Nam, Y.; Choi, S. Composite common spatial pattern for subject-to-subject transfer. *IEEE Signal Process. Lett.* **2009**, *16*, 683–686. [CrossRef]

76. Raza, H.; Prasad, G.; Li, Y.; Cecotti, H. Covariate shift-adaptation using a transductive learning model for handling non-stationarity in EEG based brain-computer interfaces. In Proceedings of the 2014 IEEE International Conference on Bioinformatics and Biomedicine (BIBM), Belfast, UK, 2–5 November 2014; pp. 230–236.

77. Yin, Z.; Wang, Y.; Liu, L.; Zhang, W.; Zhang, J. Cross-subject EEG feature selection for emotion recognition using transfer recursive feature elimination. *Front. Neurorobot.* **2017**, *11*, 19. [CrossRef]

78. Chiang, K.J.; Wei, C.S.; Nakanishi, M.; Jung, T.P. Cross-Subject Transfer Learning Improves the Practicality of Real-World Applications of Brain-Computer Interfaces. In Proceedings of the 2019 9th International IEEE/EMBS Conference on Neural Engineering (NER), San Francisco, CA, USA, 20–23 March 2019; pp. 424–427.

79. Dai, M.; Wang, S.; Zheng, D.; Na, R.; Zhang, S. Domain transfer multiple kernel boosting for classification of EEG motor imagery signals. *IEEE Access* **2019**, *7*, 49951–49960. [CrossRef]

80. Nakanishi, M.; Wang, Y.T.; Wei, C.S.; Chiang, K.J.; Jung, T.P. Facilitating Calibration in High-Speed BCI Spellers via Leveraging Cross-Device Shared Latent Responses. *IEEE Trans. Biomed. Eng.* **2019**, *67*, 1105–1113. [CrossRef] [PubMed]

81. Zhang, Y.U.; Zhou, G.; Jin, J.; Wang, X.; Cichocki, A. Frequency recognition in SSVEP-based BCI using multiset canonical correlation analysis. *Int. J. Neural Syst.* **2014**, *24*, 1450013. [CrossRef]

82. Yuan, P.; Chen, X.; Wang, Y.; Gao, X.; Gao, S. Enhancing performances of SSVEP-based brain–computer interfaces via exploiting inter-subject information. *J. Neural Eng.* **2015**, *12*, 046006. [CrossRef]

83. Waytowich, N.R.; Faller, J.; Garcia, J.O.; Vettel, J.M.; Sajda, P. Unsupervised adaptive transfer learning for Steady-State Visual Evoked Potential brain-computer interfaces. In Proceedings of the 2016 IEEE International Conference on Systems, Man, and Cybernetics (SMC), Budapest, Hungary, 9–12 October 2016; pp. 004135–004140.

84. Salami, A.; Khodabakhshi, M.B.; Moradi, M.H. Fuzzy transfer learning approach for analysing imagery BCI tasks. In Proceedings of the 2017 Artificial Intelligence and Signal Processing Conference (AISP), Shiraz, Iran, 20–27 October 2017; pp. 300–305.

85. Rodrigues, P.L.C.; Jutten, C.; Congedo, M. Riemannian Procrustes Analysis: Transfer Learning for Brain–Computer Interfaces. *IEEE Trans. Biomed. Eng.* **2018**, *66*, 2390–2401. [CrossRef] [PubMed]

86. Waytowich, N.R.; Lawhern, V.J.; Bohannon, A.W.; Ball, K.R.; Lance, B.J. Spectral Transfer Learning Using Information Geometry for a User-Independent Brain-Computer Interface. *Front. Neurosci.* **2016**, *10*, 430. [CrossRef]

87. Arvaneh, M.; Robertson, I.; Ward, T.E. Subject-to-subject adaptation to reduce calibration time in motor imagery-based brain-computer interface. In Proceedings of the 2014 36th Annual International Conference of the IEEE Engineering in Medicine and Biology Society, Chicago, IL, USA, 26–30 August 2014; pp. 6501–6504.

88. Heger, D.; Putze, F.; Herff, C.; Schultz, T. Subject-to-subject transfer for CSP based BCIs: Feature space transformation and decision-level fusion. In Proceedings of the 2013 35th Annual International Conference of the IEEE Engineering in Medicine and Biology Society (EMBC), Osaka, Japan, 3–7 July 2013; Volume 2013, pp. 5614–5617.

89. Gaur, P.; McCreadie, K.; Pachori, R.B.; Wang, H.; Prasad, G. Tangent Space Features-Based Transfer Learning Classification Model for Two-Class Motor Imagery Brain–Computer Interface. *Int. J. Neural Syst.* **2019**, *29*, 1950025. [CrossRef]

90. Zheng, W.L.; Zhang, Y.Q.; Zhu, J.Y.; Lu, B.L. Transfer components between subjects for EEG-based emotion recognition. In Proceedings of the 2015 International Conference on Affective Computing and Intelligent Interaction (ACII), Xi'an, China, 21–24 September 2015; pp. 917–922.

91. Dai, M.; Zheng, D.; Liu, S.; Zhang, P. Transfer kernel common spatial patterns for motor imagery brain-computer interface classification. *Comput. Math. Meth. Med.* **2018**, *2018*, 9871603. [CrossRef] [PubMed]

92. Zanini, P.; Congedo, M.; Jutten, C.; Said, S.; Berthoumieu, Y. Transfer learning: A Riemannian geometry framework with applications to brain–computer interfaces. *IEEE Trans. Biomed. Eng.* **2017**, *65*, 1107–1116. [CrossRef]

93. Kalunga, E.K.; Chevallier, S.; Barthélemy, Q. Transfer learning for SSVEP-based BCI using Riemannian similarities between users. In Proceedings of the 2018 26th European Signal Processing Conference (EUSIPCO), Rome, Italy, 3–7 September 2018; pp. 1685–1689.

94. Jayaram, V.; Alamgir, M.; Altun, Y.; Scholkopf, B.; Grosse-Wentrup, M. Transfer learning in brain-computer interfaces. *IEEE Comput. Intell. Mag.* **2016**, *11*, 20–31. [CrossRef]

95. García-Salinas, J.S.; Villaseñor-Pineda, L.; Reyes-García, C.A.; Torres-García, A.A. Transfer learning in imagined speech EEG-based BCIs. *Biomed. Signal Process. Control* **2019**, *50*, 151–157. [CrossRef]

96. Tu, W.; Sun, S. Transferable discriminative dimensionality reduction. In Proceedings of the 2011 IEEE 23rd International Conference on Tools with Artificial Intelligence, Boca Raton, FL, USA, 7–9 November 2011; pp. 865–868.

97. Xu, G.; Shen, X.; Chen, S.; Zong, Y.; Zhang, C.; Yue, H.; Liu, M.; Chen, F.; Che, W. A Deep Transfer Convolutional Neural Network Framework for EEG Signal Classification. *IEEE Access* **2019**, *7*, 112767–112776. [CrossRef]

98. Sakhavi, S.; Guan, C. Convolutional neural network-based transfer learning and knowledge distillation using multi-subject data in motor imagery BCI. In Proceedings of the 2017 8th International IEEE/EMBS Conference on Neural Engineering (NER), Shanghai, China, 25–28 May 2017; pp. 588–591.

99. Behncke, J.; Schirrmeister, R.T.; Volker, M.; Hammer, J.; Marusic, P.; Schulze-Bonhage, A.; Burgard, W.; Ball, T. Cross-Paradigm Pretraining of Convolutional Networks Improves Intracranial EEG Decoding. In Proceedings of the 2018 IEEE International Conference on Systems, Man, and Cybernetics (SMC), Miyazaki, Japan, 7–10 October 2018; pp. 1046–1053.

100. Tan, C.; Sun, F.; Zhang, W. Deep Transfer Learning for EEG-Based Brain Computer Interface. In Proceedings of the 2018 IEEE International Conference on Acoustics, Speech and Signal Processing (ICASSP), Calgary, AB, Canada, 15–20 April 2018; pp. 916–920.

101. Völker, M.; Schirrmeister, R.T.; Fiederer, L.D.J.; Burgard, W.; Ball, T. Deep transfer learning for error decoding from non-invasive EEG. In Proceedings of the 2018 6th International Conference on Brain-Computer Interface (BCI), GangWon, Korea, 15–17 January 2018; pp. 1–6.

102. Wu, D.; Lawhern, V.J.; Gordon, S.; Lance, B.J.; Lin, C.-T. Driver Drowsiness Estimation From EEG Signals Using Online Weighted Adaptation Regularization for Regression (OwARR). *IEEE Trans. Fuzzy Syst.* **2016**, *25*, 1522–1535. [CrossRef]

103. Fahimi, F.; Zhang, Z.; Goh, W.B.; Lee, T.S.; Ang, K.K.; Guan, C. Inter-subject transfer learning with an end-to-end deep convolutional neural network for EEG-based BCI. *J. Neural Eng.* **2019**, *16*, 026007. [CrossRef]

104. Kobler, R.J.; Scherer, R. Restricted Boltzmann Machines in Sensory Motor Rhythm Brain-Computer Interfacing: A study on inter-subject transfer and co-adaptation. In Proceedings of the 2016 IEEE International Conference on Systems, Man, and Cybernetics (SMC), Budapest, Hungary, 9–12 October 2016; pp. 000469–000474.

105. Liu, Y.; Yang, C.; Li, Z. The Application of Transfer Learning in P300 Detection. In Proceedings of the 2018 IEEE International Conference on Cyborg and Bionic Systems (CBS), Shenzhen, China, 25–27 October 2018; pp. 412–417.

106. Parvan, M.; Ghiasi, A.R.; Rezaii, T.Y.; Farzamnia, A. Transfer Learning based Motor Imagery Classification using Convolutional Neural Networks. In Proceedings of the 2019 27th Iranian Conference on Electrical Engineering (ICEE), Yazd, Iran, 30 April–2 May 2019; pp. 1825–1828.

107. Özdenizci, O.; Wang, Y.; Koike-Akino, T.; Erdoğmuş, D. Learning invariant representations from EEG via adversarial inference. *IEEE Access* **2020**, *8*, 27074–27085. [CrossRef]

108. Tan, C.; Sun, F.; Kong, T.; Fang, B.; Zhang, W. Attention-based Transfer Learning for Brain-computer Interface. In Proceedings of the ICASSP 2019–2019 IEEE International Conference on Acoustics, Speech and Signal Processing (ICASSP), Brighton, UK, 12–17 May 2019; pp. 1154–1158.

109. Albuquerque, I.; Monteiro, J.; Rosanne, O.; Tiwari, A.; Gagnon, J.F.; Falk, T.H. Cross-Subject Statistical Shift Estimation for Generalized Electroencephalography-based Mental Workload Assessment. In Proceedings of the 2019 IEEE International Conference on Systems, Man and Cybernetics (SMC), Bari, Italy, 6–9 October 2019; pp. 3647–3653.

110. Wu, H.; Li, F.; Li, Y.; Fu, B.; Shi, G.; Dong, M.; Niu, Y. A Parallel Multiscale Filter Bank Convolutional Neural Networks for Motor Imagery EEG Classification. *Front. Neurosci.* **2019**, *13*, 1275. [CrossRef]

111. Uran, A.; van Gemeren, C.; van Diepen, R.; Chavarriaga, R.; Millán, J.D. Applying Transfer Learning To Deep Learned Models for EEG Analysis. *arXiv* **2019**, arXiv:1907.01332.

112. Craik, A.; Kilicarslan, A.; Contreras-Vidal, J.L. Classification and Transfer Learning of EEG during a Kinesthetic Motor Imagery Task using Deep Convolutional Neural Networks. In Proceedings of the 2019 41st Annual International Conference of the IEEE Engineering in Medicine and Biology Society (EMBC), Berlin, Germany, 23–27 July 2019; Volume 2019, pp. 3046–3049.

113. Onishi, A.; Nakagawa, S. Comparison of Classifiers for the Transfer Learning of Affective Auditory P300-Based BCIs. In Proceedings of the 2019 41st Annual International Conference of the IEEE Engineering in Medicine and Biology Society (EMBC), Berlin, Germany, 23–27 July 2019; pp. 6766–6769.

114. Lin, Y.P. Constructing a Personalized Cross-day EEG-based Emotion-Classification Model Using Transfer Learning. *IEEE J. Biomed. Health Inf.* **2019**, *24*, 1255–1264. [CrossRef]

115. Daoud, H.; Bayoumi, M.A. Efficient epileptic seizure prediction based on deep learning. *IEEE Trans. Biomed. Circuits Syst.* **2019**, *13*, 804–813. [CrossRef]

116. Wang, Y.; Cao, J.; Wang, J.; Hu, D.; Deng, M. Epileptic Signal Classification with Deep Transfer Learning Feature on Mean Amplitude Spectrum. In Proceedings of the 2019 41st Annual International Conference of the IEEE Engineering in Medicine and Biology Society (EMBC), Berlin, Germany, 23–27 July 2019; pp. 2392–2395.

117. Fauzi, H.; Shapiai, M.I.; Khairuddin, U. Transfer Learning of BCI Using CUR Algorithm. *J. Signal Process. Syst.* **2020**, *92*, 109–121. [CrossRef]

118. Özdenizci, O.; Wang, Y.; Koike-Akino, T.; Erdoğmuş, D. Transfer learning in brain-computer interfaces with adversarial variational autoencoders. In Proceedings of the 2019 9th International IEEE/EMBS Conference on Neural Engineering (NER), San Francisco, CA, USA, 20–23 March 2019; pp. 207–210.

119. Arevalillo-Herráez, M.; Cobos, M.; Roger, S.; García-Pineda, M. Combining Inter-Subject Modeling with a Subject-Based Data Transformation to Improve Affect Recognition from EEG Signals. *Sensors* **2019**, *19*, 2999. [CrossRef]

120. Yang, S.; Yin, Z.; Wang, Y.; Zhang, W.; Wang, Y.; Zhang, J. Assessing cognitive mental workload via EEG signals and an ensemble deep learning classifier based on denoising autoencoders. *Comput. Boil. Med.* **2019**, *109*, 159–170. [CrossRef]

121. Zou, Y.; Zhao, X.; Chu, Y.; Zhao, Y.; Xu, W.; Han, J. An inter-subject model to reduce the calibration time for motion imagination-based brain-computer interface. *Med. Boil. Eng. Comput.* **2019**, *57*, 939–952. [CrossRef] [PubMed]

122. Hang, W.; Feng, W.; Du, R.; Liang, S.; Chen, Y.; Wang, Q.; Liu, X. Cross-Subject EEG Signal Recognition Using Deep Domain Adaptation Network. *IEEE Access* **2019**, *7*, 128273–128282. [CrossRef]

123. Congedo, M.; Sherlin, L. *EEG Source Analysis*; Elsevier BV: Amsterdam, The Netherlands, 2011; pp. 25–433.

124. Alamgir, M.; Grosse-Wentrup, M.; Altun, Y. Multitask learning for brain–computer interfaces. In Proceedings of the Thirteenth International Conference on Artificial Intelligence and Statistics, Sardinia, Italy, 13–15 May 2010; pp. 17–24.
125. Kang, H.; Choi, S. Bayesian common spatial patterns for multi-subject EEG classification. *Neural Netw.* **2014**, *57*, 39–50. [CrossRef]
126. Craik, A.; He, Y.; Contreras-Vidal, J.L. Deep Learning for Electroencephalogram (EEG) Classification Tasks: A Review. *J. Neural Eng.* **2019**, *16*, 031001. [CrossRef]

Publisher's Note: MDPI stays neutral with regard to jurisdictional claims in published maps and institutional affiliations.

Review

A Review on Mental Stress Assessment Methods Using EEG Signals

Rateb Katmah [1], Fares Al-Shargie [2,*], Usman Tariq [2], Fabio Babiloni [3,4], Fadwa Al-Mughairbi [5] and Hasan Al-Nashash [2]

[1] Biomedical Engineering Graduate Program, American University of Sharjah, Sharjah 26666, United Arab Emirates; b00081299@aus.edu
[2] Department of Electrical Engineering, American University of Sharjah, Sharjah 26666, United Arab Emirates; utariq@aus.edu (U.T.); hnashash@aus.edu (H.A.-N.)
[3] Department of Molecular Medicine, University of Sapienza Rome, 00185 Rome, Italy; fabio.babiloni@uniroma1.it
[4] College Computer Science and Technology, University Hangzhou Dianzi, Hangzhou 310018, China
[5] College of Medicines and Health Sciences, United Arab Emirates University, Al-Ain 15551, United Arab Emirates; f.almughairbi@uaeu.ac.ae
* Correspondence: fyahya@aus.edu

Citation: Katmah, R.; Al-Shargie, F.; Tariq, U.; Babiloni, F.; Al-Mughairbi, F.; Al-Nashash, H. A Review on Mental Stress Assessment Methods Using EEG Signals. *Sensors* **2021**, *21*, 5043. https://doi.org/10.3390/s21155043

Academic Editor: Yvonne Tran

Received: 31 May 2021
Accepted: 19 July 2021
Published: 26 July 2021

Publisher's Note: MDPI stays neutral with regard to jurisdictional claims in published maps and institutional affiliations.

Abstract: Mental stress is one of the serious factors that lead to many health problems. Scientists and physicians have developed various tools to assess the level of mental stress in its early stages. Several neuroimaging tools have been proposed in the literature to assess mental stress in the workplace. Electroencephalogram (EEG) signal is one important candidate because it contains rich information about mental states and condition. In this paper, we review the existing EEG signal analysis methods on the assessment of mental stress. The review highlights the critical differences between the research findings and argues that variations of the data analysis methods contribute to several contradictory results. The variations in results could be due to various factors including lack of standardized protocol, the brain region of interest, stressor type, experiment duration, proper EEG processing, feature extraction mechanism, and type of classifier. Therefore, the significant part related to mental stress recognition is choosing the most appropriate features. In particular, a complex and diverse range of EEG features, including time-varying, functional, and dynamic brain connections, requires integration of various methods to understand their associations with mental stress. Accordingly, the review suggests fusing the cortical activations with the connectivity network measures and deep learning approaches to improve the accuracy of mental stress level assessment.

Keywords: mental stress; EEG; data analysis; connectivity network; machine Learning

1. Introduction

Mental stress is one of the contributing factors to health problems. It is defined as the human body's response, controlled by the sympathetic nervous system (SNS) and hypothalamus–pituitary–adrenocortical axis (HPA axis), to mental, physical and emotional stimuli [1]. This expression can be used with regard to internal (personality structure) or external (dealing with problems) matters triggering various physiological and negative emotional changes [2]. Literature defined three types of stress; acute stress, episodic stress, and chronic stress [3]. Acute stress is related to short-lasting exposure and is not harmful. Episodic stress happens when the stimulus is more frequent for a limited time [4]. Meanwhile, chronic stress is the most damaging, resulting from permanent and long-standing stressors [5]. Several studies have reported that mental stress has direct physiological effects leading to several diseases including stroke, cardiovascular disease, cognitive problems, speech distinctiveness and depression [6,7]. Moreover, stress affects the human body indirectly at different levels varying between skin conditions, eating habits, inadequate sleeping and decision-making [8–10]. Thus, researchers have developed various methods to assess the stress level in its early stages to avoid the negative consequences on health and performance.

Assessment of mental stress is challenging because each individual experiences stress differently [11]. Besides, the reliability of evaluating mental stress depends on the method of assessment and analysis. Traditionally, stress is assessed using subjective methods. The most commonly used method is the self-report questionnaires [12] such as the perceived stress scale [13,14]. Many studies have established the questionnaire score and self-report rating or interview as ground truth to estimate the mental stress level. However, questionnaires are subjective and require the user's full attention. As a result, individuals are not always aware of their genuine stress levels. Hence, the procedures, such as self-report questionnaires, may result in an inaccurate stress level measurement. Furthermore, they seem to be less informative than physiological measures. Researchers have identified several physiological measurements as stress indicators such as heart rate variability (HRV), electrodermal activity (EDA), electromyogram (EMG), blood pressure, pupil diameter, salivary cortisol and salivary alpha amylase [2]. Nevertheless, physiological markers can be influenced by many factors including mental stress. Cortisol level has been reported to be affected by circadian rhythm (i.e., its concentration changes throughout the day) [15,16]. In addition, a subject's physical activity affects salivary alpha amylase level [17,18], and EDA is sensitive to skin disease and humidity [19].

Various neuroimaging techniques have been used to assess mental stress by directly or indirectly measuring the brain activity. These include functional near-infrared spectroscopy (fNIRS), electroencephalography (EEG) [20], positron emission tomography (PET) [21] and functional magnetic resonance imaging (fMRI) [22]. The EEG modality has some advantages such as high temporal resolution, low cost, and ease of use. Hence, it is the most used technique to analyze mental states including stress [23,24]. A typical EEG stress assessment method consists of two major parts: feature extraction and stress classification. There are three categories of EEG features: time-domain, frequency-domain, and synchronicity-domain features [25–27]. The time-domain features capture the temporal information using amplitude related to energy, variability, coefficient of variation, Hjorth feature, fractal dimension feature and higher-order crossing feature. On the other hand, the most used frequency-domain features are obtained from the EEG signal clinical frequency bands, delta (0.5–4 Hz), theta (4–8 Hz), alpha (8–13 Hz), beta (14–30 Hz) and gamma (30–50 Hz) [28]. These brain rhythms contain relevant information related to mental stress and other psychological disorders. The commonly used spectral EEG features include the power spectral density (PSD), differential asymmetry features, phase synchronization, phase lag index, directed transfer function and entropies [29–32]. In addition, the time-frequency features are obtained through the short-time Fourier transform (STFT), or discrete wavelet transform (DWT) [33–35]. The findings of subsequent studies on the usefulness of EEG signal analysis methods for the assessment of mental stress have been conflicting and impeding the development of further research. To resolve these difficulties, this work aims at conducting a comprehensive review of the state-of-art of the published EEG analysis methods on mental stress and to propose potential future research directions.

The rest of the paper is organized as follows. The materials and methods are described in Section 2, where the explanation of inclusion and exclusion strategy in addition to the variables of interest are reported. EEG pre-processing and the data analysis methods are presented in Section 3. Section 4 reviews the most common classifiers that have been used in quantifying stress levels. Section 5 shows the review results, including the relationship between EEG analysis methods and type of classifier and the variables can be considered in assessing mental stress. The discussion of the findings on the reviewed papers is described in Section 6. Finally, Sections 7 and 8 summarize the main challenges and conclusion of the research in stress estimation-based EEG signal.

2. Materials and Methods

2.1. Search Strategy

The Preferred Reporting Items for Systemic Reviews and Analysis (PRISMA) was used to conduct this review [36]. The following databases were searched for study publications, namely Google Scholar, PubMed, Science Direct, IEEE Xplore, and PsycINFO. The used search terms were the single terms of mental stress and EEG. This was combined with at least one of the following terms: connectivity, power spectral, coherence, entropy and classification. In addition to searching databases, the reference list for all selected articles was checked to specify any additional relevant studies that might have been overlooked during the primary search. Figure 1 shows the search strategy and identification of relevant studies.

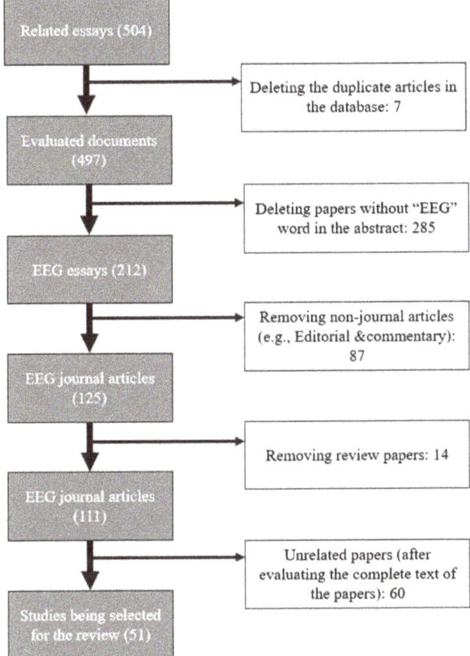

Figure 1. Flow chart of search strategy and identification of relevant studies.

2.2. Inclusion and Exclusion Strategy

Manuscripts in English and EEG experimental studies were considered in this review. In contrast, those involving animals were excluded to avoid any possible effect of cognitive impairments.

2.3. Variables of Interest

The main variables detected in each paper were (i) type of stressor, (ii) experiment duration, (iii) number of subjects who participated in the experiment, (iv) number of EEG electrodes, (v) EEG frequency bands, (vi) type of features, (vii) type of classifier, (viii) classification performance, (ix) summary of results compared before and after the stress task, and (x) comments on the findings.

3. EEG Analysis Methods

The EEG signal goes through extensive preprocessing steps to remove artifacts and noise before applying data analysis methods. Data preprocessing plays a major role in getting meaningful information about the signal. Thus, comprehensive knowledge about the types of artifacts is required. According to Jiang et al. [37], physiological artifacts are the most common artifacts that affect EEG signal. In addition, artifacts represent another vital source of biased information. The digitized EEG signal can be segmented into epochs (e.g., 2 s) for visually identifying and rejecting visible artifacts. To remove the noise and artifacts from EEG signals, researchers have utilized a variety of methods such as regression techniques, blind source separation (BSS), empirical-mode decomposition (EMD) and wavelet transform algorithms. This is in addition to Adaptive and Wiener filtering, high pass, band pass, notch filters, and independent component analysis (ICA) [38]. In fact, we still have a lack of standardization related to EEG pre-processing that can be used by all research studies.

In stress studies, the process of reviewing, cleansing, transforming, and modelling EEG signals with the aim of finding useful knowledge, informing conclusions, and assisting decision-making is known as data analysis. Several data analysis methods have been reported in the literature to analyze mental stress based on EEG signals. However, selecting an appropriate analysis method is very important to minimize the data processing cost, storage size and dimensional space. The following sub-section provides a comprehensive review of the EEG analysis methods on mental stress.

3.1. Connectivity Methods

The primary objective of EEG research is to link diverse measures of neural rhythms to functional brain states reflecting cognition, behavior, or neuropathology [39]. Each EEG signal is produced by the superposition of several brain current sources [40]. The involvement of each source varies depending on the location and orientation of the source and measuring electrodes. Several researchers have shown interest in functional or effective connectivity. However, the various forms of data used to assess functional connectivity differ in many ways, involving temporal and spatial information, as well as whether the data reflect electrical neuron activities, neuronal ensemble activities, or hemodynamics of macroscopic brain areas. Furthermore, the exact computational techniques employed to find these values vary amongst researchers, even when dealing with the same data type.

The issue is considerably more complex in the case of EEG, where numerous aspects of the signals might be linked. The information in an EEG signal comes from a complex and dense network of interconnected neurons. Hence, studying brain connectivity may provide us with a more exact model of the brain and how its various areas interact with each other. There are two types of brain connectivity: functional connectivity and effective connectivity [41]. The functional connectivity reflects the relationships between different brain regions as reflected on the temporal coherence between the networks. The various methods for determining functional connectivity may result in different conclusions depending on factors such as the strength of the interaction between neural units, type of stressor and number and location of electrodes. This can even be true for data from the same modality and even data obtained using the same task. Employing multiple interpretations of what defines functional connectivity might also lead to conflicting findings [41]. Effective connectivity, on the other hand, is the simplest circuit that describes the experimentally achieved relationship between two neurons. It explains how the neural system affects the others [42]. Effective connectivity, in contrast to the non-directional and correlative functional connectivity, assesses the directional influences between distinct brain regions [43]. As such, functional and effective connectivity measures are important in trying to understand the brain behavior under stress and non-stress conditions. There are numerous features utilized to detect this connectivity measurement, and the following is a quick description of them.

Coherence analysis aims to identify the functional connectivity and synchronization between different brain regions (several electrode sites). These mutual relations can be found by analyzing the amplitude and the phase of signal within the used EEG electrodes [44,45]. Xia et al. [44] examined coherence using multilevel stress assessment and found a significant increase for all frequency bands (except beta) at frontoparietal lobe. In addition, strong coherency for delta wave was detected in prefrontal and temporal regions at higher stress level. Meanwhile, study in [46] has shown an increased brain connectivity between interhemispheric locations in delta and theta bands, whereas the alpha and beta coherence connectivity networks spread all over the scalp. In particular, this increase in coherence level under stress in the article [46] was regarded as the brain attempting to attain redundant communication between its different regions in order to quickly process the cognitive load of the applied stressor. The full mathematical expressions of the coherence measures can be found in a previous study [47].

Magnitude Square Coherence (MSC) is another measure of functional connectivity in stress studies. A study in [48] found significant reduction in the functional connectivity from control to the stress situation in intra-hemispheric and inter-hemispheric prefrontal cortex (PFC). Meanwhile, when applying sleep deprivation as a stressor, the EEG connectivity maps show a decreased MSC for alpha band in the anterior region of scalp and increased beta coherence spread all over the scalp [49]. However, this behavior was not reproduced when dealing with Stroop color word task (SCWT) where there was only elevated beta coherence for sagittal middle regions. Darzi et al. [50] have proved that extracting MSC features with a length of 56 s achieved the highest accuracy by applying support vector machines (SVM) as a classifier compared to the directed transfer function (DTF), phase–slope Index (PSI), canonical correlation (CC), and power spectral density (PSD) techniques. Likewise, Khosrowabadi et al. [51], reported that MSC accuracy, sensitivity and specificity were greater than those obtained by Gaussian mixture models (GMM) and fractal dimension (FD) features using K-nearest neighbors (KNN) or SVM classifiers. Consequently, the most useful advantage about using coherence in analyzing EEG to quantify stress is that it cannot be affected by the amplitude oscillations for the different brain locations. However, the main drawback for coherence analysis is the high sensitivity to phase coupling and power changes [46,52]. The mathematical formulations of the employed MSC can be found in [53].

Pearson's correlation-based captures linear, time-domain dependencies among EEG signals. It could be found over a single epoch or over several epochs, and it is calculated using the Pearson's correlation coefficient, cross-covariance, and auto-covariance of EEG signals [54]. Therefore, increasing the value for the Pearson correlation coefficient from (-1) to (1) indicates intense connections between brain regions. In particular, this technique has been used by study [54] to reduce feature vectors and computational time, and to improve accuracy of SVM classifiers in detecting human stress. The main interest of such features is the high performance while reducing dimensionality of the EEG data set [55]. On the other hand, canonical correlation analysis (CCA) is useful to get information from the cross-covariance matrices in order to estimate the effect of mental stress. This is done by detecting the linear combination that achieves maximum correlation between two vectors [56]. The main advantage of using CCA is its applicability to be used with multimodal data that has different modal dimensionalities [57]. The mathematical expressions of the correlation analysis can be found in previous study [58].

Amplitude asymmetry refers to the difference in absolute amplitude that exists between the homologous electrodes positioned on the hemispheres when a stressor is applied. It is used to find the difference in the relative stimulation between brain locations [59]. Despite its high performance in estimating acute stress levels [44], this technique is influenced by HRV biofeedback [60]. The study in [61] describes the math of the asymmetry method.

Mutual information (MI) is used to detect dynamic concatenate and similarity of joint probability distribution function between two EEG signals [50,62]. Therefore, MI aims to find the statistical dependency between signals and analyze EEG with different spectral bands [63]. The MI during stress is represented by EEG connectivity maps. According to the study in [49], mutual information did not achieve significant increase in the EEG map when using Stroop task, whereas the sleep deprivation physical stressor showed widespread decreases of linear area comparing to a significant increase of nonlinear area in the anterior, central, and temporoparietal regions of head. Meanwhile, Pernice et al. [64] reported that the shared information between brain locations during relaxing state was low, whereas a significant MI increase was noted for alpha, theta and delta bands in the frontal region during mental arithmetic task. On the other hand, comparing to other connectivity measures, this technique has a short time processing and it is not restricted to real-valued variable; therefore, it could be used on several kinds of variables [65]. The mathematical formulations that describe MI can be found in [66].

Phase lag is used to detect the lag or delay between two EEG signals related to different brain regions. Xia et al. [44], has detected a significant role for the phase lag technique in discrimination between stress and control conditions for different levels. However, a study in [45] has found low accuracy for this feature compared to the other used methods such as coherence, absolute power and amplitude asymmetry. The main limitation of this technique is that it does not provide the directionality of connectivity and the volume conduction problem [67]. The mathematical expressions of the phase lag analysis can be found in previous study [68].

Phase–slope index (PSI) is a measure of phase synchronization that is not sensitive to volume conduction or common reference effects. Studies in [50] found several patterns of brain locations connectivity during the perception of external stimuli that chronic stress can change them, whereas the synchronization between left parietal and right temporal showed a decrease of 55% in the stressful subjects. Darzi et al. [50] have shown a high performance when using PSI features, whereas the results of Khosrowabadi et al. [69] achieved low accuracy for PSI comparing to DTF and PDC features. The main advantages of using PSI are overcoming the independent background activity generated between two electrodes and the ability to give meaningful information even though the nonlinear phase spectrum [69]. However, PSI may fail to correctly describe the directionality of EEG [70]. The mathematical formulations of the PSI method can be found in [71].

Partial directed coherence (PDC) is a measure used to detect the direction and weight of information flow in the frequency domain between multivariate data. Specifically, multivariate analysis will represent the stress phenomenon without loss of information of data with several variables. In particular, two directed coherences (feed-forward and feedback aspects) can be predicted from the classical coherence function using PDC. Therefore, directional flow between two channels within specific frequency involves several calculated factors such as Akaike information criterion and Granger causality [72]. Studies in [72,73] have found that when fatigue level increases due to stress, the functional coupling decreases over parietal-frontal regions while using theta, alpha and beta frequency bands. A significant form of PDC to get functional connectivity measurement is the Generalized Partial Directed Coherence (GPDC), which is used to control negative causality of the EEG multichannel analysis. Khosrowabadi et al. [69] has used GPDC features in detecting stress/non-stress cases, and they found medium and low accuracy compared to PSI and DTF features. The full mathematical expressions of the PDC measures can be found in a previous study [74].

Directed transfer function (DTF) is an effective connectivity technique used to detect the interaction patterns between neurons. Yu et al. [75] found that DTF has increased values (enhanced EEG coupling) at alpha and beta bands after applying a mental arithmetic stress task. In particular, these results lead to enhancing the flow of information from the central regions (the source of information outflow) to parietal and occipital areas for alpha and beta. According to a study [69] in quantifying stress, DTF shows the highest accuracy comparing to PSI and GPDC. However, DTF does not differentiate between directed influence of one signal to another [76], but it shows a higher performance than CCA, PSI, MSC and PSD [50]. Meanwhile, the main limitation for using DTF is its sensitivity to cortico-cortical and brain to heart functional coupling [75]. DTF mathematical expressions can be found in articles [77].

3.2. Power Spectral (Frequency Domain)

Spectral features are the characteristics obtained from the EEG signal in frequency domain. In order to get meaningful information about the EEG, it is important to check the segmentation process of EEG to get stationary signal. Thus, some of the more widely used spectral features and processing techniques are described below.

Power spectral density (PSD) pursues to find power distribution for time-domain EEG signal over frequency range and this provides significant information about cortical activation. In particular, PSD is useful in describing stochastic process of the signal and evaluating short data records [78]. There are several methods applied to estimating the PSD, for example, fast Fourier transform (FFT), Welch, Burg, Yule walker, welch method and periodogram [54]. Several studies have demonstrated the effectiveness of using PSD to estimate the level of stress. For example, study in [79], reported that mental stress decreased the EEG power spectral density in the alpha band. Likewise, the study in [20] found a significant decrease in alpha rhythm when increasing the level of stress from level 1 to level 2 (based on increasing the complexity/difficulty of the math task), and then increasing from levels 2 to level 3. In particular, the difficulty of the math task was increased from level 1 up to 3 by increasing the integer numbers and operands that were used in the math operation. Meanwhile, according to [20], the most dominant cortical structure that is involved in stress detection is the right prefrontal cortex. For detailed mathematical formulations of the PSD method, refer to [80].

Other studies utilized absolute power (AP) as an indicator of stress. The AP at a particular band is calculated by dividing the absolute value of fast Fourier transform of the EEG signal by the signal's length [81]. Meanwhile, studies in [59,82] used the relative power (RP) to check the rhythm of EEG signal by finding the ratio between the power of each band and the power of the total bands. Subhani et al. [45] and Arsalan et al. [83] found that applying AP on stress/non-stress detection shows a significant difference regarding theta EEG band (4–7 Hz) compared to other bands, whereas in the case of RP, they reported that when stress levels increased, the RP decreased [45]. Consequently, RP showed a better performance compared to the AP in spite of its sensitivity to the noise and memory recall [81]. The detailed math expressions for the AP and RP methods are identified by study [45].

Studies in [26,79] utilized powers from the wavelet transform (WT) coefficients to extract features that are highly correlated with mental stress. They found that the mean alpha rhythm power has significantly decreased from one stress level to the next higher one. Moreover, WT is an appropriate method for multi-resolution time-frequency analysis. This is done by decomposing the EEG signal into its frequency bands retaining information in both: frequency and time domain. Then, from wavelet coefficients, the average power and energy can be estimated. Even though the Fourier transform (FT) provides a frequency domain representation of the signal, the wavelet transform creates a time and frequency domain representation, providing a quick access to the localized information of the signal. In particular, since EEG signals are nonstationary, using the FT may result in tiny changes in the spectrum, and the analysis may alter depending on the duration of data. Thus, WT is preferable to FT [84]. The mathematical formulations of the employed WT can be found in [85].

Other studies used Gaussian mixtures of EEG spectrogram to detect stress by analyzing the changing of spectral density of the EEG signal related to time domain. Moreover, this data analysis method involves short-time Fourier transform (STFT) to calculate the spectrogram of the time signal. After computing spectrogram, Gaussian mixture model (GMM), which is a linear combination of Gaussian pdfs, can be estimated to find the density [51]. The obvious role of this model is extracting the symmetric and asymmetric EEG signal; however, some drawbacks of considering infinite range and symmetric nature are reported [86]. Khosrowabadi et al. [51,87] have used this technique to quantify chronic mental stress. They found that GMM has a lower accuracy than MSC, but higher than

FD features when using SVM classifier. The detailed math expressions for the Gaussian method are identified by studies [87].

The study in [88] quantified mental stress by using spectral moments (SM). SM was processed to detect three power spectral moments from each EEG segment, that are related to different root square moments with orders of zero, two and four. These moments are found depending on the phase excluded power spectrum and the EEG length. Attallah in [88] verified the effectiveness of spectral moment in differentiating stress/non-stress cases and between several stress levels and reported high accuracy for SM with a linear discriminant analysis (LDA) classifier. The full mathematical expressions of the SM method can be found in [89].

3.3. Time Domain Techniques

The most widely used temporal features in quantifying mental stress are reviewed below:

Hjorth parameters are statistical parameters used to describe the EEG signal in the time domain. The Hjorth parameters are also known as normalized slope descriptors (NSDs). They consist of activity, mobility, and complexity descriptors. Activity parameter demonstrates the signal power leading to denoting the surface of the power spectrum in the frequency domain. The mobility approximates the mean frequency, and complexity approximates the bandwidth of the signal [90,91]. These parameters depend on the time domain, but they provide information about the frequency spectrum of the EEG [92]. However, theses parameters are sensitive to noise. Besides, the Hjorth parameters need shorter computation time in getting frequency information in addition to forming a good alternative for short time Fourier transform (STFT). Oh et al. [93] found that combining Hjorth parameter with band pass filtering has a higher classification performance than the general Hjorth parameter. The mathematical formulation of the employed Hjorth parameter can be found in [94].

Other methods to estimate the complexity of EEG signals in the time domain are the entropies. For example, Shannon entropy (SE) is used to estimate EEG signal irregularity and to quantify energy distribution of power spectrum by analyzing the EEG time series. This leads us to know brain behavior during a variety of states to detect mental stress [95]. Therefore, the study in [95] found the group that had the highest stress index (high mental stress) tend to have the lowest alpha-band-entropy. Zhu et al. [96] used VR-based relaxation therapy to relieve stress by evaluating the changes in Shannon entropy. They reported that SE had an increased trend in the alpha band, before and after watching VR. Another type of entropy is the Approximate Entropy (ApEn), which is used with time series data to know the fluctuations unpredictability and the amount of the regularity. According to Wang et al. [97], the complexity of the system is responsible for determining data length when estimating the value of ApEn. Meanwhile, the study in [97] showed that mental arithmetic task induced a significant increase of ApEn at the anterior cingulate and insular cortex. The main advantage for ApEn is its ability to deal with noise and possibility to be used with stochastic and deterministic chaotic signals. Moreover, the wavelet sum of entropy was utilized by Hasan et al. [90] as a separate feature to identify the signs of stress from EEG recordings. It represents the summation of the entropy after being calculated for each wavelet band. These wavelet bands can be found as a result of dividing EEG signal onto distinct frequency bands (generally five bands) and applying discrete wavelet transform (DWPT) [90]. Finally, self-entropy (SE) is used to detect information processing within the physiological network by estimating dynamical activity of the EEG signal [19,62]. Studies [96] include the mathematical expression for all entropy kinds.

Higuchi's fractal dimension (FD) is the estimation of irregularity, complexity, and nonlinear properties of the EEG signal where high and low values of FD are related to irregular and regular waveforms, respectively [11]. Higuchi FD provides a significant analysis for stress phases by computing fractal dimension, which is useful in real-time testing for brain chaotic behavior during chronic mental stress [51]. Studies in [11,98] have shown that combining FD with statistical features outperforms spectral power features.

The recorded EEG complexity in frontal lobe has high values when using mental arithmetic stressor [98]. On the other hand, Khosrowabadi et al. [51] detected a low accuracy for Higuchi's FD comparing to GMM and MSC features for SVM and KNN classifiers. The main interest about FD is its independency of signal nature and high efficiency, but it is sensitive to noise and frequency bands and its performance will be low when it is used alone [99]. The mathematical formulations of the employed FD can be found in [100].

3.4. Statistical Features

This type of features can be found by applying standard statistical operations on the EEG signal within the time domain to quantify stress levels. Thus, statistical techniques are simple, easy to use and often complement each other [101]. Meanwhile, the most common features for EEG data analysis are the mean, skewness, kurtosis, standard deviation, shape factor, first and second difference, root mean square, and impulse factor [88,90,92,102]. Hou et al. [11] found that combining statistical features with fractal dimension and power features improved the classification accuracy of stress. Moreover, study in [103] found that the variance values are higher in rest than stress levels, whereas kurtosis showed increased values in stress conditions when moving from delta to gamma bands. On the other hand, the main drawback is related to using all these features in stress estimation, which leads to longer time processing. Furthermore, some studies utilized principal component analysis (PCA) as a conventional and statistical method for detecting samples in the EEG data of high dimension. According to Deshmukh et al. [104], the main purpose of using PCA was to reduce the dimension of the stress features before feeding into the classifier. This is done by applying features Eigen vectors on features dimensionality to get the lowest orthogonal dimensions [44]. Moreover, PCA provides information about how the investigated groups, related to stress/non-stress conditions, could be separated into principal components (PCs) space [105]. Shon et al. [92] analyzed mental stress and demonstrated that PCA has a lower accuracy (65.30%) in the process of features selection than genetic algorithm (71.76%). However, PCA limitation is the probability to fail in processing data when dealing with complicated manifold [104].

4. Classification

Stress studies have examined various types of classifiers to assess the level of mental stress. The most common and significant classifiers are SVM, LR, NB, KNN, LDA, multi-layer perceptron (MLP), convolutional neural network (CNN) and long short-term memory (LSTM). The following sections describe the implementation of the aforementioned classifiers on EEG stress studies. Table 1 summarizes the main findings of previous EEG stress studies.

SVM is a binary classification model built in feature vector to discover the hyperplane that optimizes the margin between input data classifications. Several studies used SVM to discriminate between stress levels. For example, the studies in [51,106] applied SVM to quantify two levels of stress and achieved accuracy levels of 75% and 90%, respectively. On the other hand, studies in [107] have utilized SVM to classify three levels of stress. Meanwhile, [26] combined SVM with an error-correcting output code and reported that the average classification accuracy of these mental stress levels showed a drop in value from 97.61 to 95.37 and to 91.40 with the increased stress level. Besides, Gaikwad et al. [107] had an accuracy of 72.30% in the real time by using a trained algorithm as a reference. According to Hou et al. [11], increasing the number of stress levels (from two levels up to four) declined the SVM accuracy.

Furthermore, studies in [19] have utilized LR to differentiate between stress levels. LR is a statistical model that utilizes a logistic function to represent a binary dependent variable in its most basic form, however many more advanced extensions exist. It is used to investigate the relationship between one dichotomous dependent variable and one (categorical or continuous) independent variable. Zanetti et al. [19] analyzed three mental states and the recorded accuracy by LR was 84.30%, but even though LR had some errors in detecting resting states. Meanwhile, the achieved accuracy by LR was as high as SVM and random forest classifiers when it was used with several stress states induced by arithmetic stress task [45]. Saeed et al. [108] showed that logistic regression provides a significant performance with 85.15% accuracy in stress quantification (specifically with alpha asymmetry feature) comparing to other classification techniques such as KNN, NB and MLP.

Some studies employed NB to classify stress levels. NB is a simple and fast probabilistic classifier that is used when input dimensionality is large. It is based on Bayes' theorem, which assumes that extracted features are independent to each other. Subhani et al. [45] reported that NB achieved the highest accuracy in quantifying four levels of stress with a recorded accuracy of 94.0%, 94.6%, and 91.7% for levels 1, 2, and 3, respectively. Darzi et al. [50] detected two levels of stress using NB and found that SVM has a better performance than NB even though the running time of NB is about five times shorter than SVM, therefore NB is more suitable for online tasks. Thus, NB provides fast stress quantification because no complex optimization parameters are required. Meanwhile, NB had a low accuracy in Arsalan et al. [83] when dealing with theta band of two stress levels (75%) and three stress levels (50%). Moreover, Saeed et al. [108] recorded an accuracy of 80.79% for quantifying stress by NB, whereas in [109] they showed that using low beta waves as a feature vector will reduce NB performance to get an accuracy equal to 71.4%.

Furthermore, the non-parametric learning algorithm K-NN can be involved in quantifying mental stress. The mechanism of K-NN depends on estimating the distance between neighbors and choosing the K closest neighbors. Thus, two of the critical factors to be identified are the optimal value of K and neighbors distance D [90,108]. Saeed et al. [108] used K-NN with alpha asymmetry, beta, and gamma waves as features to quantify long-term stress. They found that K-NN has an accuracy of (65.96%) when these features are combined with each other. Meanwhile, the study in [50] found that K-NN has achieved an accuracy of (90.0%) comparing to the SVM and Bayesian classifiers. The main advantage of K-NN is the low computational complexity in quantifying stress/non-stress phases when dealing with small-sized data [50,90]. However, K-NN has a drawback, which is the high sensitivity to data local structure (dimensions).

On the other hand, some studies applied LDA as a machine learning method to classify stress by finding the linear combination between EEG features. Therefore, it is difficult to apply LDA on nonlinear EEG data due to LDA's linear nature [110]. LDA was applied by Minguillon et al. [111] to quantify three levels of stress using the average relative gamma as a feature and found that, increasing the number of stress markers will enhance the value of the recorded accuracy (50.0%). Meanwhile, Vanitha et al. [112] found that LDA has the lowest accuracy (70.166%) comparing to the SVM (89.07%) and K-NN (72.67%) classifiers when detecting stress levels for students. Consequently, the main drawback of LDA is the assumptions and restrictions (linear decision boundaries) that are needed to establish this classifier [111].

Besides, MLP is a non-linear artificial neural network model that is used to map the input data into output data. It consists of multiple layers (at least three) that vary between input, output and one or more hidden layers. Since MLPs are fully connected, each layer is connected to the next one and each node will be as a neuron that uses non-linear activation function. Several studies have employed MLP to quantify mental stress. Saeed et al. [108] reported that, integrating alpha, beta and gamma features with MLP provides the highest accuracy (85.13%) compared to the one that can be achieved using a single feature. Meanwhile, Arsalan et al. [83] found that MLP outperforms both SVM and NB classifiers and gives the highest accuracy for both two-and three-class quantification of mental stress. Even though, the main drawback of MLP is the formation of over-fitting because of excessive or insufficient neurons [108].

Another example for deep networks is that of deep CNN, which is considered as a regulated MLP. It provides an alternative form to mimic the brain functionality in quantifying mental stress [23]. Comparing to the other classification algorithms, CNN needs a little pre-processing, can be used for large size nonlinear data and it provides a significant feature discrimination [113]. The main advantage of using CNN is the independence from human effort and prior knowledge. Several studies utilized CNN to analyze mental stress. For example, Jebelli et al. [114] quantified three levels of workers' stress where CNN yielded an accuracy of 79.26% that outperforms SVM's accuracy (79.12%), whereas in the study [115], CNN's accuracy was equal to 86.62%. Meanwhile, they found that the optimum network configuration to quantify workers' stress level needs two hidden layers with 83 and 23 neurons in the first and second hidden layers, respectively. Therefore, CNN facilitates the need for EEG feature extraction, which consumes time in the supervised learning algorithms [115].

5. Results

Most of the reviewed studies have reported high alpha activity during relaxation states compared to the stressful conditions. In particular, a significant increase in the spectral power is more apparent after applying stimulus. EEG gamma activity showed a varied response, but generally a relatively decreased gamma activity can be observed with both relaxed and stressful situations. Hence, gamma oscillations may not be sensitive to stress level variations. Regarding fast beta band, it has a significant positive interaction indicating stronger increase in stress phases. Furthermore, central, and parieto-temporal areas are the most affected cortical regions with alpha and slow beta. Inspection of these variations related to different frequency bands were sided by the result of having stronger interaction effects in the right hemisphere comparing to the left one. Figure 2 summarizes the classification accuracy for each of the five different frequency bands extracted from the reviewed studies.

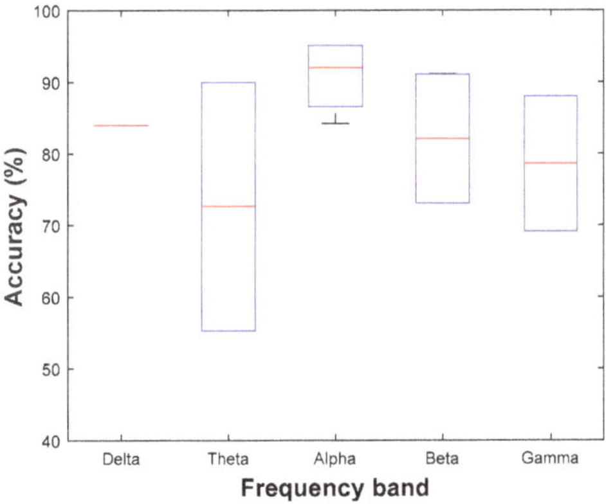

Figure 2. Classification accuracy based on EEG frequency bands.

In general, accuracy refers to the percentage of accurate predictions. A value close to 100 indicates that the classification model is performing well. As a result, features are chosen from those EEG frequency bands that improve classification accuracy. To choose the best frequency band, all possible combinations (from five frequency bands) were used. According to the discussed sections, different classifiers were used to quantify mental stress using EEG. In order to get the proper performance, there are three parameters that will be needed: accuracy, sensitivity and specificity. They have been used to identify the classifier ability in correctly distinguish between positive and negative results and to measure each one of them properly. This performance is influenced by the quality of EEG signal, processing power and the EEG feature components that are used as an input to the classifier [83,116]. Arsalan et al. [83] found that, combining MLP classifier with PSD, correlation and rational asymmetry features outperforms SVM and NB in classifying two/three levels of stress. Furthermore, combining several results for multiple sensors may provide a better classification accuracy [117]. In particular, specific features and classifiers have reached high levels of accuracy such as PSD and SVM. Meanwhile, all references that have used Montreal Imaging Stress Task (MIST) as a stressor, rely on SVM classifier except Minguillon et al. [111], which has LDA instead. Despite the achieved low accuracy (50.00%) by the article [111], EEG measurements provided shorter response time, significant cognitive information and low sensitivity to physical activity. Thus, combining EEG with physiological signals elevates the LDA accuracy up to 86.00%. On the other hand, Xia et al. [44] got an accuracy equal to 79.45% when using ECG and EEG measurements with SVM classifier in addition to high number of participants and EEG electrodes. Therefore, the selected EEG features (relative power, power ratios, amplitude asymmetry, coherence, and phase lag) have shown promising and robust results when employed with MIST stressor and SVM classifier in quantifying mental stress. Furthermore, using NB classifier and the increased number of frequency bands were the main reason of getting high accuracy by Subhani et al. [45] comparing to Xia et al. [44] that have used the same criteria.

Different stressors can be employed to generate mental stress, resulting in a variety of impacted brain regions. Students' examination periods can be used to develop long-term psychological mental stressors. According to Darzi et al. [50] long-term stress affects the functional connectivity of the temporal-parietal and the left central and temporal regions. Furthermore, for music and videos stressors, pre-frontal region of the brain has shown increased activities when using two EEG electrodes to get differences between two frontal regions [92]. Lotfan et al. [118] utilized the Trier Social Stress Test (TSST), which includes free speech and mental arithmetic task in front of an audience, to induce moderate psychosocial stress. The brain connectivity measures revealed that the two situations, including before and 20 min after the TSST exposure, produced the same levels of stress. This indicates that the persistence of stress after 20 min fades and the brain network mimics the condition before stress. Al-Shargie et al. [119] used MIST, which increased beta rhythm power and decreased alpha rhythm power in the right pre-frontal cortex (sensitive to mental stress) and this is what was estimated by fMRI studies [120,121]. Likewise, using MIST task, the ventrolateral prefrontal area (VLPFC) achieved a higher accuracy than other PFC subregions [56]. Stroop color word task affects the temporal and spatiotemporal regions where several stress levels are induced individually to each subject [11]. For Maastricht Acute Stress Test (MAST), its protocol induces a realistic stress reaction in the subjects, which leads to variation of several salient physiological features [105]. Finally, driving task shows increased cortical activities for low level of stress, but it decreases with elevated stress level and time. Hence, this test makes a drop in alpha rhythm power when moving from rest to the stress state [122]. Figures 3 and 4 compare the resulted classification accuracy of different types of EEG data analysis methods using MIST and SCWT stressors, respectively.

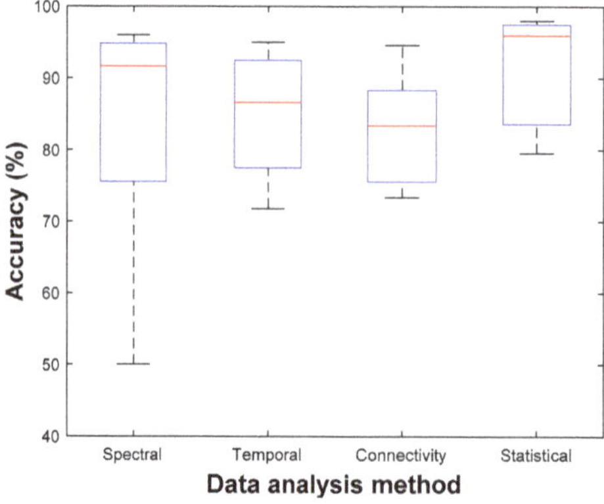

Figure 3. Classification accuracy with MIST stressor.

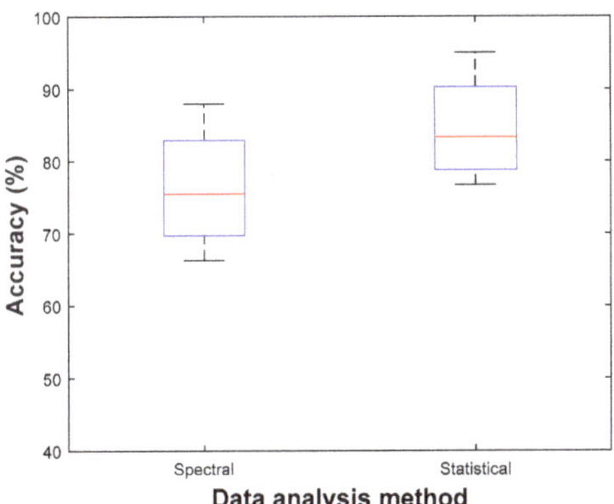

Figure 4. Classification accuracy with SCWT stressor.

Some experiments of stress detection combined more than one stressor such as arithmetic task with either Stroop test [102,123] or relaxing videos [19] and mental workload with public speaking [124]. Moreover, as discussed by studies in [19,124], employing normal four frequency bands showed accuracy levels of 83.33% and 84.30% using NB and RF classification methods, respectively. However, Ahn et al. [123] derived two frequency fields (low and high bands) and reached 77.90% accuracy by SVM whereas the accuracy of Jun et al. [102] was about 96% by three different bands (theta, alpha, and beta) with SVM classifier.

For studies that are interested in analyzing stress in normal daily life (psychological labelling), no stressors were introduced to the subjects. They used the same procedure in labelling participants and acquiring EEG data. There was an obvious variation in the treated frequency ranges. Thus, the highest accuracy was acquired when dealing with seven bands where they got 85.20% for SVM [108] comparing with the three bands 78.57% [54] and four bands 83.33% [106] that have used same classifier. Besides, the lowest performance was related to two frequency fields with 71.4% accuracy with NB classifier.

There are significant accuracies that have been achieved related to variety of stressor types. Studies in [118,124,125] used four bands but different classifiers and stimuli; for example, Lotfan et al. [118] obtained an accuracy of 92.31% with SVM and TSST stressor and noted increasing levels for another physiological measurement, which was cortisol level, whereas Masood et al. [125] detected 87.50% performance when applying CNN classifier and cognitive tasks, but Secerbegovic et al. [124] got a low value of 77.08% for SVM and mental workload test despite detecting a critical positive effect for applying EDA and ECG with the used EEG.

Another set of studies examined the temporal lobe when having stressors as a form of odor and traffic noise. They found a positive correlation between mental stress and EEG beta power rhythms [126–128]. Table 1 summarizes previous studies related to mental stress classification using EEG signal. The summary focuses on the type of techniques that are used to quantify mental stress taking into consideration the number of subjects, number of EEG channels, type of stressor, duration of the experiment, the analyzed frequency band, the extracted features, type of classifier, and the achieved performance. The summary in Table 1 orders the reviewed studies based on the type of stressor.

Table 1. Previous studies related to mental stress classification using EEG signal.

Reference Number	Year	Number of Subjects	Number of Channels	Type of Stressor	Condition Types	Brain Regions	Experiment Duration	Frequency Bands	Features	Classifier	Performance Acc: Accuracy Spec: Specificity Sen: Sensitivity	Notes
[129]	2020	25	4	MIST	Relax–Stress	PFC	24 min	4–13 Hz 8–12 Hz 13–30 Hz 25–45 Hz	Frontal asymmetry alpha, Beta and Gamma power	LDA	Acc: 85.6%	Quantifying two levels of stress had a better accuracy than three levels (85.6% vs. 58.4%).
[111]	2018	10	4	MIST	Relax–Stress–Neutral	PFC	30 min	Theta, alpha, gamma	Average relative gamma	LDA	Acc: 50.00%	Considering heart rate, skin resistance and trapezius activity features with EEG increased accuracy to 86.00%.
[44]	2018	22	128	MIST	Rest–Three stress levels	PFC	80 min	Delta, theta, alpha, beta, gamma	AP, RP, PL, coherence, relative power ratio, amplitude asymmetry	SVM	Acc: 79.54% Sen: 81.00% Spec: 78.00%	The existence of machine learning techniques provided automatic diagnosis for stress phase.
[56]	2017	25	7	MIST	Control–Stress	PFC	10 min	Delta, theta, alpha, beta	CCA	SVM	Acc: 89.80% Sen: 87.50% Spec: 92.00%	Mental stress was specific and localized to the right ventrolateral PFC.
[45]	2017	22	128	MIST	Control–Stress	Whole scalp	80 min	Delta, theta, alpha1, alpha2, beta1, beta2, beta3, gamma1, gamma2, gamma3	AP, RP, coherence, PL, Amplitude asymmetry	NB SVM	Acc: 94.6% Sen: 98.3% Spec: 93.3% Acc: 93.9% Sen: 96.7% Spec: 92.5%	Using ICA was not recommended with coherence and PL.
[119]	2016	22	7	MIST	Control–Stress	PFC	25 min	Delta, theta, alpha, beta	Mean powers	SVM	Acc: 91.7% Sen: 90.4% Spec: 93.4%	- PFC region was sensitive to mental stress. - Adding fNIRS to EEG improved the accuracy to 95.1%.
[20]	2015	12	8	MIST	Rest–Three stress levels	PFC	60 min	Delta, Theta, Alpha, Beta, Noisy gamma, Noisy signal (64–128Hz)	PSD, energy, average power	SVM	Acc: 94.00%	Increasing difficulty level reduced subject engagement with the stress task.

Table 1. *Cont.*

Reference Number	Year	Number of Subjects	Number of Channels	Type of Stressor	Condition Types	Brain Regions	Experiment Duration	Frequency Bands	Features	Classifier	Performance Acc: Accuracy Spec: Specificity Sen: Sensitivity	Notes
[79]	2015	5	7	MIST	Control–Stress	PFC	17 min	Alpha	Alpha PSD	SVM	Acc: 95.00%	Alpha rhythm and oxygenated haemoglobin had negative correlation under stress.
[130]	2021	25	7	Mental arithmetic	Control–Stress	PFC	15 min	Alpha	PSD, coherence	-	-	EEG alpha rhythmic demonstrated significant decrease in brain functional connectivity.
[88]	2020	66	19	Mental arithmetic	Relax–Stress	Temporal, Frontal, Central, Occipital, Parietal	4 min	Beta, Alpha, Theta, Delta	MDF, MFMD, SM, RMS, AR	linear SVM Cubic SVM KNN LDA	Acc: 99.70% Acc: 99.75% Acc: 99.93% Acc: 99.94%	Highest accuracy achieved using only two frontal brain electrodes.
[62]	2018	1	14	Mental arithmetic	Rest–Stress	Frontal lobe	31 min	Delta (0.5–3Hz), Theta (3–8Hz), Alpha (8–12Hz), Beta (12–25Hz)	Self-Entropy, MI, Conditional MI	RF	Acc: 97.50%	Combining EEG connectivity with cardiorespiratory features offered high accuracy.
[112]	2016	6	14	Mental arithmetic	Neutral–Three stress levels	Prefrontal, Frontal lobes	20 min	Delta, theta, alpha, beta, gamma	IMF, instantaneous frequency (using HHT)	SVM KNN LDA	Acc: 89.07% Acc: 72.67% Acc: 70.17%	The highest classification accuracy was detected in alpha band.
[23]	2011	5	19	Mental arithmetic	Relax–Stress	Whole scalp	11 min	Delta, theta, alpha, beta	Welch, Yule walker, Burg methods	ANN	91.17%	Maximum classification accuracy achieved when using burg extraction method.
[123]	2019	14	3	Mental arithmetic, SCWT	Relax–Stress	Left, right hemisphere	34 min	Low (0.04–0.15Hz) High (0.15–0.4Hz)	Normalized band power, Power asymmetry	SVM	Acc: 77.90% Spec: 72.00% Sen: 84.60%	Combining HRV with EEG increased accuracy up to 87.50%.

Table 1. *Cont.*

Reference Number	Year	Number of Subjects	Number of Channels	Type of Stressor	Condition Types	Brain Regions	Experiment Duration	Frequency Bands	Features	Classifier	Performance Acc: Accuracy Spec: Specificity Sen: Sensitivity	Notes
[19]	2019	17	14	Mental arithmetic, videos, playing games	Relax–Two stress levels	Frontal, Occipital lobes	31 min	Delta, theta, alpha, beta	Shannon entropy, mutual information, covariance, precision	RF LR	Acc: 84.30% Acc: 84.30%	Adding HRM and EDA to EEG measurements achieved better accuracy.
[102]	2016	10	14	Mental arithmetic, SCWT	Rest–Two stress levels	Frontal, Occipital lobes	18 min	Theta, alpha, beta	PSD, relative difference of alpha and beta power	SVM	Acc: 96.00%	Providing temporal sliding window with different overlapping increased the accuracy.
[131]	2020	227	62	Psychological labelling	Relax–Stress	PFC	16 min	2.5–3 Hz 24–24.5 Hz 24.5–25 Hz 26–26.5 Hz	PSD	RF	Acc: 81.33% Spec: 80.33% Sen: 82.33%	People who are more stressed, had higher spectral capacity in the left prefrontal cortex.
[108]	2020	33	5	Psychological labelling	Rest–Stress	Frontal, Temporal lobes	39–58 min	Delta, theta, alpha, beta, gamma, slow (4–13Hz), low beta (13–17Hz)	Neural oscillatory features, alpha and beta asymmetry.	SVM NB KNN LR MLP	Acc: 85.20% Acc: 80.79% Acc: 65.96% Acc: 85.15% Acc: 85.13%	- SVM was best in detecting long-term stress when used with alpha asymmetry feature.
[54]	2018	28	1	Psychological labelling	Rest–Stress	Frontal lobe	3 min	Low beta, high beta, low gamma	Neural oscillatory features	SVM	Acc: 78.57%	Using correlation-based feature subset selection method with SVM gave higher accuracy.
[109]	2017	28	1	Psychological labelling	Rest–Stress	Frontal lobe	3 min	Low beta (13–17Hz) High beta (18–30Hz)	Low beta waves, linear regression	NB	Acc: 71.4%	- NB took less computational time comparing to SVM and MLP. - Low beta can be used as feature to quantify stress.

75

Table 1. *Cont.*

Reference Number	Year	Number of Subjects	Number of Channels	Type of Stressor	Condition Types	Brain Regions	Experiment Duration	Frequency Bands	Features	Classifier	Performance Acc: Accuracy Spec: Specificity Sen: Sensitivity	Notes
[90]	2019	32	32	Music videos	Relax–Stress	Frontal lobe	-	Delta, theta, alpha, beta, gamma	Nine statistical features, Hjorth parameters, energy, standard deviation, wavelet sum of entropy, PSD.	KNN	Acc: 73.38%	- Using appropriate processing techniques enhances performance. - Final performance can be affected by comparative analysis of scalp sources.
[92]	2018	32	32	Music videos	Relax–Stress	Frontal lobe	40 min	Delta, theta, alpha, beta, gamma	Six statistical features, PSD, HOC, Hjorth parameters, Frontal Asymmetry Alpha.	KNN	Acc: 67.08%	GA-based method outperformed PCA in stress detection and gave better classification accuracy using KNN classifier.
[69]	2018	26	9	Audio-visual stimuli	Relax–Stress	PFC	10 min	Theta, alpha, beta(1,2,3,4,5)	PSI, DTF, GPDC	KNN+SVM	Acc: 90.62%	The effect of stress on brain regions connectivity depends on the subject.
[51]	2011	26	8	Exam stress	Relax–Stress	Whole scalp	6.30 min	2–32 Hz	Higuchi's fractal dimension, Gaussian mixtures of EEG spectrogram, MSCE.	KNN SVM	Acc: 90.00% Acc: 90.00%	MSCE provided a promising inter-subject validation accuracy (90%) in classifying the EEG.
[132]	2018	11	14	Working hazards and tiredness	Low stress–High stress	Frontal, Occipital lobes	Several hours	Delta, theta, alpha, low beta, beta, high beta, gamma	Seven frequency features, 17 time domain features	KNN GDA SVM	Acc: 65.80% Acc: 74.92% Acc: 75.90%	- KNN showed low accuracy because of inductive bias of KNN method.
[118]	2018	23	30	TSST	Relax–Stress	Frontal, Parietal, Temporal lobes	13 min	Delta, theta, alpha, beta	Transitivity, modularity, path length, efficiency	SVM (beta wave)	Eyes-opened Acc: 92.31% Eyes-closed Acc: 93.62%	- Salivary cortisol level increased during TSST. - Classifier accuracy arised from alpha and beta bandwidths of EEG.
[11]	2015	9	14	SCWT	Low stress–High stress	Frontal, Occipital lobes	12 min	Theta, alpha, beta	PSD, FD, Six statistical features.	SVM KNN	Acc: 85.17% Acc: 76.72%	The followed methodology provided real-time EEG-base stress recognition.

Table 1. *Cont.*

Reference Number	Year	Number of Subjects	Number of Channels	Type of Stressor	Condition Types	Brain Regions	Experiment Duration	Frequency Bands	Features	Classifier	Performance Acc: Accuracy Spec: Specificity Sen: Sensitivity	Notes
[125]	2019	24	2	SCWT	Relax–Stress	Frontal lobe	11 min	Theta, delta, alpha, beta	PSD	CNN	Acc: 87.50% Sen: 87.50% Spec: 84.00%	Adding HRM, RESP and EDA to EEG will increase stress detection accuracy.
[133]	2020	20	14	Cognitive task	Low stress– High stress	Frontal, Occipital lobes	30 min	Alpha	Discrete wavelet transform	CNN	Acc: 93.00% Sen: 0.923% Spec: 0.934%	-
[105]	2017	15	1	MAST	Relax–Stress	Forehead position	65 min	Alpha1 (8–9Hz), alpha2 (10–12Hz), Beta1 (13–17Hz), Beta2 (18–30Hz)	Attention, Meditation, PSD	SVM	Acc: 86.00% Sen: 84.00% Spec: 90.00%	Changes in the physiological features were correlated with the trend of salivary alpha amylase.
[124]	2017	9	1	Mental workload + public speaking	Three stress levels	Forehead position	-	Delta, theta, alpha, beta	Energy, mean amplitude, RMS, maximum amplitude, weighted mean frequency	SVM NB	Acc: 77.08 Acc: 83.33%	Adding ECG and EDA features to EEG features would improve accuracy.
[50]	2019	26	9	Long-term psychological task	Relax–Stress	Whole scalp	8 min	Theta, alpha1, alph2, beta1, beta2, beta3, gamma1	PSD, LI, CC, CCA, PSI, GC, DTF, MI, MSCE	SVM KNN NB	94.00% 92.00% 90.00%	Using MI and DTF provided the highest accuracy for stress detection.
[122]	2020	86	16	Driving task	Relax–Stress	Frontal, Occipital lobes	45 min	Theta, alpha, low beta, high beta, gamma	Six time-domain features, three frequency-domain features (PSD)	SVM NN RF	Acc: 90.55% Acc: 87.70% Acc: 85.00%	- Adding feature selection technique (MI, PCA, RSFS) improved classifier accuracy. - Combining SVM with MI achieved best accuracy.

Note: SCWT: Stroop Color-Word Task; MIST: Montreal Imagining Stress Task; SMO: Sequential Minimum Optimization; PL: Phase Lag; Acc: Accuracy; Sen: Sensitivity; Spec: specificity; fNIRS: Functional Near-Infrared Spectroscopy; PFC: Prefrontal Cortex; RMS: root mean square; TSST: Trier Social Stress Test; BCI: Brain–computer interface; ECOC: error-correcting output code; FD: fractal dimension; GA: genetic algorithm; NN: Neural network; RF: Random forest; RSFS: Random subset feature selection; SVC: Support vector classification; EDA: Electrodermal activity; HRM: Heart rate monitor; IMF: Intrinsic mode function; HHT: Hilbert-Huang Transform; AR: Asymmetry ration; RER: Relative energy ratio; SC: Spectral centroids; SE: Spectral entropy; ESD: Energy spectral density; FCM: Fuzzy C-mean; FKM: Fuzzy K-means; ESD: Energy spectral density; LI: Laterality index; CC: Correlation coefficient; GC: Granger causality; MAST: Maastricht Acute Stress Test; RMS: Root mean square; AR: Auto regression; EDA: Electrodermal activity; HRM: Heart rate monitor; ANN: Artificial neural network; GDA: Gaussian discernment analysis; IAPS: International Affective Picture System; ENN: Elman neural network; CNN: Convolutional Neural Network.

6. Discussion

Stress has become a growing problem in our daily lives by having a negative impact on both individuals and society. Different systems of the human body, such as the nervous, immune, cardiovascular, and gastrointestinal systems, are negatively affected by stress. This directly influences or transforms the hippocampus, a brain field, regardless of the nature of the stress. The victim's memory and decision-making capabilities are harmed as a result of this brain alteration. It also has a detrimental effect on hormone excretion, which is important for proper immune system processing. Stress also causes cardiac-arrhythmias by amplifying or decreasing heartbeats, blood pressure, and creating disturbances in the cardio-vascular system. Meanwhile, it has negative effects on the gastrointestinal (GI) system, such as decreased appetite, disruption of normal GI tract activity, and crabby-bowel-syndrome. Thus, mental stress evaluation and analysis are very important procedures that can be done to detect stress in order to prevent significant health problems. Despite the number of studies that covered this phenomenon using EEG signals, there is a lack of inclusive guidelines about the relevance between EEG feature and its extraction methods. Here, we conducted a comprehensive review on the methods of analysis of mental stress-based EEG signals. Specifically, our review focused on the type of the method used for data analysis and classification model. In particular, we found that selecting the right method of analysis is challenging because of factors variety that are exercised in the experiments. These factors include EEG sensor, sample size, stressor type, task duration, time of the day, proper EEG processing, feature extraction mechanism, number of features and type of classifier. Therefore, the significant part related to mental stress quantification is choosing the most appropriate features. Another case of concern is the large discrepancy between individuals and response to stress. For example, different stress response may be acquired for a particular subject depending on his psychology, sociality, health, and emotional state.

The methods of quantifying mental stress using EEG varies across the analysis spectrum. As previously stated, because the brain acts in networks, descriptors of network functioning will be required to completely comprehend neural processing. In this work, we provided a comprehensive review on these analysis methods. Meanwhile, we highlighted the key differences spotted between the research findings and argued that variations of the data analysis techniques could be a significant contributing factor towards several contradictory results. Besides, the extracting features that are related to brain connectivity showed a clear model of the brain and how its different regions are interacting with each other. Therefore, studying feature extraction techniques related to brain connectivity provides a clear model of the brain and how its different regions are interacting with each other.

Moreover, there is a variety in the experiment duration between the discussed references. Thirty minutes process of the study in [111] involves maximum voluntary contraction (MVC), resting state (RS), MIST training and task, three questions about self-perceived level of stress and a relaxation period. Meanwhile, the eighty minutes period of data acquisition of the studies [44,45] comes from two conditions (stress and control) where each one consists of 40 min of habituation, rest, four levels main condition, and recovery periods. In the experiment protocol of Al-Shargie et al. [20], it takes 60 min duration divided between introduction, training, resting, and the main experiment, which needs about 40 min using three levels of mental arithmetic task. Besides, the four minutes of the study in [88] depend on the experiment procedure, which includes 3 min of counting and 1 min of serial subtraction where EEG data is recorded. They have used 18 min duration in the study that involves a brief introduction, training, data recording for control and stress conditions. Saeed et al. [108] achieved an increase in beta rhythm power and a decrease in alpha rhythm power in the pre-frontal cortex with a total duration of 25 min. Consequently, to avoid the effect of time on subject's cognitive ability and the influences of circadian rhythm on stress performance, it is preferred to conduct the EEG experiment on all participants at the same time of day [44].

The task nature and sample size had a direct influence on classifier accuracy, such as the restricted duration in doing mental arithmetic tasks, which leads to low performance accuracy [45] and the varied results gained with a large number of participants with the studies in [44,45] compared to [111]. It is worth noting that the majority of studies have a limited sample size, meaning that the amount of people involved is insufficient to overcome prejudices caused by individual differences. A larger sample size is needed to ensure statistical power and to bolster our findings.

On the other hand, decreasing the number of EEG electrodes maintains real time stress detection, but could increase system mobility and ease. Therefore, using one or two frontal electrodes might be sufficient to detect stress/control phases, but in order to get their level it is better to use more electrodes as suggested [88]. As mentioned in the EEG processing section, the extracted EEG signal undergoes to several denoising processes that may eliminate the unwanted peaks and artefacts, but small remaining noise could deform the information of analyzed EEG.

Finally, the used EEG sensor to record data and measure mental stress has a huge impact on the number of channels available. The number of channels in a typical EEG system can range from 1 to 256. The 10/20 system, which governs the positioning of electrodes on the brain, is followed. The benefits of multi-channel EEG systems are that they do a better job of avoiding data loss, particularly as the sensor network expands to more channels (caused by when electrode distances grow further apart when fewer are deployed) as well as in detecting vital clinical signals. This means that medical applications need higher resolution EEG systems (larger sensor networks) to complete the task.

7. Challenges and Future Work

Most of studies induced stress in controlled environments, whereas the better method is to develop a protocol that sustains the real scenarios such as virtual reality. Furthermore, the discussed researches did not correlate the physiological changes, such as cortisol levels, with the behavioral response. Most of the reviewed studies conducted offline experiments, but we suggest developing an online system that deals with stress recognition in the real time. Moreover, one of the critical factors that influences stress assessment results is the ground truth that is needed to train the classifier by sorting subjects into stress/non-stress groups. Most of studies established this labelling by questionnaire score, psychologist interview or both of them. Nevertheless, these two methods cannot provide a direct judgment on mental stress existence because of the high dependency on participants themselves (in many cases, they expect a wrong stress situation because of subconsciousness). Unlike the used simulated experiments, a significant challenge will be faced when labelling subjects in real world tasks.

As a future work, suggesting EEG feature extraction techniques could be useful in improving stress detection such as phase synchronization and source localization. Phase synchronization is used to analyze interdependence between two-time EEG signals regardless of their amplitude. It has high sensitivity that leads to detect dynamical changes of brain functions during mental stress. While EEG is a powerful tool for measuring neuronal activity and connectivity, the lack of spatial resolution could be a drawback. EEG source localization may be used to estimate the locations of electrical activities from the scalp potential measurements. The information of localization about these active sources (depending on the recorded potential from the electrodes) provides a good diagnosis for the mental state and brain abnormalities. This method can be combined with other feature extraction techniques such as directed connectivity measures.

8. Conclusions

In this paper, we have presented a comprehensive review of EEG signal analysis methods for the assessment of mental stress. A rigorous procedure was adopted for the search strategy and identification of relevant studies. The review emphasized the major discrepancies between the research findings. It also suggests that various data processing methodologies have contributed to numerous conflicting outcomes. These various can be attributed to a number of variables, including the lack of a consistent procedure, brain regions of interest, type of stressor, duration of experiment, EEG signal processing, feature extraction technique, and the type of classifiers used. In addition, we have reported the effect of sample size bias in connectivity estimation. This problem can be solved by equalizing sample sizes between different conditions or participants, using statistical methods that explicitly account for sample size bias, or employing connectivity approaches that are not affected by sample size bias. Moreover, understanding the relationships between mental stress and the complex and diverse EEG characteristics, such as time-varying, functional, and dynamic brain connections, necessitates the integration of several data analysis methods. As a result, we propose combining the network connectivity measures with deep learning to increase the accuracy of assessing mental stress levels.

Author Contributions: Conceptualization, R.K., F.A.-S. and H.A.-N.; Methodology, R.K. and F.A.-S.; Formal Analysis, R.K. and F.A.-S.; Resource, R.K., F.A.-S., H.A.-N., U.T., F.A.-M. and F.B.; Validation, R.K., F.A.-S., H.A.-N., U.T., F.A.-M. and F.B.; Data Curation, R.K. and F.A.-S.; Writing-Original Draft Preparation, R.K., F.A.-S.; Writing-Review & Editing, R.K., F.A.-S., H.A.-N., U.T., F.A.-M. and F.B.; Supervision, F.A.-S., H.A.-N. and U.T. Funding acquisition F.A.-S., H.A.-N., U.T., F.A.-M. and F.B. All authors have read and agreed to the published version of the manuscript.

Funding: This research was funded by American University of Sharjah, FRG-20-L-E25.

Institutional Review Board Statement: Not Applicable.

Informed Consent Statement: Not Applicable.

Data Availability Statement: Not Applicable.

Acknowledgments: We would like to express our gratitude to the editor and three anonymous reviewers for their insightful comments and feedback.

Conflicts of Interest: The authors declare no conflict of interest.

References

1. Selye, H. The stress syndrome. *Am. J. Nurs.* **1965**, *65*, 97–99.
2. Giannakakis, G.; Grigoriadis, D.; Giannakaki, K.; Simantiraki, O.; Roniotis, A.; Tsiknakis, M. Review on psychological stress detection using biosignals. *IEEE Trans. Affect. Comput.* **2019**, 1–16. [CrossRef]
3. Lazarus, J. *Stress Relief & Relaxation Techniques*; McGraw Hill Professional: New York, NY, USA, 2000.
4. Bakker, J.; Pechenizkiy, M.; Sidorova, N. What's your current stress level? Detection of stress patterns from GSR sensor data. In Proceedings of the 2011 IEEE 11th International Conference on Data Mining Workshops, Vancouver, BC, Canada, 11 December 2011; pp. 573–580.
5. Colligan, T.W.; Higgins, E.M. Workplace stress: Etiology and consequences. *J. Workplace Behav. Health* **2006**, *21*, 89–97. [CrossRef]
6. Crowley, O.V.; McKinley, P.S.; Burg, M.M.; Schwartz, J.E.; Ryff, C.D.; Weinstein, M.; Seeman, T.E.; Sloan, R.P. The interactive effect of change in perceived stress and trait anxiety on vagal recovery from cognitive challenge. *Int. J. Psychophysiol.* **2011**, *82*, 225–232. [CrossRef] [PubMed]
7. O'Connor, D.B.; Thayer, J.F.; Vedhara, K. Stress and health: A review of psychobiological processes. *Annu. Rev. Psychol.* **2021**, *72*, 663–688. [CrossRef] [PubMed]
8. Thoits, P.A. Stress and health: Major findings and policy implications. *J. Health Soc. Behav.* **2010**, *51*, S41–S53. [CrossRef] [PubMed]
9. Garg, A.; Chren, M.-M.; Sands, L.P.; Matsui, M.S.; Marenus, K.D.; Feingold, K.R.; Elias, P.M. Psychological stress perturbs epidermal permeability barrier homeostasis: Implications for the pathogenesis of stress-associated skin disorders. *JAMA Dermatol.* **2001**, *137*, 53–59. [CrossRef] [PubMed]
10. C.Adam, T.; S.Epel, E. Stress, eating and the reward system. *Physiol. Behav.* **2007**, *91*, 449–458. [CrossRef] [PubMed]
11. Hou, X.; Liu, Y.; Sourina, O.; Tan, Y.R.E.; Wang, L.; Mueller-Wittig, W. EEG based stress monitoring. In Proceedings of the 2015 IEEE International Conference on Systems, Man, and Cybernetics, Hong Kong, China, 9–12 October 2015; pp. 3110–3115.
12. Holmes, T.H.; Rahe, R.H. The social readjustment rating scale. *J. Psychosom. Res.* **1967**, *11*, 213–218. [CrossRef]
13. Monroe, S.M. Modern approaches to conceptualizing and measuring human life stress. *Annu. Rev. Clin. Psychol.* **2008**, *4*, 33–52. [CrossRef]
14. Weiner, I.B.; Craighead, W.E. State-Trait anxiety inventory. *Corsini Encycl. Psychol.* **2010**, *4*, 2002.
15. Kirschbaum, C.; Hellhammer, D. Salivary cortisol in psychoneuroendocrine research: Recent developments and applications. *Psychoneuroendocrinology* **1994**, *19*, 313–333. [CrossRef]
16. Hanrahan, K.; McCarthy, A.M.; Kleiber, C.; Lutgendorf, S.; Tsalikian, E. Strategies for salivary cortisol collection and analysis in research with children. *Appl. Nurs. Res.* **2006**, *19*, 95–101. [CrossRef] [PubMed]

17. Engert, V.; Vogel, S.; Efanov, S.I.; Duchesne, A.; Corbo, V.; Ali, N.; Pruessner, J.C. Investigation into the cross-correlation of salivary cortisol and alpha-amylase responses to psychological stress. *Psychoneuroendocrinology* **2011**, *36*, 1294–1302. [CrossRef]
18. Koibuchi, E.; Suzuki, Y. Exercise upregulates salivary amylase in humans. *Exp. Ther. Med.* **2014**, *7*, 773–777. [CrossRef] [PubMed]
19. Zanetti, M.; Mizumoto, T.; Faes, L.; Fornaser, A.; De Cecco, M.; Maule, L.; Valente, M.; Nollo, G. Multilevel assessment of mental stress via network physiology paradigm using consumer wearable devices. *J. Ambient Intell. Humaniz. Comput.* **2019**, *12*, 4409–4418. [CrossRef]
20. Al-Shargie, F.; Tang, T.B.; Badruddin, N.; Kiguchi, M. Mental stress quantification using EEG signals. In Proceedings of the International Conference for Innovation in Biomedical Engineering and Life Sciences, Putrajaya, Malaysia, 6–8 December 2015; pp. 15–19.
21. Arrighi, J.A.; Burg, M.; Cohen, I.S.; Kao, A.H.; Pfau, S.; Caulin-Glaser, T.; Zaret, B.L.; Soufer, R. Myocardial blood-flow response during mental stress in patients with coronary artery disease. *Lancet* **2000**, *356*, 310–311. [CrossRef]
22. Zhang, X.; Huettel, S.A.; O'Dhaniel, A.; Guo, H.; Wang, L. Exploring common changes after acute mental stress and acute tryptophan depletion: Resting-state fMRI studies. *J. Psychiatr. Res.* **2019**, *113*, 172–180. [CrossRef]
23. Saidatul, A.; Paulraj, M.P.; Yaacob, S.; Yusnita, M.A. Analysis of EEG signals during relaxation and mental stress condition using AR modeling techniques. In Proceedings of the 2011 IEEE International Conference on Control System, Computing and Engineering, Penang, Malaysia, 25–27 November 2011; pp. 477–481.
24. Hu, B.; Peng, H.; Zhao, Q.; Hu, B.; Majoe, D.; Zheng, F.; Moore, P. Signal quality assessment model for wearable EEG sensor on prediction of mental stress. *IEEE Trans. Nanobiosci.* **2015**, *14*, 553–561.
25. Al-Shargie, F.; Tang, T.B.; Kiguchi, M. Stress assessment based on decision fusion of EEG and fNIRS signals. *IEEE Access* **2017**, *5*, 19889–19896. [CrossRef]
26. Al-Shargie, F.; Tang, T.B.; Badruddin, N.; Kiguchi, M. Towards multilevel mental stress assessment using SVM with ECOC: An EEG approach. *Med Biol. Eng. Comput.* **2018**, *56*, 125–136. [CrossRef]
27. Ehrhardt, N.M.; Fietz, J.; Kopf-Beck, J.; Kappelmann, N.; Brem, A.K. Separating EEG correlates of stress: Cognitive effort, time pressure, and social-evaluative threat. *Eur. J. Neurosci.* **2021**, 1–10. [CrossRef]
28. Kulkarni, N.; Phalle, S.; Desale, M.; Gokhale, N.; Kasture, K. A Review on EEG Based Stress Monitoring System Using Deep Learning Approach. *Mukt Shabd J.* **2020**, *9*, 1317–1325.
29. Keshmiri, S. Conditional Entropy: A Potential Digital Marker for Stress. *Entropy* **2021**, *23*, 286. [CrossRef]
30. Alyan, E.; Saad, N.M.; Kamel, N.; Yusoff, M.Z.; Zakariya, M.A.; Rahman, M.A.; Guillet, C.; Merienne, F. Frontal Electroencephalogram Alpha Asymmetry during Mental Stress Related to Workplace Noise. *Sensors* **2021**, *21*, 1968. [CrossRef] [PubMed]
31. Chae, J.; Hwang, S.; Seo, W.; Kang, Y. Relationship between rework of engineering drawing tasks and stress level measured from physiological signals. *Autom. Constr.* **2021**, *124*, 103560. [CrossRef]
32. Al-Shargie, F.; Tariq, U.; Babiloni, F.; Al-Nashash, H. Cognitive Vigilance Enhancement Using Audio Stimulation of Pure Tone at 250 Hz. *IEEE Access* **2021**, *9*, 22955–22970. [CrossRef]
33. Movahed, R.A.; Jahromi, G.P.; Shahyad, S.; Meftahi, G.H. A Major Depressive Disorder Classification Framework based on EEG Signals using Statistical, Spectral, Wavelet, Functional Connectivity, and Nonlinear Analysis. *J. Neurosci. Methods* **2021**, *358*, 109–209. [CrossRef]
34. Cohen, L. *Time-Frequency Analysis*; Prentice Hall: Englewood Cliffs, NJ, USA, 1995; Volume 778.
35. Boonyakitanont, P.; Lek-Uthai, A.; Chomtho, K.; Songsiri, J. A review of feature extraction and performance evaluation in epileptic seizure detection using EEG. *Biomed. Signal Process. Control* **2020**, *57*, 101702. [CrossRef]
36. Moher, D. Corrigendum to: Preferred Reporting Items For Systematic Reviews And Meta-Analyses: The PRISMA Statement International Journal of Surgery. *Int. J. Surg.* **2010**, *8*, 336–341. [CrossRef]
37. Jiang, X.; Bian, G.-B.; Tian, Z. Removal of Artifacts from EEG Signals: A Review. *Sensors* **2019**, *19*, 987. [CrossRef] [PubMed]
38. Mumtaz, W.; Rasheed, S.; Irfan, A. Review of challenges associated with the EEG artifact removal methods. *Biomed. Signal Process. Control* **2021**, *68*, 1–13. [CrossRef]
39. Srinivasan, R.; Winter, W.R.; Ding, J.; Nunez, P.L. EEG and MEG coherence: Measures of functional connectivity at distinct spatial scales of neocortical dynamics. *J. Neurosci. Methods* **2007**, *166*, 41–52. [CrossRef] [PubMed]
40. Nunez, P.L.; Srinivasan, R. *Electric Fields of the Brain: The Neurophysics of EEG*; Oxford University Press: Oxford, MI, USA, 2006.
41. Horwitz, B. The Elusive Concept of Brain Connectivity. *Neuroimage* **2003**, *19*, 466–470. [CrossRef]
42. Friston, K.J. Functional and effective connectivity in neuroimaging: A synthesis. *Hum. Brain Mapp.* **1994**, *2*, 56–78. [CrossRef]
43. Friston, K.J. Functional and effective connectivity: A review. *Brain Connect.* **2011**, *1*, 13–36. [CrossRef]
44. Xia, L.; Malik, A.S.; Subhani, A.R. A physiological signal-based method for early mental-stress detection. *Biomed. Signal Process. Control* **2018**, *46*, 18–32. [CrossRef]
45. Subhani, A.R.; Mumtaz, W.; Saad, M.N.B.M.; Kamel, N.; Malik, A.S. Machine learning framework for the detection of mental stress at multiple levels. *IEEE Access* **2017**, *5*, 13545–13556. [CrossRef]
46. Subhani, A.R.; Malik, A.S.; Kamil, N.; Saad, M.N.M. Difference in brain dynamics during arithmetic task performed in stress and control conditions. In Proceedings of the 2016 IEEE EMBS Conference on Biomedical Engineering and Sciences (IECBES), Kuala Lumpur, Malaysia, 4–8 December 2016; pp. 695–698.
47. Thatcher, R.; Krause, P.; Hrybyk, M. Cortico-cortical associations and EEG coherence: A two-compartment model. *Electroencephalogr. Clin. Neurophysiol.* **1986**, *64*, 123–143. [CrossRef]

48. Al-Shargie, F. Early Detection of Mental Stress Using Advanced Neuroimaging and Artificial Intelligence. *arXiv* **2019**, arXiv:1903.08511.

49. Alonso, J.; Romero, S.; Ballester, M.; Antonijoan, R.; Mañanas, M. Stress assessment based on EEG univariate features and functional connectivity measures. *Physiol. Meas.* **2015**, *36*, 1351. [CrossRef]

50. Darzi, A.; Azami, H.; Khosrowabadi, R. Brain functional connectivity changes in long-term mental stress. *J. Neurodev. Cogn.* **2019**, *1*, 16–41.

51. Khosrowabadi, R.; Quek, C.; Ang, K.K.; Tung, S.W.; Heijnen, M. A Brain-Computer Interface for classifying EEG correlates of chronic mental stress. In Proceedings of the 2011 International Joint Conference on Neural Networks, San Jose, CA, USA, 31 July–5 August 2011; pp. 757–762.

52. Harada, H.; Shiraishi, K.; Kato, T.; Soda, T. Coherence analysis of EEG changes during odour stimulation in humans. *J. Laryngol. Otol.* **1996**, *110*, 652–656. [CrossRef] [PubMed]

53. Carter, G.; Knapp, C.; Nuttall, A. Estimation of the magnitude-squared coherence function via overlapped fast Fourier transform processing. *IEEE Trans. Audio Electroacoust.* **1973**, *21*, 337–344. [CrossRef]

54. Umar Saeed, S.M.; Anwar, S.M.; Majid, M.; Awais, M.; Alnowami, M. Selection of neural oscillatory features for human stress classification with single channel eeg headset. *Biomed Res. Int.* **2018**, *2018*, 1049257. [CrossRef] [PubMed]

55. Hall, M.A. Correlation-based feature selection of discrete and numeric class machine learning. In Proceedings of the Seventeenth International Conference on Machine Learning, Standord, CA, USA, 29 June–2 July 2000; pp. 1–10.

56. Al-Shargie, F.; Tang, T.B.; Kiguchi, M. Assessment of mental stress effects on prefrontal cortical activities using canonical correlation analysis: An fNIRS-EEG study. *Biomed. Opt. Express* **2017**, *8*, 2583–2598. [CrossRef]

57. Ha, U.; Kim, C.; Lee, Y.; Kim, H.; Roh, T.; Yoo, H.-J. A multimodal stress monitoring system with canonical correlation analysis. In Proceedings of the 2015 37th Annual International Conference of the IEEE Engineering in Medicine and Biology Society (EMBC), Milan, Italy, 25–29 August 2015; pp. 1263–1266.

58. Gibbons, J.D.; Chakraborti, S. *Nonparametric Statistical Inference: Revised and Expanded*; CRC Press: Boca Raton, FL, USA, 2014.

59. Vuppalapati, C.; Raghu, N.; Veluru, P.; Khursheed, S. A system to detect mental stress using machine learning and mobile development. In Proceedings of the 2018 International Conference on Machine Learning and Cybernetics (ICMLC), Chengdu, China, 15–18 July 2018; pp. 161–166.

60. Dziembowska, I.; Izdebski, P.; Rasmus, A.; Brudny, J.; Grzelczak, M.; Cysewski, P. Effects of heart rate variability biofeedback on EEG alpha asymmetry and anxiety symptoms in male athletes: A pilot study. *Appl. Psychophysiol. Biofeedback* **2016**, *41*, 141–150. [CrossRef]

61. Thatcher, R.; Cantor, D. Electrophysiological techniques in the assessment of malnutrition. In *Malnutrition and Behavior: Critical Assessment of Key Issues: An International Symposium at a Distance, 1982–1983/Josef Brozek, Beat Schurch, Editors*; Nestle Foundation: Lausanne, Switzerland, 1984; pp. 116–136.

62. Zanetti, M.; Faes, L.; De Cecco, M.; Fornaser, A.; Valente, M.; Guandalini, G.; Nollo, G. Assessment of mental stress through the analysis of physiological signals acquired from wearable devices. In Proceedings of the Italian Forum of Ambient Assisted Living, Lecce, Italy, 2–4 July 2018; pp. 243–256.

63. Subhani, A.R.; Mumtaz, W.; Kamil, N.; Saad, N.M.; Nandagopal, N.; Malik, A.S. MRMR based feature selection for the classification of stress using EEG. In Proceedings of the 2017 Eleventh International Conference on Sensing Technology (ICST), Sydney, Australia, 4–6 December 2017; pp. 1–4.

64. Pernice, R.; Zanetti, M.; Nollo, G.; De Cecco, M.; Busacca, A.; Faes, L. Mutual Information Analysis of Brain-Body Interactions during different Levels of Mental stress. In Proceedings of the 2019 41st Annual International Conference of the IEEE Engineering in Medicine and Biology Society (EMBC), Berlin, Germany, 23–27 July 2019; pp. 6176–6179.

65. Barraza, N.; Moro, S.; Ferreyra, M.; de la Peña, A. Mutual information and sensitivity analysis for feature selection in customer targeting: A comparative study. *J. Inf. Sci.* **2019**, *45*, 53–67. [CrossRef]

66. Kraskov, A.; Stögbauer, H.; Grassberger, P. Estimating mutual information. *Phys. Rev. E* **2004**, *69*, 066138. [CrossRef] [PubMed]

67. Al-Shargie, F.; Tariq, U.; Alex, M.; Mir, H.; Al-Nashash, H. Emotion recognition based on fusion of local cortical activations and dynamic functional networks connectivity: An EEG study. *IEEE Access* **2019**, *7*, 143550–143562. [CrossRef]

68. Collura, T.F. Towards a coherent view of brain connectivity. *J. Neurother.* **2008**, *12*, 99–110. [CrossRef]

69. Khosrowabadi, R. Stress and Perception of Emotional Stimuli: Long-term Stress Rewiring the Brain. *Basic Clin. Neurosci.* **2018**, *9*, 107. [CrossRef] [PubMed]

70. Bastos, A.M.; Schoffelen, J.-M. A tutorial review of functional connectivity analysis methods and their interpretational pitfalls. *Front. Syst. Neurosci.* **2016**, *9*, 175. [CrossRef] [PubMed]

71. Stam, C.J.; Nolte, G.; Daffertshofer, A. Phase lag index: Assessment of functional connectivity from multi channel EEG and MEG with diminished bias from common sources. *Hum. Brain Mapp.* **2007**, *28*, 1178–1193. [CrossRef] [PubMed]

72. Al-Shargie, F.M.; Hassanin, O.; Tariq, U.; Al-Nashash, H. EEG-Based Semantic Vigilance Level Classification Using Directed Connectivity Patterns and Graph Theory Analysis. *IEEE Access* **2020**, *8*, 115941–115956. [CrossRef]

73. Al-Shargie, F.; Tariq, U.; Hassanin, O.; Mir, H.; Babiloni, F.; Al-Nashash, H. Brain Connectivity Analysis Under Semantic Vigilance and Enhanced Mental States. *Brain Sci.* **2019**, *9*, 363. [CrossRef]

74. Baccalá, L.A.; Sameshima, K. Partial directed coherence: A new concept in neural structure determination. *Biol. Cybern.* **2001**, *84*, 463–474. [CrossRef] [PubMed]

75. Yu, X.; Zhang, J. Estimating the cortex and autonomic nervous activity during a mental arithmetic task. *Biomed. Signal Process. Control* **2012**, *7*, 303–308. [CrossRef]

76. Djordjević, Z.; Jovanović, A.; Perović, A. Brain connectivity measure—The direct transfer function—Advantages and weak points. In Proceedings of the 2012 IEEE 10th Jubilee International Symposium on Intelligent Systems and Informatics, Subotica, Serbia, 20–22 September 2012; pp. 93–97.

77. Kamiński, M.; Ding, M.; Truccolo, W.A.; Bressler, S.L. Evaluating causal relations in neural systems: Granger causality, directed transfer function and statistical assessment of significance. *Biol. Cybern.* **2001**, *85*, 145–157. [CrossRef]

78. Ameera, A.; Saidatul, A.; Ibrahim, Z. Analysis of EEG Spectrum Bands Using Power Spectral Density for Pleasure and Displeasure State. In Proceedings of the IOP Conference Series: Materials Science and Engineering, Kazimierz Dolny, Poland, 21–23 November 2019; p. 012030.

79. Al-shargie, F.; Tang, T.B.; Badruddin, N.; Kiguchi, M. Simultaneous measurement of EEG-fNIRS in classifying and localizing brain activation to mental stress. In Proceedings of the 2015 IEEE International Conference on Signal and Image Processing Applications (ICSIPA), Kuala Lumpur, Malaysia, 19–21 October 2015; pp. 282–286.

80. Cooley, J.W.; Tukey, J.W. An algorithm for the machine calculation of complex Fourier series. *Math. Comput.* **1965**, *19*, 297–301. [CrossRef]

81. Park, K.S.; Choi, H.; Lee, K.J.; Lee, J.Y.; An, K.O.; Kim, E.J. Patterns of electroencephalography (EEG) change against stress through noise and memorization test. *Int. J. Med. Med Sci.* **2011**, *3*, 381–389.

82. Zhao, C.; Zhao, M.; Liu, J.; Zheng, C. Electroencephalogram and electrocardiograph assessment of mental fatigue in a driving simulator. *Accid. Anal. Prev.* **2012**, *45*, 83–90. [CrossRef] [PubMed]

83. Arsalan, A.; Majid, M.; Butt, A.R.; Anwar, S.M. Classification of perceived mental stress using a commercially available EEG headband. *IEEE J. Biomed. Health Inform.* **2019**, *23*, 2257–2264. [CrossRef] [PubMed]

84. Akin, M. Comparison of wavelet transform and FFT methods in the analysis of EEG signals. *J. Med Syst.* **2002**, *26*, 241–247. [CrossRef] [PubMed]

85. Khushaba, R.N.; Kodagoda, S.; Lal, S.; Dissanayake, G. Driver drowsiness classification using fuzzy wavelet-packet-based feature-extraction algorithm. *IEEE Trans. Biomed. Eng.* **2010**, *58*, 121–131. [CrossRef]

86. Krishna, N.M.; Sekaran, K.; Vamsi, A.V.N.; Ghantasala, G.P.; Chandana, P.; Kadry, S.; Blažauskas, T.; Damaševičius, R. An efficient mixture model approach in brain-machine interface systems for extracting the psychological status of mentally impaired persons using EEG signals. *IEEE Access* **2019**, *7*, 77905–77914. [CrossRef]

87. Khosrowabadi, R.; bin Abdul Rahman, A.W. Classification of EEG correlates on emotion using features from Gaussian mixtures of EEG spectrogram. In Proceedings of the 3rd International Conference on Information and Communication Technology for the Moslem World (ICT4M) 2010, Jakarta, Indonesia, 13–14 December 2010; pp. E102–E107.

88. Attallah, O. An Effective Mental Stress State Detection and Evaluation System Using Minimum Number of Frontal Brain Electrodes. *Diagnostics* **2020**, *10*, 292. [CrossRef] [PubMed]

89. Benoit, C.; Royer, E.; Poussigue, G. The spectral moments method. *J. Phys. Condens. Matter* **1992**, *4*, 3125. [CrossRef]

90. Hasan, M.J.; Kim, J.M. A Hybrid Feature Pool-Based Emotional Stress State Detection Algorithm Using EEG Signals. *Brain Sci* **2019**, *9*, 376. [CrossRef]

91. Sharma, N.; Gedeon, T. Modeling stress recognition in typical virtual environments. In Proceedings of the 2013 7th International Conference on Pervasive Computing Technologies for Healthcare and Workshops, Venice, Italy, 5–8 May 2013; pp. 17–24.

92. Shon, D.; Im, K.; Park, J.-H.; Lim, D.-S.; Jang, B.; Kim, J.-M. Emotional stress state detection using genetic algorithm-based feature selection on EEG signals. *Int. J. Environ. Res. Public Health* **2018**, *15*, 1–11. [CrossRef]

93. Oh, S.-H.; Lee, Y.-R.; Kim, H.-N. A novel EEG feature extraction method using Hjorth parameter. *Int. J. Electron. Electr. Eng.* **2014**, *2*, 106–110. [CrossRef]

94. Hjorth, B. EEG analysis based on time domain properties. *Electroencephalogr. Clin. Neurophysiol.* **1970**, *29*, 306–310. [CrossRef]

95. Sulaiman, N.; Taib, M.N.; Lias, S.; Murat, Z.H.; Aris, S.A.M.; Mustafa, M.; Rashid, N.A. Development of EEG-based stress index. In Proceedings of the 2012 International Conference on Biomedical Engineering (ICoBE), Penang, Malaysia, 27–28 February 2012; pp. 461–466.

96. Zhu, L.; Tian, X.; Xu, X.; Shu, L. Design and evaluation of the mental relaxation VR scenes using forehead EEG features. In Proceedings of the 2019 IEEE MTT-S International Microwave Biomedical Conference (IMBioC), Penang, Malaysia, 27–28 February 2019; pp. 1–4.

97. Wang, X.; Liu, B.; Xie, L.; Yu, X.; Li, M.; Zhang, J. Cerebral and neural regulation of cardiovascular activity during mental stress. *Biomed. Eng. Online* **2016**, *15*, 335–347. [CrossRef]

98. Wang, Q.; Sourina, O. Real-time mental arithmetic task recognition from EEG signals. *IEEE Trans. Neural Syst. Rehabil. Eng.* **2013**, *21*, 225–232. [CrossRef]

99. Kesić, S.; Spasić, S.Z. Application of Higuchi's fractal dimension from basic to clinical neurophysiology: A review. *Comput. Methods Programs Biomed.* **2016**, *133*, 55–70. [CrossRef]

100. Higuchi, T. Approach to an irregular time series on the basis of the fractal theory. *Phys. D Nonlinear Phenom.* **1988**, *31*, 277–283. [CrossRef]

101. Motamedi-Fakhra, S.; Moshrefi-Torbatia, M.; Hill, M.; Hill, C.M.; White, P.R. Signal processing techniques applied to human sleep EEG signals—A review. *Biomed. Signal Process. Control* **2014**, *10*, 21–33. [CrossRef]

102. Jun, G.; Smitha, K.G. EEG based stress level identification. In Proceedings of the 2016 IEEE International Conference on Systems, Man, and Cybernetics (SMC), Budapest, Hungary, 9–12 October 2016; pp. 003270–003274.

103. Korde, K.S.; Paikrao, P.; Jadhav, N. Analysis of EEG Signals and Biomedical Changes Due to Meditation on Brain by Using ICA for Feature Extraction. In Proceedings of the 2018 Second International Conference on Intelligent Computing and Control Systems (ICICCS), Madurai, India, 14–15 June 2018; pp. 1479–1484.

104. Deshmukh, R.; Deshmukh, M. Mental Stress Level Classification: A Review. *Int. J. Comput. Appl.* **2014**, *1*, 15–18.

105. Betti, S.; Lova, R.M.; Rovini, E.; Acerbi, G.; Santarelli, L.; Cabiati, M.; Del Ry, S.; Cavallo, F. Evaluation of an integrated system of wearable physiological sensors for stress monitoring in working environments by using biological markers. *IEEE Trans. Biomed. Eng.* **2017**, *65*, 1748–1758.

106. Sani, M.; Norhazman, H.; Omar, H.; Zaini, N.; Ghani, S. Support vector machine for classification of stress subjects using EEG signals. In Proceedings of the 2014 IEEE Conference on Systems, Process and Control (ICSPC 2014), Kuala Lumpur, Malaysia, 12–14 December 2014; pp. 127–131.

107. Gaikwad, P.; Paithane, A. Novel approach for stress recognition using EEG signal by SVM classifier. In Proceedings of the 2017 International Conference on Computing Methodologies and Communication (ICCMC), Erode, India, 18–19 July 2017; pp. 967–971.

108. Saeed, S.M.U.; Anwar, S.M.; Khalid, H.; Majid, M.; Bagci, A.U. EEG based Classification of Long-term Stress Using Psychological Labeling. *Sensors* **2020**, *20*, 1886. [CrossRef]

109. Saeed, S.M.U.; Anwar, S.M.; Majid, M. Quantification of human stress using commercially available single channel EEG Headset. *Ieice Trans. Inf. Syst.* **2017**, *100*, 2241–2244. [CrossRef]

110. Subhani, A.R.; Xia, L.; Malik, A.S. EEG signals to measure mental stress. In Proceedings of the 2nd International Conference on Behavioral, Cognitive and Psychological Sciences, Bandos, Maldives, 25–27 November 2011; pp. 84–88.

111. Minguillon, J.; Perez, E.; Lopez-Gordo, M.A.; Pelayo, F.; Sanchez-Carrion, M.J. Portable system for real-time detection of stress level. *Sensors* **2018**, *18*, 2504. [CrossRef] [PubMed]

112. Vanitha, V.; Krishnan, P. Real time stress detection system based on EEG signals. *Biomed Res. Int.* **2016**, *27*, 271–275.

113. Najafabadi, M.M.; Villanustre, F.; Khoshgoftaar, T.M.; Seliya, N.; Wald, R.; Muharemagic, E. Deep learning applications and challenges in big data analytics. *J. Big Data* **2015**, *2*, 1–21. [CrossRef]

114. Jebelli, H.; Habibnezhad, M.; Khalili, M.M.; Fardhosseini, M.S.; Lee, S. Multi-Level Assessment of Occupational Stress in the Field Using a Wearable EEG Headset. In Proceedings of the Construction Research Congress 2020: Safety, Workforce, and Education, Tempe, AZ, USA, 8–10 March 2020; pp. 140–148.

115. Jebelli, H.; Khalili, M.M.; Lee, S. Mobile EEG-based workers' stress recognition by applying deep neural network. In *Advances in Informatics and Computing in Civil and Construction Engineering*; Springer: Berlin/Heidelberg, Germany, 2019; pp. 173–180.

116. Ling, C.; Goins, H.; Ntuen, A.; Li, R. EEG signal analysis for human workload classification. In Proceedings of the IEEE SoutheastCon 2001 (Cat. No. 01CH37208), Clemson, SC, USA, 1 April 2001; pp. 123–130.

117. Calibo, T.K.; Blanco, J.A.; Firebaugh, S.L. Cognitive stress recognition. In Proceedings of the 2013 IEEE International Instrumentation and Measurement Technology Conference (I2MTC), Minneapolis, MN, USA, 6–9 May 2013; pp. 1471–1475.

118. Lotfan, S.; Shahyad, S.; Khosrowabadi, R.; Mohammadi, A.; Hatef, B. Support vector machine classification of brain states exposed to social stress test using EEG-based brain network measures. *Biocybern. Biomed. Eng.* **2019**, *39*, 199–213. [CrossRef]

119. Al-Shargie, F.; Kiguchi, M.; Badruddin, N.; Dass, S.C.; Hani, A.F.M.; Tang, T.B. Mental stress assessment using simultaneous measurement of EEG and fNIRS. *Biomed. Opt. Express* **2016**, *7*, 3882–3898. [CrossRef] [PubMed]

120. Dedovic, K.; Renwick, R.; Mahani, N.K.; Engert, V.; Lupien, S.J.; Pruessner, J.C. The Montreal Imaging Stress Task: Using functional imaging to investigate the effects of perceiving and processing psychosocial stress in the human brain. *J. Psychiatry Neurosci.* **2005**, *30*, 319.

121. Dedovic, K.; D'Aguiar, C.; Pruessner, J.C. What stress does to your brain: A review of neuroimaging studies. *Can. J. Psychiatry* **2009**, *54*, 6–15. [CrossRef]

122. Halim, Z.; Rehan, M. On identification of driving-induced stress using electroencephalogram signals: A framework based on wearable safety-critical scheme and machine learning. *Inf. Fusion* **2020**, *53*, 66–79. [CrossRef]

123. Ahn, J.W.; Ku, Y.; Kim, H.C. A novel wearable EEG and ECG recording system for stress assessment. *Sensors* **2019**, *19*, 1991. [CrossRef]

124. Secerbegovic, A.; Ibric, S.; Nisic, J.; Suljanovic, N.; Mujcic, A. Mental workload vs. stress differentiation using single-channel EEG. In *CMBEBIH 2017*; Springer: Berlin/Heidelberg, Germany, 2017; pp. 511–515.

125. Masood, K.; Alghamdi, M.A. Modeling Mental Stress Using a Deep Learning Framework. *IEEE Access* **2019**, *7*, 68446–68454. [CrossRef]

126. Choi, Y.; Kim, M.; Chun, C. Measurement of occupants' stress based on electroencephalograms (EEG) in twelve combined environments. *Build. Environ.* **2015**, *88*, 65–72. [CrossRef]

127. Puterman, E.; O'Donovan, A.; Adler, N.E.; Tomiyama, A.J.; Kemeny, M.; Wolkowitz, O.M.; Epel, E. Physical activity moderates stressor-induced rumination on cortisol reactivity. *Psychosom. Med.* **2011**, *73*, 604. [CrossRef]

128. Tran, Y.; Thuraisingham, R.; Wijesuriya, N.; Nguyen, H.; Craig, A. Detecting neural changes during stress and fatigue effectively: A comparison of spectral analysis and sample entropy. In Proceedings of the 2007 3rd International IEEE/EMBS Conference on Neural Engineering, Kohala Coast, HI, USA, 2–5 May 2007; pp. 350–353.

129. Zhang, Y.; Wang, Q.; Chin, Z.Y.; Ang, K.K. Investigating different stress-relief methods using Electroencephalogram (EEG). In Proceedings of the 2020 42nd Annual International Conference of the IEEE Engineering in Medicine & Biology Society (EMBC), Montreal, QC, Canada, 20–24 July 2020; pp. 2999–3002.
130. Al-Shargie, F. Prefrontal cortex functional connectivity based on simultaneous record of electrical and hemodynamic responses associated with mental stress. *arXiv* **2021**, arXiv:2103.04636.
131. Baumgartl, H.; Fezer, E.; Buettner, R. Two-level classification of chronic stress using machine learning on resting-state EEG recordings. In Proceedings of the 25th Americas Conference on Information Systems, Cancún, Mexico, 15–17 August 2019; pp. 1–10.
132. Jebelli, H.; Hwang, S.; Lee, S. EEG-based workers' stress recognition at construction sites. *Autom. Constr.* **2018**, *93*, 315–324. [CrossRef]
133. Devi, D.; Sophia, S.; Janani, A.A.; Karpagam, M. Brain wave based cognitive state prediction for monitoring health care conditions. *Mater. Today Proc.* **2020**, 1–7. [CrossRef]

Article

EEG Mental Stress Assessment Using Hybrid Multi-Domain Feature Sets of Functional Connectivity Network and Time-Frequency Features

Ala Hag [1], Dini Handayani [1,*], Thulasyammal Pillai [1], Teddy Mantoro [2], Mun Hou Kit [3] and Fares Al-Shargie [4]

[1] School of Computer Science & Engineering, Taylor's University, Jalan Taylors, Subang Jaya 47500, Malaysia; alaahmedyahyahag@sd.taylors.edu.my (A.H.); thulasyammal.ramiahpillai@taylors.edu.my (T.P.)
[2] Faculty of Engineering and Technology, Sampoerna University, Jakarta 12780, Indonesia; teddy.mantoro@sampoernauniversity.ac.id
[3] Department of Mechatronic and Biomedical Engineering, Universiti Tunku Abdul Rahman, Bandar Sungai Long, Kajang 43000, Malaysia; munhk@utar.edu.my
[4] Department of Electrical Engineering, American University of Sharjah, Sharjah 26666, United Arab Emirates; fyahya@aus.edu
* Correspondence: dinioktarina.dwihandayani@taylors.edu.my

Citation: Hag, A.; Handayani, D.; Pillai, T.; Mantoro, T.; Kit, M.H.; Al-Shargie, F. EEG Mental Stress Assessment Using Hybrid Multi-Domain Feature Sets of Functional Connectivity Network and Time-Frequency Features. *Sensors* 2021, 21, 6300. https://doi.org/10.3390/s21186300

Academic Editors: Ruben Pauwels and Steve Ling

Received: 9 July 2021
Accepted: 16 September 2021
Published: 20 September 2021

Publisher's Note: MDPI stays neutral with regard to jurisdictional claims in published maps and institutional affiliations.

Abstract: Exposure to mental stress for long period leads to serious accidents and health problems. To avoid negative consequences on health and safety, it is very important to detect mental stress at its early stages, i.e., when it is still limited to acute or episodic stress. In this study, we developed an experimental protocol to induce two different levels of stress by utilizing a mental arithmetic task with time pressure and negative feedback as the stressors. We assessed the levels of stress on 22 healthy subjects using frontal electroencephalogram (EEG) signals, salivary alpha-amylase level (AAL), and multiple machine learning (ML) classifiers. The EEG signals were analyzed using a fusion of functional connectivity networks estimated by the Phase Locking Value (PLV) and temporal and spectral domain features. A total of 210 different features were extracted from all domains. Only the optimum multi-domain features were used for classification. We then quantified stress levels using statistical analysis and seven ML classifiers. Our result showed that the AAL level was significantly increased ($p < 0.01$) under stress condition in all subjects. Likewise, the functional connectivity network demonstrated a significant decrease under stress, $p < 0.05$. Moreover, we achieved the highest stress classification accuracy of 93.2% using the Support Vector Machine (SVM) classifier. Other classifiers produced relatively similar results.

Keywords: mental stress; electroencephalography; feature extraction; functional connectivity network; time-frequency features; machine learning

1. Introduction

Mental stress has become a catchphrase nowadays, affecting almost everyone, due to the increasing demands in the workplace, life burdens, changing lifestyles, and technological interventions. The long-term effects of stress not only impact health issues, such as heart disease, obesity, diabetes, stroke, and depression [1–3], but have economic consequences too. The economic losses can reach up to billions of dollars [4]. Thus, researchers are trying to detect mental stress at its early stage to prevent it from becoming chronic. The evaluation of human psychological stress usually performed using subjective or objective measurement methods. The subjective stress assessment methods used psychological assessment approaches, such as a clinical interview and psychological-based questionnaires, such as the Trier Social Stress Test (TSST) [5,6], Perceived Stress Scale (PSS) [7–9], State-Trait Anxiety Inventory (STAI), and Hospital Anxiety and Depression Scale (HADS) [10].

The drawback of subjective methods is that it is subjective to the user's reported answers, and it only describe the current state of the subject's stress level. Recent studies have focused on an objective method of physiological assessment, which gives the individuals the freedom to assess their mental stress states without the expert's intervention and drives more reliable evaluation [11–13]. The physiological assessment depends on the body's reactions towards stress, such as facial expressions [5], blink rate [14], pupil dilation [15], eye gaze, and voice intonation [16]. Several studies have reported that for anyone diagnosed with stress, their body had shown different symptoms by changing the normal activities of these bio-markers: catecholamine, cortisol level, and alpha-amylase enzyme [11,12]. Additionally, during stress, the frontal area of the brain showed high activity of glucose metabolism and blood flow [9,17]. Therefore, several studies utilized different modalities such as functional brain imaging (i.e., functional magnetic resonance imaging (fMRI) and electroencephalography (EEG)) technologies to identify the brain regions and fluctuation of brain activities affected by stress [12,17,18]. The prefrontal cortex (PFC) was the common area that appeared to be sensitive to stress exposure. Moreover, there is an evidence of changes in the autonomic nervous system's (ANS) activities under stress [19]. Consequently, physiological features of stress from the ANS can be seen as subtle changes in heart rate (HR), heart rate variability (HRV) [6], respiration [20], skin conductance [21], and blood pressure [22]. Currently, the focus is on brain activities as, according to the latest work in neuroscience, it is the main target organ of mental stress due to its responsibility to distinguish between different situations' contexts (i.e., stressful and threatening or not) [23]. The brain activities are usually analyzed by several tools, such as EEG [24], fMRI [25], positron emission tomography (PET) [26] and other neuroimaging modalities. EEG is a measurement tool that depicts the electrical activities on the brain's surface. Compared to other modalities, EEG provides high temporal resolution to detect the time variance of changes in the brain's state [27], is easy to setup, and is commercially available at a lower cost.

To measure stress in real life, researchers used a different approach to stimulate stress in laboratory settings. Several validated stress inducement methods were established such as mental arithmetical tasks [13], negative feedback and time pressure [28], public speaking [29], and noise manifested by music [30]. For the sake of validity and increasing the accuracy of detection, researchers, in many cases, employed a combination of one or more modalities with EEG, such as skin conductance [31], functional near-infrared spectroscopy (fNIRS) [13], and electrocardiography (ECG) [11]. Even though applying additional modulation with EEG improved the accuracy, it is not suitable for home-based applications due to the knowledge base required, lengthy setup time, and the expensive and inconvenient usage for wearable devices. Therefore, current studies suggest that the enhancement of EEG signals could be accomplished by obtaining the optimal features from specific regions of the brain related to the task.

For the aim of finding the relative EEG markers that explain mental stress and increase its detection rate, several studies employed different types of features from the time domain, frequency domain, and time-frequency domain [8,32–36], and several machine learning algorithms have been used to predict the mental stress state, such as SVM [37], K-Nearest Neighbors(KNN) [29,38], LR [1], Feed-Forward Neural Network (FF-NN) [30], Naive Bayes(NB) [9,38], and Random Forest(RF) [39]. In the literature, non-invasive EEG-based stress studies suggested that bio-markers (i.e., alpha, beta, and gamma) in specific brain areas could reveal the mental stress state [18,40,41]. However, no consensus has been reached about the particular established EEG patterns/features that differentiate stress levels, see review [36]. In studies [8,29,42], different frequency band features have been demonstrated to classify stress tasks. The low beta was considered as one approach to recognize mental stress [9]. Similarly, alpha rhythm power at the right PFC was shown to be more discriminative to stress and rest states [28,36]. Another study in [11] showed that the PFC relative gamma power (RG) was more discriminative between stress levels than alpha asymmetry.

Current researchers acknowledge that multi-domain features and multi-channel analyses are required to create an effective information feature space in which a good interpreter can eventually produce effective alarms for the current mental state. As a result, studies presented by Attallah [34] and Hasan [43] revealed that hybrid feature sets from various domains (time domain, frequency domain, and time-frequency domain) may enhance the overall classification of EEG emotion analysis. To the best of our knowledge, no study on fusing such domains with functional connectivity network features has been done. In contrast to the majority of cortical activation features, which focus on a single channel feature, functional connectivity features look for relationships and interactions across different regions of the brain (inter-channel relations). This knowledge aids in a better understanding of how the brain functions and could offer more accurate representations of mental stress states. To address the aforementioned limitation, this study aims to investigate the fusion of functional connectivity network features with cortical activation features from the time and frequency domains to detect mental stress in order to aid in the development of wearable devices. In contrast to prior research, our objective is to combine single-channel features with inter-join channel features (connectivity). Thus, we employ a well-established clinical assessment method using salivary alpha amylase (cortisol measure) to enhance the labeling of the task given. We then propose an objective method based on the machine learning framework to classify stress levels with a minimum number of EEG channels. We present a novel methodology to identify mental stress by investigating the statistical difference between stress and rest conditions. The proposed method is analyzed and evaluated using seven classifiers [36], namely: KNN, RF, Logistic Regression (LR), SVM, classification and regression Decision Tree (CART), Linear Discrimination Analysis (LDA), and NB. The accuracy, precision, recall, and F-score matrices were used to evaluate the performance of the classifiers.

The following section summarizes our contributions in this work.

1. Developing an experimental protocol to induce two levels of mental stress (stress/rest or control) in a short time, which is important for real-life application.
2. A multi-domain feature set is proposed by fusing features from the time domain, frequency domain, and functional connectivity networks.
3. A feature selection method was implemented to select the most discriminative feature sets.
4. The performance of the proposed method is tested and evaluated using seven machine learning classifiers.

The rest of the paper is organized as follows: Section 2 describes the dataset, protocol setup, and data annotation. Section 3 explains the detailed methodology. In Section 4, the classification evaluation method is presented. In Section 5, results and analysis are discussed in detail. In Section 6, a discussion of the results is provided, and the study's conclusions are presented in Section 7.

2. Dataset and Materials

2.1. Participants

In this dataset, the total number of participants was 22 subjects (aged 26 ± 4 with head size of 56 ± 2 cm). All subjects were male right-handed healthy adults having the same culture and background (undergraduate students). The participants were asked about their medical condition to fit the experiment eligibility. Smokers and drug users were excluded due to their effect on the sympathetic nervous system. Moreover, participants must have no history of any physical or mental health problems. Several rules had been imposed on them before starting the experiment. For example, no eating or drinking two hours before the experiment and no physical activity occurred [13]. The experiment time was chosen between 4.00 and 5.30 p.m. to reduce the circadian rhythm's effects on cortisol collection. The experiment protocol was approved by the institute review board of University Teknologi Petronas.

2.2. Stress EEG Measurement and Protocol

The experiment task protocol was based on the Montreal Imaging Stress Task (MIST), which was described in detail in [44]. The task was created using MATLAB and presented using a Graphical User Interface (GUI). It involved a mental arithmetic task (MA) using simple calculation of two-digit integers (ranging from 0 to 100) with operands restricted to $+, -$, and (/ or *) (example $99/3 - 76 + 51$). The answer for each question was displayed in the GUI using a numerical order ranging from '0' to '9', and participants were trained to select the correct answer with a mouse click. The experiment task was performed in three subsequent phases: preparation, rest condition, and stress condition. Each phase is described in detail below. In the preparation phase, participants were given five minutes to practice the MA task, and the average time taken to answer the questions was recorded for each participant, which would later be utilized as a time constraint to induce stress.

In the stress phase, a cap of EEG electrodes was placed on the frontal region of each participant's scalp, and simultaneous measurement was performed while the participant solved the arithmetic within a time limit (derived based on a 10% reduction from the average time recorded during the preparation phase). Additionally, the average peer performance was displayed on the screen as a real-time performance indicator of subject's performance compared to other participants. Notification of a negative message for each response that exceeds the time limit or gets the answer wrong, i.e., a message of "Incorrect", or "Time's up" being flashed on the screen. The negative feedback was intended to add more stress to the participants.

In the rest phase, the participant was instructed to keep calm and relax while looking at the fixation cross presented at a computer monitor. The presentation of the stress and rest states was in a block design. There was a total of five blocks in each of the stress and rest conditions, as shown in Figure 1. For every block, an arithmetic task popped up for 30 s to induce stress, followed by 20 s of rest. During the 30 s of the stress task, multiple mathematical questions were displayed on the computer monitor based on participants' response time in answering each question. For the 20 s of a rest condition, the participant looked at the fixation cross in the computer screen as a visual cue for the trial onset.

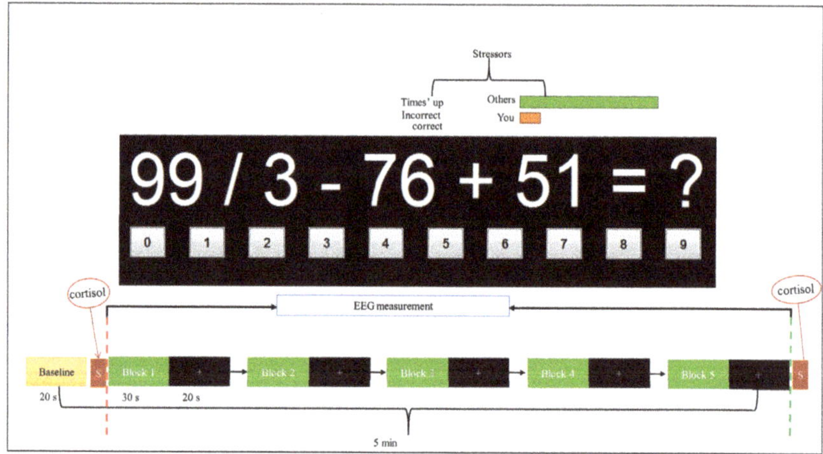

Figure 1. Experiment block design. A total of five active blocks for each task with salivary alpha amylase (SAA) cortisol was collected before and after the stress task and presented by the letter S with a red background. For each block, arithmetic tasks are given for the 30 s followed by 20 s of rest. The red dashed line marks the start of the task, and the green dashed line marks the end of the task (the marking is done at every block).

The EEG signals were recorded using the Discovery 24E system (BrainMaster Technologies Inc, Bedford, OH, USA). The system was equipped with 7-electrodes (Fp1, Fp2, F7, F3, F4, Fz, F) placed on the prefrontal cortex, as shown in Figure 2. The EEG electrodes were referenced to the earlobe electrodes (A1 and A2). The placement of the EEG electrodes was based on the 10–20 system and was sampled at 256 Hz.

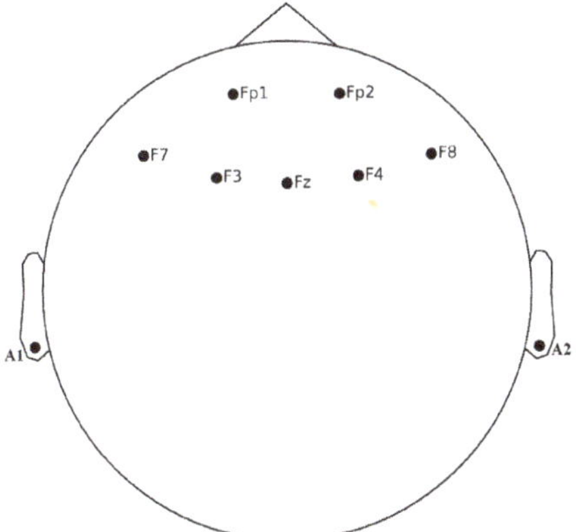

Figure 2. EEG Channels' Position on Scalp.

2.3. Dataset Labelling

The EEG signal has been labeled for each subject based on the cortisol of the salivary amylase level (AAL). During the experiment, two samples of AAL were obtained, as shown in Figure 1. The first AAL sample was collected before starting the experiment task (stress/rest condition) as a baseline. The first AAL result was supposed to show the initial state of the subject as not stressed; otherwise, the subject would be removed from the study. The second AAL sample was collected at the end of the experiment. The data annotation/labeling of the EEG signal was based on the cortisol level; medically, cortisol levels greater than 60 micrograms per decilitre (mcg/dL) indicate that the subject is stressed, while those between 30 and 60 (mcg/dL) are labeled as working brain condition, and those less than 30 (mcg/dL) are labeled as the rest state [44,45].

3. EEG Base Mental Stress Analysis Method

This section describes the proposed methodology process for implementing the stress detection method, namely signal preprocessing, feature extraction and selection, and classification, as shown in Figure 3.

3.1. Signal Preprocessing

The raw EEG signals were preprocessed using Python and an external MNE package [46]. The raw EEG signals were band-pass filtered using a finite impulse response (FIR) filter with 1 Hz to 35 Hz bandwidth. Since we only measured the frontal cortex, the EEG data were re-referenced to the average reference as suggested by [47]. Consequently, the noise caused by 50/60 Hz of line power was omitted. Furthermore, Fast-ICA has been used to eliminate the associated noise caused by eye blinking called electrooculogram (EOG) artifacts under 4 Hz, muscle artifacts (EMG) with frequency beyond 30 Hz, and heart rate [48]. Fast-ICA has the significant ability of denoising ocular artifacts (OAs) that exist in low frequencies less than 16 Hz, therefore delineating the overlapping frequency bands [49]. The EEG signals were segmented into 1000 ms EEG epochs relative to the target task. The selection of 1000 ms or 256 EEG data points was due to its stationarity at an epoch size of >256 for experiments involved in event-related potential. This number of data points is appropriate to show the stationarity of EEG signals and have been reported in previous EEG studies with a comparable data point [50–52]. The baseline was extracted and omitted using the full length of each epoch. Then, all EEG epochs were visually double-checked to eliminate data segments contaminated with noise. Lastly, we identify from the clean EEG signals two mental states (stress and rest). The first 20 s of rest from the first block were considered for the rest state, and another 20 s from the last stress block (block 5) were considered for the stress condition. The two states were labeled based on the results obtained from the cortisol data collection of AAL.

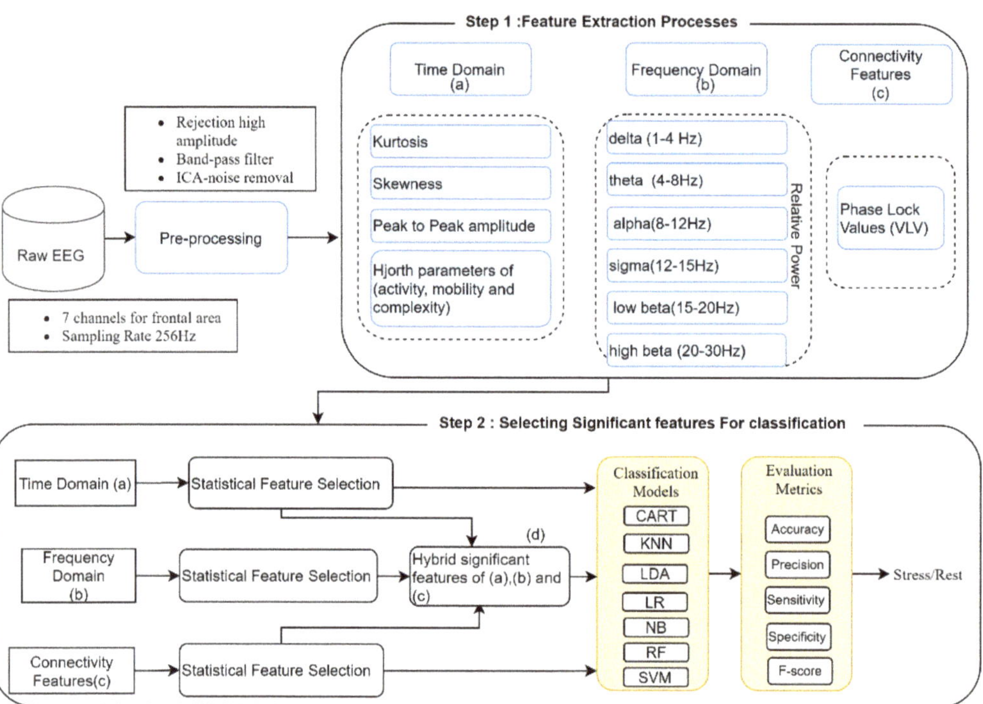

Figure 3. Proposed ML methodology flow chart for mental stress state recognition.

3.2. Feature Extraction

Feature extraction is a crucial step in analyzing and classifying EEG signals [43]. Because the EEG signal is a non-stationary and time-varying signal, choosing an appropriate technique to extract useful features that reflect brain activity is critical for reducing dimensional space, improving processing performance, and increasing the detection rate. EEG features can be broadly categorized into single-channel features and multi-channel features. The majority of the existing features are computed on a single channel that involves temporal and or spatial information from a specific brain region, e.g., statistical features, frequency-domain features, e.g., PSD. A few multi-channel features are computed to reflect the relationships between different brain regions, e.g., brain connectivity features. The EEG signal comes from a complex of interconnected brain neurons. Hence, the fusion of the brain connectivity with cortical activation features may provide us with a more exact model of the brain and how its various areas interact with each other. In this paper, both cortical activation features (single-channel features) and functional connectivity network features (multi-channel features) have been employed.

In particular, twelve (single-channel) features were extracted from both the time domain and frequency domain of the cleaned EEG signals for each of the seven channels (Fp1, Fp2, F7, F3, F4, Fz, and F8) located at the prefrontal and frontal region of the brain. Those EEG features were six features per EEG channel from the time domain: kurtosis, peak-to-peak amplitude, skewness, and Hjorth parameters of activity, complexity, and mobility. Likewise, six features per EEG channels were extracted from the frequency domain of relative powers for frequency bands: delta δ, theta θ, alpha α, sigma σ, low beta β, and high beta β. Additionally, a total of 126 features (multi-channel features) were extracted from the connectivity network of all channels. The EEG signal's length used for feature extraction methods was 40 s (20 s stress, 20 s rest), segmented by the epoch of 1 s, which results in a total of 40 segments, and each segment consists of 210 features per subject for both stress and rest tasks. Table 1 shows the summary of dataset content.

Table 1. A summary of Dataset Structure Content.

Name	Array Shape	Array Content
Data	$40 \times 7 \times 256$	Trails $\times$ channels $\times$ samples (256 Hz $\times$ 1 s)
Label	40×2	Trail $\times$ label (stress, rest)

Each of the domain's features was explained in detail in the following subsections.

3.2.1. Time-Domain Features (TDFs)

The TDFs were calculated from the cleaned EEG signals at each epoch. The TDFs are also called statistical features widely used in the classification of EEG signals to measure the irregularity of signal amplitude in the time domain. Therefore, several studies employed TDFs in emotion [29] and stress classification [37,39]. In this paper, six statistical features were extracted from the time domain, namely: kurtosis, peak amplitude, skewness, and Hjorth parameters of activity, complexity, and mobility. Each of these features was extracted from each channel per subject. The full details of these parameters are given below.

Kurtosis: is the measure of the relative flatness of an EEG signal distribution per segment (epoch), and it is calculated using the equation.

$$Kurtosis = \frac{\frac{1}{T}\sum_{t=1}^{T}(x(t) - \mu)^4}{\sigma^4} \tag{1}$$

where T is the number of epochs , $x(t)$ is time-series sample points, and μ, σ are the mean and standard deviation of the signal.

Skewness: measures the distribution difference between the mean and the median for each variable of epochs.

$$Skewness = \frac{\frac{1}{T}\sum_{t=1}^{T}(x(t) - \mu)^3}{\sigma^3} \tag{2}$$

Peak-to-peak amplitude (ptp_amp): the change between the peak of the highest amplitude value and the lowest amplitude value among the various time windows.

Hjorth parameters: three features of Hjorth Parameters (TDHPs), namely activity, mobility, and complexity of the signal, are extracted, which are useful for the quantitative evaluation of an EEG signal and can be expressed as:

- Hjorth Activity: The activity measure represents the signal power and measures the variance of a time function using the equation.

$$Activity = var(x(t)) \tag{3}$$

where $x(i)$ represents the signal on time.

- Hjorth Mobility: mobility represents the mean frequency or the proportion of the standard deviation of the signal and is denoted by:

$$Mobility = \sqrt{\frac{var(\frac{dy(t)}{dt})}{Activity(y(t))}} \tag{4}$$

where mobility represents the square root of the variance of the first derivative of the signal $x(t)$ divided by the activity.

- Hjorth complexity: the complexity parameter gives an estimate of the bandwidth of the signal, which indicates the similarity of the shape of the signal to a pure sine wave.

$$Complexity = \sqrt{\frac{Mobility(\frac{dy(t)}{dt})}{Mobility(y(t))}} \tag{5}$$

All these extracted features were then fed as an input to the classifiers.

3.2.2. Frequency-Domain Features (FDFs)

In the frequency domain, the multitaper method is used to estimate the power spectral density (PSD) because it provides a more robust spectral estimation than the classical methods and Welch's periodograms [53]. Compared to Welch's approach, the multitaper method does not need to identify a window duration because it computes the periodogram on the whole signal and provides a high-frequency resolution and low variance [53].

Multitaper spectrum estimation (MSE): a Nonparametric method used to estimate PSD from a combination of multiple orthogonal tapers (or "windows"). MSE aims to recover the information lost when using a single taper and offers significant performance gains over a nonparametric single taper. The estimator is the average of the K direct spectral estimators, each acting on the whole data record (rather than on a signal segment, as happens in the Welch method) and applying different tapers. Each (partial) estimator is computed by:

$$\hat{S}_k = \left| \sum_{i=1}^{N} h_{i,k} X_{i+l-1} e^{-2j\Pi ft\Delta t} \right|^2 \tag{6}$$

Let $x(t)$, for $t = 0, 1,..., N\ 1$, be a zero-mean time series with unit sampling and spectral density $S(f)$, Δt is the sampling interval, $h_{i,k}$ is the kth data taper, and the bandwidth for Δt is 1 s.

The final estimator is computed as:

$$\hat{S}_k = \frac{1}{k} \sum_{k=0}^{k-1} \hat{S}_k(f) \tag{7}$$

where K is equal to $2NW - 1$, and 2 W is the normalized bandwidth of the tapers.

The relative power (RP) of six frequency bands: delta (1–4 Hz), theta (4–8 Hz), alpha (8–12 Hz), sigma (12–15 Hz), low beta (15–20 Hz), and high beta (20–30 Hz) were computed from the MSE of PSD. The RP is expressed by divided the specific power band over the total power of all bands and calculated as below:

$$RP = \frac{power(selected\_band)}{power(total\_bands)} * 100 \tag{8}$$

The RP features were then used as an input to the classifiers.

3.2.3. Functional Connectivity Network

The functional connectivity network is generated by measuring the connection between electrode pairs in each frequency band using Phase Locking Value (PLV). The PLV technique, similar to the conventional coherence method, computes the correlation between two pairs of EEG channels in distinct frequency bands. The PLV is an effective measure of brain functional signals due to its ability to quantify locking between the phases of the signals from two distinct electrodes and does not depend on the assumption of stationary signals [54]. Therefore, PLV was proved to be a valid method to investigate task-induced changes in the long-range synchronization of neural activity from EEG data [55]. To calculate the phase-locking value, we extract the instantaneous phase $\phi_i^a(t)$ of the analytical signal $x_i^a(t)$ of the time series $x_i(t)$.

Then, for each pair (i, j) of EEG channels, we compute the modulus of the time-averaged phase difference projection onto the unit circle and computed it in Equation (9):

$$VLP_{ij} = \left| \frac{1}{T} \sum_t e^{i(\phi_i^a(t) - \phi_j^a(t))} \right| \tag{9}$$

where N is the total number of trials in time series, and ϕ_i and ϕ_j are the instantaneous phase values at trial index n. The PLV values range between [0, 1], with 0 indicating no phase synchronization and 1 indicating that there is a fixed relative relationship between the two signals in all trials. Because PLV employs undirected measurement for all electrodes, it is known as symmetric measure (PLV($k1$, $k2$) = PLV($k2$, $k1$)).Thus, the direct connection is ignored, and the total number of connections between the EEG channels is measured using:

$$N = \frac{k(k-1)}{2} \tag{10}$$

where k is the total number of channels. In this paper, the total extracted connectivity features are 126 features (21 features × 6 bands) since we are using only 7 EEG channels. However, we only utilized the PLV with a phase-difference distribution that was significantly different from zero using t-test feature selection at $p < 0.05$.

3.3. Hybrid Features of Time, Frequency Domain and Connectivity Features

Fusion information from cortical activation (Time and Frequency Domain) and connectivity features might complement each other, giving us a more accurate representation of the brain and how its various regions interact. In this paper, the total features extracted from multi-domain features were 210 features (42 features from time, 42 features from frequency domain, and 126 features from PLV connectivity features), resulting in high-dimensional feature space. Therefore, the significant-features-based channels from the time domain, frequency domain, and connectivity features were identified using a statistical t-test with a 95% confident interval and $p = 0.05$ level of significance. Thus, the most significant features from each domain were fused to form a new subset of the significant features from the time domain, frequency domain, and connectivity network. The total significant-features-based channels were 42 (15 from the time domain, 20 from the frequency domain, and 64 from connectivity features) and used as a new fusion feature set to subsequent classifiers.

4. Classification

To classify and evaluate stress levels, three scenarios have been conducted. First, individual feature of the selected channels within each domain was considered as a bio-marker and evaluated separately (i.e., Hjorth complexity, Hjorth mobility, relative alpha, . . . , etc., see Tables 3–5). In the second scenario , we utilized the features from the selected channels in each domain as a feature vector and classified them separately (i.e., see Figures 8–10). Meanwhile, we fused the features of the selected channels from the three domains: time domain, frequency domain, and connectivity network features into a single feature vector and used them as an input to the ML classifiers (see Figure 11). Several machine learning algorithms have been used for EEG signal analysis to train and predict the features extracted from target EEG tasks. In this paper, seven classifiers, namely LR, RF,LDA, KNN, SVM, DT, and NB, were employed to evaluate the model performance of mental stress recognition based on the scenarios mentioned and provide the researchers with useful information about the effective classifier to be considered in future work. Table 2 shows the classifier's tuning parameters utilized. More details about the utilized classifiers can be found in our previous study [56]. The extracted features are split into 80% for training and 20% for testing. In each classifier, an independent subject test with 5-fold cross-validation was performed.

Table 2. Default Parameters for Classification Techniques.

No.	Classifier	Default Value
1	SVM	C = 1.0, Kernal = Radial Basis Function (RBF), 1.0×10^{-3}
2	KNN	K = 5, distance function = euclidean distance
3	NB	variance_smoothing = 1×10^{-9}
4	RF	n_estimators = 100 trees, criterion = 'gini'
5	DT	criterion = 'gini'
6	LR	penalty = 'l2', *, tolerance = 0.0001, C = 1.0
7	LDA	solver = 'Singular value decomposition (svd)', tolerance = 0.0001

The proposed model's performance has been evaluated using seven classifiers with 5-fold cross-validation and a four-measure matrix. These include accuracy, precision, sensitivity, and F-measure. The equations below show the mathematical formulation for each prediction. The results obtained from the confusion matrix has:

- True Positives (Tp): The number of labels correctly identified as stress conditions.
- True Negatives (Tn): The number of labels correctly identified as a rest condition.
- False Positives (Fp): The number of labels incorrectly identified as stress.
- False Negatives (Fn): The number of labels incorrectly identified as rest.

$$Accuracy = \frac{Tp + Tn}{Tp + Tn + Fp + Fn} \tag{11}$$

$$Precision = \frac{Tp}{Tp + Fp} \tag{12}$$

$$Sensitivity = \frac{Tp}{Tp + Fn} \tag{13}$$

$$Specificity = \frac{Tn}{Tn + Fp} \tag{14}$$

$$F\text{-}measure = 2\frac{Precision * Sensitivity}{Precision + Sensitivity} \tag{15}$$

Accuracy denotes the measurement of how many correct predictions were made in the whole dataset in two-class problems, i.e., stress and rest conditions. Precision indicates the correct measure of a positive prediction. Meanwhile, sensitivity refers to the completeness measure of a classifier, measuring the number of true stress conditions that get predicted over whole stress labels in the dataset. Specificity measures the proportion of rest conditions that are correctly identified. The F-measure was used to evaluate the detection result using both sensitivity and precision.

5. Result and Analysis

5.1. Statistical Analysis

The stress inducement using an arithmetic task under time pressure and negative feedback for 22 subjects was evaluated using a salivary alpha-amylase level and EEG. The stress inducement reported higher levels of salivary alpha-amylase in stress (M = 93.64, SD = 13.99) (KIU/L) compared rest condition (M = 24.45, SD = 4.44) (KIU/L), as shown in Figure 4. The increase in the alpha-amylase level from rest condition to stress condition was significant with a mean $p < 0.0001$. This also correlates with our previous results [13], which revealed a significant difference in alpha-amylase level between the two conditions across all subjects. Therefore, time pressure and negative feedback prove to be reliable for stress induction in the lab.

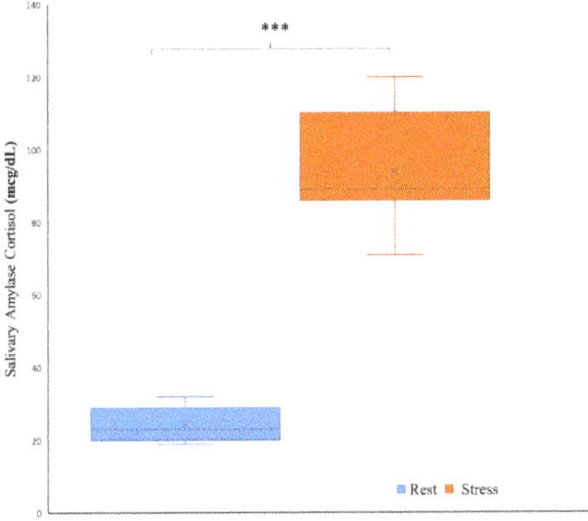

Figure 4. The mean and standard deviation of the salivary amylase cortisol measured by (mcg/dL) for rest and stress conditions.

In EEG signal analysis, an independent-sample *t*-test was conducted to compare stress and rest for each feature-based electrode. The star symbols are used in topographic maps to show the significant electrodes per feature. For time-domain features, Figure 5a shows the mean and standard deviation of the Hjorth complexity, Hjorth mobility, Hjorth activity, kurtosis, peak-to-peak amplitude (*ptp_amp*), and skewness of EEG signals at 1–30 Hz for stress and rest conditions, which were taken by averaging all subjects' data for each condition.

The placement of EEG electrodes is coordinated based on the international 10–20 system, as shown in Figure 2. The means of Hjorth complexity, Hjorth activity, and *ptp_amp* were decreasing from rest condition to stress condition when subjects were exposed to mathematical stressor tasks and increased from rest to stress conditions in Hjorth activity and skewness. This variation in different parameters indicates a further decrease in complexity and *ptp_amp* from rest to stress conditions but a high increase in the mobility component in the signal. The significant electrodes for time-domain features are shown in Figure 5b, where the color scale represents statistical differences based on *t*-test values.

The total number of features for each channel is six, giving a total of (7 channels × 6 features) 42 features in the time domain. However, only 15 significant features in the time domain that discriminate the rest and stress conditions were selected and used in this study. The topographic T-map for both complexity and mobility features shows the same significant channels of 'Fp1', 'F3', and 'F4'; significant Hjorth activity channels were 'F7' and 'F3'. Finally, *ptp_amp* has four significant channels: 'F7', 'F3', 'Fz', and 'F8'. Note that skewness and kurtosis were also selected as a useful feature even though there was only one channel, 'Fp4' and 'Fp1', respectively, for each one, showing a statistically significant difference between the two conditions. This means that the skewness and kurtosis features can offer additional discriminative information between the two conditions. Figure 6 shows the frequency changes on the brain with respect to the stressor test using scale colors of the power distribution of PSD. A statistical analysis of the averaged normalized relative power of the frequency bands ($delta, theta, alpha, sigma, low\_beta, high\_beta$) was carried out to demonstrate the difference between rest and stress states. The topographic T-map shows the significant electrodes corresponding to each band with '*' star symbols and the color scale of the T-map. Out of 42 features (7 channels * 6 relative power bands) in the frequency domains, only 20 features were selected as significant features based on *t*-test values for the experiment task of rest and stress conditions.

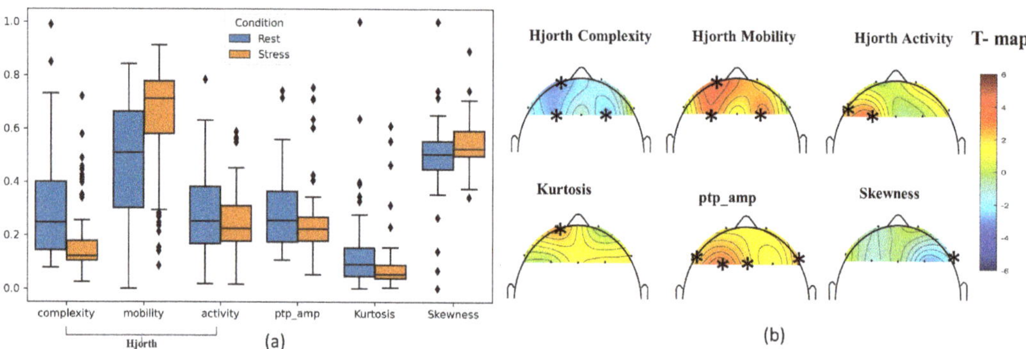

Figure 5. (**a**) The mean and standard deviation using scatter for time-domain features of the Hjorth complexity, Hjorth mobility, Hjorth activity, kurtosis, peak-to-peak amplitude (ptp_amp), and skewness of EEG signals at 1–30 Hz for stress and rest conditions. The difference between stress and rest is shown using T-maps in (**b**). The star (*) symbols denote statistically significant electrodes using topographic maps (two-sample *t*-test; $p < 0.01$, Bonferroni correction).

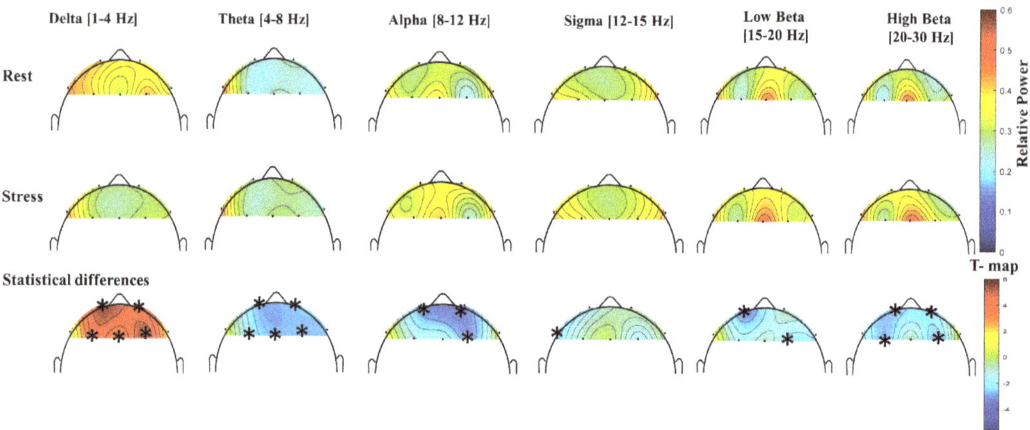

Figure 6. The mean topographic maps for relative bands power of delta, theta, alpha, sigma, low beta, and high beta at 1–30 Hz for rest and stress conditions. The difference between stress and rest relative powers are shown using T-maps. The star (*) symbols denote to the significant electrodes related to specific feature (two-sample *t*-test; $p < 0.01$, Bonferroni correction).

High beta β(20–30 Hz) showed a significant decrease from the rest condition to the stress condition across all participants with ($p < 0.001$). Consequently, a noticeable significant decrease in the alpha α relative power (8–12 Hz) was found in the right cortex of the frontal area for the mathematical stressor tests from the rest condition to the stress condition. Likewise, theta θ (4–8 Hz) relative power indicated a slightly significant increase in stress conditions compared to rest conditions. The overall statistical analysis of the average relative power was shown to be discriminative among stress and rest conditions in all significant electrodes with $p < 0.0001$ in alpha and high beta and $p < 0.001$ in delta and sigma (12–15 Hz) with p-value 0.05. Interestingly, the prefrontal right cortex channels ('Fp1', 'Fp2', and 'F4') were shown to be more significant in most relative power bands to distinguish rest and stress conditions.

Similarly, Figure 7 shows the functional connectivity network features of PLV, which measures the changes (increase/decrease) in the connectivity network between two pairs of channels. The PLV was extracted from six frequency bands of both tasks (rest/stress). The significant channels denoted either an increase or decrease (*+/*−) in the connectivity network measurements from the rest condition to the stress condition. From Figure 7, it can be seen that significant functional connectivity networks in delta and alpha decreased from the rest condition to the stress condition. On the other hand, high beta shows an increase in the connectivity network from the rest condition to the stress conditions. Other bands showed increases and decreases in the connectivity network between different brain regions. The significant discrimination connectivity features between stress and rest conditions were selected using a t-test and fused with other significant features from the time and frequency domains.

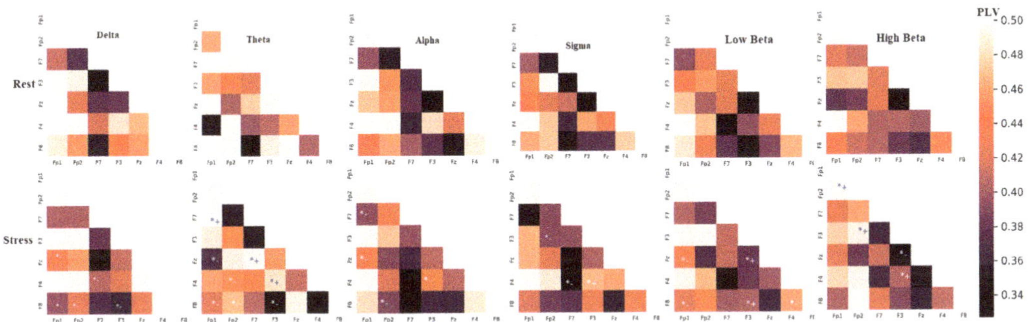

Figure 7. The PLV connectivity network among EEG channel pairs over all trails for rest and stress condition. The star (*) symbol denotes the significant connections between electrodes selected by the t-test.

5.2. Classification Results

The overall classification performance results in terms of the average accuracy/ precision/recall/f-score and standard deviations of the proposed methods with the types of classifiers are presented in Table 3 for time-domain features, in Table 4 for frequency-domain features, and Table 5 for connectivity network features. Those average accuracies were evaluated for the seven classification algorithms with respect to the number of channels selected by the t-test. From the time-domain features in Table 3, we obtain the following significant findings:

- The best classification accuracy was obtained by using the peak-to-peak amplitude feature with four channels ('F7', 'F3', 'Fz', and 'F8') of the frontal region with a mean accuracy of 79.4% using Random Forest and 76.1% using SVM. Meanwhile, the rest of the classifiers achieved an average accuracy of 75% for *ptp_amp*.
- The Hjorth parameters of complexity, mobility, and activity achieved a result of an average of 69.1%, 71.5%, and 71.8% using KNN, SVM, and NB, respectively. The significant channels of Hjorth complexity and mobility were located in the prefrontal

cortex ('Fp1', 'F3', and 'F4'), while Hjorth activity had only two significant channels that were selected from the left frontal cortex of ('F7' and 'F3').

- Features with a low number of channels tends to achieve low accuracy due to low spatial resolution. Kurtosis and skewness got one significant channel for each and obtained a maximum average accuracy of 55% and 56%, respectively, for 'Fp1' and 'F4'.

Table 3. The average accuracies and standard deviations of significant time-domain features from the selected channels with the seven classifiers.

Features	Sig. Channels ($p < 0.001$)	Performance	CART	KNN	LDA	LR	NB	RF	SVM
Hjorth complexity	'Fp1', 'F3', 'F4'	accuracy	0.644 ± 0.146	0.691 ± 0.119	0.671 ± 0.103	0.659 ± 0.118	0.682 ± 0.114	0.671 ± 0.145	0.677 ± 0.124
		precision	0.653 ± 0.151	0.715 ± 0.118	0.680 ± 0.109	0.672 ± 0.126	0.696 ± 0.116	0.673 ± 0.15	0.694 ± 0.128
		recall	0.650 ± 0.149	0.701 ± 0.117	0.675 ± 0.107	0.668 ± 0.123	0.689 ± 0.117	0.672 ± 0.146	0.688 ± 0.127
		f1-score	0.641 ± 0.149	0.688 ± 0.119	0.667 ± 0.108	0.656 ± 0.119	0.678 ± 0.118	0.667 ± 0.148	0.675 ± 0.125
Hjorth mobility	'Fp1', 'F3', 'F4'	accuracy	0.706 ± 0.127	0.702 ± 0.158	0.705 ± 0.122	0.516 ± 0.072	0.709 ± 0.129	0.712 ± 0.146	0.715 ± 0.140
		precision	0.711 ± 0.128	0.721 ± 0.159	0.713 ± 0.114	0.519 ± 0.081	0.715 ± 0.126	0.718 ± 0.140	0.729 ± 0.135
		recall	0.704 ± 0.124	0.711 ± 0.155	0.711 ± 0.117	0.519 ± 0.078	0.711 ± 0.124	0.715 ± 0.136	0.725 ± 0.137
		f1-score	0.703 ± 0.128	0.698 ± 0.160	0.702 ± 0.123	0.609 ± 0.039	0.707 ± 0.128	0.708 ± 0.147	0.710 ± 0.144
Hjorth activity	'F7', 'F3'	accuracy	0.541 ± 0.000	0.706 ± 0.184	0.655 ± 0.173	0.511 ± 0.001	0.718 ± 0.168	0.540 ± 0.120	0.688 ± 0.117
		precision	0.301 ± 0.000	0.714 ± 0.189	0.659 ± 0.179	0.431 ± 0.000	0.727 ± 0.164	0.387 ± 0.032	0.699 ± 0.183
		recall	0.371 ± 0.000	0.705 ± 0.183	0.656 ± 0.177	0.372 ± 0.000	0.723 ± 0.166	0.394 ± 0.013	0.691 ± 0.177
		f1-score	0.321 ± 0.000	0.699 ± 0.190	0.648 ± 0.180	0.321 ± 0.000	0.708 ± 0.177	0.386 ± 0.024	0.681 ± 0.176
Kurtosis	'Fp1'	accuracy	0.549 ± 0.118	0.531 ± 0.089	0.515 ± 0.130	0.517 ± 0.129	0.516 ± 0.136	0.549 ± 0.118	0.518 ± 0.136
		precision	0.551 ± 0.117	0.539 ± 0.107	0.539 ± 0.163	0.541 ± 0.163	0.511 ± 0.200	0.551 ± 0.117	0.524 ± 0.184
		recall	0.549 ± 0.115	0.536 ± 0.100	0.526 ± 0.139	0.526 ± 0.139	0.523 ± 0.147	0.549 ± 0.115	0.521 ± 0.153
		f1-score	0.544 ± 0.119	0.524 ± 0.095	0.504 ± 0.131	0.509 ± 0.129	0.484 ± 0.151	0.544 ± 0.119	0.498 ± 0.153
PTP_AMP	'F7', 'F3', 'Fz', 'F8'	accuracy	0.758 ± 0.127	0.735 ± 0.146	0.754 ± 0.126	0.401 ± 0.000	0.745 ± 0.152	0.794 ± 0.122	0.761 ± 0.145
		precision	0.764 ± 0.125	0.745 ± 0.144	0.767 ± 0.120	0.301 ± 0.000	0.752 ± 0.154	0.798 ± 0.119	0.766 ± 0.146
		recall	0.759 ± 0.125	0.738 ± 0.141	0.761 ± 0.121	0.371 ± 0.000	0.745 ± 0.162	0.796 ± 0.121	0.766 ± 0.146
		f1-score	0.754 ± 0.126	0.732 ± 0.146	0.752 ± 0.127	0.321 ± 0.000	0.734 ± 0.162	0.790 ± 0.123	0.756 ± 0.146
Skewness	'F4'	accuracy	0.561 ± 0.108	0.538 ± 0.101	0.490 ± 0.141	0.467 ± 0.119	0.490 ± 0.120	0.561 ± 0.108	0.483 ± 0.125
		precision	0.566 ± 0.112	0.546 ± 0.104	0.493 ± 0.148	0.467 ± 0.132	0.503 ± 0.153	0.566 ± 0.112	0.492 ± 0.154
		recall	0.564 ± 0.110	0.546 ± 0.101	0.491 ± 0.145	0.467 ± 0.125	0.491 ± 0.128	0.564 ± 0.110	0.493 ± 0.134
		f1-score	0.559 ± 0.107	0.533 ± 0.103	0.488 ± 0.140	0.461 ± 0.124	0.480 ± 0.118	0.559 ± 0.107	0.471 ± 0.134

Furthermore, significant findings from frequency-domain features in Table 4 were elaborated below.

- The highest average accuracy achieved by the relative power of the high beta band (20–30 Hz) with significant channels 'Fp1', 'Fp2', 'F3', and 'F4' was 73% accuracy with KNN and 71% with both RF and SVM.
- For the lower frequencies of delta (1–4 Hz) and theta (4–8 Hz), the significant selected channels were located in the prefrontal and middle frontal cortex area of the scalp— 'Fp1', 'Fp2', 'F3', and 'F4'. Both achieved an average accuracy of 68% using SVM and KNN, respectively.
- The lowest accuracy obtained in frequency-band features was 52.4% from sigma relative power with only one channel of 'F7'.
- Likewise, the average accuracy of alpha relative power and low betas were 63.4% and 64.5% with KNN and LDA, respectively.

The overall observation of significant selected channels for both time-domain and frequency-domain features was observed in 'Fp1', 'Fp2', 'F3', and 'F4'.

Table 4. The summary of average accuracies and standard deviations of significant frequency-domain features from the selected channels with the seven classifiers.

Band	Sig. Channels ($p < 0.001$)	Performance	CART	KNN	LDA	LR	NB	RF	SVM
Delta	'Fp1', 'Fp2', 'F3', 'Fz', 'F4'	accuracy	0.631 ± 0.127	0.659 ± 0.152	0.680 ± 0.123	0.488 ± 0.069	0.672 ± 0.128	0.675 ± 0.113	0.682 ± 0.144
		precision	0.639 ± 0.122	0.679 ± 0.150	0.692 ± 0.120	0.488 ± 0.076	0.682 ± 0.128	0.686 ± 0.102	0.694 ± 0.145
		recall	0.635 ± 0.121	0.674 ± 0.146	0.689 ± 0.121	0.495 ± 0.066	0.679 ± 0.129	0.682 ± 0.104	0.690 ± 0.145
		f1-score	0.627 ± 0.127	0.652 ± 0.152	0.679 ± 0.123	0.481 ± 0.079	0.672 ± 0.128	0.673 ± 0.113	0.677 ± 0.145
Theta	'Fp1', 'Fp2', 'F3', 'Fz', 'F4'	accuracy	0.614 ± 0.088	0.619 ± 0.087	0.679 ± 0.142	0.473 ± 0.059	0.655 ± 0.118	0.643 ± 0.107	0.656 ± 0.135
		precision	0.614 ± 0.091	0.623 ± 0.100	0.683 ± 0.146	0.471 ± 0.064	0.662 ± 0.132	0.646 ± 0.107	0.659 ± 0.140
		recall	0.614 ± 0.091	0.616 ± 0.095	0.671 ± 0.138	0.475 ± 0.060	0.652 ± 0.119	0.645 ± 0.104	0.655 ± 0.140
		f1-score	0.611 ± 0.091	0.607 ± 0.097	0.669 ± 0.138	0.464 ± 0.068	0.647 ± 0.122	0.640 ± 0.103	0.648 ± 0.136
Alpha	'Fp1', 'Fp2', 'F4'	accuracy	0.622 ± 0.107	0.634 ± 0.149	0.634 ± 0.089	0.481 ± 0.070	0.619 ± 0.123	0.606 ± 0.096	0.622 ± 0.137
		precision	0.627 ± 0.108	0.646 ± 0.153	0.640 ± 0.096	0.477 ± 0.076	0.622 ± 0.126	0.608 ± 0.094	0.630 ± 0.141
		recall	0.622 ± 0.101	0.639 ± 0.144	0.633 ± 0.091	0.477 ± 0.076	0.615 ± 0.124	0.607 ± 0.093	0.626 ± 0.136
		f1-score	0.618 ± 0.105	0.628 ± 0.151	0.630 ± 0.091	0.474 ± 0.073	0.611 ± 0.122	0.602 ± 0.095	0.618 ± 0.136
Sigma	'F7'	accuracy	0.524 ± 0.101	0.485 ± 0.153	0.467 ± 0.143	0.422 ± 0.083	0.520 ± 0.145	0.524 ± 0.101	0.494 ± 0.166
		precision	0.532 ± 0.121	0.491 ± 0.161	0.470 ± 0.147	0.422 ± 0.091	0.522 ± 0.153	0.532 ± 0.120	0.501 ± 0.181
		recall	0.528 ± 0.111	0.490 ± 0.156	0.473 ± 0.145	0.422 ± 0.087	0.519 ± 0.150	0.528 ± 0.111	0.499 ± 0.175
		f1-score	0.521 ± 0.103	0.475 ± 0.156	0.463 ± 0.144	0.415 ± 0.085	0.510 ± 0.151	0.521 ± 0.103	0.480 ± 0.169
Low Beta	'Fp1', 'F4'	accuracy	0.608 ± 0.081	0.597 ± 0.134	0.646 ± 0.087	0.494 ± 0.050	0.626 ± 0.123	0.574 ± 0.100	0.617 ± 0.104
		precision	0.612 ± 0.082	0.601 ± 0.136	0.649 ± 0.091	0.489 ± 0.060	0.635 ± 0.124	0.580 ± 0.101	0.624 ± 0.106
		recall	0.610 ± 0.082	0.598 ± 0.136	0.647 ± 0.090	0.492 ± 0.055	0.630 ± 0.120	0.578 ± 0.101	0.618 ± 0.101
		f1-score	0.605 ± 0.081	0.592 ± 0.134	0.643 ± 0.091	0.488 ± 0.060	0.626 ± 0.123	0.571 ± 0.100	0.613 ± 0.103
High Beta	'Fp1', 'Fp2', 'F3', 'F4'	accuracy	0.658 ± 0.120	0.729 ± 0.122	0.726 ± 0.133	0.505 ± 0.069	0.734 ± 0.136	0.713 ± 0.118	0.714 ± 0.115
		precision	0.660 ± 0.105	0.736 ± 0.125	0.733 ± 0.137	0.505 ± 0.071	0.735 ± 0.135	0.716 ± 0.115	0.718 ± 0.109
		recall	0.656 ± 0.104	0.731 ± 0.119	0.726 ± 0.129	0.509 ± 0.065	0.732 ± 0.134	0.713 ± 0.112	0.716 ± 0.111
		f1-score	0.651 ± 0.111	0.727 ± 0.124	0.723 ± 0.132	0.499 ± 0.078	0.731 ± 0.137	0.707 ± 0.119	0.711 ± 0.116

Table 5 presents the classification performance of each significant PLV of the connectivity frequency bands. The highest accuracy achieved by PLV bands were 0.752 ± 0.144, 0.734 ± 0.145, and 0.719 ± 0.177 for PLV's of delta, high beta, and alpha, respectively, using LDA. The rest of the PLV bands got an average accuracy of 0.65 ± 0.12.

Table 5. The summary of average accuracies and standard deviations of PLV's significant connectivity network features from each band.

Features	Performance	CART	KNN	LDA	LR	NB	RF	SVM
Delta_PLV	accuracy	0.678 ± 0.107	0.684 ± 0.141	0.752 ± 0.144	0.657 ± 0.161	0.718 ± 0.166	0.717 ± 0.150	0.724 ± 0.179
	precision	0.678 ± 0.111	0.691 ± 0.153	0.757 ± 0.142	0.661 ± 0.161	0.720 ± 0.162	0.719 ± 0.152	0.731 ± 0.182
	recall	0.677 ± 0.109	0.681 ± 0.143	0.753 ± 0.140	0.658 ± 0.160	0.717 ± 0.162	0.713 ± 0.146	0.725 ± 0.177
	f1-score	0.674 ± 0.108	0.678 ± 0.143	0.749 ± 0.145	0.653 ± 0.161	0.712 ± 0.165	0.711 ± 0.148	0.717 ± 0.181
Theta_PLV	accuracy	0.606 ± 0.172	0.681 ± 0.137	0.683 ± 0.129	0.631 ± 0.163	0.683 ± 0.145	0.629 ± 0.170	0.651 ± 0.179
	precision	0.614 ± 0.172	0.681 ± 0.139	0.687 ± 0.130	0.631 ± 0.168	0.690 ± 0.144	0.636 ± 0.174	0.651 ± 0.180
	recall	0.612 ± 0.172	0.674 ± 0.136	0.685 ± 0.128	0.627 ± 0.164	0.681 ± 0.146	0.631 ± 0.169	0.649 ± 0.178
	f1-score	0.604 ± 0.174	0.673 ± 0.137	0.681 ± 0.129	0.624 ± 0.164	0.676 ± 0.145	0.627 ± 0.169	0.646 ± 0.178
Alpha_PLV	accuracy	0.686 ± 0.140	0.662 ± 0.148	0.719 ± 0.177	0.657 ± 0.180	0.710 ± 0.162	0.699 ± 0.142	0.691 ± 0.143
	precision	0.686 ± 0.139	0.664 ± 0.154	0.719 ± 0.182	0.659 ± 0.181	0.709 ± 0.161	0.701 ± 0.142	0.696 ± 0.146
	recall	0.686 ± 0.139	0.657 ± 0.149	0.713 ± 0.181	0.658 ± 0.181	0.706 ± 0.160	0.703 ± 0.145	0.687 ± 0.143
	f1-score	0.682 ± 0.139	0.653 ± 0.152	0.711 ± 0.178	0.654 ± 0.179	0.707 ± 0.159	0.698 ± 0.143	0.686 ± 0.142
Sigma_PLV	accuracy	0.613 ± 0.124	0.675 ± 0.125	0.659 ± 0.125	0.591 ± 0.120	0.677 ± 0.117	0.629 ± 0.136	0.656 ± 0.131
	precision	0.619 ± 0.125	0.680 ± 0.126	0.662 ± 0.126	0.597 ± 0.122	0.681 ± 0.117	0.632 ± 0.137	0.659 ± 0.133
	recall	0.616 ± 0.122	0.675 ± 0.121	0.657 ± 0.122	0.595 ± 0.121	0.675 ± 0.114	0.631 ± 0.136	0.656 ± 0.130
	f1-score	0.611 ± 0.122	0.670 ± 0.124	0.653 ± 0.124	0.589 ± 0.121	0.672 ± 0.118	0.626 ± 0.136	0.655 ± 0.130
L_beta_PLV	accuracy	0.663 ± 0.148	0.628 ± 0.134	0.679 ± 0.163	0.595 ± 0.132	0.655 ± 0.127	0.642 ± 0.137	0.641 ± 0.183
	precision	0.663 ± 0.144	0.635 ± 0.145	0.686 ± 0.162	0.604 ± 0.137	0.659 ± 0.126	0.641 ± 0.137	0.642 ± 0.186
	recall	0.663 ± 0.144	0.631 ± 0.139	0.683 ± 0.161	0.603 ± 0.132	0.655 ± 0.123	0.643 ± 0.136	0.640 ± 0.186
	f1-score	0.659 ± 0.149	0.624 ± 0.136	0.675 ± 0.165	0.592 ± 0.135	0.653 ± 0.126	0.636 ± 0.140	0.636 ± 0.183
H_beta_PLV	accuracy	0.714 ± 0.143	0.680 ± 0.133	0.734 ± 0.145	0.675 ± 0.134	0.696 ± 0.141	0.719 ± 0.126	0.698 ± 0.137
	precision	0.715 ± 0.146	0.699 ± 0.136	0.738 ± 0.150	0.684 ± 0.137	0.701 ± 0.141	0.723 ± 0.129	0.708 ± 0.140
	recall	0.715 ± 0.147	0.688 ± 0.131	0.734 ± 0.149	0.680 ± 0.133	0.696 ± 0.143	0.722 ± 0.127	0.702 ± 0.136
	f1-score	0.710 ± 0.145	0.677 ± 0.132	0.732 ± 0.147	0.672 ± 0.131	0.691 ± 0.140	0.716 ± 0.126	0.696 ± 0.136

We further classified each subset of significant features of the time domain, frequency domain, and connectivity features as feature vectors and passed them to the classifiers. Figure 8 shows the average accuracy of 15 significant time-domain features and achieved a high accuracy of 81.4% and 80% using RF and SVM, respectively, while other classifiers achieved an average of 76.4%. Figure 9 represents the average accuracies of 20 significant features of relative powers in the frequency domain. The highest accuracy of 80% was obtained by SVM, and 74% was the average accuracy of the other classifiers. Similarly, Figure 10 shows the results of 29 significant features from the connectivity network of PLV, and the average performance accuracy obtained was 88% with SVM and RF, while the rest of the classifiers achieved an average of 84%. Meanwhile, Figure 11 presents the average classification result of 64 hybrid significant features from the time domain, frequency domain, and functional connectivity network, as shown in Tables 3–5.

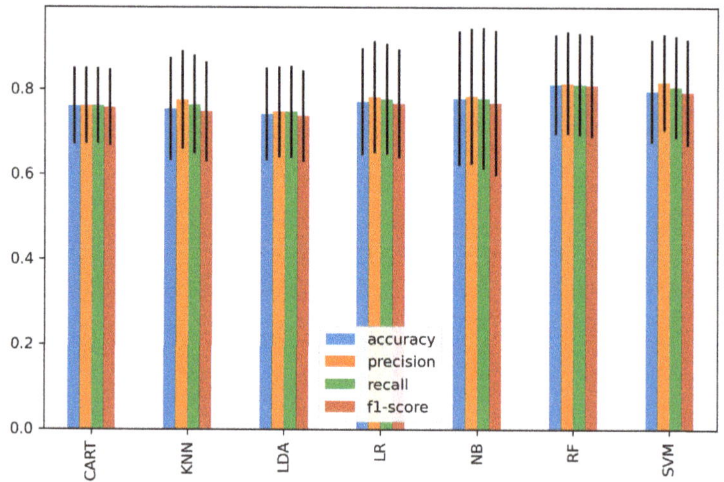

Figure 8. The average classification performance and standard deviation σ of 15 significant features of the time domain. The vertical line indicates σ.

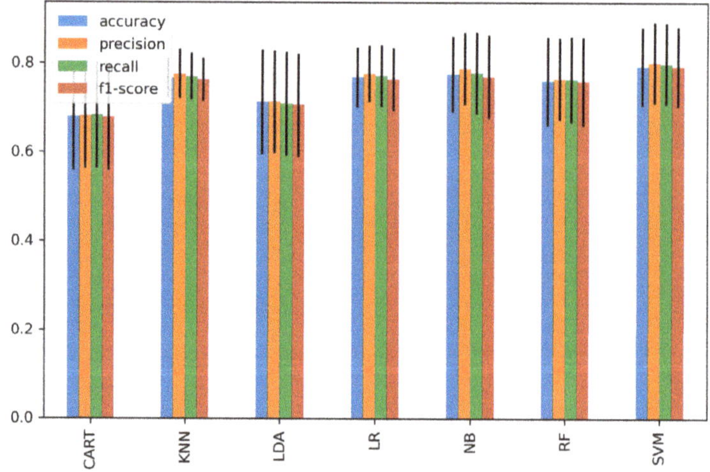

Figure 9. The average classification performance and standard deviation of 20 significant features from the frequency domain.

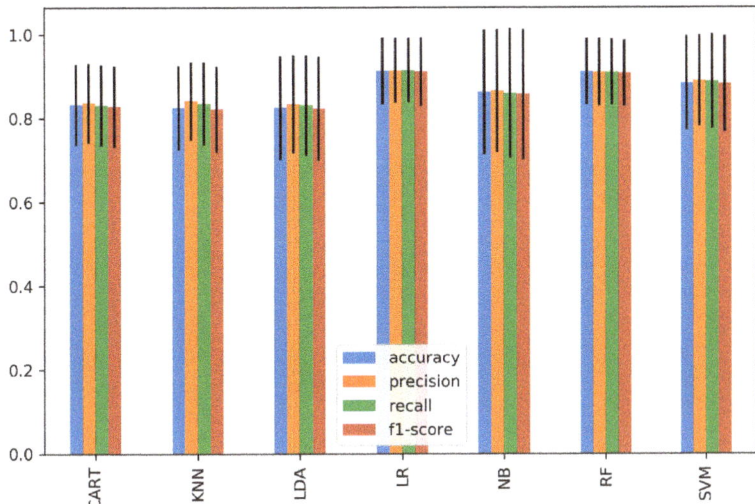

Figure 10. The average classification performance and standard deviation of all significant connectivity network features of PLV.

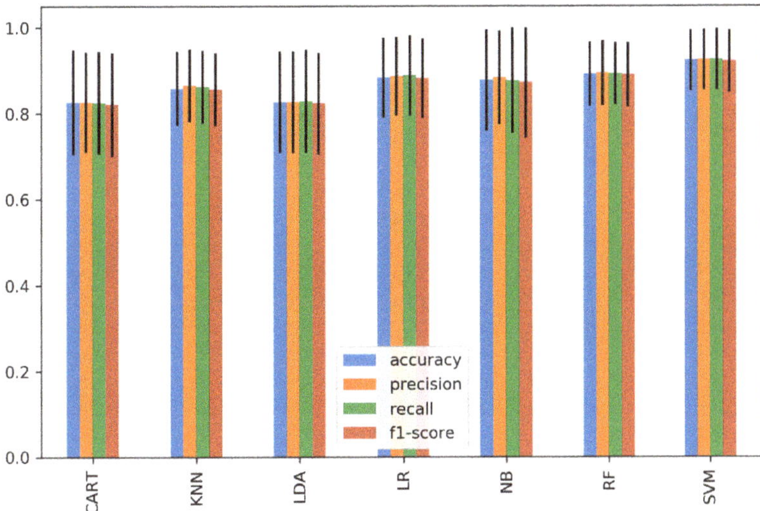

Figure 11. The average classification performance and standard deviation of hybrid features consisting of 42 significant features from the time domain, frequency domain, and connectivity network.

Figure 12 demonstrates the comparison of classification accuracy of each feature's subset domain (time-domain feature, frequency domain, connectivity network feature) as well as after their fusion. In summary, these results show that SVM achieved the best classification performance when fusing connectivity features with cortical connectivity features, scoring 93.2%, 92.4%, 92.5%, and 92.1% for accuracy, precision, recall, and f1-score, respectively. Overall, fusing the multi-domain feature set from cortical and connectivity features improves the classification performance by 13% compared to a single subset domain alone.

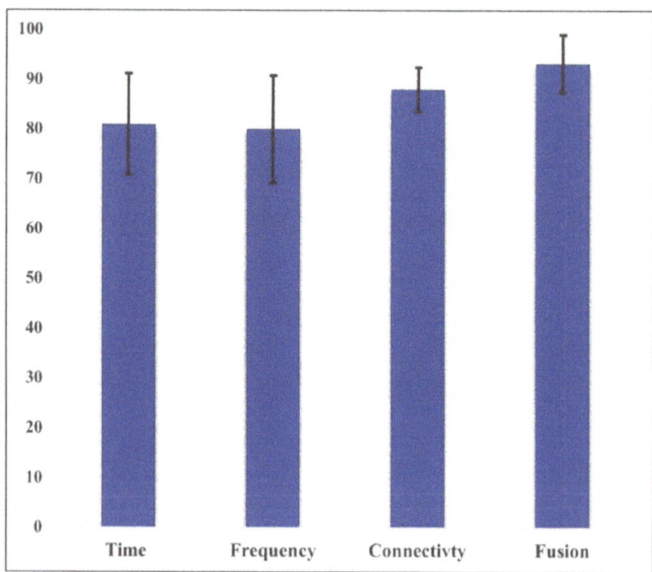

Figure 12. A summary comparison between the average accuracy of single subset domain features (time domain, frequency domain, connectivity network features) and the fusion of all three using SVM.

6. Discussion

This study has presented a methodological approach based on the fusion of multi-domain EEG features and ML for the sake of mental stress classification. To the best of our knowledge, this is the first EEG study on stress that fused functional connectivity features with temporal and spectral features. For this aim, an experimental paradigm based on MIST was designed to induce mental stress and rest conditions using mathematical task with time pressure and negative feedback. For the comparison between the two conditions, a valid objective measurement using the alpha-amylase level (AAL) was collected from the saliva of each subject under both conditions (rest/stress) and quantitatively analyzed, as shown in Figure 4. We found that induced stress revealed a significant difference in AAL between the rest and stress conditions across all subjects, with a considerable increase in AAL from the rest condition to the stress condition. This study correlates to prior findings [13,57] of utilizing arithmetical tasks to induce mental stress in the laboratory setting.

Compared to previous stress detection methods, the main contributions of the proposed method are exploiting the different feature extraction methods and analyzing the significant corresponding channels. For the identification of mental stress in EEG, three scenarios for feature analysis were conducted:

The first scenario analyzed features of the time domain, frequency domain, and functional connectivity network separately, as shown in Tables 3 and 4 and Figure 7. Prior to the analysis, only significant channels were selected for classification. The selection of significant channels in all types of features was based on a statistical t-test at $p < 0.05$. The second scenario was based on combining the significant features within the time domain, frequency domain, and connectivity features of PLV to form subset feature vectors for classification (Figures 8 and 9). The third scenario was based on the fusion of the significant features from all domains (time domain, frequency domain, and connectivity network features) to form a single hybrid subset feature vector for subsequent classifiers.

In particular, for the temporal features of Hjorth complexity, Hjorth activity, ptp_amp, and kurtosis, we found a significant decrease from the rest condition to the stress condition. The decrease in the temporal activities within stress conditions indicates that participants experience difficulties in engaging with the MA task. In fact, the greater the complexity value is, the more active the brain is. Previous studies have found that higher complexity meant increased behavioral performance [58,59]. In line with that, the decreased complexity in our study is a sign of decreased behavioral performance (accuracy of detection) due to stress. It should be noted, however, that the decrease in brain activity/complexity in our study was localized to a certain brain region. For example, when the temporal features of Hjorth complexity are considered, the left frontal region at 'Fp1' and 'F3' is highly reduced under stress. This reduction is consistent with the previous emotion study that utilized videos to induce negative emotions in the participants [60]. Likewise, when the complexity and skewness are considered, the right frontal region at 'FP2' and 'F4' is highly reduced. This is also consistent with our previous studies that utilized simple arithmetic tasks with time pressure to induce stress [13,27].

On the other hand, we found that the relative EEG power in theta, alpha, sigma, and beta showed a significant increase from the rest condition to the stress condition at a particular region of the brain. Considering all of the relative powers together, we found that the right hemisphere was highly sensitive to stress exposure. This confirms that negative emotions are induced under stress. This is in line with previous studies that showed when the stress level increased, the alpha power increased across the frontal cortex [44,57,61,62]. Likewise, the increase in relative beta power in our study is also consistent with previous studies that utilized driving and public speaking as stressors in their studies [39,63]. These findings demonstrate the potential of using temporal, spectral, and connectivity features in finding patterns associated with mental stress, as demonstrated in our previous studies on stress and control states [64,65].

We further analyze the classification accuracy of stress based on the first scenario using CAR, KNN, LDA, LR, NB, RF, and SVM classifiers. The temporal features of the peak-to-peak amplitude of the significant channels—'F7', 'F3', 'Fz', and 'F8'—showed the highest classification accuracy of 79.4% using SVM. Meanwhile, the frequency-domain features of the high beta band (20–30 Hz) at significant channels of 'Fp1', 'Fp2', 'F3', and 'F4' achieved the highest classification accuracy of 73% using KNN.

Meanwhile, in the second scenario (domain feature subset analysis), we observed a 2%, 7%, and 13% improvement in classification accuracy in the time domain, frequency domain, and connectivity features, respectively, when compared to the first scenario. Particularly, the high accuracies of 81.4%, 80% and 88% were achieved when using the significant feature subset of the 15 time-domain features, 20 frequency-domain features, and 29 connectivity network features, respectively. It is noteworthy that the subset of connectivity features outperformed other domains in classifying mental stress. Our findings are consistent with prior functional connectivity research, which has shown that functional connection is more reflective of the mental task performance [66].

Additionally, in the third scenario, fusing significant features of the time domain, frequency domain, and connectivity features of PLV (a total of 64 features) improved the overall accuracy of detecting the rest/stress condition with the highest accuracy of 93.2% obtained using SVM. As expected, the improvements in the classification performance support the hypothesis that fusing multi-domain features may provide complementary information for better stress detection. In general, the proposed method of selecting a significant EEG-channel-based feature yield a total reduction of feature space of almost 60%, with 64 significant features out of 210 features.

The overall classifiers' performance depends significantly on relevant EEG features and the selected channel related to the given task. Previous studies have found that using a large number of EEG channels could provide high resolution and improve accuracy; however, they have inherited issues, such as cost and applicability, particularly outside laboratories. Subhani [57] discusses the identification of stress using 19 EEG channels with features of absolute power, relative power (RP), coherence, phase lag, and amplitude asymmetry, and they reported high accuracy of 94.58%; yet, high dimensionality existed when the 190-feature vector was used. This high accuracy could be interpreted as the result of using a high spatial resolution of 19 EEG channels. However, in this study, the maximum accuracy was 93.2% with the 64-feature vector. Although the accuracy was slightly lower than the one reported by [57,63], this study efficiently reduced feature space with high performance.

This study confirmed that the fusion of temporal, spectral, and connectivity features significantly improved mental stress classification accuracy. Although the study was informative for the mental stress classification, it had a few limitations. First, we only reported the results of EEG feature extraction at an epoch length of one second, which corresponds to 256 EEG data points. Future studies may consider reporting the results of epochs with more than one second, i.e., in the range of one to ten seconds. Second, this study was constructed on fusion functional connections using PLV with cortical features. Other connectivity features, such as the Phase Lock Index (PLI), Partial Directed Coherence (PDC), and Directed Transfer Function (DTF) [54], were not included in the study. Third, while we conducted statistical analysis with a *t*-test to select the brain regions relevant to mental stress in this study, different methods for detecting mental stress using feature-based channel selection (e.g., swarm intelligence) should be considered in future work to reduce the high dimensionality and select the optimal feature set. Finally, throughout the experiment, we found that the SVM outperformed other classifiers in terms of classification performance using the selected hyperparameters, but using algorithm optimization for finding optimal parameters could improve the overall performance.

In essence, the proposed framework empirically proved the possibility of having significant channels corresponding to each feature while eliminating the redundancy and ignoring un-relevant channels. Then, it could be suggested that EEG signals have the potential to be reliable for identifying stress for home-based applications with an optimal number of channels and the relevant features. However, multiple methods for detecting mental stress using feature-based channel selection should be considered in future work.

7. Conclusions

This paper aims to find feature sets that would distinguish the stress and non-stress conditions using seven frontal EEG channels. EEG's features from the time domain, frequency domain, functional connectivity network, and all three fused together were investigated. Seven classifiers were used to evaluate the performance of each feature set before and after fusion. The highest accuracy of 93.2% was achieved using hybrid features with the SVM classifier. In comparison, the evaluation performance of the time domain, frequency domain, and connectivity feature subsets were 81.4%, 80%, and 88% respectively. The results demonstrated that the proposed method of fusing the connectivity network with temporal and spectral features was capable of detecting mental stress state with high classification performance. The overall results support developing a real-time system for stress measurement and analysis.

Author Contributions: Conceptualization, A.H. and D.H.; methodology, A.H., D.H. and F.A.-S.; software, A.H.; formal analysis, A.H. and F.A.-S.; resources, A.H., D.H., T.M., F.A.-S., T.P. and M.H.K.; data curation, F.A.-S.; writing original draft preparation, A.H.; writing-review and editing, A.H., D.H., F.A.-S., T.M., T.P. and M.H.K.; supervision, D.H., T.M. and T.P.; funding acquisition, D.H.; validation, A.H., D.H., T.M., F.A.-S., T.P. and M.H.K. All authors have read and agreed to the published version of the manuscript.

Funding: This research was supported by the Fundamental Research Grant Scheme (FRGS) funded by the Ministry of Higher Education (Grant code: FRGS/1/2019/ICT02/TAYLOR/03/1).

Institutional Review Board Statement: The study was conducted according to the guidelines of the Declaration of Helsinki, and approved by the Institutional Review Board of Universiti Teknologi Petronas.

Informed Consent Statement: Informed consent was obtained from all subjects involved in the study.

Data Availability Statement: Raw EEG data can be obtained by writing formal email to Fares Al-Shargie.

Acknowledgments: The authors would like to acknowledge the support received from Taylor's University, Malaysia, through its Ph.D. Scholarship programs and the support from the Fundamental Research Grant Scheme (FRGS) funded by the Ministry of Higher Education (Grant code: FRGS/1/2019/ICT02/TAYLOR/03/1). The authors would like to thank the Universiti Teknologi Petronas for sharing the dataset.

Conflicts of Interest: The authors declare no conflict of interest.

References

1. Asif, A.; Majid, M.; Anwar, S.M. Human stress classification using EEG signals in response to music tracks. *Comput. Biol. Med.* **2019**, *107*, 182–196. [CrossRef]
2. Can, Y.S.; Arnrich, B.; Ersoy, C. Stress Detection in Daily Life Scenarios Using Smart Phones and Wearable Sensors: A Survey. *J. Biomed. Inform.* **2019**, *92*, 103139. [CrossRef]
3. Song, H.; Fang, F.; Arnberg, F.K.; Mataix-Cols, D.; De La Cruz, L.F.; Almqvist, C.; Fall, K.; Lichtenstein, P.; Thorgeirsson, G.; Valdimarsdóttir, U.A. Stress related disorders and risk of cardiovascular disease: Population based, sibling controlled cohort study. *BMJ* **2019**, *365*, 1–10. [CrossRef] [PubMed]
4. Blanding, M. Workplace Stress Responsible for up to $190B in Annual U.S. Healthcare Costs. In *HBS Working Knowledge*; Forbes: Jersey City, NJ, USA, 2015. Available online: https://www.forbes.com/sites/hbsworkingknowledge/2015/01/26/workplace-stress-responsible-for-up-to-190-billion-in-annual-u-s-heathcare-costs/?sh=59952af2235a (accessed on 9 July 2021).
5. Daudelin-Peltier, C.; Forget, H.; Blais, C.; Deschênes, A.; Fiset, D. The effect of acute social stress on the recognition of facial expression of emotions. *Sci. Rep.* **2017**, *7*, 1036. [CrossRef] [PubMed]
6. Pereira, T.; Almeida, P.R.; Cunha, J.P.; Aguiar, A. Heart rate variability metrics for fine-grained stress level assessment. *Comput. Methods Programs Biomed.* **2017**, *148*, 71–80. [CrossRef]
7. Cohen, S.; Kamarck, T.; Mermelstein, R. A Global Measure of Perceived Stress. *J. Health Soc. Behav.* **1983**, *24*, 385. [CrossRef]
8. Nagar, P.; Sethia, D. Brain Mapping Based Stress Identification Using Portable EEG Based Device. In Proceedings of the 2019 11th International Conference on Communication Systems & Networks (COMSNETS), Bengaluru, India, 7–11 January 2019; Volume 2061, pp. 601–606.
9. Sanay, M.U.S.; Syed, M.A.; Majid, M. Quantification of Human Stress Using Commercially Available Single Channel EEG Headset. *IEICE Trans. Inf. Syst.* **2017**, *100*, 2241–2244. [CrossRef]
10. Zigmond, A.S.; Snaith, R.P. The Hospital Anxiety and Depression Scale. *Br. J. Psychiatry* **1983**, *134*, 382–389. [CrossRef] [PubMed]
11. Minguillon, J.; Lopez-Gordo, M.A.; Pelayo, F. Stress Assessment by Prefrontal Relative Gamma. *Front. Comput. Neurosci.* **2016**, *10*, 1–9. [CrossRef]
12. Poole, K.L.; Schmidt, L.A. Frontal brain delta-beta correlation, salivary cortisol, and social anxiety in children. *J. Child Psychol. Psychiatry Allied Discip.* **2019**, *60*, 646–654. [CrossRef]
13. Al-Shargie, F.; Tang, T.B.; Kiguchi, M. Assessment of mental stress effects on prefrontal cortical activities using canonical correlation analysis: An fNIRS-EEG study. *Biomed. Opt. Express* **2017**, *8*, 2583. [CrossRef] [PubMed]
14. Gowrisankaran, S.; Nahar, N.K.; Hayes, J.R.; Sheedy, J.E. Asthenopia and Blink Rate Under Visual and Cognitive Loads. *Optom. Vis. Sci.* **2012**, *89*, 97–104. [CrossRef] [PubMed]
15. Barreto, A.; Zhai, J.; Adjouadi, M. Non-intrusive Physiological Monitoring for Automated Stress Detection in Human-Computer Interaction. In *Human–Computer Interaction*; Springer: Berlin/Heidelberg, Germany, 2007; pp. 29–38. [CrossRef]
16. Davis, M.S. Voice Stress Analysis. In *The Concise Dictionary of Crime and Justice*; SAGE Publications, Inc.: Thousand Oaks, CA, USA, 2012. [CrossRef]
17. Giannakakis, G.; Grigoriadis, D.; Giannakaki, K.; Simantiraki, O.; Roniotis, A.; Tsiknakis, M. Review on psychological stress detection using biosignals. *IEEE Trans. Affect. Comput.* **2019**, 1. [CrossRef]
18. Wang, Y.; Veluvolu, K.C. Evolutionary algorithm based feature optimization for multi-channel EEG classification. *Front. Neurosci.* **2017**, *11*, 1–14. [CrossRef]
19. Jiang, D.; Lu, Y.N.; Yu, M.A.; Yuanyuan, W. Robust sleep stage classification with single-channel EEG signals using multimodal decomposition and HMM-based refinement. *Expert Syst. Appl.* **2019**, *31*, 188–203. [CrossRef]
20. Jung, Y.; Yoon, Y.I. Multi-level assessment model for wellness service based on human mental stress level. *Multimed. Tools Appl.* **2017**, *76*, 11305–11317. [CrossRef]
21. Hosseini, S.A.; Khalilzadeh, M.A.; Naghibi-Sistani, M.B.; Homam, S.M. Emotional stress recognition using a new fusion link between electroencephalogram and peripheral signals. *Iran. J. Neurol.* **2015**, *14*, 142–151. [PubMed]
22. Xu, Q.; Nwe, T.L.; Guan, C. Cluster-Based Analysis for Personalized Stress. *IEEE J. Biomed. Health Inform.* **2015**, *19*, 275–281. [CrossRef]
23. Dedovic, K.; Renwick, R.; Mahani, N.K.; Engert, V.; Lupien, S.J.; Pruessner, J.C. The Montreal Imaging Stress Task: Using functional imaging to investigate the effects of perceiving and processing psychosocial stress in the human brain. *J. Psychiatry Neurosci.* **2005**, *30*, 319.

24. Ullah, I.; Hussain, M.; Qazi, E.U.H.; Aboalsamh, H. An automated system for epilepsy detection using EEG brain signals based on deep learning approach. *Expert Syst. Appl.* **2018**, *107*, 61–71. [CrossRef]

25. Dubois, J.; Adolphs, R. Building a Science of Individual Differences from fMRI. *Feature Rev.* **2016**, *20*, 425–443. [CrossRef] [PubMed]

26. Bartenstein, P. PET in neuroscience. *Nuklearmedizin* **2004**, *43*, 33–42. [CrossRef] [PubMed]

27. Al-Shargie, F.; Tang, T.B.; Badruddin, N.; Kiguchi, M. Simultaneous measurement of EEG-fNIRS in classifying and localizing brain activation to mental stress. In Proceedings of the IEEE 2015 International Conference on Signal and Image Processing Applications ICSIPA, Kuala Lumpur, Malaysia, 19–21 October 2015; pp. 282–286. [CrossRef]

28. Al-Shargie, F.; Tang, T.B.; Badruddin, N.; Kiguchi, M. Towards multilevel mental stress assessment using SVM with ECOC: An EEG approach. *Med. Biol. Eng. Comput.* **2018**, *56*, 125–136. [CrossRef]

29. Shon, D.; Im, K.; Park, J.H.; Lim, D.S.; Jang, B.; Kim, J.M. Emotional Stress State Detection Using Genetic Algorithm-Based Feature Selection on EEG Signals. *Int. J. Environ. Res. Public Health* **2018**, *15*, 2461. [CrossRef]

30. Mahajan, R. Emotion Recognition via EEG Using Neural Network Classifier. *Adv. Intell. Syst. Comput.* **2018**, *583*, 429–438. [CrossRef]

31. Sriramprakash, S.; Prasanna, V.D.; Murthy, O.V. Stress Detection in Working People. *Procedia Comput. Sci.* **2017**, *115*, 359–366. [CrossRef]

32. Zhang, S.; Li, B.; Chen, X.; Lu, T.; Zhao, C.; Ren, J.; Ji, X. Research on the Method of Evaluating Psychological Stress by EEG. *IOP Conf. Ser. Earth Environ. Sci.* **2019**, *310*, 042033. [CrossRef]

33. Keshmiri, S. Conditional Entropy: A Potential Digital Marker for Stress. *Entropy* **2021**, *23*, 286. [CrossRef] [PubMed]

34. Attallah, O. An effective mental stress state detection and evaluation system using minimum number of frontal brain electrodes. *Diagnostics* **2020**, *10*, 292. [CrossRef]

35. Saeed, S.M.U.; Anwar, S.M.; Khalid, H.; Majid, M.; Bagci, U. EEG Based Classification of Long-Term Stress Using Psychological Labeling. *Sensors* **2020**, *20*, 1886. [CrossRef] [PubMed]

36. Katmah, R.; Al-Shargie, F.; Tariq, U.; Babiloni, F.; Al-Mughairbi, F.; Al-Nashash, H. A Review on Mental Stress Assessment Methods Using EEG Signals. *Sensors* **2021**, *21*, 5043. [CrossRef]

37. Jebelli, H.; Hwang, S.; Lee, S.H. EEG-based workers' stress recognition at construction sites. *Autom. Constr.* **2018**, *93*, 315–324. [CrossRef]

38. Ahuja, R.; Banga, A. Mental stress detection in university students using machine learning algorithms. *Procedia Comput. Sci.* **2019**, *152*, 349–353. [CrossRef]

39. Halim, Z.; Rehan, M. On identification of driving-induced stress using electroencephalogram signals: A framework based on wearable safety-critical scheme and machine learning. *Inf. Fusion* **2020**, *53*, 66–79. [CrossRef]

40. Al-Nafjan, A.; Hosny, M.; Al-Ohali, Y.; Al-Wabil, A. Review and Classification of Emotion Recognition Based on EEG Brain-Computer Interface System Research: A Systematic Review. *Appl. Sci.* **2017**, *7*, 1239. [CrossRef]

41. Chen, J.; Abbod, M.; Shieh, J.S. Pain and Stress Detection Using Wearable Sensors and Devices—A Review. *Sensors* **2021**, *21*, 1030. [CrossRef]

42. Zhou, B.; Wu, X.; Ruan, J.; LV, Z.; Zhang, L. How many channels are suitable for independent component analysis in motor imagery brain-computer interface. *Biomed. Signal Process. Control* **2019**, *50*, 103–120. [CrossRef]

43. Hasan, M.J.; Kim, J.M. A Hybrid Feature Pool-Based Emotional Stress State Detection Algorithm Using EEG Signals. *Brain Sci.* **2019**, *9*, 376. [CrossRef]

44. Al-Shargie, F.; Kiguchi, M.; Badruddin, N.; Dass, S.C.; Hani, A.F.M.; Tang, T.B. Mental stress assessment using simultaneous measurement of EEG and fNIRS. *Biomed. Opt. Express* **2016**, *7*, 3882. [CrossRef] [PubMed]

45. Peng, H.T.; Savage, E.; Vartanian, O.; Smith, S.; Rhind, S.G.; Tenn, C.; Bjamason, S. Performance Evaluation of a Salivary Amylase Biosensor for Stress Assessment in Military Field Research. *J. Clin. Lab. Anal.* **2016**, *30*, 223–230. [CrossRef]

46. Gramfort, A.; Luessi, M.; Larson, E.; Engemann, D.A.; Strohmeier, D.; Garcia, S.; Claude, U.; Brodbeck, C.; Goj, R.; Jas, M.; et al. MEG and EEG data analysis with MNE-Python. *Front. Neurosci.* **2013**, *7*, 1–13. [CrossRef] [PubMed]

47. Trujillo, L.T.; Stanfield, C.T.; Vela, R.D. The Effect of Electroencephalogram (EEG) Reference Choice on Information-Theoretic Measures of the Complexity and Integration of EEG Signals. *Front. Neurosci.* **2017**, *11*, 425. [CrossRef]

48. Ingle, R.; Oimbe, S.; Kehri, V.; Awale, R.N. Classification of EEG Signals during Meditation and Controlled State Using PCA, ICA, LDA and Support Vector Machines. *Int. J. Pure Appl. Math.* **2018**, *118*, 3179–3190.

49. Li, X.; Hu, B.; Sun, S.; Cai, H. EEG-based mild depressive detection using feature selection methods and classifiers. *Comput. Methods Programs Biomed.* **2016**, *136*, 151–161. [CrossRef]

50. Blanco, S.; Garcia, H.; Quiroga, R.; Romanelli, L.; Rosso, O. Stationarity of the EEG series. *IEEE Eng. Med. Biol. Mag.* **1995**, *14*, 395–399. [CrossRef]

51. Seraj, E.; Sameni, R. Robust electroencephalogram phase estimation with applications in brain-computer interface systems. *Physiol. Meas.* **2017**, *38*, 501–523. [CrossRef] [PubMed]

52. Fraschini, M.; Demuru, M.; Crobe, A.; Marrosu, F.; Stam, C.J.; Hillebrand, A. The effect of epoch length on estimated EEG functional connectivity and brain network organisation. *J. Neural Eng.* **2016**, *13*, 036015. [CrossRef] [PubMed]

53. Prerau, M.J.; Brown, R.E.; Bianchi, M.T.; Ellenbogen, J.M.; Purdon, P.L. Sleep neurophysiological dynamics through the lens of multitaper spectral analysis. *Physiology* **2017**, *32*, 60–92. [CrossRef]

54. Al-Shargie, F.; Tariq, U.; Alex, M.; Mir, H.; Al-Nashash, H. Emotion Recognition Based on Fusion of Local Cortical Activations and Dynamic Functional Networks Connectivity: An EEG Study. *IEEE Access* **2019**, *7*, 143550–143562. [CrossRef]

55. Gong, A.; Liu, J.; Lu, L.; Wu, G.; Jiang, C.; Fu, Y. Characteristic differences between the brain networks of high-level shooting athletes and non-athletes calculated using the phase-locking value algorithm. *Biomed. Signal Process. Control* **2019**, *51*, 128–137. [CrossRef]

56. Al-Shargie, F.M.; Hassanin, O.; Tariq, U.; Al-Nashash, H. EEG-Based Semantic Vigilance Level Classification Using Directed Connectivity Patterns and Graph Theory Analysis. *IEEE Access* **2020**, *8*, 115941–115956. [CrossRef]

57. Subhani, A.R.; Mumtaz, W.; Naufal, M.; Mohamed, B.I.N.; Kamel, N.; Malik, A.S. Machine Learning Framework for the Detection of Mental Stress at Multiple Levels. *IEEE Access* **2017**, *5*, 13545–13556. [CrossRef]

58. Zhang, R.; Xu, P.; Chen, R.; Li, F.; Guo, L.; Li, P.; Zhang, T.; Yao, D. Predicting Inter-session Performance of SMR-Based Brain–Computer Interface Using the Spectral Entropy of Resting-State EEG. *Brain Topogr.* **2015**, *28*, 680–690. [CrossRef] [PubMed]

59. Tian, Y.; Zhang, H.; Xu, W.; Zhang, H.; Yang, L.; Zheng, S.; Shi, Y. Spectral entropy can predict changes of working memory performance reduced by short-time training in the delayed-match-to-sample task. *Front. Hum. Neurosci.* **2017**, *11*, 1–12. [CrossRef]

60. Chakladar, D.D.; Chakraborty, S. EEG based emotion classification using "correlation Based Subset Selection". *Biol. Inspired Cogn. Archit.* **2018**, *24*, 98–106. [CrossRef]

61. Al-shargie, F.; Tang, T.B.; Badruddin, N.; Dass, S.C.; Kiguchi, M. Mental stress assessment based on feature level fusion of fNIRS and EEG signals. In Proceedings of the 2016 6th International Conference on Intelligent and Advanced Systems (ICIAS), Kuala Lumpur, Malaysia, 15–17 August 2016; pp. 1–5. [CrossRef]

62. Al-shargie, F.M.; Tang, T.B.; Badruddin, N.; Kiguchi, M. Mental Stress Quantification Using EEG Signals. In *International Conference for Innovation in Biomedical Engineering and Life Sciences*; Ibrahim, F., Usman, J., Mohktar, M.S., Ahmad, M.Y., Eds.; Springer: Singapore, 2016; pp. 15–19.

63. Arsalan, A.; Majid, M.; Butt, A.R.; Anwar, S.M. Classification of Perceived Mental Stress Using A Commercially Available EEG Headband. *IEEE J. Biomed. Health Inform.* **2019**, *23*, 2257–2264. [CrossRef] [PubMed]

64. Al-Shargie, F.; Tang, T.B.; Kiguchi, M. Stress Assessment Based on Decision Fusion of EEG and fNIRS Signals. *IEEE Access* **2017**, *5*, 19889–19896. [CrossRef]

65. Al-Shargie, F. Prefrontal cortex functional connectivity based on simultaneous record of electrical and hemodynamic responses associated with mental stress. *arXiv* **2021**, arXiv:2103.04636. preprint.

66. Al-Shargie, F.; Tariq, U.; Hassanin, O.; Mir, H.; Babiloni, F.; Al-Nashash, H. Brain connectivity analysis under semantic vigilance and enhanced mental states. *Brain Sci.* **2019**, *9*, 363. [CrossRef] [PubMed]

 *sensors*

Article

Functional Connectivity and Frequency Power Alterations during P300 Task as a Result of Amyotrophic Lateral Sclerosis

Claudia X. Perez-Ortiz [1], Jose L. Gordillo [1], Omar Mendoza-Montoya [1,*], Javier M. Antelis [1], Ricardo Caraza [2] and Hector R. Martinez [2]

[1] Escuela de Ingeniería y Ciencias, Tecnologico de Monterrey, Monterrey 64849, Mexico; c.xochitl96@gmail.com (C.X.P.-O.); JLGordillo@Tec.mx (J.L.G.); mauricio.antelis@tec.mx (J.M.A.)

[2] Escuela de Medicina y Ciencias de la Salud, Tecnologico de Monterrey, Monterrey 64849, Mexico; rcaraza@tec.mx (R.C.); hector.ramon@tec.mx (H.R.M.)

* Correspondence: omendoza83@tec.mx

Abstract: Amyotrophic Lateral Sclerosis (ALS) is one of the most aggressive neurodegenerative diseases and is now recognized as a multisystem network disorder with impaired connectivity. Further research for the understanding of the nature of its cognitive affections is necessary to monitor and detect the disease, so this work provides insight into the neural alterations occurring in ALS patients during a cognitive task (P300 oddball paradigm) by measuring connectivity and the power and latency of the frequency-specific EEG activity of 12 ALS patients and 16 healthy subjects recorded during the use of a P300-based BCI to command a robotic arm. For ALS patients, in comparison to Controls, the results ($p < 0.05$) were: an increment in latency of the peak ERP in the Delta range (OZ) and Alpha range (PO7), and a decreased power in the Beta band among most electrodes; connectivity alterations among all bands, especially in the Alpha band between PO7 and the channels above the motor cortex. The evolution observed over months of an advanced-state patient backs up these findings. These results were used to compute connectivity- and power-based features to discriminate between ALS and Control groups using Support Vector Machine (SVM). Cross-validation achieved a 100% in specificity and 75% in sensitivity, with an overall 89% success.

Keywords: ALS; EEG; classifier; neural; connectivity; frequency-specific; BCI

Citation: Perez-Ortiz, C.X.; Gordillo, J.L.; Mendoza-Montoya, O.; Antelis, J.M.; Caraza, R.; Martinez, H.R. Functional Connectivity and Frequency Power Alterations during P300 Task as a Result of Amyotrophic Lateral Sclerosis. *Sensors* **2021**, *21*, 6801. https://doi.org/10.3390/s21206801

Academic Editor: Yvonne Tran

Received: 1 August 2021
Accepted: 26 September 2021
Published: 13 October 2021

Publisher's Note: MDPI stays neutral with regard to jurisdictional claims in published maps and institutional affiliations.

1. Introduction

Amyotrophic Lateral Sclerosis (ALS) is one of the most aggressive neurodegenerative diseases causing the patient to lose the ability to move their muscles; it affects and kills upper and lower motor neurons [1]. Being a complex disease, the specific nature of the affectations is still unknown [1]. There is no particular detection method for ALS; thus, the detection procedure involves taking several tests to discard other diseases; this may be due to the lack of a biological marker or biomarker for ALS [1]. To understand ALS, more global holistic approaches have been undertaken. New research has indicated neurodegeneration in nonmotor areas too [2], and the disease has also been recognized as a multisystem network disorder characterized by impaired connectivity (a measure of how synchronized two brain regions are) [3–7]. Correlations have been found between changes in connectivity and cognitive scores from a neuropsychological battery (cognitive tests) [7]. Frontotemporal dementia [8] and cognitive disability [9] have been linked with ALS.

The use of different technologies gives some insight; for example, Positron Emission Tomography (PET) and Magnetic Resonance Imaging (MRI) studies have shown deterioration in motor cortex regions [3]. Alterations and differences in ALS are primarily found in functional Magnetic Resonance Imaging (fMRI) or similar costly procedures [3]. Correlations have been found, in studies with ALS patients, between Electro-encephalography (EEG) rhythms and MRI and transcranial magnetic stimulation (TMS) findings [10], and between fMRIs and EEG [11] in the past, suggesting that other neuroimaging findings could be replicated with EEG. EEG offers an understanding of the activity happening on the cerebral cortex originating from neural activity. It is also a portable, noninvasive brain imaging sensor that obtains cerebral information in real-time and generates responses [12]. This type of neuroimaging has been used as a detection method for neurodegenerative diseases. For example, analyzing the sharpness of the brain signals is commonly used to detect epilepsy [13]. Thus, EEG studies can show how the activity in an ALS brain is changing or degenerating, offering specialists further comprehension of a disease whose evolutive nature is still unknown. Even though some studies have been performed on ALS patients, they have usually focused on rest-state activity [5,7]. The potential use of these types of findings as a biomarker to detect ALS has even been assessed with a result of 100% in specificity but only 58% in sensitivity [4].

Testing during a cognitive activity offers additional information of a mental state. The P300 oddball paradigm tests cognitive activity through EEG. The oddball test is the presentation of repetitive stimuli randomly interrupted by a different stimulus. P300 is an Event-Related Potential (ERP) elucidated 300 ms after a stimulus. It is a signal that arises as a response of the brain to an external stimulus [14]. The visual P300 task involves communication between the parietal region and the frontal region [15], the frontal one being one of the most affected areas for ALS individuals [3]. P300 is commonly used in Brain–Computer Interfaces (BCIs) as its response is repetitive and detectable through the cognitive task. This quality permits P300 to act as a detection method. Additionally, the magnitude and latency of the P300 peak have shown a solid ability to detect other diseases such as Alzheimer's, which is linked to a decreased peak and higher latency [16], and many other neurological disorders [14].

More recent studies have demonstrated the cognitive impairment of ALS individuals, especially with delayed latencies in P300 peaks [17–19]. These studies could offer a more robust understanding of the EEG if analyzed in the frequency domain as it offers a significant correlation to neural oscillations. Differences have been found in the functional connectivity and amplitude of Alpha and Beta frequency bands (9–13 and 14–30 Hz, respectively), suggesting a frequency-specific reduction in patterns in functional connectivity and amplitude [11,20], but a longitudinal analysis of the biomarkers found during the EEG studies among the same individuals has not been performed, in other words, showing that the advance of the disease has not been assessed until now.

The present study provides insight into the neural alterations or changes occurring in ALS patients during a cognitive task (P300 oddball paradigm) by measuring EEG activity. The objective was to find the neural alterations, define these alterations as biomarkers, analyze their change in a longitudinal study in ALS patients, and classify between groups with them. Three numeric values were calculated (variables) from the time–frequency power signals to measure the EEG activity: the value of the peak power and its latency, and the connectivity between electrodes (a value between 0 and 1) for each frequency band. These variables were tested for a significant difference between ALS and Control groups; those who were found different were used to train a classifier that separates both groups and were further analyzed.

2. Materials and Methods

Finding neural alterations in ALS patients with respect to healthy subjects is a method of obtaining insight into the mechanism that ALS follows. It could also serve as a detection aid to confirm or diagnose ALS. Different tests were made with the EEG data of 12 ALS patients and 16 healthy subjects to find these alterations. The data came from a 16 electrode-P300-based BCI system that was previously designed to aid ALS patients with communication with the outer world, such as moving a robotic hand orthosis [21].

The tests made to the data had the objective of obtaining a numeric value from the data, such as the time in milliseconds where the P300 peak was found or the power of that peak. The connectivity analysis calculated the connectivity (values from 0 to 1) between one electrode and the others. These three tests were made on the data decomposed in spectral power. A point-wise graphical analysis helped to analyze the 3D images from the time–frequency–power charts. Finally, the numeric values were taken as variables. The significantly different variables were further analyzed and some variables were selected as biomarkers. The biomarkers were used to train an SVM classifier or to observe an ALS patient's evolution over time.

Finally, the numeric values were taken as variables. The significantly different variables were selected as biomarkers, and the biomarkers were used to train an SVM classifier.

2.1. P300-BCI System

The data used for this analysis were taken from the training stage of a P300-BCI designed for ALS patients. A P300-based BCI was previously developed with the purpose of assisting ALS patients to control a Hand of Hope robotic arm (Rehab-Robotics Company, China). Muscle movement loss is a common ALS symptom, hence deteriorating the original pathway for muscle movement; the objective of this BCI is to generate an alternative path for hand-muscles movement, as shown in Figure 1a. Instead of the brain moving the muscles of the hand, a WiFi-controlled hand orthosis forces the movement onto the hand [21].

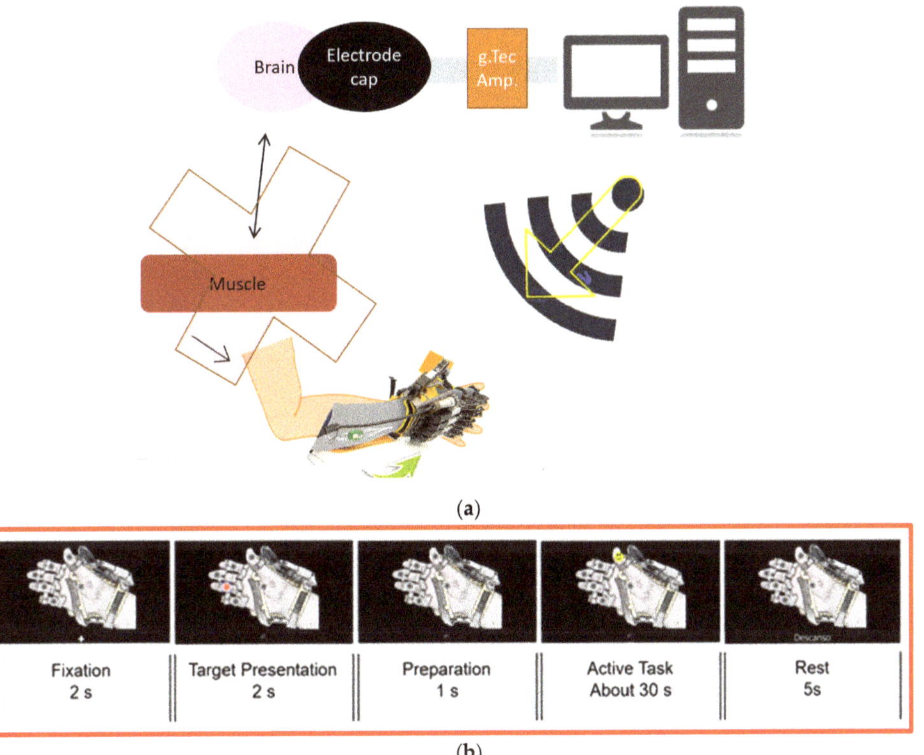

Figure 1. *Cont.*

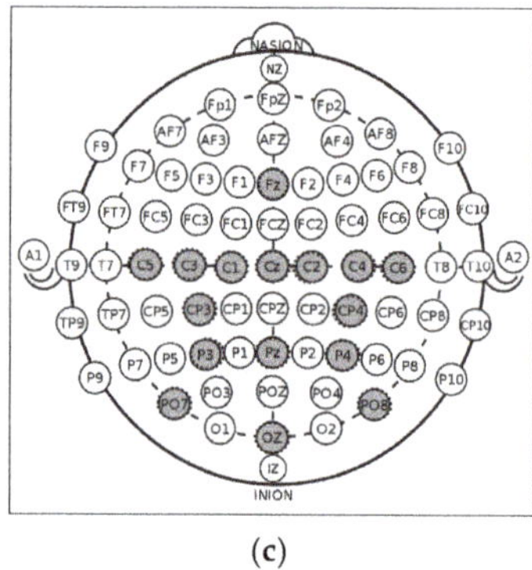

(**c**)

Figure 1. (**a**) Layout of the BCI (Brain Computer Interface) system whose purpose is to generate an alternative pathway for muscle movement. (**b**) Graphical interface of the P300-BCI in its different stages used to control a robotic hand-orthosis [21]. (**c**) Position of electrodes used in this P300-BCI according to 10–20 system.

For this test, subjects are comfortably seated in front of an LCD monitor in a silent room where only the testers and subject are present. An EEG cap is placed on the subject's head, and gel is applied to the 16 selected electrodes (shown in grey in Figure 1a) and on the reference (which is attached to the right ear) by the testers. The Graphical User Interface (GUI) is composed of a picture of an open hand wearing the hand orthosis with one grey dot on top of each finger and another dot on the hand's palm, and on the bottom, it has a rectangle for different instructions, depending on the stage. The task is based on the P300 oddball paradigm. The happy faces represent the oddball stimuli that will cause the P300 response or the ERP. This is the response the BCI is looking for to detect intentions. The first stage of the P300 experiments consists of training or calibration of the BCI system. With this training, the algorithm learns the subject's P300 characteristics so that they can be identified later.

The Training stage is composed of 8 blocks of training. Each block is composed of 5 stages, and the GUI's state in each stage is shown in Figure 1b. First, in the Fixation stage, where the subjects must get ready for the experiment, a cross appears on the GUI's rectangle for 2 s. Then, in the Target Presentation stage, one finger or the whole hand is indicated in the rectangle. The indicated finger shows which of the grey dots to observe in the Active Task stage, or if the whole hand appears indicated; then, the finger in the palm must be observed in the Active Task stage. Then, a second of Preparation is given, for the subject to prepare to begin the Active Task. Then, the Active Task stage begins. A happy face begins flashing in one dot at a time and the participant is asked to count in their head the number of happy faces that appear in the indicated spot while observing only the indicated dot. The face appears in a dot for 75 ms and then all dots are grey for 75 ms, repeating this pattern until the face appears in the indicated spot about 32 times. Finally, a 5 s rest is given to the subject, in preparation for the next block [21].

The Free Validation stage comes after the Training stage, where the subject is indicated to choose any desired dot, and as soon as the BCI detects the subject's desired dot, it is colored red. This is repeated about 3 or 4 total times for the user to see that the BCI is following their instructions. Then, the Online Validation stage begins. This stage is very similar in procedure to the Training stage, except the BCI's objective is to detect the dot the user is indicated to concentrate on. This block is repeated once per dot. Finally, the robotic arm is attached to the subject's left hand by the testers and the Free Validation and Online Validations are repeated with it on. With the robotic arm attached, when the desired dot is detected by the BCI, the corresponding finger is contracted by the motors in the Hand of Hope. When the palm dot is detected, all the fingers are contracted. The results of these BCIs have already been reported in another paper [21].

The data from the Training stage were selected as it is the stage where more trials are available; with more trials, the signal-to-noise ratio is increased. To extract the trials, the moment where the happy faces or stimuli appear was extracted. The period from 300 ms before each stimulus until 700 ms after it was extracted for each trial. Only the trials where the happy face was located on the indicated (by the BCI) dot were taken into consideration for this analysis, as these trials were where the ERPs were present. A total of 264 trials were initially used for each patient.

2.2. Data Acquisition

The data were recorded by 16 monopolar electrodes positioned according to the 10–20 international system at positions in FZ, CZ, PZ, OZ, C1, C2, C3, C4, C5, C6, CP3, CP4, P3, P4, PO7, and PO8, as shown in Figure 1c, with the reference placed on the right earlobe and ground electrode at AFz. The electrodes were selected by the designers of the BCI with the objective of covering the motor cortex and the sites commonly used in P300 BCIs [21]. The signals were amplified using a g.USBamp amplifier (a g.GAMMASYS active wet electrode arrangement and a g.USBamp amplifier provided by g.tec medical engineering GmbH, Schiedlberg, Austria). The sampling rate was set at 256 Hz. The computer processed the EEG signals, displayed the GUI, synchronized and displayed stimuli, and sent control messages to the robotic arm.

2.3. Subjects

The users of the experimental protocol were divided into two groups: ALS patients and the Control group. The ALS group contained 12 patients with Bulbar or Spinal ALS with mild to advanced levels of hand atrophy, six women and six men whose age had a mean of 59 ± 7. Additionally, 4 ALS patients went through the training more than once (with a minimum of three months between tests). These older training data were used to observe the evolution of selected variables. Only the oldest training data of each patient were used in the variable's extractions. The ALS group was recruited from the patients attending the TecSalud ALS Multidisciplinary Clinic [21]. The Control group consisted of 16 healthy subjects, eight women and eight men whose ages were 33 ± 15.

2.4. Analysis

The trials obtained from the training stage were the basis of the analysis. The steps for the presented analysis are shown in Figure 2. These signals were preprocessed to obtain the most information out of them, and then two different studies were performed on each individual's data: power and connectivity analysis. Latency and amplitude variables were extracted from the power analysis and the connectivity value between two electrodes was extracted from the connectivity analysis. All the variables obtained were compared between ALS and Control groups to select the best ones. Finally, the selected variables were observed over time in the ALS individuals and were used to train a classification model.

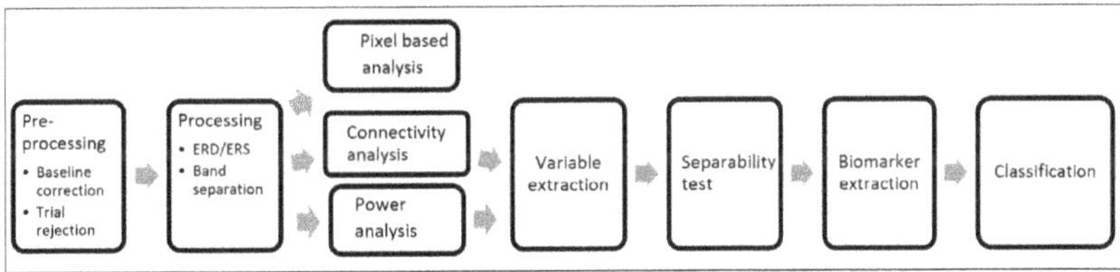

Figure 2. Processing stages of information.

2.5. Pre-Processing

A DC baseline correction was performed by averaging the activity from 200 ms before the P300 stimulus to time 0 (the specific time of the stimulus) and subtracting this value from every time point. An automatic trial rejection was performed based on three parameters: first, the maximum peak-to-peak value after the stimulus >200, as the brain signals being observed were between -100 and 100 mV [14]; then, the standard deviation of the trial after stimulus <50, and the noise to signal ratio >0.7. If any of these conditions is true, the trial is rejected.

2.6. Power Analysis

For each trial, the EEG signal was extracted from 300 milliseconds (ms) pre-stimulus to 700 ms post-stimulus. It was decomposed via complex Morlet wavelet convolution with a set of wavelets ranging from 0.5 to 40 Hz and a variable number of cycles from 2 to 10. For each user and each electrode, the percentage change with a pre-stimulus base was calculated, so the power could be compared among frequencies and time.

The power, *pow*, at each time point, *t*, is obtained by squaring the voltage, *v*, at each corresponding time point, as shown in Formula (1). The baseline interval selected was between 200 ms pre-stimulus and 0 ms (the moment of the stimulus). The value of the baseline, *R*, is the average power in this interval for each frequency band, *f*, as shown in Formula (2). Then, to obtain the activity, *A*, the voltage at each time point, $v(t)$, is squared to obtain the power, $pow(t)$, and an average is performed among all the trials, *tr* (for each subject), as shown in Formula (3). Finally, the power percentage change is obtained as shown in Formula (4). In this case, this was performed for each 0.75 Hz in the plot. Anything over 0 is considered a power increase or Event-Related Synchronization, and everything between below 0 is considered a power decrease or Event-Related Desynchronization [22]. The final powers were divided by neural bands (Delta [0.5–3 Hz], Theta [4–8 Hz], Alpha [9–13 Hz], and Beta [14–30 Hz]), and an average was performed among the frequencies of each band.

$$pow(t) = v(t)^2 \tag{1}$$

$$R(f) = \frac{\sum_{i=ix}^{iy} pow(i,f)}{tp} \tag{2}$$

$$A(f,t) = \frac{\sum_{x=1}^{tr} pow(i,f,t)}{tr} \tag{3}$$

$$\text{power \% change} = \frac{(A-R)}{R} * 100 \tag{4}$$

2.7. Point-Wise Analysis

Complementary to the maximum values power analysis, a point-wise analysis was performed over the 3D graphs whose axes were time, frequency, and spectral power percentage change. Permutation-based statistics were performed to determine areas of significant difference. For this analysis, we assumed that there were no significant areas on the map. A null-hypothesis map was made by shuffling subjects among groups, taking the mean of each new group, and subtracting one map from another. Point by point (pixel by pixel), this operation was permuted 1000 times to create a distribution. Finally, the observed value was compared to the distribution obtained. The points whose observed value had a *p*-value below 0.05 were considered significant.

2.8. Connectivity Analysis

Connectivity is a measure used to determine the oscillatory synchronization that exists between two brain regions, represented by electrodes. InterSite Phase Clustering (*ISPC*) is a connectivity measure that relies on the phase of the signals to determine the degree of connectivity. This theory is based on the concept that for two regions to be synchronized, they must be sending information and reading it at its maximum excitation point.

To calculate the *ISPC* values of each subject between signals from electrodes *x* and *y*, first, the signal is converted to an analytic signal; then, the analytic signal, as, is divided in frequency bands, *f*. The signal is then divided among the frequencies of each band, resulting in four signals, one for each band. Then, the signal's instantaneous phases, $ph(t,f)$, are calculated at each frequency band and each instant of time, t. An average of the differences (at each instant of time) between the two signals is obtained, as shown in Formula (5), and this is the *ISPC*. If the signals are synchronized, the value should be close to 1. The *ISPC* was calculated between every electrode and the other 15 electrodes.

$$ISPC_{x,y}(f) = \overline{ph(t,f,x) - ph(t,f,y)} \tag{5}$$

Formula (5): How *ISPC* is obtained.

2.9. Variables Extraction

Variables from the power and connectivity analysis were extracted. All variables mentioned were extracted for each subject and patient. From the power analysis, the values of the P300 peak and its latency were extracted. To compute these, the maximum value and its latency between 200 and 650 ms post-stimulus in the time–frequency data were calculated. From the connectivity, the *ISPC* value was used. In addition, 16 electrodes, 4 neural bands, and two conditions (magnitude and latency of P300 peak) resulted in 128 power variables; 16 electrodes compared with 15 electrodes in 4 neural bands resulted in 960 connectivity variables. All variables added to a total of 1088 total variables for each of the 28 subjects and patients.

2.10. Separability Test

To find the variables that might be of interest among both groups (ALS and Control), a Wilcoxon rank-sum test (a nonparametric test that contrasts two samples in order to determine if they come from equally distributed populations) was performed for each variable. Only the variables significantly different were further observed ($p < 0.05$).

Two approaches were made with the selected variables. First, we analyzed how the selected variables were represented in the original data. Second, we observed where patients' older training data (the four of them that we have) stood among these selected data.

2.11. Multiple Comparisons Correction

When multiple tests are being performed, a multiple-tests correction is needed. The reason is that in a normal distribution, we are expected to obtain some results that seem significantly different but occur because of chance. For the tests performed, two types of multiple-tests correction were performed. For the extracted variables analysis (maximum power, maximum power latency, and connectivity), a False Discovery Rate (FDR) was performed. For the graphical power analysis, an Extreme Point Correction was performed.

2.11.1. FDR Correction

The False Discovery Rate (FDR) is a method to fix the p-value when testing multiple comparisons, as in this case. The basis of this method is to compute the p-value of all nonsignificant results and use this value as a cutoff to make everything above it significant and below it nonsignificant, adjusting the p-value. While this is a graphical method, it can be performed through a mathematical approach. In this approach, first, all the p-values must be sorted from smallest to largest and ranked. The last value (top rank) is kept. The next largest value is the smaller between the previously adjusted p-value and the result of Formula (6) (where $p(r)$ is the p-value of rank r of the current p-value, n_p is the number of p-values, and r is the rank r), and so on, until the smallest p-value.

$$\text{adjusted } p\text{-value}_r = p(r) * \frac{n_p}{r} \tag{6}$$

Formula (6): Adjusted p-value

After performing this correction, the significance of all the discoveries was above 0.05, with the least value being 0.056, corresponding to the power results.

2.11.2. Extreme Point Correction

When performing graphical analysis, FDR is not the correct approach, as the number of points (or pixels) in the image affects the parameters of FDR, and intuitively this makes no sense, as a significant cumulus of points (or pixels) should remain significant independently of the resolution of the image. For this reason, an extreme point correction was performed. In this correction, the assumption is made that there are no significant areas on the map. A null-hypothesis map was made by shuffling subjects among groups, taking the mean of each new group, and subtracting one map from another. In this new graph, the least value and maximum values were extracted. The process was repeated 1000 times to create a bi-modal distribution. Then, everything above 0.025 and below 0.975 was considered not statistically important or different. Only the points outside of this area were still considered statistically different.

3. Results

In this section, the results are presented and described. Additional figures were used to examine the distribution of groups variables.

3.1. Power: Magnitude

The four bands were examined, each of which contained 16 electrodes. As a result, 64 variables were obtained for magnitude for each patient or Control.

A significant difference was found only in the beta band. All of the ten variables that were found to have statistical differences were found in the same band. The electrodes that were found to be different between both groups are shown in Table 1. They were in the locations FZ, C2, CP3, and PO7 ($p < 0.05$), and C1, CZ, C4, P3, PZ, and P4 ($p < 0.01$). The locations of these specific electrodes are shown in Figure 3a.

Table 1. Variables of power magnitude with significant difference resulting from a ranksum test (Wilcoxon for independent groups) between ALS group and Control group.

ELECTRODE	p-Value
BETA	
FZ	0.0244
C1	0.0013
CZ	0.0028
C2	0.0114
C4	0.0028
CP3	0.0434
P3	0.005
PZ	0.0066
P4	0.0043
PO7	0.0148

3.1.1. Distribution of Power Variables

In Figure 3b, the distribution of the power magnitude variables of each group, Control and ALS, can be seen. The distributions of electrodes found to be statistically different can be seen to have different interquartile ranges. The ALS median is below the Control median, showing a decrease in magnitude for the ALS group or a decreased activity.

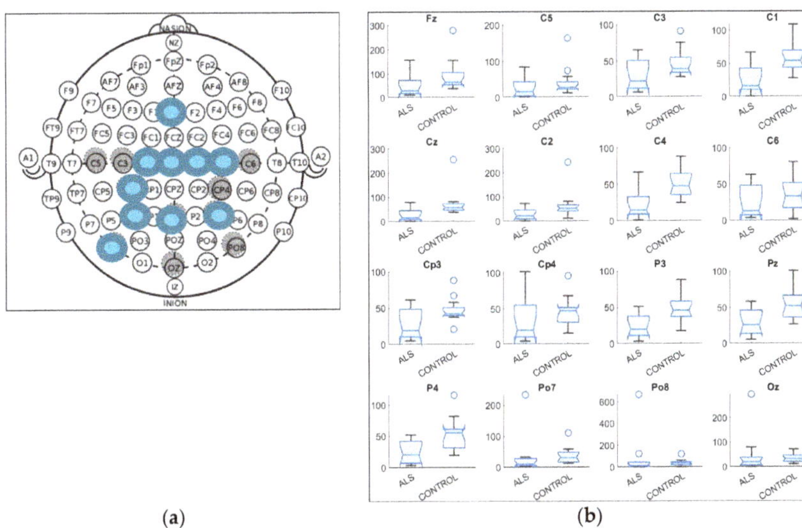

(a)   (b)

Figure 3. (**a**) Electrodes' locations found to be statistically different in power magnitude variables between Amyotrophic Lateral Sclerosis (ALS) ALS and Control group; (**b**) distribution of power magnitude variables for each electrode of ALS group and Control group.

3.1.2. ERPs

The ERP of the Beta band was computed to observe the differences remarked in the statistical analysis. As the ERPs show, in the Beta band, a peak can be seen at about 500 ms after the stimuli. The mean ERPs of channels located in the Central area (e.g., C1, CZ, C2, and C4) show opposite phases for both groups around 450 ms after the stimuli. This area is significantly different, as shown in Figure 4. In electrodes located on the Parietal area, the ALS group has lower peaks in P3, PZ, and P4. All of these variables (e.g., the peak value of spectral power in electrode CZ in the Beta band) are found amongst selected variables. For instance, in Figure 4, the ERP of channel CZ in the Beta band is shown. The Control average shows a positive peak around 400 ms, while the ALS average shows a negative deflection in this same time point. A negative peak where a positive peak should appear is usually due to an overactivation in another cerebral region. In contrast, in Figure 4, the PZ electrode, Beta band, both the Control and ALS group have a positive peak around 400 ms. However, the peak magnitude corresponding to the Control group is almost double that the ALS peak. This indicates a reduced activity in the beta band for the ALS group.

These results can also be seen in the time–frequency charts. In Figure 5c,d, we can see channel PZ for the Control and ALS group. There is a clear difference in the magnitudes in all frequencies around 400 ms, especially around 20 Hz. The magnitude can be seen much higher for the Control group. In addition, in Figure 5a,b, the differences between groups are noticeable—the ALS group shows a spectral power decrement, where the Control group shows an increase.

117

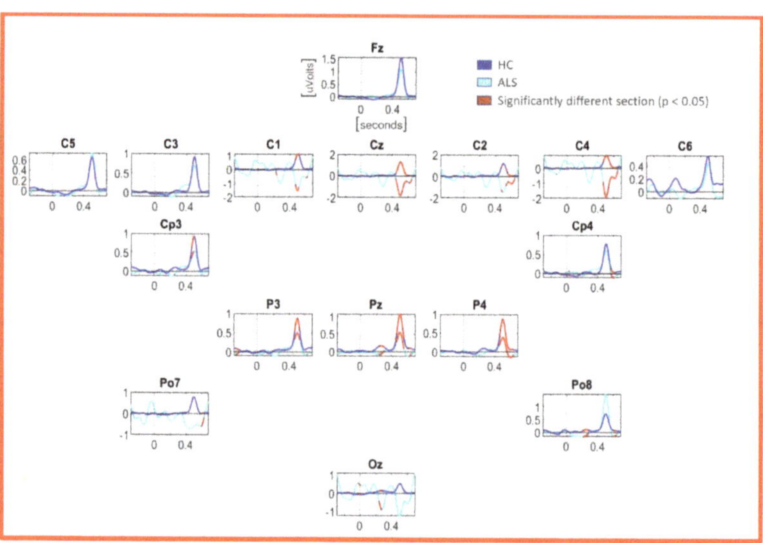

Figure 4. Mean ERPs of all 16 channels in Beta band. In cyan, the ALS group mean is observed, while blue represents the Control group. The segments that are red are significantly different with $p < 0.05$. The *x*-axis represents milliseconds and *y*-axis microvolts.

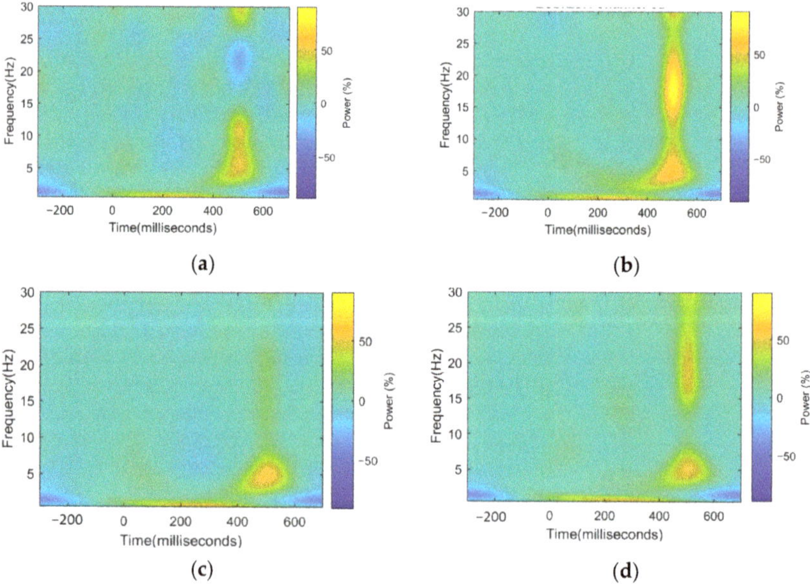

Figure 5. (**a**) Mean spectral power percent change of electrode CZ for ALS group; (**b**) mean spectral power percent change of electrode CZ for Control group; the Control group has a power increment and the ALS group has a power decrement; (**c**) mean spectral power percent change in electrode PZ for ALS group; (**d**) mean spectral power percent change in electrode PZ for Control group.

3.1.3. FDR Correction

An FDR correction was needed, as mentioned before. All 1088 variables were no longer significant after FDR correction with $p > 0.05$. Power results had the lowest value after the FDR correction with a *p*-value of 0.056. To avoid false negatives (error type 2), further tests were performed.

3.1.4. Point-Wise Analysis

After performing the point-wise analysis and the extreme pixel correction, the channels that were found to have an area of significant difference were C1, CZ, C2, C4, and OZ. In the central band, the effect seemed to dissipate as the channel became more distant from the central (Z or CZ) zone, which is on the center of the scalp in the central band, and underactivation was occurring for the ALS group. Another channel that had a significant difference, but was barely observed in previous results, was in channel OZ, almost at the same latency, as shown in the Figure 6.

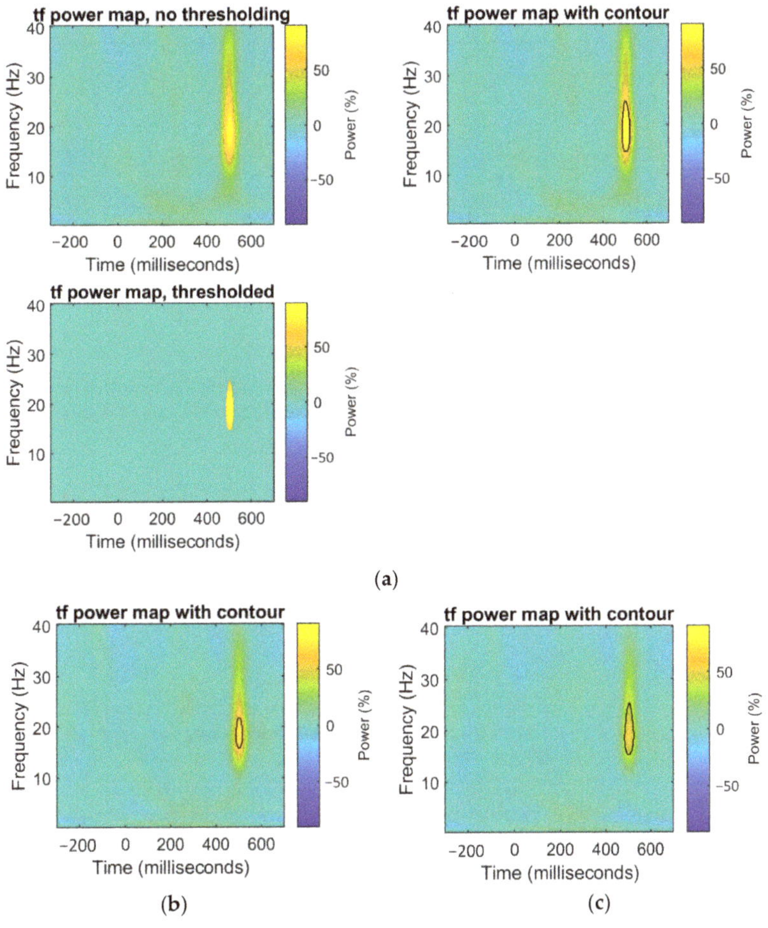

Figure 6. (**a**) Channel CZ. Time–frequency power map (up left). Time–frequency power map with the significant area highlighted (up right). Time–frequency power map displaying only significantly different area (down left); (**b**) Channel C2. Time–frequency power map with significant area highlighted; (**c**) Channel C4. Time–frequency power map with the significant area highlighted.

3.2. Power: Latency

The four bands were examined, each of which contained 16 electrodes. As a result, 64 variables were obtained for magnitude for each patient or Control.

From the 64 variables for latency, only two were found to be statistically different, with $p < 0.05$, shown in Table 2. These two variables were in two different bands, Delta and Alpha. The variable in the Delta band was located in OZ ($p < 0.01$), and the one in the Alpha band was situated in PO7 ($p < 0.01$). The locations of these two electrodes are shown in Figure 7a.

Table 2. Variables of power latency with significant difference resulting from a ranksum test (Wilcoxon for independent groups) between ALS group and Control group.

ELECTRODE	p-Value
DELTA	
OZ	0.015
ALPHA	
PO7	0.011

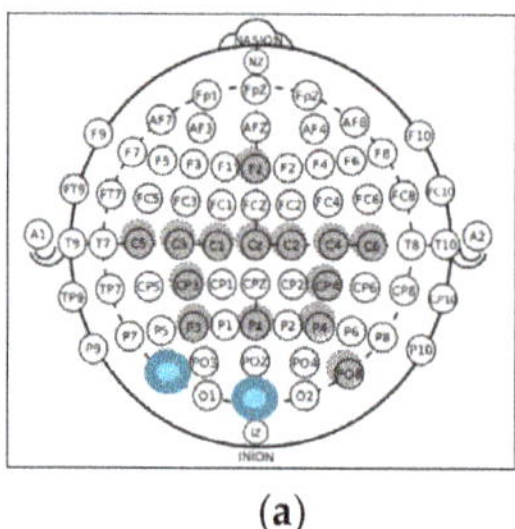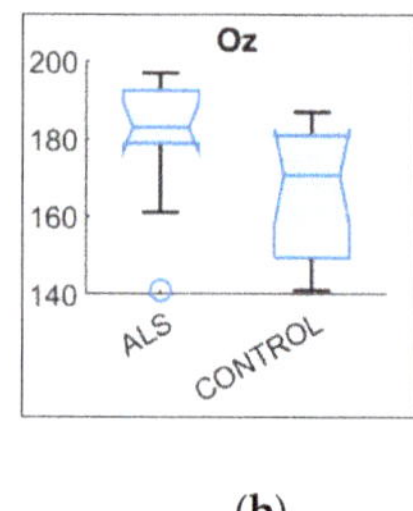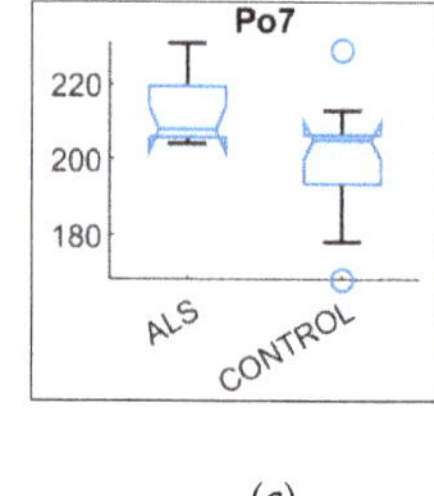

(a)   (b)   (c)

Figure 7. (**a**) Electrodes' locations found to be statistically different in power latency variables between ALS and Control group; (**b**) distribution of latencies for ALS and Control group in location OZ for the Theta band; (**c**) distribution of latencies for ALS and Control group in location PO7 for the Alpha band.

The power latency had the least variables. Only two electrodes in one band each were found to have statistical differences. In Figure 7b, the distribution of latencies is shown for the ALS and Control group in location OZ for the Theta band. The ALS group showed a higher latency than the Control. In Figure 7c, the Alpha band was shown in location PO7. A higher latency was seen for ALS as well.

3.3. Connectivity

For the connectivity analysis, all 16 electrodes were compared with all the other 15 electrodes to find their connectivity. This was performed among four bands, resulting in 960 variables for each patient. All of the 960 groups underwent the statistical analysis, and only nine variables were found to be statistically different, with $p < 0.05$. Six of these variables were in the same band, in the same electrode.

The selected variable pairs, in this case, were located in CZ—C4 in the Delta band, PO8—OZ in the Theta band, and FZ-CP4 in the Beta band ($p < 0.05$); these three pairs are shown in Figure 8a. The other six pairs were located in the Alpha band and were all between the PO7 and another electrode ($p < 0.05$). The other pairs were C3, C1, CZ, C2, C3, and CP4, as shown in Figure 8b. This information can be seen in Table 3.

The connectivity variables were mainly focused on the Alpha band between electrode PO7 and other electrodes in the Central area. In Figure 8c, we can see the distributions of connectivity of channels CP4 and PO7. It is clear that in the CP4 electrodes, both groups were difficult to distinguish. In contrast, in PO7, the green group (ALS) was below the Control group in almost every electrode.

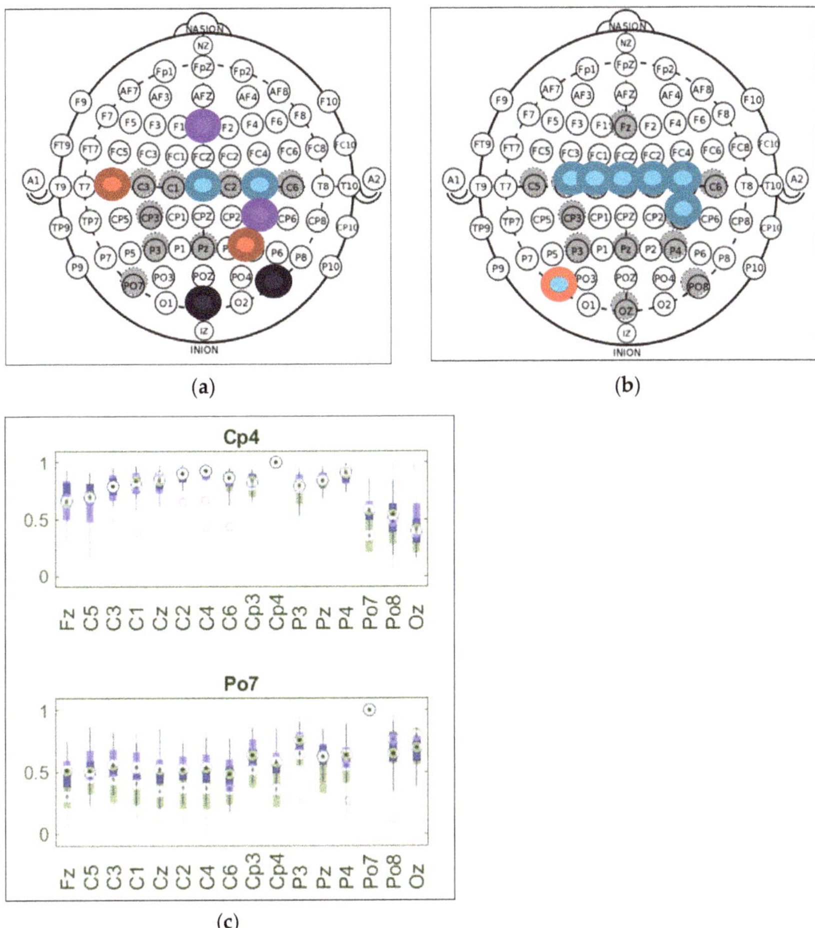

Figure 8. (a) Electrodes' locations found to be statistically different in connectivity variables between ALS and Control group for bands Delta, Theta, and Beta; (**b**) electrodes' locations found to be statistically different in connectivity variables between ALS and Control group for Beta band; (c) *ISPC* of all subjects and patients of electrodes PO7 (bottom) and CP4 (top) with all channels. Control is in blue and ALS in green. Most connectivity values' significant differences were found between the PO7 electrode and the electrodes above the motor cortex area, CZ, C1 C2, C3, C4, and CP4. This can be seen in the PO7 chart, where almost all ALS connectivities are below the Control group.

Table 3. Variables of connectivity with significant difference resulting from a ranksum test (Wilcoxon for independent groups) between ALS group and Control group.

ELECTRODE PAIR	*p*-Value
DELTA	
CZ—C4	0.0484
THETA	
PO8—OZ	0.0274
ALPHA	
PO7—C3	0.0484
PO7—C1	0.0274
PO7—CZ	0.0484
PO7—C2	0.0346
PO7—C4	0.0484
PO7—CP4	0.0308
BETA	
FZ—CP4	0.0434

This indicates a decreased connectivity for the ALS group.

In Figure 9a–c, the following electrode pairs are seen, PO7 and CP4, PO7 and C2, and FZ and CP4, respectively. In the first two-electrode pair, a statistical difference was found in the Alpha band, which is marked in blue in Figure 9a,b. In this area, we can see a negative deflection for the ALS group. The Alpha occipital cycles are thought to be activated during temporal integration in visual perception [23].

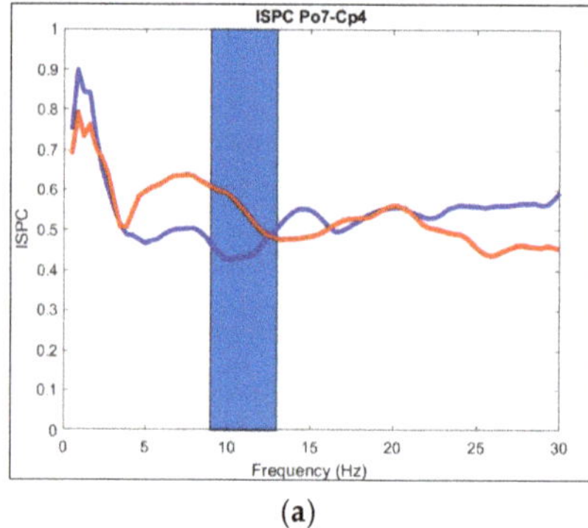

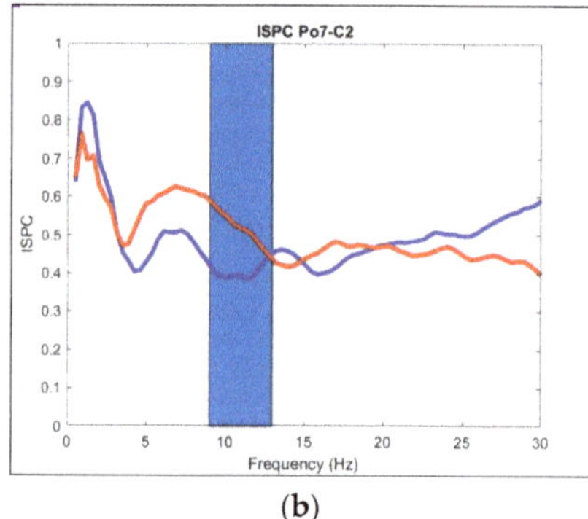

Figure 9. *Cont.*

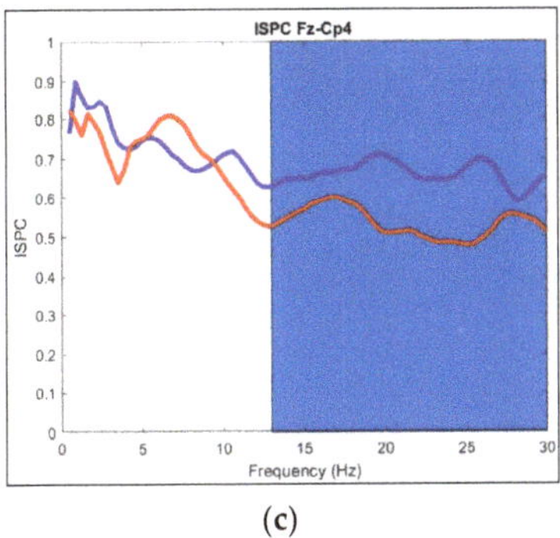

(c)

Figure 9. (**a**) Average InterSite Phase Clustering (*ISPC*) between electrodes PO7 and CP4 of Control (red) and ALS (blue). Alpha band is shown in blue. In the Alpha band, the ALS group is below the Control group; (**b**) Average *ISPC* between electrodes PO7 and C2 of Control (red) and ALS (blue). Alpha band is shown in blue. In the Alpha band, the ALS group is below the Control group; (**c**) Average *ISPC* between electrodes FZ and CP4 of Control (red) and ALS (blue). Beta band is shown in blue. In the Beta band, the ALS group is above the Control group.

In Figure 9c in the beta band, we can see that the Control group was below the ALS group in all of the bands, indicating augmented connectivity for the ALS group.

3.4. Classification Model

To select the variables for the classifier, first, a fourfold was performed to extract 25% of participants of each group for testing. With the remaining 75%, a threefold was performed to select the most important variables. To pick them, in each fold, it was determined which variables were significantly different between groups. This process was in a loop and was repeated 25 times. The variables selected the most times among all the repetitions were considered for the classification testing. The chosen variable was power in CZ in the Beta band. A simple linear SVM was trained with this variable, and with a leave-one-out cross-validation, a 100% in specificity and 75% in sensitivity were achieved, with an overall 89% success classifying individuals into each group. The reason of the effectiveness of this classification is shown in Figure 10a, where the ALS and control groups showed different population densities. The probability density function of the Healthy Control (HC) group and ALS group was estimated with the nonparametric kernel density estimation method. The ALS group seemed to have a bimodal distribution with one of its modes being inside the HC group.

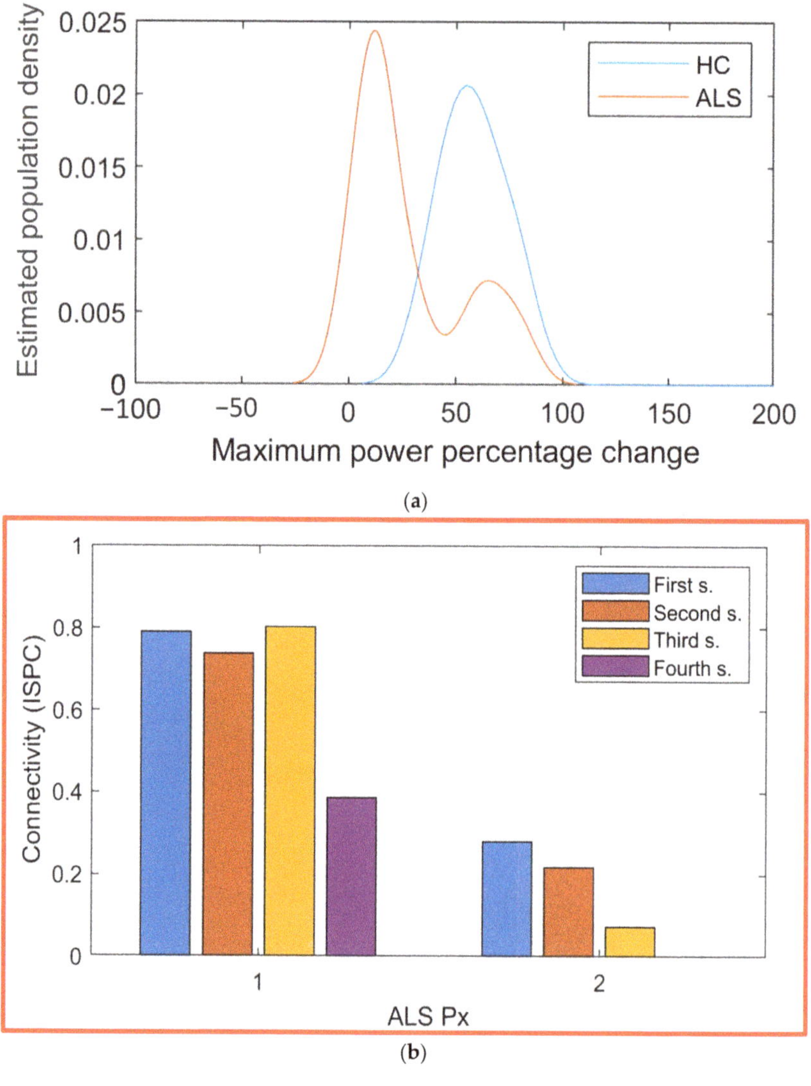

Figure 10. (**a**) Estimated population density of maximum power for ALS group and Healthy Control; (**b**) connectivity (*ISPC*) value between electrodes CZ and PO7 in the Alpha range for four different sessions for Px1 and three different sessions for Px2. Each session was three months apart.

3.5. Evolution of Patients

Data were gathered of two ALS patients at different times, with three months between each session. In Figure 10b, a graph that contains the values of the *ISPC* between Cz and PO7 in the Alpha range can be appreciated. The values calculated for three sessions for patient 2 and four sessions for patient 1 can be seen. For patient 2, a reduction in connectivity was seen as the disease advanced. Patient 1 had a similar result, but a connectivity increase was seen in the third session.

4. Discussion

The main objective of this study was to understand the underlying cognitive neural alterations that affect people with ALS as an aid for the detection or to monitor the disease's evolution. In the present work, alterations were found for ALS patients' EEG data during a cognitive test, a P300 task, in comparison to HC subjects. These alterations for the ALS group were a decreased activity in the Beta band in electrode locations FZ, CZ, and PZ; around them, an augmented latency in frequency bands Delta (OZ) and Alpha (PO7); and variations in connectivity among all frequency bands, but an especially reduced band-specific connectivity in the Alpha band between channel PO7 and channels above the motor cortex (CZ, C1, C2, C3, C4, and CP4). The tracking of connectivity values in two ALS patients indicated an Alpha-related connectivity decrease between channels PO7 and CZ. Finally, data from CZ were used to classify individuals between both groups; cross-validation achieved a 100% in specificity and 75% in sensitivity, with an overall 89% success.

Decreased activity in the Beta band in electrodes over the sensorimotor band was found. Beta band activity in the sensory-motor band had been found to be important for accurate motor performance in healthy individuals [24]. This also supports what has been reported for ALS individuals [10,23–25]. Motor system degeneration in ALS individuals has been linked to a decrease in the Beta band [25]. This strengthens the theory that CZ could work as a biomarker to monitor ALS. The only problem is that those studies were made during a motor task, not a cognitive task. On the cognitive side, beta oscillations are also traditionally associated with sensorimotor processing [22]. This indicated a sensorimotor processing dysfunction for ALS individuals. Additionally, the Beta band is associated with attention, so it is expected to be of interest. A reduction in P300 power is usually seen in older patients, but this effect is typically present in the PZ electrode [26]. However, in the graphical analysis, the difference was detectable in CZ but not in PZ, giving a strong indication that the effect is not due to age but ALS degeneration. Moreover, when reperforming the graphical analysis only with subjects older than 50 in the control group, the region in the CZ channel was still present.

Connectivity results indicated overactivation in the Beta band and underactivation in the Alpha band for ALS individuals. The difference in the connectivity maps could be clearly seen, and it was noticeable how these changes in connectivity were band-specific. A decrement in Alpha connectivity was so evident that an apparent valley in connectivity was located on this band between electrodes PO7 and CP4. Alpha band oscillations has been linked with a top-down control of the temporal resolution of visual perception [27]. More research is needed, and additional tests must be performed to study more deeply the connectivity in ALS patients; this could mean applying additional filtering to the signals, such as a Laplace, or calculating another type of connectivity between electrodes, such as power correlation. Additionally, the evolution of the two ALS patients whose results were available at different time points indicates an overall decrement in the connectivity between PO7 and CZ in the Alpha band, strongly indicating a connectivity degeneration in the Alpha band for ALS individuals. Alterations in connectivity in the Alpha band have been found in ALS patients [6,28] and also in the Beta band [5]; yet, these results were found in the rest-state. Beta band connectivity has also been found to be essential for accurate motor performance in healthy individuals [24]. This may indicate an over-effort from ALS individuals, but as the connectivity results in the Beta band were only significant in one pair of electrodes, this is hard to generalize. A reduction in connectivity has been linked to a decrease in cognition for patients with Multiple Sclerosis [29]. Rates of ALS-related impairment are noted to be related to the disease stage. Cognitive deficit is more frequent with more severe ALS stages [30]. This strengthens the theory that connectivity could serve as a tool to monitor the disease's advances in cognitive atrophy, specifically.

Most of the previous papers that have studied either connectivity or signal amplitudes did so in the resting-state, not during an active task [6,7,23,28], or during a motor task [10,24,25]. The few that have studied it during a P300 task did not analyze connectivity and amplitude simultaneously in a spectral analysis [11]. The selected variable for classification was the peak power in the Beta band in CZ. Achieving a 100% in sensitivity indicates that no false positives were present, which strengthens the possibility of using this characteristic as a potential biomarker to track ALS degeneration. Classifiers for ALS individuals have been performed, but they do not usually analyze the signals [31]. Using cognitive alterations in the Beta band to classify between ALS and Control groups has been performed with success but using a generic BCI (BCI2000) [20]. The signals used in this work came from a P300-based BCI for ALS patients [21], so understanding the cognitive neural alterations occurring could help improve the BCI's performance as its performance will most likely be different for a CLIS ALS patient than for a control subject, or even a more moderate ALS case. A follow-up of an ALS patient's potential biomarker has not been performed to our knowledge. For the magenta patient in Figure 10b, a gradual decrement was seen for the peak power in the Beta band in CZ. Yet, this was not the case for the other ALS patients represented. This may be because the magenta patient had a more severe case of atrophy, but without further research, this is mere speculation. The results were achieved with a simple univariate Support Vector Machine (SVM) classification model. The only three patients that were classified as Controls had either not very good signals or had taken the P300 evaluation previously, which may alter the results.

In conclusion, the ALS group seemed to have a statistically important difference in power and connectivity during the P300 task, a cognitive task, the two most important being in power magnitude in the Beta band and connectivity in the Alpha band. All 21 variables had $p > 0.05$ after FDR correction. Yet, the evidence implied that some of these results may be false negatives. This evidence is the fact that the Cz, C2, and C4 electrodes showed significant differences in the same region (Beta band around 500 ms after stimuli) in the time–frequency power maps, and that the power values of electrode Cz in the Beta band had a good overall performance in classifying correctly between ALS and HC groups. This is also the case for the connectivity decrease for the ALS group in the Alpha band, as Figure 9 clearly shows a frequency band-related decrease, and the connectivity value decreased in ALS patients (in the same electrode pair) as the disease advanced. The variables selected by the analysis did not seem to be random but had a correlation with what other researchers have found, strengthening the theory that they could serve as biomarkers for ALS. The SVM model that resulted from the classification between ALS patients and Control subjects had very promising results. All Control subjects were classified correctly, which means that a false ALS diagnosis would not occur. Clearly, this was a simple bivariate model. A much more complex model may be obtainable. The potential of these variables as ALS biomarkers that could aid detection or monitor the advance of the disease is noticeable and must be further studied. Finally, this was an exploratory research whose objective was to find areas that potentially need further examination. The hypothesis for the underactivations presented by the ALS group is a general neural degeneration; more studies are needed to localize the degeneration site and the level of its affectations. The next stage is to test these results by observing these specific variables with more electrodes in these areas.

Author Contributions: Conceptualization, C.X.P.-O., J.L.G., O.M.-M., J.M.A., R.C., and H.R.M.; funding acquisition, J.M.A.; investigation, C.X.P.-O., J.L.G., O.M.-M., J.M.A., R.C., and H.R.M.; project administration J.L.G., O.M.-M., J.M.A., R.C., and H.R.M.; resources, R.C., and H.R.M.; supervision, C.X.P.-O., J.L.G., O.M.-M., J.M.A., R.C., and H.R.M.; validation, R.C.; writing—original draft, C.X.P.-O.; writing—review and editing, C.X.P.-O., J.L.G., O.M.-M., J.M.A., R.C., and H.R.M. All authors have read and agreed to the published version of the manuscript.

Funding: This research has been funded by the National Council of Science and Technology of Mexico (CONACyT) through grant PN2015-873, and the UIC-TEC Seed Funding Program 2021–2022.

Institutional Review Board Statement: This study followed the ethical principles of the World Medical Association (WMA) Declaration of Helsinki (WMA, 2013).

Informed Consent Statement: Informed consent was obtained from all subjects involved in the study.

Data Availability Statement: The datasets generated and analyzed for this study are available upon request to the corresponding author.

Acknowledgments: The authors would like to acknowledge the support received from the Amyotrophic Lateral Sclerosis Clinics from the Neurology and Neuroscience Institute, TecSalud Hellion Hospital, CONACyT and the Robotics Research group.

Conflicts of Interest: The authors declare no conflict of interest.

References

1. Zarei, S.; Carr, K.; Reiley, L.; Diaz, K.; Guerra, O.; Altamirano, P.F.; Pagani, W.; Lodin, D.; OrOZco, G.; Chinea, A. A comprehensive review of amyotrophic lateral sclerosis. *Surg. Neurol. Int.* **2015**, *6*, 171. [CrossRef]
2. Kellmeyer, P.; Grosse-Wentrup, M.; Schulze-Bonhage, A.; Ziemann, U.; Ball, T. Electrophysiological correlates of neurodegeneration in motor and non-motor brain regions in amyotrophic lateral sclerosis—Implications for brain-computer interfacing. *J. Neural Eng.* **2018**, *15*, 041003. [CrossRef]
3. Chiò, A.; Pagani, M.; Agosta, F.; Calvo, A.; Cistaro, A.; Filippi, M. Neuroimaging in amyotrophic lateral sclerosis: Insights into structural and functional changes. *Lancet Neurol.* **2014**, *13*, 1228–1240. [CrossRef]
4. Iyer, P.M.; Egan, C.; Pinto-Grau, M.; Burke, T.; Elamin, M.; Nasseroleslami, B.; Pender, N.; Lalor, E.C.; Hardiman, O. Functional connectivity changes in resting-state EEG as potential biomarker for Amyotrophic Lateral Sclerosis. *PLoS ONE* **2015**, *10*, e0128682. [CrossRef]
5. Fraschini, M.; Demuru, M.; Hillebrand, A.; Cuccu, L.; Porcu, S.; Di Stefano, F.; Puligheddu, M.; Floris, G.; Borghero, G.; Marrosu, F. EEG functional network topology is associated with disability in patients with amyotrophic lateral sclerosis. *Nat. Publ. Gr.* **2016**, *6*, 1–7. [CrossRef]
6. Fraschini, M.; Lai, M.; Demuru, M.; Puligheddu, M.; Floris, G.; Borghero, G.; Marrosu, F. Functional brain connectivity analysis in amyotrophic lateral sclerosis: An EEG source-space study. *Biomed. Phys. Eng. Express* **2018**, *4*, 037004. [CrossRef]
7. Dukic, S.; McMackin, R.; Buxo, T.; Fasano, A.; Chipika, R.; Pinto-Grau, M.; Costello, E.; Schuster, C.; Hammond, M.; Heverin, M.; et al. Patterned functional network disruption in amyotrophic lateral sclerosis. *Hum. Brain Mapp.* **2019**, *40*, 4827–4842. [CrossRef]
8. Lomen-Hoerth, C.; Murphy, J.; Langmore, S.; Kramer, J.H.; Olney, R.K.; Miller, B. Are amyotrophic lateral sclerosis patients cognitively normal? *Neurology* **2003**, *60*, 1094–1097. [CrossRef]
9. Abe, K.; Fujimura, H.; Toyooka, K.; Sakoda, S.; Yorifuji, S.; Yanagihara, T. Cognitive function in amyotrophic lateral sclerosis. *J. Neurol. Sci.* **1997**, *148*, 95–100. [CrossRef]
10. Riva, N.; Falini, A.; Inuggi, A.; Gonzalez-Rosa, J.J.; Amadio, S.; Cerri, F.; Fazio, R.; Del Carro, U.; Comola, M.; Comi, G.; et al. Cortical activation to voluntary movement in amyotrophic lateral sclerosis is related to corticospinal damage: Electrophysiological evidence. *Clin. Neurophysiol.* **2012**, *123*, 1586–1592. [CrossRef]
11. Borgheai, S.B.; Deligani, R.J.; McLinden, J.; Zisk, A.; Hosni, S.I.; Abtahi, M.; Mankodiya, K.; Shahriari, Y. Multimodal exploration of non-motor neural functions in ALS patients using simultaneous EEG-fNIRS recording. *J. Neural Eng.* **2019**, *16*, 1–25. [CrossRef] [PubMed]
12. Teplan, M. Fundamentals of EEG measurement. *Meas. Sci. Rev.* **2002**, *2*, 1–11.
13. Ramele, R.; Villar, A.J.; Santos, J.M. EEG waveform analysis of p300 ERP with applications to brain computer interfaces. *Brain Sci.* **2018**, *81*, 199. [CrossRef] [PubMed]
14. Picton, T.W. The P300 wave of the human event-related potential. *J. Clin. Neurophysiol.* **1992**, *9*, 456–479. [CrossRef]
15. Bledowski, C.; Prvulovic, D.; Hoechstetter, K.; Scherg, M.; Wibral, M.; Goebel, R.; Linden, D.E.J. Behavioral/Systems/Cognitive Localizing P300 Generators in Visual Target and Distractor Processing: A Combined Event-Related Potential and Functional Magnetic Resonance Imaging Study. *J. Neurosci.* **2004**, *24*, 9353–9360. [CrossRef] [PubMed]
16. Pedroso, R.V.; Fraga, F.J.; Corazza, D.I.; Andreatto, C.A.A.; Coelho, F.G.D.M.; Costa, J.L.R.; Santos-Galduróz, R.F. P300 latency and amplitude in Alzheimer's disease: A systematic review. *Braz. J. Otorhinolaryngol.* **2012**, *78*, 126–132. [CrossRef]
17. McCane, L.M.; Heckman, S.M.; McFarland, D.J.; Townsend, G.; Mak, J.N.; Sellers, E.W.; Zeitlin, D.; Tenteromano, L.M.; Wolpaw, J.R.; Vaughan, T.M. P300-based brain-computer interface (BCI) event-related potentials (ERPs): People with amyotrophic lateral sclerosis (ALS) vs. age-matched Controls. *Clin. Neurophysiol.* **2015**, *126*, 2124–2131. [CrossRef]
18. Ogawa, T.; Tanaka, H.; Hirata, K. Cognitive deficits in amyotrophic lateral sclerosis evaluated by event-related potentials. *Clin. Neurophysiol.* **2009**, *120*, 659–664. [CrossRef]
19. Paulus, K.S.; Magnano, I.; Piras, M.R.; Solinas, M.A.; Solinas, G.; Sau, G.F.; Aiello, I. Visual and auditory event-related potentials in sporadic amyotrophic lateral sclerosis. *Clin. Neurophysiol.* **2002**, *113*, 853–861. [CrossRef]
20. Rupom, A.I.; Patwary, A.B. P300 Speller Based ALS Detection Using Daubechies Wavelet Transform in Electroencephalograph. In Proceedings of the 2nd International Conference on Electrical, Computer and Communication Engineering, ECCE 2019, Cox's Bazar, Bangladesh, 7–9 February 2019; Institute of Electrical and Electronics Engineers Inc.: Piscataway, NJ, USA, 2019.
21. Delijorge, J.; MendOZa-montoya, O.; Gordillo, J.L.; Caraza, R.; Martinez, H.R.; Antelis, J.M. Evaluation of a P300-Based Brain-Machine Interface for a Robotic Hand-Orthosis Control. *Front. Neurosci.* **2020**, *14*, 1–16. [CrossRef]
22. Pfurtscheller, G.; Lopes Da Silva, F.H. Event-related EEG/MEG synchronization and desynchronization: Basic principles. *Clin. Neurophysiol.* **1999**, *110*, 1842–1857. [CrossRef]

23. Maruyama, Y.; Yoshimura, N.; Rana, A.; Malekshahi, A.; Tonin, A.; Jaramillo-Gonzalez, A.; Birbaumer, N.; Chaudhary, U. Electroencephalography of completely locked-in state patients with amyotrophic lateral sclerosis. *Neurosci. Res.* **2021**, *162*, 45–51. [CrossRef]

24. Chung, J.W.; Ofori, E.; Misra, G.; Hess, C.W.; Vaillancourt, D.E. Beta-band Activity and Connectivity in Sensorimotor and Parietal Cortex are Important for Accurate Motor Performance HHS Public Access. *Neuroimage* **2017**, *144*, 164–173. [CrossRef] [PubMed]

25. Proudfoot, M.; Rohenkohl, G.; Quinn, A.; Colclough, G.L.; Wuu, J.; Talbot, K.; Woolrich, M.W.; Benatar, M.; Nobre, A.C.; Turner, M.R. Altered Cortical Beta-Band Oscillations Reflect Motor System Degeneration in Amyotrophic Lateral Sclerosis. *Hum. Brain Mapp.* **2016**, *38*, 237–254. [CrossRef] [PubMed]

26. Dinteren, R.; Arns, M.; Jongsma, M.L.A. Kessels, RPC P300 development across the lifespan: A systematic review and meta-analysis. *PLoS ONE* **2014**, *9*, e87347. [CrossRef]

27. Wutz, A.; Melcher, D.; Samaha, J. Frequency modulation of neural oscillations according to visual task demands. *Proc. Natl. Acad. Sci. USA* **2018**, *115*, 1346–1351. [CrossRef]

28. Iyer, P.M. Mismatch negativity as an indicator of cognitive sub-Domain Dysfunction in amyotrophic lateral sclerosis. *Front. Neurol.* **2017**, *8*, 395. [CrossRef] [PubMed]

29. d'Ambrosio, A.; Valsasina, P.; Gallo, A.; De Stefano, N.; Pareto, D.; Barkhof, F.; Ciccarelli, O.; Enzinger, C.; Tedeschi, G.; Stromillo, M.L.; et al. Reduced dynamics of functional connectivity and cognitive impairment in multiple sclerosis. *Mult. Scler. J.* **2020**, *26*, 476–488. [CrossRef]

30. Crockford, C.; Newton, J.; Lonergan, K.; Chiwera, T.; Booth, T.; Chandran, S.; Colville, S.; Heverin, M.; Mays, I.; Pal, S.; et al. ALS-specific cognitive and behavior changes associated with advancing disease stage in ALS. *Neurology* **2018**, *91*, E1370–E1380. [CrossRef]

31. Poletti, B.; Carelli, L.; Solca, F.; Lafronza, A.; Pedroli, E.; Faini, A.; Zago, S.; Ticozzi, N.; Meriggi, P.; Cipresso, P.; et al. Cognitive assessment in Amyotrophic Lateral Sclerosis by means of P300-Brain Computer Interface: A preliminary study. *Amyotroph. Lateral Scler. Front. Degener.* **2016**, *17*, 473–481. [CrossRef]

 MDPI

Article

Complex Pearson Correlation Coefficient for EEG Connectivity Analysis

Zoran Šverko [1,2], Miroslav Vrankić [1], Saša Vlahinić [1] and Peter Rogelj [2,*]

[1] Faculty of Engineering, Department of Automation and Electronics, University of Rijeka, 51000 Rijeka, Croatia; zsverko@riteh.hr (Z.Š.); miroslav.vrankic@riteh.hr (M.V.); sasa.vlahinic@riteh.hr (S.V.)

[2] Faculty of Mathematics, Natural Sciences and Information Technologies, University of Primorska, 6000 Koper, Slovenia

* Correspondence: peter.rogelj@upr.si

Abstract: In the background of all human thinking—acting and reacting are sets of connections between different neurons or groups of neurons. We studied and evaluated these connections using electroencephalography (*EEG*) brain signals. In this paper, we propose the use of the complex Pearson correlation coefficient (*CPCC*), which provides information on connectivity with and without consideration of the volume conduction effect. Although the Pearson correlation coefficient is a widely accepted measure of the statistical relationships between random variables and the relationships between signals, it is not being used for *EEG* data analysis. Its meaning for *EEG* is not straightforward and rarely well understood. In this work, we compare it to the most commonly used undirected connectivity analysis methods, which are phase locking value (*PLV*) and weighted phase lag index (*wPLI*). First, the relationship between the measures is shown analytically. Then, it is illustrated by a practical comparison using synthetic and real *EEG* data. The relationships between the observed connectivity measures are described in terms of the correlation values between them, which are, for the absolute values of CPCC and PLV, not lower that 0.97, and for the imaginary component of CPCC and wPLI—not lower than 0.92, for all observed frequency bands. Results show that the *CPCC* includes information of both other measures balanced in a single complex-numbered index.

Keywords: *EEG*; functional connectivity; phase locking value; weighted phase lag index; complex Pearson correlation coefficients

Citation: Šverko, Z.; Vrankić, M.; Vlahinić, S.; Rogelj, P. Complex Pearson Correlation Coefficient for EEG Connectivity Analysis. *Sensors* **2022**, *22*, 1477. https://doi.org/10.3390/s22041477

Academic Editor: Yvonne Tran

Received: 21 January 2022
Accepted: 11 February 2022
Published: 14 February 2022

Publisher's Note: MDPI stays neutral with regard to jurisdictional claims in published maps and institutional affiliations.

1. Introduction

A human brain contains on average about 100 billion (10^{11}) neurons connected by about 100 trillion (10^{14}) synapses. The neurons are anatomically organized in different spatial regions and functionally interact over different time points [1]. In this work, electroencephalography (*EEG*) was used to record neuron activity. *EEG* is an electrophysiological monitoring method for observing neurophysiological changes related to postsynaptic activity in the neocortex, i.e., a method for recording the electrical activity of the brain [2]. Monitoring brain activity using this method provides high temporal resolution. This property makes *EEG* one of the most suitable monitoring methods for non-invasive detection of neurons' interactions inside the brain and, consequently, for detection of information transmission within the same brain regions and between different brain regions [3]. Brain connectivity analysis is generally divided into two types: structural and functional. Tracking the direction of fibers between different brain regions or within a brain region is called structural connectivity analysis [4]. The most suitable recording methods for determining structural connectivity are magnetic resonance imaging (*MRI*) [5] and diffusion tensor imaging (*DTI*) [6]. On the other hand, functional connectivity analysis can be defined as an analysis of the amount of information transmitted between brain regions or within a brain region. This type of connectivity analysis is usually divided into two groups: undirected and directed. Undirected connectivity measures evaluate the degree of connectivity, while directed connectivity measures evaluate the degree and direction of connectivity between observed brain regions. In this paper, we focus on undirected connectivity measures. Different cognitive tasks require different information flows within a brain area or between different brain areas. This is due to the fact that neuronal oscillations are background mechanisms essential for dynamic cooperation in the brain [7–12].

The most suitable methods for monitoring brain activity to determine functional connectivity are magnetoencephalography (*MEG*) and electroencephalography (*EEG*) due to their good temporal resolutions [13].

Different types of measures can be used to determine functional connectivity, such as phase synchronization, generalized synchronization measures, linear temporal correlation, etc. [14–16]. In this paper, we focus on undirected phase synchronization measures. The most often used measures are the phase locking value (*PLV*) [17,18] and the weighted phase lag index (*wPLI*) [19]. The main difference between these two measures is the ability to avoid the effect of volume conduction.

The *PLV* index is based on phase differences of signals from two *EEG* channels. For a set of N time points, it calculates an average of N unit vectors that represent the phase difference between the signals of both channels. The *PLV* value of zero represents no connection between the observed signals' regions and the maximum *PLV* value of one represents a perfect connection. Although very widely used, a drawback of the *PLV* measure is the tendency to be biased towards higher values due to volume conduction [17].

The phase lag index (*PLI*) was designed as a solution to avoid the misinterpretation of volume conduction as a connectivity component [17]. Volume conduction reflects in the appearance of signal components with phase differences closer to 0 or $\pm\pi$. *PLI* avoids them by only considering the number of samples with positive and negative phase differences. Only if the number of samples in one group, i.e., positive or negative, is predominant then *PLI* gets value close to one. This cancels out the components with phase angle distributions centered at 0 and $\pm\pi$.

The extended version of the *PLI* is the weighted phase lag index (*wPLI* [19]). The *wPLI* measure adds weighting of samples by the imaginary component of the cross-spectral density. Because the real component of cross spectral density is not considered, samples where the phase differences are close to 0 or $\pm\pi$ have no contribution to the connectivity estimation and signal components that may arise due to volume conduction have no influence.

There are several other undirected connectivity measures, such as coherence, imaginary part of coherence, mutual information, etc., but for the purposes of this article, we limit our analysis to only those two most common ones, i.e., *PLV* and *wPLI*. They complement one another, providing connectivity estimation with and without consideration of the volume conduction effect. As an alternative, we propose a complex Pearson correlation coefficient (*CPCC*), which in a single unique measure provides information of both connectivity components.

The rest of the article is structured in the following way. In Section 2, we propose the complex Pearson correlation coefficient as a novel measure of undirected channel connectivity, review the *PLV* and *wPLI* measures, and analytically show their relationships to *CPCC*. In section three, the relationship is demonstrated with practical experiments, using synthetic and real *EEG* signals. We end the paper with a discussion and conclusion.

2. Methods

In this section, we define the proposed complex Pearson correlation measure (*CPCC*) and show its analytical relationship with *PLV* and *wPLI* connectivity measures.

2.1. Complex Pearson Correlation Coefficient as a Measure of Undirected Connectivity

Various types of complex correlation calculations are used in the literature [20], and in different research fields, such as geophysics [21], radar systems [22], optics [23], etc. In this section, we propose the use of complex Pearson correlation coefficient for *EEG* connectivity analysis.

Pearson's linear correlation coefficient (r) is the most commonly used linear correlation coefficient. It is a statistical measure of the degree to which variables change their values in relation to each other, or in other words, expresses the level to which two variables are linearly related. It is defined as follows:

$$r(x_1, x_2) = \frac{\sum_{n=1}^{N}(x_{1,n} - \overline{x_1})(x_{2,n}^* - \overline{x_2^*})}{\sqrt{\sum_{n=1}^{N}(x_{1,n} - \overline{x_1})^2 \sum_{n=1}^{N}(x_{2,n}^* - \overline{x_2^*})^2}}. \tag{1}$$

Here, N is the number of samples, x_1 and x_2, are the series being analyzed, $\overline{\{.\}}$ represents mean values of observed series, and $\{.\}^*$ is the complex conjugate operator (if the values in series are complex). The resulting r ranges from -1 (indicating perfect negative correlation) to $+1$ (indicating perfect positive correlation). A zero value is an indicator of no linear signal relationship. Assuming that *EEG* signals for the analysis should be pre-filtered, which removes *DC* signal components, the equation can be simplified:

$$r(x_1, x_2) = \frac{\sum_{n=1}^{N} x_{1,n} \cdot x_{2,n}^*}{\sqrt{\sum_{n=1}^{N} |x_{1,n}|^2} \cdot \sqrt{\sum_{n=1}^{N} |x_{2,n}|^2}}. \tag{2}$$

The numerator in the Equation (2) can be understood as a time-averaged temporal estimation of sample relationship, while the denominator is a weighting factor to obtain the desired range from -1 to 1. Let us first focus to the numerator. For two oscillatory signals represented as series of real values, the temporal relationship estimation is also an oscillatory signal. Consequently, the temporal contribution of a single time step does not have any direct meaning and at least one period of signal samples need to be averaged to become informative. To improve temporal meaningfulness of the estimation, analytic signal representations can be used instead of real valued ones. Analytic signal sample is a complex number that adds an imaginary part indicating the oscillatory nature of the signal to its existent real valued part. Thus, in addition to a real signal value, the analytic signal sample includes the information of signal instantaneous amplitude and instantaneous phase, which can be represented as a vector in a complex plane. Because basic sinusoidal oscillatory signal keeps the instantaneous amplitude constant over time while its instantaneous phase increases linearly, these vectors are also called phasors. For two phase-locked signals the phase difference is constant and the numerator of Equation (2) gets constant over time, too. Its real value represents the dot product of the phasors, while its imaginary part equals the size of the cross product. Altogether the product of two phasors of two analytic signals is analogous to the cross spectral density for the stationary or quasi-stationary signals. The denominator of Equation (2), needed for scaling, relates to the power of both signals. The final result when using analytic signals is the complex Pearson correlation coefficient (*CPCC*):

$$CPCC = r(x_{a,1}, x_{a,2}), \tag{3}$$

where x_a denotes analytic signals.

The analytic signal representation is defined only for narrow frequency band signals. In such cases analytic signals can be computed from real ones by adding imaginary part equal to the Hilbert transform (*HT*) of the original signal:

$$x_a(t) = x(t) + iHT(x(t)), \tag{4}$$

where *HT* (*x(t)*) represents the Hilbert transform of x (real signal) and $x_a(t)$ is an analytic signal, as explained in [17]. With *HT* we obtain a phasor influenced by all the frequencies in the observed narrow band. Phasors can also be obtained using the discrete Fourier transform, where one phasor presents each frequency component, but only stationary, without temporal dimension. *HT* provides this additional temporal perspective, which enables analysis of non-stationary signals. Because *EEG* signals are non-stationary, in this paper we limit to the analysis of their narrow band pre-filtered components with the analytic form obtained using *HT*.

Connectivity measures estimate the relationship between two signals and this can be performed using the *CPCC*. The in-phase signals have high real *CPCC* part and zero imaginary part. On the other hand, imaginary component represents the relationship between signals with the phase lag of $\pm\pi/2$. Thus, the connectivity of two brain regions can be estimated considering both parts of the complex *CPCC* value for the corresponding *EEG* signals, by computing its absolute value (*absCPCC*). In such a case the obtained value should be related to the *PLV* value. When the volume conduction effect needs to be avoided, only the imaginary component shall be used (*imCPCC*). Such estimation is expected to be related to the *wPLI* value.

2.2. Phase Locking Value PLV and Its Relation to CPCC

Phase-locking value (*PLV*) is calculated based on the phase differences of the two analytical signals [17,18]:

$$PLV_{x_1,x_2} = \left| \frac{1}{N} \sum_{n=1}^{N} e^{i(\Delta\phi_{x_{1,n},x_{2,n}})} \right|. \tag{5}$$

In Equation (5), $\Delta\phi$ represents the phase difference and N represents the number of samples. The instantaneous phase difference is defined as:

$$\Delta\phi_{x_{1,n},x_{2,n}} = \phi_{x_{1,n}} - \phi_{x_{2,n}}, \tag{6}$$

where $\phi_{x_{1,n}}$ and $\phi_{x_{2,n}}$ stand for the phase angles at n-th sample. In order to obtain instantaneous phases, analytical signals need to be computed, using (*HT*). Computation of *PLV* can be visualized by creating a set of N unit vectors corresponding to N time samples, see Figure 1. Phase angles of those vectors are equal to phase differences between the two *EEG* signals for samples from 1 to N. All the N unit vectors representing phase differences are averaged to obtain *PLV*.

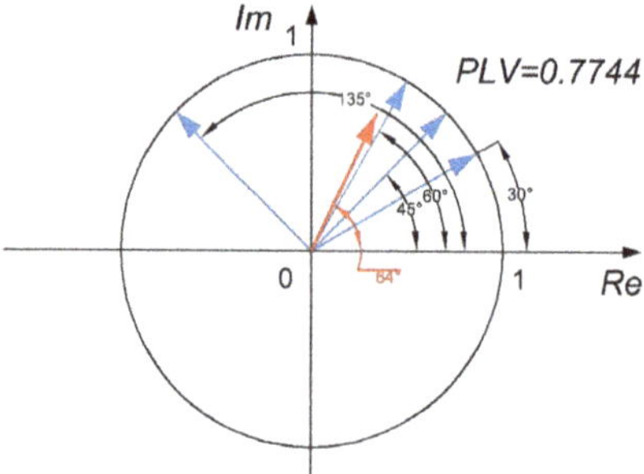

Figure 1. Visualization of averaging used in calculation of the *PLV*. The *PLV* is computed from unit vectors representing instantaneous phase differences.

The high value of *PLV* is obtained when the vectors are well clustered, which means that the phase difference between the two *EEG* channels is mostly constant for all the time samples On the other hand, when the phase difference between the two channels is changing with time, the unit vectors are scattered, which results in low *PLV* value.

The lack of this measure is its tendency to falsely over-estimate the connectivity level due to the volume conduction. The reason is that the volume conduction enables a signal from a single source to be measured on both *EEG* electrodes under consideration, which results in a zero-phase difference over a larger time interval, leading to a larger *PLV* value.

In order to prove our assumption that the absolute value of the complex Pearson correlation is related to the *PLV* index, the Equation (2) can be rewritten in the following way:

$$absCPCC_{x_1,x_2} = |r(x_1,x_2)| = \frac{|\sum_{n=1}^{N} A_{x_{1,n}} \cdot A_{x_{2,n}} \cdot e^{i\Delta\phi_{x_{1,n},x_{2,n}}}|}{\sqrt{\sum_{n=1}^{N} A_{x_{1,n}}^2} \cdot \sqrt{\sum_{n=1}^{N} A_{x_{2,n}}^2}}, \tag{7}$$

where A_x represents the instantaneous amplitude of a complex signal.

Comparing Equations (5) and (7), we can see that the *PLV* is related to the *absCPCC*, but scales the contributions of instantaneous phases with instantaneous amplitudes:

$$\frac{|\sum_{n=1}^{N} A_{x_{1,n}} \cdot A_{x_{2,n}} \cdot e^{i\Delta\phi_{x_{1,n},x_{2,n}}}|}{\sqrt{\sum_{n=1}^{N} A_{x_{1,n}}^2} \cdot \sqrt{\sum_{n=1}^{N} A_{x_{2,n}}^2}} \Diamond \neq \left| \frac{1}{N} \sum_{n=1}^{N} e^{i(\Delta\phi_{x_{1,n},x_{2,n}})} \right|, \tag{8}$$

absCPCC is therefore a weighted version of *PLV*.

2.3. Weighted Phase Lag Index wPLI and Its Relation to CPCC

The *PLI* and *wPLI* measures of connectivity address the volume conduction problem. Let us first present the *PLI* measure, as a transitional step towards a more refined weighted *PLI* measure (*wPLI*). The *PLI* is defined as [17]:

$$PLI_{x_1,x_2} = \left| \frac{1}{N} \sum_{n=1}^{N} sign(Im(S_{x_{1,n},x_{2,n}})) \right|, \tag{9}$$

where N is the number of samples. In the original definition [24], $S_{x_{1,n},x_{2,n}}$ is the cross-spectral density of the observed signals defined by Fourier transform. In [17], *PLI* was defined using analytical signals obtained by HT and the cross-spectral density is defined as:

$$S_{x_{1,n},x_{2,n}} = |A_{x_{1,n}}| \cdot |A_{x_{2,n}}| e^{i(\phi_{x_{1,n}} - \phi_{x_{2,n}})}, \tag{10}$$

where $A_{x_{1,n}}$ and $A_{x_{2,n}}$ are the instantaneous amplitudes of the observed signals x_1 and x_2 at sample n. Based on Equation (9) and as shown in , Computation of *PLI* is illustrated in Figure 2. All unit vectors that represent phase differences are first divided into two subsets: those with positive, and those with negative imaginary part. Then, the difference of subsets' sizes is divided by the number of all vectors N, and its absolute value equals the *PLI*.

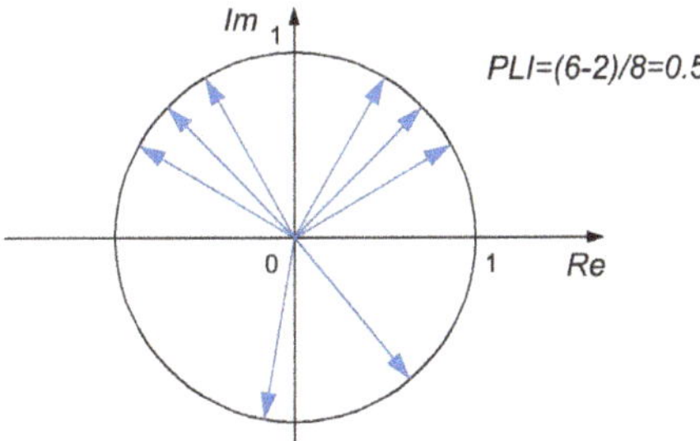

Figure 2. Visualization of averaging used in calculation of the *PLI*. The *PLI* is computed from unit vectors representing instantaneous phase differences.

Therefore, if there is a predominant positive or negative phase difference throughout the observed time interval, then the obtained value of *PLI* will be close or equal to 1. On the contrary, *PLI* which equals 0 is obtained when half of the phase differences are negative and the other half of them are positive.

The weighted phase lag index (*wPLI*) is an improved version of the phase lag index connectivity measure. The unit vectors of phase differences from *PLI* are now scaled with instantaneous amplitudes of both signals [19]. In other words, *wPLI* is obtained by weighting *PLI* with the imaginary part of the cross spectral density:

$$wPLI_{x_1,x_2} = \left| \frac{\frac{1}{N}\sum_{n=1}^{N} |Im(S_{x_{1,n},x_{2,n}})| sign(Im(S_{x_{1,n},x_{2,n}}))}{\frac{1}{N}\sum_{n=1}^{N} |Im(S_{x_{1,n},x_{2,n}})|} \right|. \tag{11}$$

By expressing the cross spectral density from Equation (10) using the complex conjugate operator, the *wPLI* can be rewritten as follows:

$$S_{x_{1,n},x_{2,n}} = x_{1,n} \cdot x_{2,n}^*. \tag{12}$$

$$wPLI_{x_1,x_2} = \left| \frac{\frac{1}{N}\sum_{n=1}^{N} |Im(x_{1,n} \cdot x_{2,n}^*)| sign(Im(x_{1,n} \cdot x_{2,n}^*))}{\frac{1}{N}\sum_{n=1}^{N} |Im(x_{1,n} \cdot x_{2,n}^*)|} \right|, \tag{13}$$

which can be further simplified as:

$$wPLI_{x_1,x_2} = \frac{|\sum_{n=1}^{N} Im(x_{1,n} \cdot x_{2,n}^*)|}{\sum_{n=1}^{N} |Im(x_{1,n} \cdot x_{2,n}^*)|}. \tag{14}$$

Now we can show its relationship to the *CPCC* or more specifically to its imaginary part, denoted *imCPCC*:

$$
\begin{aligned}
imCPCC_{x_1,x_2} &= |Im[r(x_1,x_2)]| \\
&= \frac{|Im[\sum_{n=1}^{N} x_{1,n} \cdot x_{2,n}^*]|}{\sqrt{\sum_{n=1}^{N}|x_{1,n}|^2} \cdot \sqrt{\sum_{n=1}^{N}|x_{2,n}|^2}} \\
&= \frac{|\sum_{n=1}^{N} Im(x_{1,n} \cdot x_{2,n}^*)|}{\sqrt{\sum_{n=1}^{N}|x_{1,n}|^2} \cdot \sqrt{\sum_{n=1}^{N}|x_{2,n}|^2}}
\end{aligned}
\tag{15}
$$

Comparing Equations (14) and (15) we see that both measures, *wPLI* and *imCPCC*, are based on the imaginary part of the cross spectral density *S* in the numerator, and differ only in scaling in the denominator. The *wPLI* is scaled using the imaginary part of *S* only, while *imCPCC* with the power of both signals.

2.4. Connectivity Estimation Based on Phase Difference Histograms

In this section, we explain how connectivity reflects in phase difference histograms, to illustrate the connectivity measures. Although statistical properties of the phase difference distribution can clearly indicate phase locking of the signals [25], in practice, connectivity measures are rarely explained in these terms. We will use it to gain better insight into real connectivity between signals, particularly for the cases where values of connectivity measures are the highest or most different between each other.

Let us first assume no volume conduction is present. When two signals are not connected, they change independently and the phase differences are uniformly distributed. Connectivity between two brain regions reflect in more expressed phase differences of corresponding signals. The higher the connectivity, the more pronounced the extreme gets, and the standard deviation of the distribution gets lower, see Figure 3a. The "red" distribution reflects the highest connectivity and its standard deviation is the lowest. On the other hand, the "orange" distribution has the highest standard deviation and reflects the lowest connectivity. The mean value of the phase distribution equals the average phase difference, and can have an arbitrary value in the $[-\pi, \pi]$ range, and it does not depend on the connectivity level.

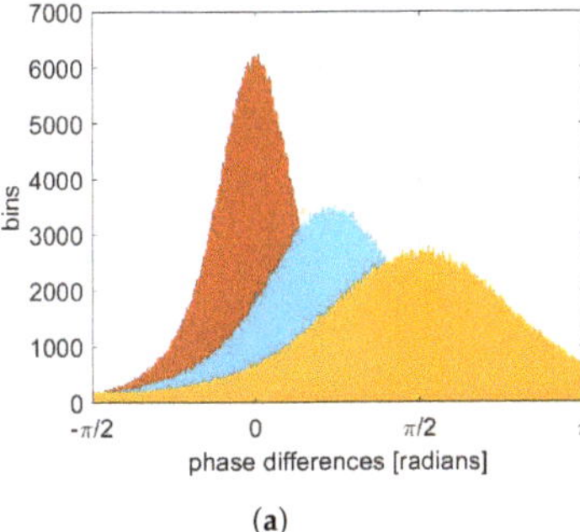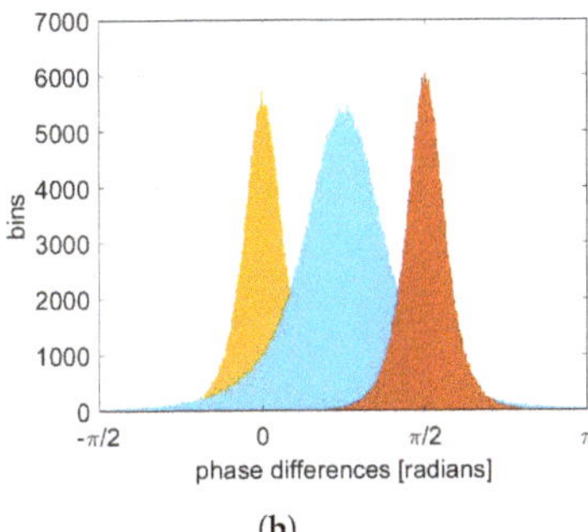

(a)   (b)

Figure 3. The relationship between connectivity and phase differences distributions. When volume conduction is not considered (**a**) higher connectivity reflects in lower variance, while the mean value is irrelevant. In the presence of volume conduction (**b**), it reflects in higher values for phase differences close to 0 or π, which, therefore, do not (necessarily) indicate connectivity. Connectivity level is expressed with colors; red is the highest and yellow is the lowest.

If volume conduction is present, certain signal components are included in both of the signals under consideration. These signal components have a phase difference of 0 or $\pm\pi$, but due to noise and signal interference, instantaneous phase differences spread around these values. These values therefore do not (necessarily) imply higher connectivity. In the example in Figure 3b, we can expect that distributions with peaks closer to $0 \pm k\pi$ are more likely to reflect volume conduction and not connectivity. The estimated connectivity is therefore the highest when the value at 0 and $\pm\pi$ is the lowest and the variance the smallest.

3. Results

In this section, we compare the proposed *CPCC* measures with *PLV* and *wPLI* using synthetic signals and real-life signals from freely available datasets.

3.1. Synthetic Signals from the MRC Brain Network Dynamics Unit (University of Oxford)

In the first experiment, we generated synthetic signals following Mäkinen et al. [26]. The *EEG* data we generated contained 31 channels from 973 trials, which were concatenated into a single large signal. This suited our particular purpose, as we analyzed general brain connectivity independent of specific brain events.

We computed connectivity using the proposed and established methods for different frequency bands. The connectivity matrices representing the estimated connectivity for each electrode pair and for all four measures are shown in Figure 4. We can clearly see strong visual similarities between the proposed measures and the most commonly used measures, i.e., between *PLV* and *absCPCC*, as well as *wPLI* and *imCPCC*.

To better compare the connectivity measures, see Figure 5, with scatter plots for measure pairs *PLV* to *absCPCC* and *wPLI* to *imCPCC*. Each dot in a scatter plot represents one electrode pair. The color of the dots depends on the relative density of the dots in the graph. There are also two lines shown, where the black one represents identity while the cyan one the best linear fit. High correlation is evident for both connectivity measure pairs, while the scaling differences depend on the frequency band, most evidently for the *PLV* to *absCPCC* pair. There are some electrode pairs that deviate slightly from the general linear relationship while the overall correlation of the measures seem to be high.

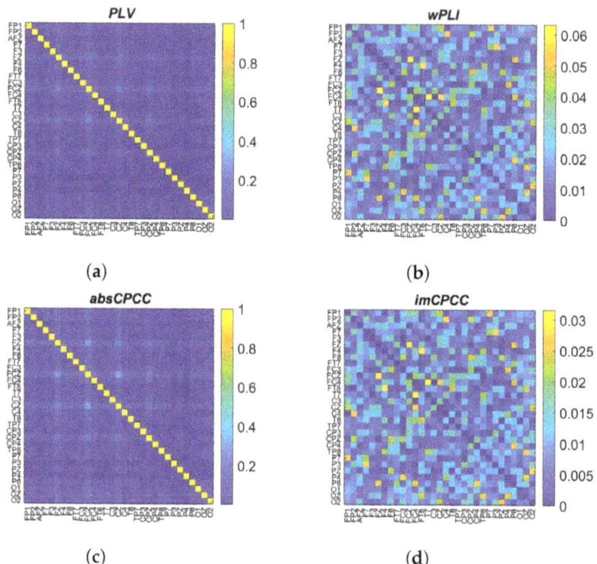

Figure 4. Connectivity matrices obtained with *PLV* (**a**), *wPLI* (**b**), *absCPCC* (**c**), and *imCPCC* (**d**) for signals generated with [26], for 8–13 Hz frequency band.

To evaluate the relationship between the measures, evident from Figure 5, we computed their correlation. In addition to frequency bands shown in Figure 5, 0.5–4 Hz, 4–8 Hz and 8–13 Hz, we computed it for 13–18, 18–30, and 35–45 Hz frequency bands. For all the frequency bands and both pairs, i.e., *PLV* to *absCPCC*, and *wPLI* to *imCPCC*; the obtained correlation equaled 0.99, proving the close to perfect relationship between the measures.

Finally, we selected electrode pairs with the highest connectivity values and the highest ratio between them. Their phase difference distributions are shown in Figure 6. As expected, the same electrode pair (16–11) had the highest *PLV* and *absCPCC* values. The corresponding phase distribution was centered at the phase angle 0, indicating the possibility of volume conduction. Similarly, one electrode pair (14–12) had the highest in both *wPLI* and *imCPCC* values. The corresponding phase distribution has a less pronounced peak off the center. The ratios between *PLV* and *wPLI*, as well as *absCPCC* and *imCPCC* values, were the highest when the later ones equalled 0 and the histogram was centered (electrode pair 15–12). The highest and *wPLI* to *PLV* ratio and *imCPCC* to *absCPCC* ratio were obtained when *absCPCC* equaled *imCPCC*.

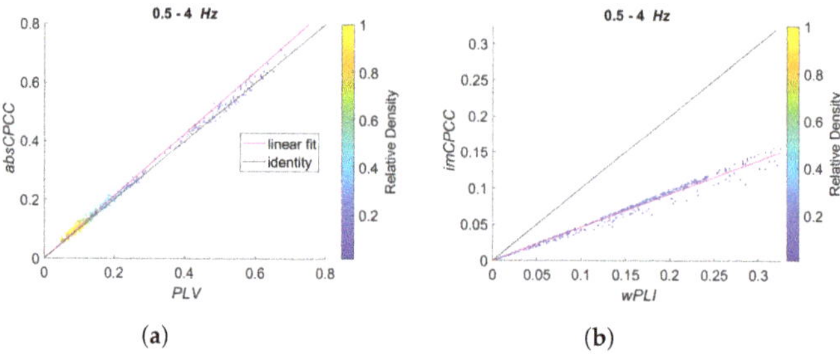

Figure 5. *Cont.*

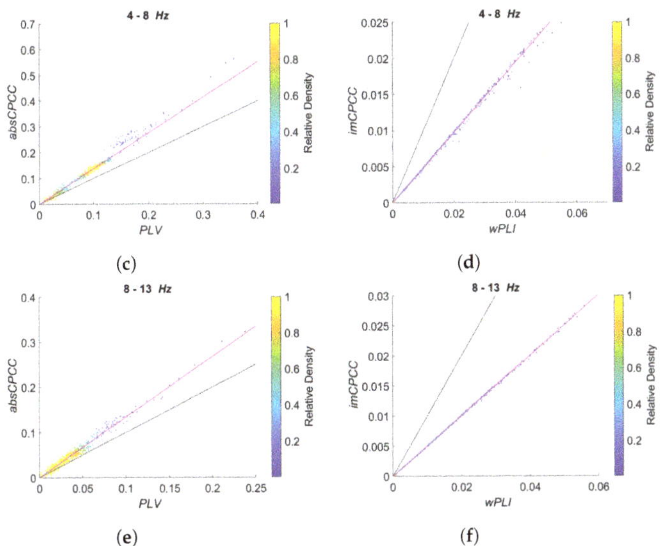

Figure 5. Scatter plots of *absCPCC* to *PLV* relationship (**left**) and *imCPCC* to *wPLI* relationship (**right**). Each dot represents one electrode pair. Dots are colored according to their relative density. The black line represents identity, while the cyan one is the best linear fit. Rows correspond to different frequency bands: (**a**,**b**) 0.5–4 Hz; (**c**,**d**) 4–8 Hz; (**e**,**f**) 8–13 Hz.

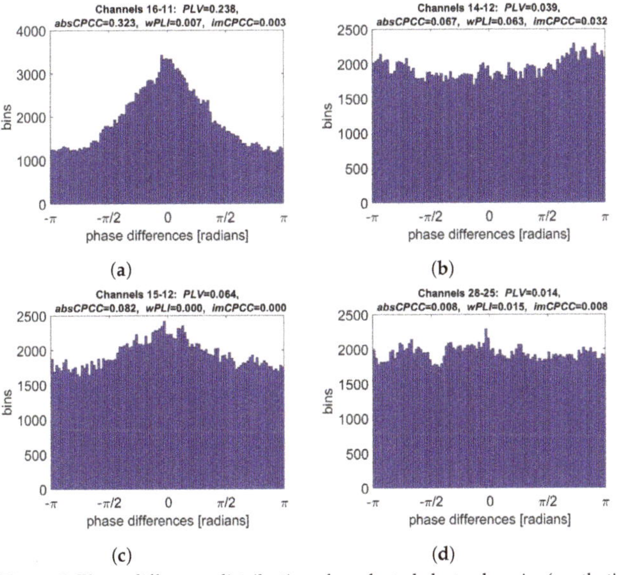

Figure 6. Phase difference distributions for selected electrode pairs (synthetic signals [26]). Shown are the distributions corresponding to the highest: (**a**) *PLV* and *absCPCC* values, (**b**) *wPLI* and *imCPCC* values, (**c**) ratio between *absCPCC* and *imCPCC* values, (**d**) ratio between *imCPCC* and *absCPCC* values.

3.2. Synthetic Signals Generated with the Kuramoto Model

The second set of synthetic signals to test connectivity estimation methods was generated using the Kuramoto model according to [27]. The reason for using it is that the relationship between electrodes is defined and known in advance. Twenty-four signals (channels/electrodes) were generated. The signals form three groups, from 1 to 8, 9 to 16, and 17 to 24. They are composed of two signal components, where the first component is synchronized between all electrodes in a group and the second component is not synchronized and gives a more realistic variability to the signal set. The signals from 17 to 24 are composed of the first components only and due to the high coupling factor ($K = 1000$), these signals are synchronized very quickly. As a result of the fast synchronization, these signals are in phase and can be observed as an example of high volume conduction.

Figure 7 shows the connectivity matrices of the observed connectivity measures. It is visible that the signals are connected within groups and much less between the groups. The random nature of the generation process could lead to synchronous signals even between different groups. Looking at the third group of signals, we also see a tendency for the *PLV* and *absCPCC* to include volume conduction as an acceptable contribution to the connectivity, while *wPLI* and *imCPCC* avoid this component.

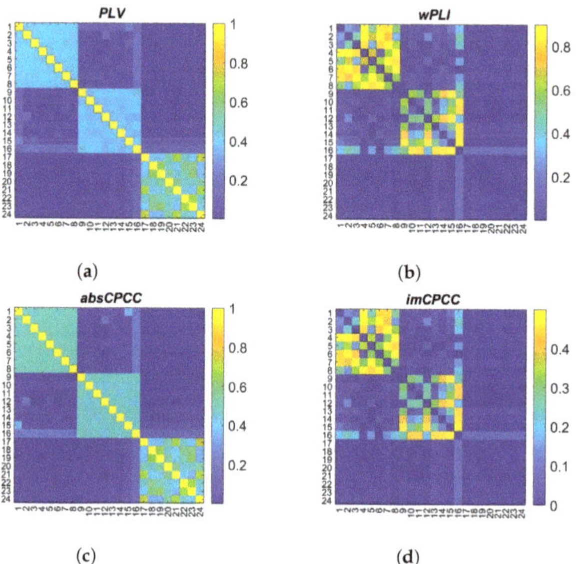

(**a**)   (**b**)

(**c**)   (**d**)

Figure 7. Connectivity matrices obtained with *PLV*, *wPLI*, *absCPCC*, and *imCPCC* for signals generated with the Kuramoto model [27].

In Figure 8 is a scatter plot showing the relationship between *absCPCC* and *PLV* values (a) and between the *imCPCC* and *wPLI* values (b). A strong linear relationship is evident for both connectivity estimation method pairs, while the scaling is different.

The correlation between *PLV* and *absCPCC* values as well as the correlation between *wPLI* and *imCPCC* equals 0.99, indicating strong similarities between these measures.

The phase difference distributions for signal pairs with the highest connectivity values and the highest ratio between them are shown in Figure 9. The phase difference distribution corresponding to the highest *PLV* and *absCPCC* values was narrow and centered around 0. It was obtained for two signals from the last group (24–17), which modeled volume conduction. The highest *wPLI* and *imCPCC* values were obtained for signals from the first group (7–3), with wide distribution centered at $\pi/2$ radians. The ratio between the *absCPCC* and *imCPCC* values was the highest when the latter one equaled 0 and the histogram was centered, indicating possible volume conduction (signal pair 19–18). The highest *imCPCC* to *absCPCC* ratio was, again, obtained when values for both measures were equal, with a clear peak of the distribution at $\pi/2$ radians.

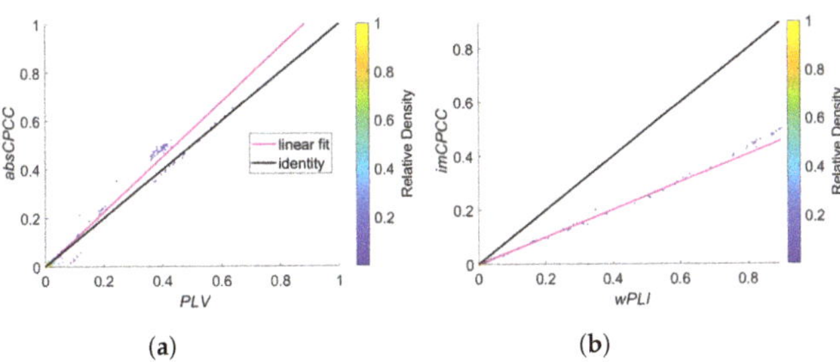

Figure 8. Relationships between *absCPCC* and *PLV* (**a**), and *wPLI* and *imCPCC* (**b**), shown as a scatter plot of values for all signal pairs where signals were generated with the Kuramoto model [27]. The black line represent the identity while the cyan line shows the best linear fit. The colors of the dots represent the relative density of the connectivity values.

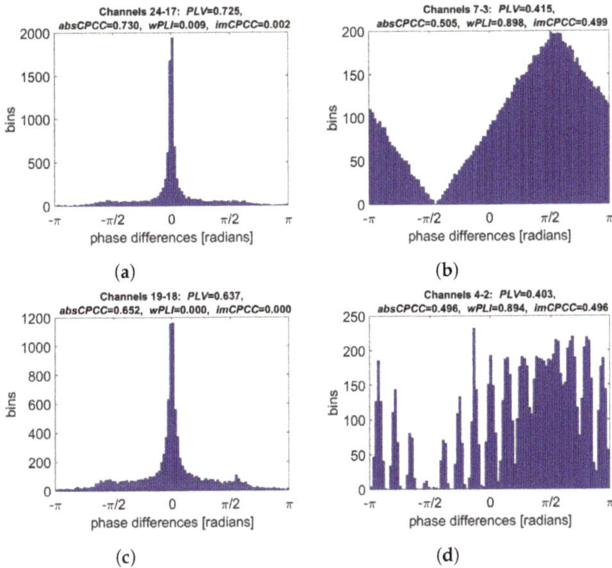

Figure 9. Phase difference distributions for selected synthetic signal pairs generated using the Kuramoto model [27]. Shown are the distributions corresponding to the highest: (**a**) *PLV* and *absCPCC* values, (**b**) *wPLI* and *imCPCC* values, (**c**) ratio between *absCPCC* and *imCPCC* values, (**d**) ratio between *imCPCC* and *absCPCC* values.

3.3. Real-Life Signals

For testing on real-life data, we used the SPIS Resting State Dataset [28], a multimodal dataset with *EEG* and forehead *EOG* signals. In our analysis, we used only *EEG* signals from the "eyes closed" (*EC*) and "eyes open" (*EO*) states with a duration of 2.5 min, using 256 Hz sampling rate.

Offline preprocessing of the EEG signal was performed in the following sequence of steps:

1. The raw brain activity data were imported into *MATLAB* using the *EEGLAB* toolbox;
2. Electrode positions (also called channel locations) were defined in the software;
3. The data were referenced to average;
4. The data were filtered with a band pass filter limited to 0.5 and 45 Hz;
5. Automatic spectral-based channel suppression ($z = 5$) was performed using the *EEGLAB "pop rejchan"* function;

6. Artifacts were removed using the *ICLabel* plugin for *EEGLAB* (thresholds for removing components were less than or equal to 0.05 for brain activity and greater than or equal to 0.9 for artifacts);
7. The data were re-referenced to average;
8. Sub-bands of the *EEG* signal were extracted (delta 0.5–4 Hz, theta 4–8 Hz, alpha 8–13 Hz, low beta 13–18 Hz, high beta 18–30 Hz, gamma 35–45 Hz).

Figure 10 shows the connectivity matrices for *PLV*, *absCPCC*, *wPLI*, and *imCPCC* for both conditions (*EC* and *EO*) in the alpha band (8–13 Hz). The similarity between *PLV* and *absCPCC*, as well as between *wPLI* and *imCPCC*, is visible, although with some evident differences, mainly between the latter two. Although the patterns are similar, the color scaling (based on the highest value) is different.

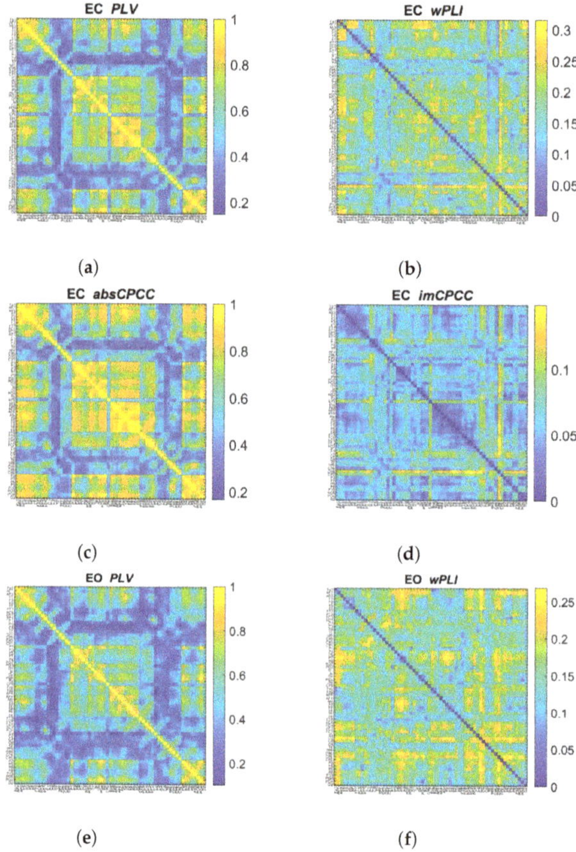

(a)   (b)

(c)   (d)

(e)   (f)

Figure 10. *Cont.*

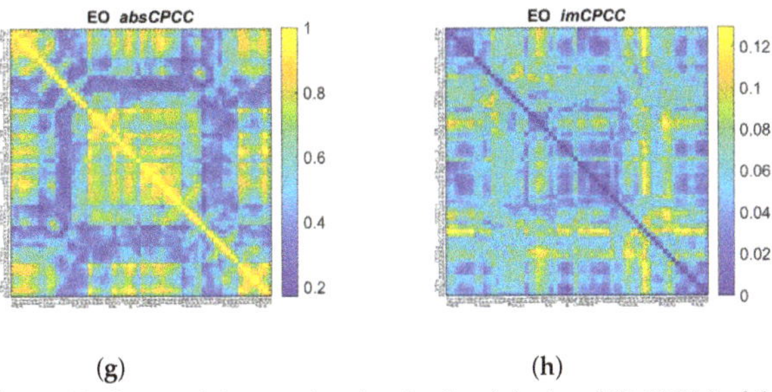

(g)   (h)

Figure 10. Connectivity matrices for the the alpha band (8–13 Hz) of the real-life signals [28] for eyes closed (EC) and eyes open (EO) states, computed with *PLV*, *wPLI*, *absCPCC* (**g**), and *imCPCC* (**h**).

Figures 11 and 12 show the relationship between *absCPCC* and *PLV* values (left) and between the *imCPCC* and *wPLI* values (right) for all the frequency ranges. Figure 11 shows the signals recorded with eyes closed (EC state), while Figure 12 is for eyes open (EO state). We can see that *absCPCC* and *PLV* as well as *imCPCC* and *wPLI* are positively correlated in all frequency bands. However, we have to be aware that real-life signals include multiple signal components with different amplitudes, while the scaling is common for the whole sequence. This makes the results more scattered in *PLV-absCPCC* and *wPLI-imCPCC* distributions. The relationship between *absCPCC* and *PLV* is evident, but with visible deviation from being perfectly linear due to reduced scaling difference for high connectivity values. The relationship between *imCPCC* and *wPLI* also deviates from linear, with evident range of scaling differences. The relationships do not seem to be dependent on the EC/EO state.

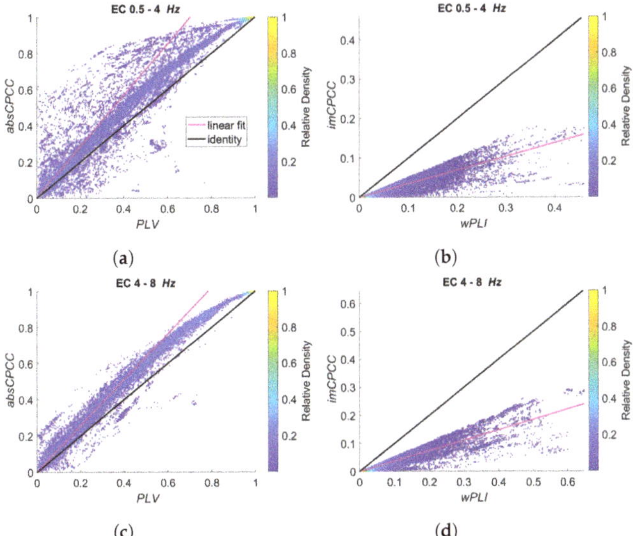

(a)   (b)

(c)   (d)

Figure 11. *Cont.*

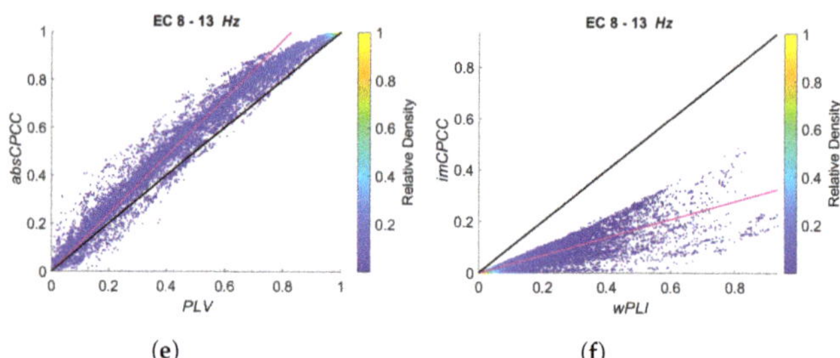

Figure 11. Scatter plots of the *absCPCC* to *PLV* relationship (**left**) and the *imCPCC* to *wPLI* relationship (**right**), for all electrode pairs and for *10* test subjects (*EC* state). The black line represents the identity, while the cyan line shows the best linear fit. The colors of the dots represent the relative density of the connectivity values. Each row is shown for a different frequency band: (**a**,**b**) 0.5–4 Hz; (**c**,**d**) 4–8 Hz; (**e**,**f**) 8–13 Hz.

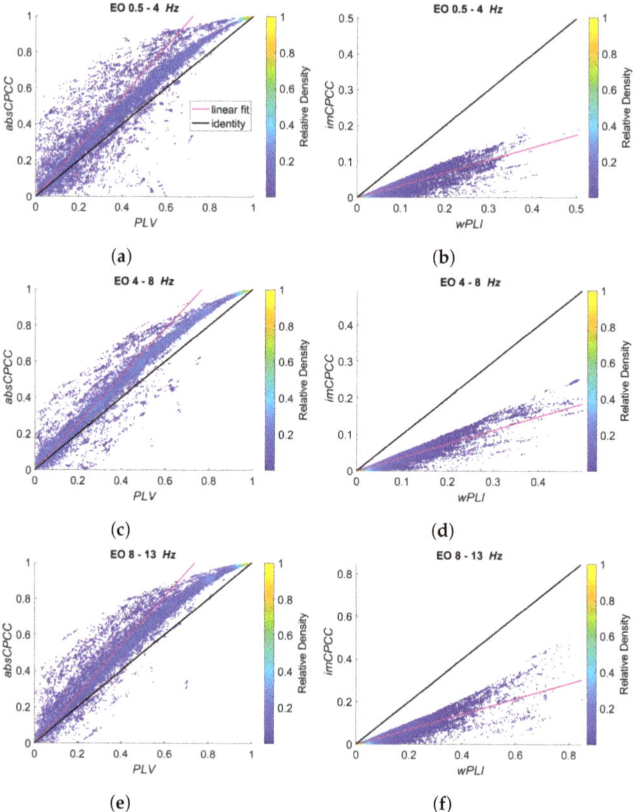

Figure 12. Scatter plots of the *absCPCC* to *PLV* relationship (**left**) and the *imCPCC* to *wPLI* relationship (**right**), for all electrode pairs and for 10 test subjects (*EO* state). The black line represents the identity, while the cyan line shows the best linear fit. The colors of the dots represent the relative density of the connectivity values. Each row is shown for a different frequency band: (**a**,**b**) 0.5–4 Hz; (**c**,**d**) 4–8 Hz; (**e**,**f**) 8–13 Hz.

To enumerate the linearity of relationships between the connectivity measures, we show the correlation values between them in Table 1.

Table 1. Correlation values between compared connectivity measures (real-life signal). Here, r_{abs} and r_{im} denote $r(absCPCC, PLV)$ and $r(imCPCC, wPLI)$ respectively.

Frequency	State-EC		State-EO	
(Hz)	r_{abs}	r_{im}	r_{abs}	r_{im}
0.5–4	0.93	0.86	0.94	0.89
4–8	0.98	0.91	0.97	0.94
8–13	0.98	0.86	0.97	0.91
13–18	0.99	0.94	0.99	0.96
18–30	0.98	0.96	0.99	0.96
35–45	0.96	0.95	0.99	0.95

The correlation between *absCPCC* and *PLV* connectivity measures is high for all frequency bands and both states (*EC* and *EO*), with an average of 0.97. Only slightly lower values are obtained for the correlation between *imCPCC* and *wPLI*, with an average of 0.92. The corresponding p-values for the alternative hypothesis that measures are not correlated are all smaller than 0.0001 and, thus, well below the significance level of 0.05, which means that the hypothesis of the correlation between the absCPCC and PLV and between imCPCC and wPLI is proven for all frequency bands and both states.

Phase difference distributions for real-life signals for electrode pairs with the highest connectivity values and the highest ratios between them are shown in Figure 13. The phase difference distributions corresponding to the highest *PLV* and *absCPCC* values are narrow and centered around 0. The highest *wPLI* and *imCPCC* values are obtained when the distribution that is wide, slightly asymmetric, and centered at non-zero phase difference. The ratio between *absCPCC* and *imCPCC* values is the highest when the later one equal 0 and the histogram is centered at $\pm\pi$ radians. The highest *imCPCC* to *absCPCC* ratio is, again, obtained when their values are equal and the histogram is not symmetric around 0.

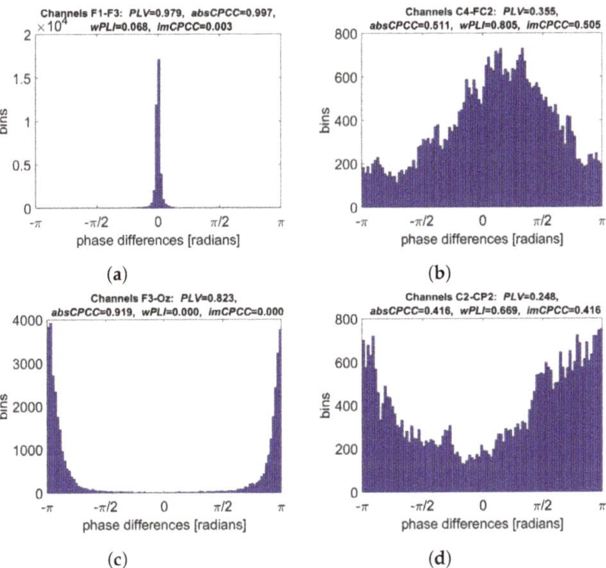

Figure 13. Phase difference distributions for selected electrode pairs (real-life signals [28]). For each distribution, all four measures are calculated. The figure shows the electrode pair with the highest: (**a**) *PLV* and *absCPCC* values, (**b**) *wPLI*, and *imCPCC* values, (**c**) ratio between *absCPCC* and *imCPCC* values, (**d**) ratio between *imCPCC* and *absCPCC* values.

4. Discussion and Conclusions

In this paper, we shed new light on the Pearson correlation for the *EEG* connectivity analysis. We introduced the complex correlation (CPCC) as a measure of brain connectivity. We compared it to the (currently) most widely used brain connectivity measures, i.e., *PLV* [17,18] and *wPLI* [19]. The correlation coefficient (CC) has been used before, but only between the real signals, not the analytic ones, and it was shown that it does not represent the optimal metric to estimate functional interactions [29]. It equals the real component of CPCC, while we showed the importance of the absolute value and the imaginary component of CPCC.

We showed that the imaginary part of the complex Pearson correlation (*imCPCC*) is closely related to the *wPLI* measure and that the absolute value of the complex Pearson correlation *absCPCC* is closely related to *PLV*. The relationships are proven analytically and numerically, on two types of synthetic signals [26,27] and on real-life *EEG* signals [28]. Analytically, the differences are only in the denominators that are normalizing the measures to the $[0,1]$ interval. Numerically, high correlations between the results obtained with related measures are shown. For synthetic signals, the correlation level is for all frequency bands equal to 0.99. The scaling differences are evident for *absCPCC* to *PLV* relationship, and differ for different frequency bands. The connectivity results for real-life signals show more differences and the measures are less correlated, but still with an average correlation of 0.97 for *absCPCC* to *PLV* relationship and 0.92 for the *imCPCC* to *wPLI* relationship. Real-life signals consist of more components, which originate in different sources, are related through different neural paths, and include different (although similar) frequencies. Even when limiting frequency bandwidths, they are more information-rich than simulated signals. Connectivity could be understood as a portion of signal components that affects two distinct electrode signals, but can vary with time. All of this reflects in more complex phase difference histograms and more complex scatter-plotted relationships between the measures.

Based on the results shown in this paper, and the fact that connectivity measures are currently typically analyzed relatively, we conclude that *PLV* can be replaced with *absCPCC* and *wPLI* with *imCPCC*. Moreover, the *absCPCC* and *imCPCC* measures are defined as two components of the the same *CPCC* measure and are, therefore, related, while *PLV* and *wPLI* are not. This enables comparison of the connectivity components that may be affected by volume conduction and those which are certainly not. The imaginary component of CPCC can only be lower or equal to the absolute CPCC value due to the excluded real CPCC component, which depends on signal components that may result from volume conduction. It can be expected that the true connectivity may also yield in-phase signals of two electrodes, which makes them indistinguishable from volume conduction. Similar to wPLI, imCPCC scales the components, such that ones more likely arise from volume conduction have lower influence. Thus, the estimated connectivity deviates from the true one, but CPCC provides the upper and lower boundary, with absCPCC and imCPCC respectively. Neuroscientists sometimes have a practice of calculating both the *PLV* and *wPLI* (or *PLI*) and then interpreting the results [29–32] because the *PLV* method, unlike the *wPLI*, does not take into account the influence of volume conduction. With the proposed *CPCC* measures, they can get additional information, related to the ratio of both connectivity components. As such, the CPCC measure could be used for various neurology related studies. Such studies include the EEG-based brain mechanism of sleep stages, which is important for sleep quality assessment and disease diagnosis [33]. By averaging over the trial set, the proposed measures could also be used as a solution to improve the prediction results of the phases of the synchronization and desynchronization tasks [34]. The potential application of CPCC also lies in the assessment of mental stress levels using functional connectivity as a parameter [35,36] and in the diagnosis dyslexia [37]. In addition, the proposed measures could be used as parameters for the evaluation of simulated EEG data based on the theory of functional connectivity of the brain [38].

The next valuable property of the proposed *CPCC* measure is that it can be computed as a summation of temporal sample contributions. This enables the measure to reveal temporal changes of connectivity and opens a new direction for further research. This can be especially useful for analyzing human brain networks in auditory and visual tasks [39,40] and is also promising for assessing motor skills [41]. This also allows us to observe changes in the organization of brain network connectivity over time using well-known measures from complex network graph theory [42].

The *CPCC* measure has an advantage in the computational complexity. In our experiments the computation of *absCPCC* and *imCPCC* was 65% to 179% faster than the computations of *PLV* and *wPLI*.

Finally, the computation of the correlation of two analytic signals is easy to implement and already available in most of the statistical and signal analysis tools.

Following the above discussion—we can state that the newly proposed *CPCC* connectivity measure, with *absCPCC* and *imCPCC* as its components, could replace *PLV* and *wPLI* measures, accelerate the computation of brain connectivity, and provide further information about brain processes. The data, code, and instructions for replicating the study presented in this article are freely available at https://github.com/zsverko/Code_CPCC.git (accessed on 10 January 2022).

Author Contributions: Conceptualization, S.V. and P.R.; methodology, S.V., P.R. and Z.Š.; software, Z.Š. and P.R.; validation, Z.Š. and P.R.; formal analysis, Z.Š.; investigation, Z.Š.; resources, M.V. and P.R.; data curation, Z.Š.; writing—original draft preparation, Z.Š. and P.R.; writing—review and editing, Z.Š., P.R., S.V. and M.V.; visualization, Z.Š.; supervision, P.R., S.V. and M.V.; project administration, Z.Š.; funding acquisition, Z.Š., S.V. and M.V. All authors have read and agreed to the published version of the manuscript.

Funding: This research received no external funding.

Institutional Review Board Statement: Not applicable.

Informed Consent Statement: Not applicable.

Data Availability Statement: The data, code, and instructions are freely available at https://github.com/zsverko/Code_CPCC.git (accessed on 10 January 2022).

Conflicts of Interest: The authors declare no conflict of interest.

Abbreviations

The following abbreviations are used in this manuscript:

absCPCC	absolute value of complex Pearson correlation coefficient
CPCC	complex Pearson correlation coefficient
DTI	diffusion tensor imaging
EC	eyes closed
EEG	electroencephalography
EO	eyes open
HT	Hilbert transform
imCPCC	imaginary component of complex Pearson correlation coefficient
MEG	magnetoencephalography
MRI	magnetic resonance imaging
PLV	phase locking value
PLI	phase lag index
wPLI	weighted phase lag index

References

1. Fornito, A.; Zalesky, A.; Bullmore, E. *Fundamentals of Brain Network Analysis*, 1st ed.; Academic Press: London, UK, 2016; pp. 1–3.
2. Sanei, S.; Chambers, J.A. *EEG Signal Processing*, 1st ed.; John Wiley & Sons: Hoboken, NJ, USA, 2016; pp. 1–8.
3. Olejarczyk, E.; Marzetti, L.; Pizzella, V.; Zappasodi, F. Comparison of connectivity analyses for resting state EEG data. *J. Neural Eng.* **2017**, *14*, 036017. [CrossRef] [PubMed]

4.   Koch, M.A.; Norris, D.G.; Hund-Georgiadis, M. An investigation of functional and anatomical connectivity using magnetic resonance imaging. *Neuroimage* **2002**, *16*, 241–250. [CrossRef] [PubMed]

5.   Rose, S.; Pannek, K.; Bell, C.; Baumann, F.; Hutchinson, N.; Coulthard, A.; McCombe, P.; Henderson, R. Direct evidence of intra- and interhemispheric corticomotor network degeneration in amyotrophic lateral sclerosis: An automated MRI structural connectivity study. *Neuroimage* **2012**, *59*, 2661–2669. [CrossRef] [PubMed]

6.   Lim, S.; Han, C.E.; Uhlhaas, P.J.; Kaiser, M. Preferential detachment during human brain development: Age-and sex-specific structural connectivity in diffusion tensor imaging (DTI) data. *Cereb. Cortex* **2015**, *25*, 1477–1489. [CrossRef]

7.   Singer, W. Neuronal synchrony: A versatile code for the definition of relations? *Neuron* **1999**, *24*, 49–65. [CrossRef]

8.   Varela, F.; Lachaux, J.P.; Rodriguez, E.; Martinerie, J. The brainweb: Phase synchronization and large-scale integration. *Nat. Rev. Neurosci.* **2001**, *2*, 229–239. [CrossRef]

9.   Fries, P. A mechanism for cognitive dynamics: Neuronal communication through neuronal coherence. *Trends Cogn. Sci.* **2005**, *9*, 474–480. [CrossRef]

10.   Fries, P. Rhythms for cognition: Communication through coherence. *Neuron* **2015**, *88*, 220–235. [CrossRef]

11.   Siegel, M.; Donner, T.H.; Engel, A.K. Spectral fingerprints of large-scale neuronal interactions. *Nat. Rev. Neurosci.* **2012**, *13*, 121–134. [CrossRef]

12.   Bastos, A.M.; Schoffelen, J.M. A tutorial review of functional connectivity analysis methods and their interpretational pitfalls. *Front. Syst. Neurosci.* **2016**, *9*, 175. [CrossRef]

13.   Sakkalis, V. Review of advanced techniques for the estimation of brain connectivity measured with EEG/MEG. *Comput. Biol. Med.* **2011**, *41*, 1110–1117. [CrossRef]

14.   Hwa, R.C.; Ferree, T.C. Scaling properties of fluctuations in the human electroencephalogram. *Phys. Rev. E* **2002**, *66*, 021901. [CrossRef]

15.   Li, L.; Zhang, J.X.; Jiang, T. Visual working memory load-related changes in neural activity and functional connectivity. *PLoS ONE* **2011**, *6*, e22357. [CrossRef]

16.   Stam, C.J.; Van Dijk, B.W. Synchronization likelihood: An unbiased measure of generalized synchronization in multivariate data sets. *Phys. D* **2002**, *163*, 236–251. [CrossRef]

17.   Stam, C.J.; Nolte, G.; Daffertshofer, A. Phase lag index: Assessment of functional connectivity from multi channel EEG and MEG with diminished bias from common sources. *Hum. Brain Mapp.* **2007**, *28*, 1178–1193. [CrossRef]

18.   Lachaux, J.P.; Rodriguez, E.; Martinerie, J.; Varela, F.J. Measuring phase synchrony in brain signals. *Hum. Brain Mapp.* **1999**, *8*, 194–208. [CrossRef]

19.   Vinck, M.; Oostenveld, R.; Van Wingerden, M.; Battaglia, F.; Pennartz, C.M. An improved index of phase-synchronization for electrophysiological data in the presence of volume-conduction, noise and sample-size bias. *Neuroimage* **2011**, *55*, 1548–1565. [CrossRef]

20.   Schreier, P.J. A unifying discussion of correlation analysis for complex random vectors. *IEEE Trans. Signal Process* **2008**, *56*, 1327–1336. [CrossRef]

21.   Ma, G.; Li, L. Depth and structural index estimation of 2D magnetic source using correlation coefficient of analytic signal. *J. Appl. Geophy.* **2013**, *91*, 9–13. [CrossRef]

22.   Fuhrmann, D.R.; San Antonio, G. Transmit beamforming for MIMO radar systems using signal cross-correlation. *IEEE Trans. Aerosp. Electron. Syst.* **2008**, *44*, 171–186. [CrossRef]

23.   Makita, S.; Kurokawa, K.; Hong, Y.J.; Miura, M.; Yasuno, Y. Noise-immune complex correlation for optical coherence angiography based on standard and Jones matrix optical coherence tomography. *Biomed. Opt. Express* **2016**, *7*, 1525–1548. [CrossRef]

24.   Welch, P. The use of fast Fourier transform for the estimation of power spectra: A method based on time averaging over short, modified periodograms. *IEEE Trans. Audio Electroacoust.* **1967**, *15*, 70–73. [CrossRef]

25.   Tass, P.; Rosenblum, M.G.; Weule, J.; Kurths, J.; Pikovsky, A.; Volkmann, J.; Schnitzler, A.; Freund H.J. Detection of n:m Phase Locking from Noisy Data: Application to Magnetoencephalography. *Phys. Rev. Lett.* **1998**, *81*, 3291–3294. [CrossRef]

26.   Mäkinen, V.; Tiitinen, H.; May, P. Auditory event-related responses are generated independently of ongoing brain activity. *NeuroImage* **2005**, *24*, 961–968. [CrossRef]

27.   Šverko, Z.; Sajovic, J.; Drevenšek, G.; Vlahinić, S.; Rogelj, P. Generation of Oscillatory Synthetic Signal Simulating Brain Network Dynamics. In Proceedings of the 2021 44th International Convention on Information, Communication and Electronic Technology (MIPRO), MEET—Microelectronics, Electronics and Electronic Technology, Opatija, Croatia, 27 September–1 October 2021.

28.   Torkamani Azar, M.; Kanik, S.D.; Aydin, S.; Cetin, M. Prediction of reaction time and vigilance variability from spatio-spectral features of resting-state EEG in a long sustained attention task. *IEEE J. Biomed. Health Inform.* **2020**, *24*, 2550–2558. [CrossRef]

29.   Fraschini, M.; Pani, S.M.; Didaci, L.; Marcialis, G.L. Robustness of functional connectivity metrics for EEG-based personal identification over task-induced intra-class and inter-class variations. *Pattern Recognit. Lett.* **2019**, *125*, 49–54. [CrossRef]

30.   Hardmeier, M.; Hatz, F.; Bousleiman, H.; Schindler, C.; Stam, C.J.; Fuhr, P. Reproducibility of functional connectivity and graph measures based on the phase lag index (PLI) and weighted phase lag index (wPLI) derived from high resolution EEG. *PLoS ONE* **2014**, *9*, e108648. [CrossRef]

31.   Hu, L.; Zhang, Z. *EEG Signal Processing and Feature Extraction*; Springer: Singapore, 2019; pp. 241–266.

32.   Sarmukadam, K.; Bitsika, V.; Sharpley, C.F.; McMillan, M.M.; Agnew, L.L. Comparing different EEG connectivity methods in young males with ASD. *Behav. Brain Res.* **2020**, *383*, 112482. [CrossRef]

33. Huang, H.; Zhang, J.; Zhu, L.; Tang, J.; Lin, G.; Kong, W.; Lei, X.; Zhu, L. EEG-Based Sleep Staging Analysis with Functional Connectivity. *Sensors* **2021**, *21*, 1988. [CrossRef]

34. Caicedo-Acosta, J.; Castaño, G.A.; Acosta-Medina, C.; Alvarez-Meza, A.; Castellanos-Dominguez, G. Deep Neural Regression Prediction of Motor Imagery Skills Using EEG Functional Connectivity Indicators. *Sensors* **2021**, *21*, 1932. [CrossRef]

35. Hag, A.; Handayani, D.; Pillai, T.; Mantoro, T.; Kit, M.H.; Al-Shargie, F. EEG Mental Stress Assessment Using Hybrid Multi-Domain Feature Sets of Functional Connectivity Network and Time-Frequency Features. *Sensors* **2021**, *21*, 6300. [CrossRef] [PubMed]

36. Katmah, R.; Al-Shargie, F.; Tariq, U.; Babiloni, F.; Al-Mughairbi, F.; Al-Nashash, H. A Review on Mental Stress Assessment Methods Using EEG Signals. *Sensors* **2021**, *21*, 5043. [CrossRef]

37. Formoso, M.A.; Ortiz, A.; Martinez-Murcia, F.J.; Gallego, N.; Luque, J.L. Detecting Phase-Synchrony Connectivity Anomalies in EEG Signals. Application to Dyslexia Diagnosis. *Sensors* **2021**, *21*, 7061. [CrossRef] [PubMed]

38. Anzolin, A.; Toppi, J.; Petti, M.; Cincotti, F.; Astolfi, L. SEED-G: Simulated EEG Data Generator for Testing Connectivity Algorithms. *Sensors* **2021**, *21*, 3632. [CrossRef] [PubMed]

39. Chikara, R.K.; Ko, L.-W. Modulation of the Visual to Auditory Human Inhibitory Brain Network: An EEG Dipole Source Localization Study. *Brain Sci.* **2019**, *9*, 216. [CrossRef] [PubMed]

40. Perez-Ortiz, C.X.; Gordillo, J.L.; Mendoza-Montoya, O.; Antelis, J.M.; Caraza, R.; Martinez, H.R. Functional Connectivity and Frequency Power Alterations during P300 Task as a Result of Amyotrophic Lateral Sclerosis. *Sensors* **2021**, *21*, 6801. [CrossRef] [PubMed]

41. García-Murillo, D.G.; Alvarez-Meza, A.; Castellanos-Dominguez, G. Single-Trial Kernel-Based Functional Connectivity for Enhanced Feature Extraction in Motor-Related Tasks. *Sensors* **2021**, *21*, 2750. [CrossRef]

42. Vecchio, F.; Pappalettera, C.; Miraglia, F.; Alù, F.; Orticoni, A.; Judica, E.; Cotelli, M.; Pistoia, F.; Rossini, P.M. Graph Theory on Brain Cortical Sources in Parkinson's Disease: The Analysis of 'Small World' Organization from EEG. *Sensors* **2021**, *21*, 7266. [CrossRef]

 *sensors*

Article

An Exploratory EEG Analysis on the Effects of Virtual Reality in People with Neuropathic Pain Following Spinal Cord Injury

Yvonne Tran [1,*,†], Philip Austin [2,†], Charles Lo [3], Ashley Craig [4,5], James W. Middleton [4,5], Paul J. Wrigley [4,6] and Philip Siddall [2,4]

1. Department of Linguistics, Macquarie University Hearing, Macquarie University, Sydney, NSW 2109, Australia
2. Department of Pain Management, HammondCare, Greenwich Hospital Greenwich, Sydney, NSW 2065, Australia; paustin@hammond.com.au (P.A.); phil.siddall@sydney.edu.au (P.S.)
3. Management Disciplinary Group, Wentworth Institute of Higher Education, Surrey Hills, NSW 2010, Australia; charles.lo@win.edu.au
4. Sydney Medical School-Northern, Faculty of Medicine and Health, The University of Sydney, Sydney, NSW 2006, Australia; a.craig@sydney.edu.au (A.C.); james.middleton@sydney.edu.au (J.W.M.); paul.wrigley@sydney.edu.au (P.J.W.)
5. John Walsh Centre for Rehabilitation Research, Kolling Institute, Northern Sydney Local Health District, St Leonards, NSW 2065, Australia
6. Pain Management Research Institute, Kolling Institute, Northern Sydney Local Health District, St Leonards, NSW 2065, Australia
* Correspondence: yvonne.tran@mq.edu.au
† These authors contributed equally to this work.

Citation: Tran, Y.; Austin, P.; Lo, C.; Craig, A.; Middleton, J.W.; Wrigley, P.J.; Siddall, P. An Exploratory EEG Analysis on the Effects of Virtual Reality in People with Neuropathic Pain Following Spinal Cord Injury. *Sensors* **2022**, *22*, 2629. https://doi.org/10.3390/s22072629

Academic Editor: Juan Pablo Martínez

Received: 10 March 2022
Accepted: 27 March 2022
Published: 29 March 2022

Publisher's Note: MDPI stays neutral with regard to jurisdictional claims in published maps and institutional affiliations.

Abstract: Neuropathic pain in people with spinal cord injury is thought to be due to altered central neuronal activity. A novel therapeutic intervention using virtual reality (VR) head-mounted devices was investigated in this study for pain relief. Given the potential links to neuronal activity, the aim of the current study was to determine whether use of VR was associated with corresponding changes in electroencephalography (EEG) patterns linked to the presence of neuropathic pain. Using a within-subject, randomised cross-over pilot trial, we compared EEG activity for three conditions: no task eyes open state, 2D screen task and 3D VR task. We found an increase in delta activity in frontal regions for 3D VR with a decrease in theta activity. There was also a consistent decrease in relative alpha band (8–12 Hz) and an increase in low gamma (30–45 Hz) power during 2D screen and 3D VR corresponding, with reduced self-reported pain. Using the nonlinear and non-oscillatory method of extracting fractal dimensions, we found increases in brain complexity during 2D screen and 3D VR. We successfully classified the 3D VR condition from 2D screen and eyes opened no task conditions with an overall accuracy of 80.3%. The findings in this study have implications for using VR applications as a therapeutic intervention for neuropathic pain in people with spinal cord injury.

Keywords: EEG; brain activity; virtual reality; neuropathic pain; spinal cord injury; fractal dimension

1. Introduction

Spinal cord injury (SCI) is a life-changing event that causes not only a debilitating loss of sensorimotor and autonomic functions but is also associated with numerous secondary conditions. One prevalent secondary condition is chronic pain, with research showing over 50% of patients reporting more than one pain type and the pain often described as unrelenting and excruciating [1,2]. For people with SCI, neuropathic pain (NP) has been reported to be as common as musculoskeletal pain [2]. At the injury level, SCI NP is thought to result from altered central neuronal activity, with hyperexcitable neurones having exaggerated responses to stimuli; however, below the level of injury, the mechanisms are less clear [3]. The neurophysiological responses are thought to generate abnormal pain impulses back to the brain. SCI also leads to the reorganisation of the primary somatosensory cortex, which is associated with abnormal patterns of firing in the cortex and thalamus, known as thalamocortical dysrhythmia (TCD) [4], and is proposed as a mechanism underlying the generation of neuropathic pain and other neurological symptoms [5,6].

Given these complex mechanisms, involving structural and functional changes in central pain pathways at multiple levels of the neuroaxis, current treatments provide only partial and often unsatisfactory pain relief [7]. As such, alternative therapeutic approaches, such as virtual reality (VR), are now being examined [8] where advancement in technology offers an alternative treatment for a number of medical and psychological conditions and procedures [9–11]. VR is a simulated creation of a 3D environment using computer technology [12]. Current VR systems include head-mounted devices (HMD) with 3D-enabled glasses, noise-cancelling headphones for sound and head and/or body-tracking sensors in addition to devices such as joysticks and data gloves [13]. Together, this forms a realistic multisensory experience that surrounds the user, generating strong feelings of "presence", a subjective sensation of being in another place [14].

Several pilot studies using a variety of 3D HMD and 2D screen-based VR applications have shown a reduction in NP in people with SCI pain in over two-thirds of participants [15–17]. Such encouraging findings suggest that VR may be an effective, accessible, and inexpensive method of reducing NP in both the long and short term. Recent evidence suggests that, compared to 2D VR, 3D VR technologies are more realistic and vivid [18], where the three-dimensional perception of an image or video is considered more immersive where users feel completely involved [19].

Although clinical studies in people with SCI-related NP have shown promise for the effectiveness of VR, the neural mechanism underlying the positive response to VR is unknown. In previous studies, there is evidence for neural mechanisms underlying VR immersion. From electroencephalography (EEG) studies, task-related differences in EEG alpha activity and coherence were correlated with spatial presence [20]. Frontal-midline theta activity increases were found from different levels of immersion in VR applications [21].

Given that the brain activity of SCI people with NP has been found to be associated with resting-state EEG [6,22,23], we were interested in examining underlying brain activity changes in persons with SCI and NP during VR intervention. Current studies have demonstrated brain activity markers for NP, specifically, increases in theta- and beta-wave frequencies and reduced alpha-wave frequencies in EEG signals, thought to be associated with TCD [6,22,23]. Another study from Vuckovic and colleagues showed that these EEG frequency changes can be used to identify patients with SCI who are at risk of developing NP before physical symptoms appear [24]. These EEG markers of NP have also been shown to be reversible following treatments to reduce SCI pain, where, for example, Hasan and colleagues showed significant reductions in beta- and theta-wave frequencies following biofeedback treatment [22]. Additionally, recent pilot studies investigating EEG and VR encouragingly show (a) decreases in beta-wave frequencies in response to VR in people with anxiety [23], and (b) in a case report, increases in alpha-wave frequency during phantom limb pain relief in people with brachial plexus injury during VR [25].

Thus, the aim of the current study was to determine whether use of VR is associated with corresponding changes in EEG patterns linked to the presence of neuropathic pain. We hypothesised that using a 3D VR application would be associated with a shift of EEG activity from a TCD brain wave pattern towards a reduced TCD state and thus a reduction in the severity of NP. We examined brain activity in three states, a resting eyes-open state with no task (EO-no task), using a 2D screen-based VR (2D screen) as an active control and during immersive 3D HMD VR (3D VR).

2. Materials and Methods

2.1. Study Design

We used a randomised cross-over study design for this exploratory study. This involved two sequential VR interventions and a baseline measure using within participant comparisons. Baseline measure was taken for EEG comparisons involving an eyes-open condition, whereby participants were asked to remain still and focus on the middle of a blank computer screen. There were two VR interventions, one utilising an immersive 3D VR and one with 2D screen applications using the same virtual environment. Seventeen adults with SCI and known NP were recruited using convenience sampling. We randomly allocated the type of VR intervention used first and second using sequentially numbered, opaque sealed envelopes. As it was important to show parity in describing both interventions, a script using neutral language was prepared. This study was registered by the Australia New Zealand Clinical Trials Registry, number ACTRN12618000959279, in May 2018, and further detail on the exploratory trial can be found in Austin et al., 2020 [26].

2.2. Participants

Participants were recruited from both a database of participants with SCI as well as through clinical contact. The inclusion criteria were adult males with SCI of longer than 12 months duration, lesion at C6 level or below, a confirmed diagnosis of NP (>6 months), reported neuropathic pain over the previous week prior to attending interventions, and stable pharmacological or no pharmacological treatment for at least four weeks. We limited the study to male participants only as they account for the majority of new SCI cases (up to 80%) and because of potential gender differences in pain reporting and medication use [27]. The exclusion criteria were the presence of other pain types that were more prominent during the time of the interventions, a SCI level higher than C6, presence of brain injury, or other neurological diagnosis.

2.3. Study Schedule

All participants attended the intervention on one occasion. To account for circadian influences on wakefulness in the brain activity of people with SCI, all participants were asked to attend at 11 a.m. Baseline pain intensity measures were taken with an 11-point numerical pain rating scale (NPRS). We examined average, worst, least and current NP intensities. Current NP intensity were taken immediately after the intervention and used for the analysis. As we used a cross-over design, we included a washout period in the experimental design. This was implemented to reduce any potential carryover effect that may be from the effects of the first intervention. The washout period separated the two intervention periods. Washout periods need to be at least five times the half-life of a given treatment [28], so a 60 min washout period was used. The hour-long washout was calculated from reports that pain significantly reduces immediately after VR exposures but returns to baseline levels at 10 min after VR exposure [29]. The cross-over was counterbalanced so that exposure to both interventions were equal. The entire study took place in a temperature-controlled room maintained at 25 °C. Details on the intervention protocol can be found in Austin et al. (2020) [26]. The height of the bench for the screens was modified for wheelchair access and adjusted appropriately for each person. Participants were required to report any headset discomfort and cyber-sickness (includes symptoms of nausea, vomiting, headache, vertigo and fatigue) prior to, during or after using the 3D HMD VR device.

2.4. 3D HDM VR Device and Task

The Oculus Rift® headset is commercially available, inexpensive and commonly used for VR studies in medical research [30]. The screen sampling rate was 80 Hz. For the VR task in this study, participants viewed a 3D VR experience called Nature Trek® that includes nine nature environments all containing many types of animals and calming music. Prior to use, participants were instructed on the use of a hand-held joystick to move around an alpine meadow environment and make full use of the 360° scene. The VR application was standardised across the group. The VR headset was calibrated for participants' eyesight in addition to advice on motion sickness prevention during VR such as reducing the speed of their character and/or reducing head movement.

2.5. 2D Screen Application

The same Nature Trek® application was run on a 17.3-inch Alienware® laptop screen with the participant seated in the same position. This allowed for a reliable comparison between the effects of 3D VR and 2D screen experiences. The screen sampling rate was 60 Hz.

2.6. Self-Reported Pain Measures

We used the numerical pain rating scale (NPRS) to investigate the effects of 3D HMD VR and 2D screen applications on SCI NP. The NPRS was completed at three time points to gather pain information for baseline and VR interventions. Participants completed the 11-point NPRS after each intervention and reported levels of pain intensity immediately after the intervention, as well as reporting their average pain intensity during each intervention and lowest pain intensity during each intervention. The 11-point NPRS is a reliable and valid measure used across many pain populations [31]. Changes in pain intensity from these VR interventions has been reported in our previous feasibility study (see Austin et al. [26]).

2.7. EEG Recording and Preprocessing

Thirty-two EEG channels were measured using the EmotivPro® system over the entire cortex, following the International 10–20 Montage System. EEG was recorded using the EmotivPRO® software and was sampled at a 256 Hz sample rate with left and right ear (A1, A2) references. Once fitted, the Oculus Rift VR HMD system was placed over the top of the EEG cap (Figure 1). Two minutes of baseline EEG was taken. During baseline, participants were instructed to remain still and focus on the middle of a blank computer screen to avoid eye movement. They were asked to fixate on a printed cross placed in the middle of the screen. The VR interventions were each 15 min in length.

EEG pre-processing was conducted in the following steps.

1. The EEG signals were re-sampled to 128 Hz from EmotivePRO®
2. EEG signals were imported into EEGLab
3. All channels were transformed to the average reference in EEGlab.
4. EEG signals were filtered with a 0.1 Hz high-pass filter.
5. EEG signals were visually inspected so that 20 s segments relatively free of major artifact could be extracted from the three conditions EO-no task, 2D screen and 3D VR. EEG from EO-no task was taken at approximately the 1 min point, after participant's EEG had settled and both 2D screen and 3D VR were taken at approximately the 5 min mark during the immersive task, where the first 20 s of relatively clean EEG segment could be obtained.
6. With the 20 s EEG segment, EEG artifact from blinks, eye saccades, lateral eye movements and cardiac signal components were analysed using independent component analysis (ICA) using EEGLab [32]. Artifact components were visually inspected and manually removed. Fast frequency and EMG (electromyography) noise was then removed using the Automatic artifact removal (AAR) plugin for EEGLab [33].

2.8. Spectral Analysis

To determine the corresponding changes in EEG patterns linked to the use of VR in those with neuropathic pain, we started with an examination of EEG spectral activity using a robust power spectral estimation method following Melman and Victor (2016), which utilises a multi-taper method [34]. We used this method to ensure that any noise from the VR system does not influence the spectral EEG activity, as it is more resistant to transient artifacts. Spectral analysis quantifies the amount of oscillatory activity in the different frequencies, and we examined the relative power for the widely accepted frequency bands: the delta (1–4 Hz), theta (4–8 Hz), alpha (8–12 Hz), beta (12–30 Hz) and low gamma (30–45 Hz) waves. Relative power for the spectral bands was calculated as the power of each given band divided by the sum power from 1 to 45 Hz. To ensure that signals were not affected by the VR headsets (that is, electrode sites where the VR device band sat on top) and representative of the regions of the scalp, we chose nine channels that covered the frontal region (F3, Fz, F4), central region (C3, Cz, C4) and parietal region (P3, Pz, P4). The conditions were baseline eyes open (EO), 2D screen and 3D VR.

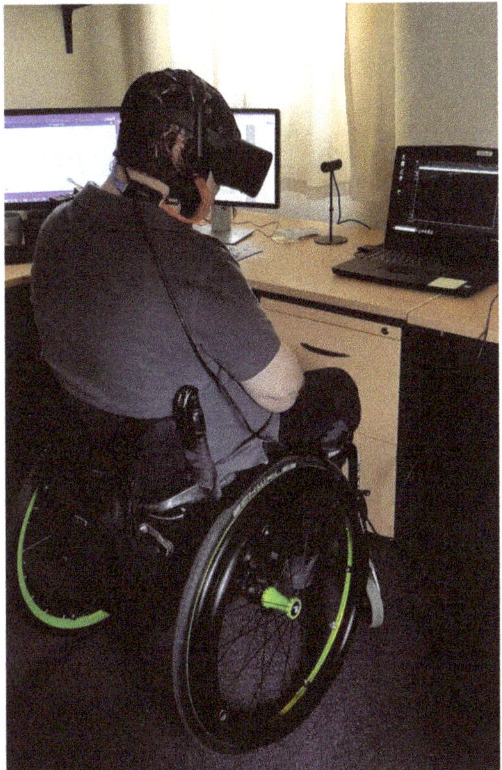

Figure 1. Participant with EEG and 3D HMD VR set up.

2.9. Fractal Dimension Analysis

We also explored a nonlinear and non-oscillatory EEG feature using Fractal Dimension (FD). Whereas spectral analysis explored oscillatory markers and different frequency bands, using FD, we could examine non-oscillatory markers for the different immersion levels from VR interventions. The FD of an EEG signal measures its complexity, that is, the amount of irregularity within the time series. We explored the FD of data from individual EEG channels using a method used most with EEG signals that was introduced by Higuchi (1988) [35]. The Higuchi's FD is a straightforward method that can be applied to time series data in order to extract the fractal dimension.

Suppose we have a time series:

$$X(i)(i = 1, \ldots . N)$$

From this, the length of the curve $L_m(k)$, for $m = 1, \ldots . k$ can be defined as follows:

$$L_m(k) = \frac{1}{k}\{(\sum_{i=1}^{[\frac{N-m}{k}]} |X(m+ik) - X(m+(i-1)k|)\frac{N-1}{k[\frac{N-m}{k}]}\}$$

where the square brackets [] denotes Gauss' notation, both m and k are integers, k indicates the discrete time interval and m indicates the initial time value.

The length of the curve for the time interval k is then defined as:

$$L(k) = \frac{1}{k} \sum_{m=1}^{k} L_m(k)$$

If the curve contains fractal properties, then $L(k)$ is proportional to k^{-D}, where D is the fractal dimension. The value of the time interval is varied from $k = 1, 2, 3$ up to k_{max}.

A log–log plot of $L(k)$ against k will give a straight line with slope $-D$.

From numerical analysis, a choice of $k_{max} = 6$ was found to sufficiently estimate the slope. The FD for this study was calculated over time in a sliding non-overlapping window with a fixed length.

2.10. Artificial Neural Network Analysis and Evaluation Metrics

We used the artificial neural network (ANN) analysis from the SPSS v27 toolbox (SPSS Inc., Chicago, IL, USA) to demonstrate whether neural differences existed in brain activity between three immersion conditions that are distinct and can be classified using ANN models. We used a multilayer perceptron (MLP) ANN model with three-layer feedforward back propagation. EEG spectral and FD data were randomly divided into training (70%) and testing (30%) sets. A hyperbolic tangent function was used for the hidden and output layer. A gradient descent was used to estimate the synaptic weights. The initial learning rate was set as 0.4 with a momentum of 0.9. We performed these models for binary classifications. For performance of the classifications, we used the receiver operating characteristics (ROC) analysis as a measure of predictive accuracy. We also used well-known performance indicators sensitivity or true positive rate (TPR), specificity or true negative rate (TNR) and accuracy, obtained from the testing sample. These were calculated as follows:

$$Sensitivity = \frac{TP}{TP + FN}$$

$$Specificity = \frac{TN}{TN + FP}$$

$$Accuracy = \frac{TP + TN}{TP + TN + FP + FN}$$

3. Results

One participant was excluded from all analysis due to reporting no pain over the previous week before the study. Another participant was excluded from EEG analysis as they had poor and corrupted EEG signals. All participants reported no cybersickness following the VR interventions. Table 1 shows the participants' demographic characteristics including age, duration in years since SCI, level and extent of SCI, pain consistency and prescribed pain medication.

3.1. Evidence of Improvement in Pain Scores from Participating in the Tasks

The mean pain intensity scores over the week prior to their attendance, during and after 2D screen and 3D VR interventions were examined. Repeated measures ANOVA showed overall significant differences in pain ratings for all three conditions (pre-task, 2D screen and 3D VR), $F (2, 14) = 46.6$, $p < 0.001$. Post hoc analysis using the Bonferroni test showed significant reductions within participants from pre-task to interventions ($p < 0.001$), with mean (95% CI) pain ratings of 4.9 (4.1–5.8) for pre-task, 3.4 (2.4–4.4) for 2D screen and 1.9 (1.0–2.9) for 3D VR.

Table 1. Participant demographic characteristics (*n* = 15).

Characteristics		*n* (%)
Age	Mean (SD)	56.0 (13.1)
Years in pain since injury	Mean (SD)	20.0 (12.3)
ASIA impairment grade	A	10 (66.7)
	B	2 (13.3)
	C	0 (0.0)
	D	3 (20.0)
Injury level	Paraplegia	13 (86.7)
	Tetraplegia	2 (13.3)
Pain consistency	Constant	12 (80.0)
	Intermittent	3 (20.0)
Pain medication	Multiple	4 (26.7)
	Single	6 (40)
	None	5 (33.3)

3.2. Regional Differences in Relative Power for the Three Conditions

Repeated measures MANOVAs were conducted to test for differences in the spectral relative power of the three conditions for the three regions, frontal (F3, Fz, F4), central (C3, Cz, C4) and parietal (P3, Pz, P4). Tables 2–6 shows the relative power (Mean (SE)) breakdowns for each of the frequency bands from the nine sites in the three conditions. Univariate main effects from repeated-measures ANOVAs were used to determine any statistical differences between the three conditions. For the frontal region, EEG differences were found in the relative delta (Wilks' Lambda = 0.52, F (6, 52) = 3.34, *p* = 0.007, η^2_P = 0.28), theta (Wilks' Lambda = 0.56, F (6, 52) = 2.92, *p* = 0.016, η^2_P = 0.25 alpha (Wilks' Lambda = 0.47, F (6, 52) = 4.0, *p* = 0.002, η^2_P = 0.31) and gamma (Wilks' Lambda = 0.58, F (6, 52) = 2.72, *p* = 0.023, η^2_P = 0.24 frequencies. Post hoc test using Bonferroni found significant increases in delta activity in the F3 and F4 sites with the 3D VR condition. A significant reduction in frontal theta was found in the Fz site, whereas a reduction in alpha activity was found in F3 and F4, and was greatest during 3D VR. There were statistical differences in the beta band, but significant increases in gamma activity were found in the F3 site with greatest increase for the 2D screen condition. For the central region, relative power difference was found for the theta frequency only (Wilks' Lambda = 0.57, F (6, 52) = 2.77, *p* = 0.020, η^2_P = 0.24), with significant decreases for the VR interventions compared with resting EO. The alpha frequency band did not show an overall significant difference between the three conditions, despite significant differences in the univariate main effects. In the parietal region, relative power differences were significant for the alpha (Wilks' Lambda= 0.58, F (6, 52) = 2.75, *p* = 0.021, η^2_P = 0.24), theta (Wilks' Lambda = 0.54, F (6, 52) = 3.09, *p* = 0.012, η^2_P = 0.26) and gamma (Wilks' Lambda = 0.46, F (6, 52) = 4.09, *p* = 0.002, η^2_P = 0.32) frequency bands. There were significant reductions in relative theta and alpha power for the 3D VR condition. Gamma activity increases were also found to be significantly greater for the 2D screen condition.

Table 2. Relative EEG delta power for the three conditions EO-no task, 2D screen and 3D VR.

	Three Conditions			Main Effect between Conditions		
Channel	EO-No Task Mean (SE)	2D Screen Mean (SE)	3D VR Mean (SE)	F (2, 28)	*p*	η^2_P
F3	0.370 (0.018)	0.338 (0.011)	0.422 (0.010)	8.852	0.001	0.387
F4	0.358 (0.018)	0.355 (0.019)	0.397 (0.015)	2.674	0.087	0.160
Fz	0.366 (0.019)	0.349 (0.013)	0.365 (0.015)	0.373	0.692	0.026
C3	0.325 (0.016)	0.341 (0.017)	0.374 (0.019)	2.628	0.090	0.158
C4	0.329 (0.023)	0.342 (0.014)	0.352 (0.018)	0.539	0.589	0.037
Cz	0.348 (0.014)	0.363 (0.018)	0.375 (0.018)	0.769	0.473	0.052

Table 2. *Cont.*

Channel	Three Conditions			Main Effect between Conditions		
	EO-No Task Mean (SE)	**2D Screen Mean (SE)**	**3D VR Mean (SE)**	**F (2, 28)**	*p*	$\eta^2{}_P$
P3	0.346 (0.013)	0.326 (0.011)	0.359 (0.019)	1.487	0.243	0.096
P4	0.322 (0.013)	0.344 (0.013)	0.338 (0.019)	0.525	0.597	0.036
Pz	0.352 (0.015)	0.365 (0.015)	0.388 (0.016)	2.051	0.147	0.128

SE = Standard error, $\eta^2{}_P$ = Partial eta squared.

Table 3. Relative EEG theta power for the three conditions EO-no task, 2D screen and 3D VR.

Channel	Three Conditions			Main Effect between Conditions		
	EO-No Task Mean (SD)	**2D Screen Mean (SD)**	**3D VR Mean (SD)**	**F (2, 28)**	*p*	$\eta^2{}_P$
F3	0.169 (0.009)	0.166 (0.010)	0.132 (0.013)	8.031	0.002	0.365
F4	0.180 (0.013)	0.159 (0.012)	0.149 (0.013)	2.507	0.100	0.152
Fz	0.192 (0.011)	0.165 (0.012)	0.155 (0.011)	3.165	0.058	0.184
C3	0.191 (0.010)	0.167 (0.013)	0.145 (0.011)	5.135	0.013	0.268
C4	0.191 (0.012)	0.152 (0.012)	0.172 (0.011)	4.261	0.024	0.233
Cz	0.196 (0.008)	0.173 (0.014)	0.172 (0.013)	2.214	0.128	0.137
P3	0.203 (0.009)	0.172 (0.009)	0.164 (0.012)	8.059	0.002	0.365
P4	0.201 (0.011)	0.158 (0.011)	0.148 (0.012)	10.484	<0.001	0.428
Pz	0.180 (0.007)	0.147 (0.011)	0.148 (0.014)	4.274	0.024	0.234

SE = Standard error, $\eta^2{}_P$ = Partial eta squared.

Table 4. Relative EEG alpha power for the three conditions EO-no task, 2D screen and 3D VR.

Channel	Three Conditions			Main Effect between Conditions		
	EO-No Task Mean (SD)	**2D Screen Mean (SD)**	**3D VR Mean (SD)**	**F (2, 28)**	*p*	$\eta^2{}_P$
F3	0.126 (0.013)	0.111 (0.009)	0.074 (0.006)	10.625	<0.001	0.431
F4	0.115 (0.009)	0.107 (0.014)	0.081 (0.008)	6.952	0.004	0.332
Fz	0.114 (0.009)	0.098 (0.009)	0.088 (0.003)	3.056	0.063	0.179
C3	0.139 (0.015)	0.109 (0.011)	0.090 (0.008)	4.490	0.020	0.243
C4	0.146 (0.018)	0.117 (0.012)	0.106 (0.010)	4.764	0.017	0.254
Cz	0.132 (0.013)	0.109 (0.012)	0.096 (0.008)	4.520	0.020	0.244
P3	0.128 (0.012)	0.112 (0.010)	0.102 (0.012)	1.501	0.240	0.097
P4	0.144 (0.014)	0.101 (0.009)	0.100 (0.007)	5.897	0.007	0.296
Pz	0.127 (0.010)	0.099 (0.008)	0.089 (0.011)	6.715	0.004	0.324

SE = Standard error, $\eta^2{}_P$ = Partial eta squared.

Table 5. Relative EEG beta power for the three conditions EO-no task, 2D screen and 3D VR.

Channel	Three Conditions			Main Effect between Conditions		
	EO-No Task Mean (SD)	**2D Screen Mean (SD)**	**3D VR Mean (SD)**	**F (2, 28)**	*p*	$\eta^2{}_P$
F3	0.157 (0.012)	0.175 (0.012)	0.138 (0.007)	3.583	0.041	0.204
F4	0.166 (0.013)	0.175 (0.012)	0.148 (0.013)	1.818	0.181	0.115
Fz	0.157 (0.013)	0.184 (0.012)	0.160 (0.013)	1.835	0.178	0.116
C3	0.186 (0.011)	0.188 (0.009)	0.174 (0.016)	0.598	0.557	0.041
C4	0.174 (0.012)	0.180 (0.010)	0.183 (0.014)	0.300	0.743	0.021
Cz	0.166 (0.008)	0.173 (0.013)	0.161 (0.013)	0.325	0.725	0.023
P3	0.158 (0.008)	0.190 (0.008)	0.175 (0.012)	4.022	0.029	0.223
P4	0.181 (0.009)	0.181 (0.011)	0.199 (0.021)	0.673	0.518	0.046
Pz	0.169 (0.010)	0.179 (0.011)	0.150 (0.012)	2.251	0.124	0.139

SE = Standard error, $\eta^2{}_P$ = Partial eta squared.

Table 6. Relative EEG gamma power for the three conditions EO-no task, 2D screen and 3D VR.

Channel	Three Conditions			Main Effect between Conditions		
	EO-No Task Mean (SD)	2D Screen Mean (SD)	3D VR Mean (SD)	F (2, 28)	p	$\eta^2{}_P$
F3	0.058 (0.007)	0.085 (0.009)	0.061 (0.005)	7.584	0.002	0.351
F4	0.066 (0.008)	0.082 (0.008)	0.066 (0.007)	1.953	0.161	0.122
Fz	0.060 (0.008)	0.082 (0.007)	0.073 (0.008)	2.798	0.078	0.167
C3	0.061 (0.005)	0.082 (0.008)	0.074 (0.007)	3.904	0.032	0.218
C4	0.061 (0.009)	0.076 (0.008)	0.077 (0.010)	1.323	0.283	0.086
Cz	0.056 (0.005)	0.073 (0.10)	0.066 (0.007)	1.884	0.171	0.119
P3	0.060 (0.006)	0.089 (0.007)	0.078 (0.009)	9.533	<0.001	0.405
P4	0.059 (0.006)	0.085 (0.009)	0.086 (0.009)	7.017	0.003	0.334
Pz	0.064 (0.007)	0.080 (0.009)	0.068 (0.008)	2.108	0.140	0.131

SE = Standard error, $\eta^2{}_P$ = Partial eta squared.

Figure 2 shows the overall EEG power spectrum for the three conditions in three EEG channels Fz, Cz and Pz, representative of the frontal, central and parietal regions. An increase in delta activity occurred for the VR interventions compared with baseline EO condition. The greatest reduction in the theta band was observed for the 3D VR condition. The reduction in the alpha frequency band was gradual, with the greatest reduction during the 3D VR intervention. Increases at higher frequencies occurred in the gamma frequencies (30–45 Hz). This can be seen with 2D screen and 3D VR, with greater increases during the 2D screen task.

3.3. Regional Differences in Higuchi's FD for the Three Conditions

Repeated measures MANOVA also was conducted to test for differences in FD for three regions, frontal (F3, Fz, F4), central (C3, Cz, C4) and parietal (P3, Pz, P4). There were no significant differences in the FD for the three conditions in the frontal region. There were significant FD differences in brain activity for both the central region (Wilks' Lambda = 0.21, F (6, 9) = 5.6, p = 0.011) and the parietal region (Wilks' Lambda = 0.27, F (6,9) = 4.1, p = 0.028). Post hoc analysis using Bonferonni found differences between the eyes-open with both 2D screen and 3D VR, but not between 2D screen and 3D VR. The mean percent change for 2D screen from baseline EO was 4.73%, and the mean percent change for 3D VR from baseline EO was 5.08% (See Table 7 for summary statistics). Figure 3 shows the FD by EEG channel for each of the three conditions. Compared with the eyes-open task, both 2D screen and 3D VR conditions displayed raised FD.

3.4. Performance of ANN model for Classifying 3D VR Using EEG Activity

The final three-layer model consisted of 6-6-2 feedforward back propagation. Table 8 shows the performance of the ANN model for each binary classification explored. For the classification of 3D VR against both EO-no task and 2D screen, we obtained an overall accuracy of 80.3%. However, sensitivity for this model was low at only 43.8%. The highest sensitivity was between 3D VR with EO-no task at 78%. The differences in the sensitivity, specificity and accuracy demonstrate that the neural activity during 3D VR was distinct and can be classified.

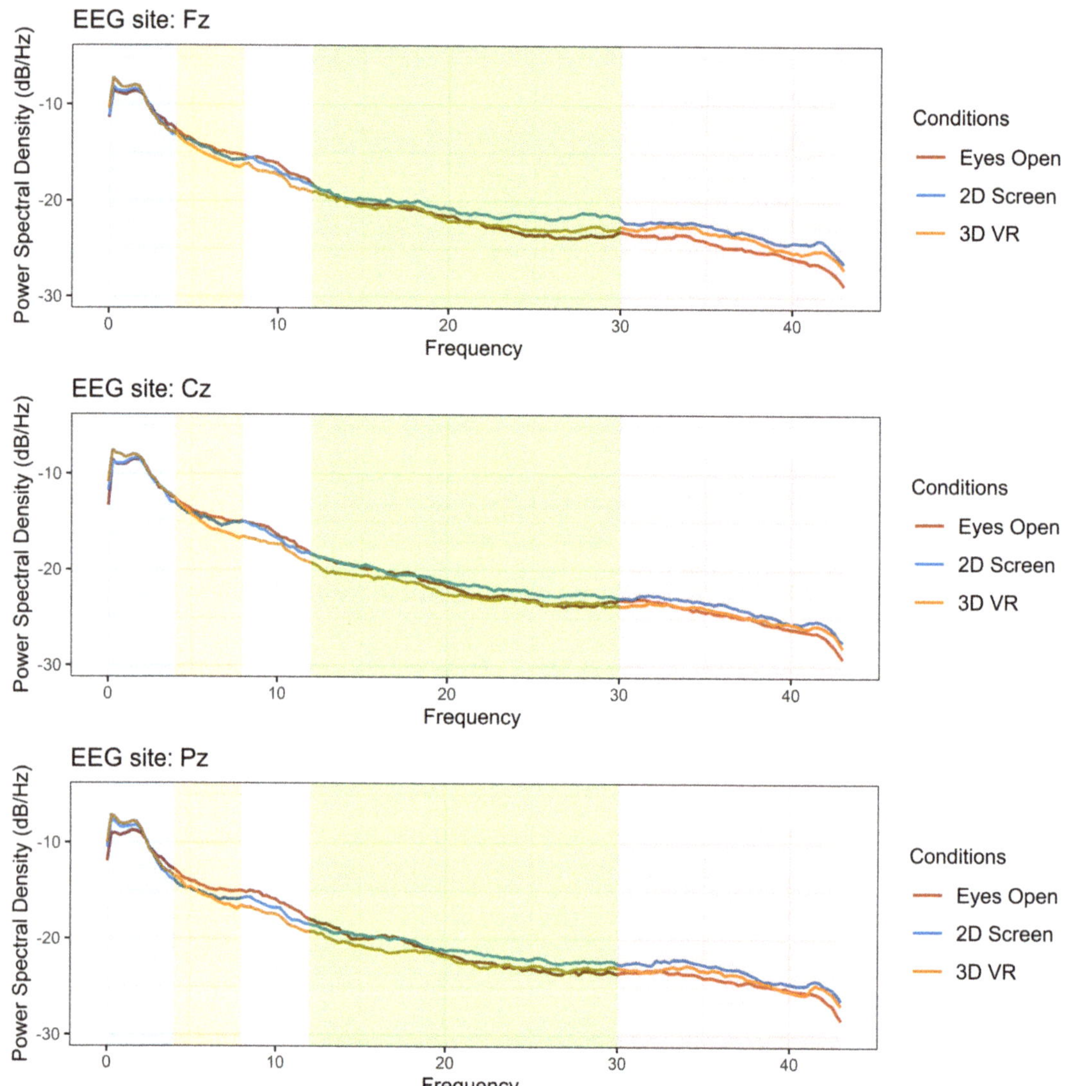

Figure 2. EEG average power spectrum for *n* = 15 participant during eyes-open task (red line), 2D screen (blue line) and 3D VR (orange line) in the Fz, Cz and Pz channels. Blue shade = Delta (1–4 Hz), Yellow shade = Theta (4–8 Hz), Gray shade = Alpha (8–12 Hz), Green shade = Beta (12–30 Hz) and Pink shade = Gamma (30–45 Hz).

Table 7. Fractal dimensions in nine EEG sites for the three conditions EO-no task, 2D screen and 3D VR.

Channel	Three Conditions			% Change from EO		Main Effect	
	EO-No Task Mean (SD)	2D Screen Mean (SD)	3D VR Mean (SD)	2D screen	3D VR	F (2, 13)	p
F3	1.58 (0.09)	1.66 (0.11)	1.65 (0.12)	5.06	4.43	3.25	0.07
F4	1.61 (0.11)	1.65 (0.12)	1.66 (0.10)	2.48	3.11	1.45	0.27
Fz	1.56 (0.08)	1.67 (0.09)	1.66 (0.09)	7.05	6.41	9.87	0.002
C3	1.57 (0.06)	1.65 (0.09)	1.67 (0.08)	5.10	6.37	18.13	<0.001
C4	1.58 (0.10)	1.64 (0.09)	1.65 (0.09)	3.80	4.43	4.81	0.03
Cz	1.56 (0.09)	1.63 (0.11)	1.62 (0.09)	4.49	3.84	4.62	0.03
P3	1.58 (0.10)	1.64 (0.09)	1.65 (0.09)	3.80	4.43	4.81	0.03
P4	1.57 (0.11)	1.66 (0.09)	1.67 (0.09)	5.73	6.37	8.39	0.005
Pz	1.57 (0.09)	1.65 (0.10)	1.67 (0.12)	5.10	6.37	10.20	0.002

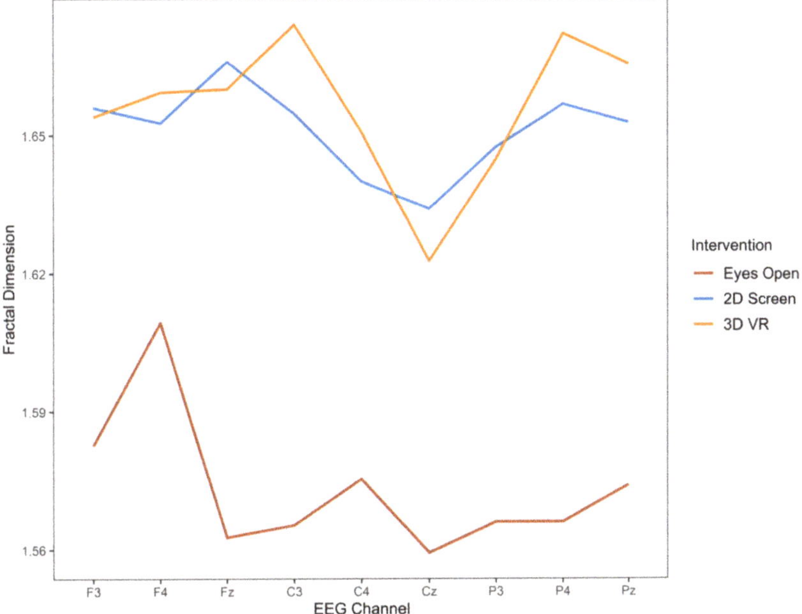

Figure 3. The fractal dimension from EEG activity during three conditions, no-eyes-open task (red), 2D screen task (blue) and 3D VR task (orange).

Table 8. Performance of binary classifications using ANN models.

Binary Classification	Sensitivity (%)	Specificity (%)	Accuracy (%)	ROC
3D VR vs. 2D screen + EO	43.8	94.1	80.3	0.767
3D VR vs. EO	78.0	79.5	78.8	0.877
3D VR vs. 2D screen	68.3	73.8	71.1	0.744
2D screen vs. EO	66.7	76.7	72.4	0.730

4. Discussion

The occurrence of NP in people with a SCI is thought to have a neural basis that is both complex and multilevel. Current available treatments have only provided partial and often unsatisfactory pain relief [36]. In this study, we explored a novel alternative therapeutic approach of utilising VR to reduce NP in people with SCI while examining the underlying brain activity during the interventions. As reported in our previous paper [26], our findings showed a significant reduction in self-reported pain intensity ratings in participants with SCI for their NP during VR interventions. The reduction in pain intensity was greatest during the 3D VR task. This demonstrates that VR interventions can be viable alternative therapeutic interventions for NP in persons with SCI.

We then examined whether there were any corresponding neural changes during the VR intervention. The results from this study found a reduction in EEG theta in the frontal and parietal regions. The decrease in relative theta was greatest for the 3D VR intervention. A significant and consistent reduction in relative EEG alpha frequency band was found in almost all EEG sites for 3D VR and 2D screen. The full EEG spectrum averaged for all participants shows the reduction to be gradual in the frontal and parietal regions and based on the level of immersion, such that reduced alpha power was greatest during 3D VR. These EEG changes for the three conditions partially supports our hypothesis that EEG changes will shift in the direction towards reduced TCD with decreased theta activity and increased alpha activity. A reduction in theta activity was observed during VR interventions; however, rather than an increase in alpha activity to counter the TCD, we found a further decrease in alpha frequency power during VR application. Similarly, Jensen et al. (2013) confirmed TCD EEG patterns in chronic pain for SCI (increased theta and reduced alpha), but they also found significant associations between pain severity and EEG alpha wave activity, with higher alpha activity associated with increased levels of pain [37]. They concluded that successful pain suppression may be associated with decreased frontal alpha activity, and this was confirmed in the current study.

The reduction in EEG alpha power, with increases in delta and low gamma activity, may also be associated with the "distraction" or immersion effects of VR applications with 3D VR thought to have the greatest degree of immersion [38]. Lim et al. 2019 found alpha waves to decrease during concentration and immersion [39]. Similarly, in a study with cancer pain and VR immersion, it was low frequency power in theta and alpha frequency ranges that were found to decrease during VR meditation task compared to their pre-condition [40]. Other EEG changes such as frontal delta activity increases has also been found to be associated with concentration during cognitive tasks [41], and it is thought that this link is moderated by motivation [42]. We found increases in frontal delta activity, and this only occurred during the 3D VR condition and not during the 2D screen. Low gamma activity increases were highest for the 2D screen condition in the frontal region but were the same as in the 3D VR in the parietal region. Gamma activity has often been linked to cognitive function or processes [43]. Increases in gamma activity are thought to be related to perception [44].

We were also interested in non-oscillatory EEG markers for NP and effects of VR interventions. FD is a nonlinear measure for complexity in brain signals. The FD of EEG signals have repeatedly been shown to be of a lower value for people suffering from brain disorders compared to healthy individuals [45]. Anderson and colleagues (2021) found able-bodied participants to have higher FD compared to SCI participants with neuropathic pain and used FD as a diagnostic marker for NP [45]. Foss and colleagues (2006) were able to differentiate between different pain states from FD values. They found FD to be lowest for thermal pain and greatest for back pain [46]. Using Higuchi's FD, we found significant increases in FD for the central and parietal regions during 2D screen and 3D VR compared with the EO-no task. However, there were no distinguishable differences between the FD for 2D screen and 3D VR. The increase in FD for 2D screen and 3D VR may be showing changes in neural signals, demonstrating a normalisation of the affected thalamocortical system. Higher FD values generally correspond to higher signal complexity, and a reduction in FD may indicate a loss of neural efficiency, as previously found in people with Alzheimer's disease [47,48].

Although we did not find significant differences for 2D screen and 3D VR in FD, we were able to distinguish the two brain activity states using ANN. Using both oscillatory and non-oscillatory measures for our feature set, classification of 2D screen against 3D

VR had an accuracy of 68.3%. Classifying 3D VR from both 2D screen and EO-no task gave an overall accuracy of 80.3%; however, this was largely from the high-specificity (true negative) result. Sensitivity was only at 43.8%. This low sensitivity was probably due to the gradual changes from level of immersion between baseline EO to 2D VR to 3D VR, making it difficult to detect 3D VR from a mixture of immersion levels. Classifying 3DVR from baseline EO had highest predictive accuracy with an ROC of 0.877. The results match the findings from the spectral bands, in that the gradual changes in immersion levels are reflected with the sensitivity analysis. Sensitivity was highest between baseline EO and 3D VR. Sensitivity between baseline EO with 2D screen and 3D VR with 2D screen were similar at 66.7% and 68.6%, respectively. The results from the ANN models indicate that brain activity during 3D VR immersion is distinct and can be classified; the ROC shows reasonable predictive accuracy.

5. Strengths and Limitations

The findings in this study have implications for using VR applications as a therapeutic intervention for NP in people with SCI and our understanding of the mechanisms responsible for VR-associated pain relief. Both changes in alpha and theta wave activity have been demonstrated in association with SCI-related NP. The significant reduction in self-reported pain intensity after the intervention was found to correspond to significant changes in EEG brain activity. The changes seem to suggest that two pathways may be occurring during VR intervention. There is some evidence that VR-associated pain relief is associated with a remediation or reversal of TCD through reductions in theta activity, but there is also evidence for a possible attention-related mechanism involving alpha activity, delta activity in the frontal cortex, and low gamma activity in the parietal region. The strengths in this study include the use of both oscillatory and non-oscillatory methods in EEG signal processing to understand the underlying neural mechanisms during VR immersion. As to limitations, we applied a common average reference strategy for the EEG signals to first remove any common noise. However, common average referencing can introduce bias and lower amplitudes of the signal when the coverage of electrodes is not dense enough. Our study only utilized 32 channels, and the recommended electrode density is at least 64 channels (Nunez, 2006). As such, there may have been bias in the amplitudes of the signals in this study. Additionally, as this study is a preliminary exploratory examination, we did not examine post-VR session effects, and we are unable to determine if the effects on pain reduction are able to persist longer than the VR session. However, a more recent study using the same VR distraction protocols in people with cancer pain did show pain relief remained for up to 20 min after the VR sessions [49]. Encouragingly, recent work also shows that more frequent use of cognitive-based VR applications over weeks or months in people with NP and phantom limb pain in combination with other non-pharmacological therapies offers both long-term relief in pain intensity and decreases in pain-related behaviours [50,51]. Future research should be conducted with a larger sample with focus on longer-term outcomes to test for cumulative effects from VR interventions. A larger study should also examine different baseline pain intensity levels to test whether this intervention can be used in people with high levels of pain.

Author Contributions: P.S., A.C., J.W.M., P.J.W. and Y.T. initiated the project partnership, conceptualized the project and obtained funding. P.A. contributed to the design of study protocols, screening and recruitment of subjects, administering questionnaires, organizing clinic visits, performance of VR, collection of EEG signals. Y.T. and P.A. drafted the first draft of the paper. Y.T. conducted the EEG analysis and statistical analysis. C.L. conducted the data wrangling and setup of EEG data for analysis. Y.T., P.A., C.L., A.C., J.W.M., P.J.W. and P.S. all made substantial contributions to the interpretation of data. All authors have read and agreed to the published version of the manuscript.

Funding: This study funded by the Australian and New Zealand College of Anaesthetists, reference 19/002.

Institutional Review Board Statement: The study was conducted in accordance with the Declaration of Helsinki and approved by The Northern Sydney Local Health District Research Ethics Committee, reference RESP/18/133.

Informed Consent Statement: Informed consent was obtained from all subjects involved in the study.

Data Availability Statement: The datasets generated and/or analysed during the current study are available from the corresponding author on reasonable request.

Conflicts of Interest: The authors declare no conflict of interest.

References

1. Masri, R.; Keller, A. Chronic pain following spinal cord injury. *Adv. Exp. Med. Biol.* **2012**, *760*, 74–88. [PubMed]
2. Kim, H.Y.; Lee, H.J.; Kim, T.L.; Kim, E.; Ham, D.; Lee, J.; Kim, T.; Shin, J.W.; Son, M.; Sung, J.H.; et al. Prevalence and Characteristics of Neuropathic Pain in Patients With Spinal Cord Injury Referred to a Rehabilitation Center. *Ann. Rehabil. Med.* **2020**, *44*, 438–449. [CrossRef] [PubMed]
3. Hadjipavlou, G.; Cortese, A.M.; Ramaswamy, B. Spinal cord injury and chronic pain. *BJA Educ.* **2016**, *16*, 264–268. [CrossRef]
4. Gerke, M.B.; Duggan, A.W.; Xu, L.; Siddall, P.J. Thalamic neuronal activity in rats with mechanical allodynia following contusive spinal cord injury. *Neuroscience* **2003**, *117*, 715–722. [CrossRef]
5. Sarnthein, J.; Stern, J.; Aufenberg, C.; Rousson, V.; Jeanmonod, D. Increased EEG power and slowed dominant frequency in patients with neurogenic pain. *Brain* **2006**, *129 Pt 1*, 55–64. [CrossRef]
6. Llinás, R.R.; Ribary, U.; Jeanmonod, D.; Kronberg, E.; Mitra, P.P. Thalamocortical dysrhythmia: A neurological and neuropsychiatric syndrome characterized by magnetoencephalography. *Proc. Natl. Acad. Sci. USA* **1999**, *96*, 15222–15227. [CrossRef]
7. Jensen, T.S.; Madsen, C.S.; Finnerup, N.B. Pharmacology and treatment of neuropathic pains. *Curr. Opin. Neurol.* **2009**, *22*, 467–474. [CrossRef]
8. Austin, P.D.; Siddall, P.J. Virtual reality for the treatment of neuropathic pain in people with spinal cord injuries: A scoping review. *J. Spinal Cord Med.* **2021**, *44*, 8–18. [CrossRef]
9. Park, M.J.; Kim, D.J.; Lee, U.; Na, E.J.; Jeon, H.J. A Literature Overview of Virtual Reality (VR) in Treatment of Psychiatric Disorders: Recent Advances and Limitations. *Front. Psychiatry* **2019**, *10*, 505. [CrossRef]
10. Oing, T.; Prescott, J. Implementations of Virtual Reality for Anxiety-Related Disorders: Systematic Review. *JMIR Serious Games* **2018**, *6*, e10965. [CrossRef]
11. Niki, K.; Okamoto, Y.; Maeda, I.; Mori, I.; Ishii, R.; Matsuda, Y.; Takagi, T.; Uejima, E. A Novel Palliative Care Approach Using Virtual Reality for Improving Various Symptoms of Terminal Cancer Patients: A Preliminary Prospective, Multicenter Study. *J. Palliat. Med.* **2019**, *22*, 702–707. [CrossRef]
12. Pourmand, A.; Davis, S.; Marchak, A.; Whiteside, T.; Sikka, N. Virtual Reality as a Clinical Tool for Pain Management. *Curr. Pain Headache Rep.* **2018**, *22*, 53. [CrossRef]
13. Li, L.; Yu, F.; Shi, D.; Shi, J.; Tian, Z.; Yang, J.; Wang, X.; Jiang, Q. Application of virtual reality technology in clinical medicine. *Am. J. Transl. Res.* **2017**, *9*, 3867–3880.
14. Servotte, J.-C.; Goosse, M.; Campbell, S.H.; Dardenne, N.; Pilote, B.; Simoneau, I.L.; Guillaume, M.; Bragard, I.; Ghuysen, A. Virtual Reality Experience: Immersion, Sense of Presence, and Cybersickness. *Clin. Simul. Nurs.* **2020**, *38*, 35–43. [CrossRef]
15. Eick, J.; Richardson, E.J. Cortical activation during visual illusory walking in persons with spinal cord injury: A pilot study. *Arch. Phys. Med. Rehabil.* **2015**, *96*, 750–753. [CrossRef]
16. Jordan, M.; Richardson, E.J. Effects of Virtual Walking Treatment on Spinal Cord Injury-Related Neuropathic Pain: Pilot Results and Trends Related to Location of Pain and at-level Neuronal Hypersensitivity. *Am. J. Phys. Med. Rehabil.* **2016**, *95*, 390–396. [CrossRef]
17. Villiger, M.; Estévez, N.; Hepp-Reymond, M.C.; Kiper, D.; Kollias, S.S.; Eng, K.; Hotz-Boendermaker, S. Enhanced activation of motor execution networks using action observation combined with imagination of lower limb movements. *PLoS ONE* **2013**, *8*, e72403. [CrossRef]
18. Rooney, B.; Hennessy, E. Actually in the Cinema: A Field Study Comparing Real 3D and 2D Movie Patrons' Attention, Emotion, and Film Satisfaction. *Media Psychol.* **2013**, *16*, 441–460. [CrossRef]
19. Roettl, J.; Terlutter, R. The same video game in 2D, 3D or virtual reality—How does technology impact game evaluation and brand placements? *PLoS ONE* **2018**, *13*, e0200724.
20. Kober, S.E.; Kurzmann, J.; Neuper, C. Cortical correlate of spatial presence in 2D and 3D interactive virtual reality: An EEG study. *Int. J. Psychophysiol.* **2012**, *83*, 365–374. [CrossRef]
21. Slobounov, S.M.; Ray, W.; Johnson, B.; Slobounov, E.; Newell, K.M. Modulation of cortical activity in 2D versus 3D virtual reality environments: An EEG study. *Int. J. Psychophysiol.* **2015**, *95*, 254–260. [CrossRef]
22. Hasan, M.A.; Fraser, M.; Conway, B.A.; Allan, D.B.; Vučković, A. Reversed cortical over-activity during movement imagination following neurofeedback treatment for central neuropathic pain. *Clin. Neurophysiol.* **2016**, *127*, 3118–3127. [CrossRef]
23. Tarrant, J.; Viczko, J.; Cope, H. Virtual Reality for Anxiety Reduction Demonstrated by Quantitative EEG: A Pilot Study. *Front. Psychol.* **2018**, *9*, 1280. [CrossRef]

24. Vuckovic, A.; Gallardo, V.J.F.; Jarjees, M.; Fraser, M.; Purcell, M. Prediction of central neuropathic pain in spinal cord injury based on EEG classifier. *Clin. Neurophysiol.* **2018**, *129*, 1605–1617. [CrossRef]

25. Osumi, M.; Sano, Y.; Ichinose, A.; Wake, N.; Yozu, A.; Kumagaya, S.I.; Kuniyoshi, Y.; Morioka, S.; Sumitani, M. Direct evidence of EEG coherence in alleviating phantom limb pain by virtual referred sensation: Case report. *Neurocase* **2020**, *26*, 55–59. [CrossRef]

26. Austin, P.D.; Craig, A.; Middleton, J.W.; Tran, Y.; Costa, D.S.J.; Wrigley, P.J.; Siddall, P.J. The short-term effects of head-mounted virtual-reality on neuropathic pain intensity in people with spinal cord injury pain: A randomised cross-over pilot study. *Spinal Cord* **2021**, *59*, 738–746. [CrossRef]

27. Norrbrink Budh, C.; Lund, I.; Hultling, C.; Levi, R.; Werhagen, L.; Ertzgaard, P.; Lundeberg, T. Gender related differences in pain in spinal cord injured individuals. *Spinal Cord* **2003**, *41*, 122–128. [CrossRef]

28. Evans, S.R. Clinical trial structures. *J. Exp. Stroke Transl. Med.* **2010**, *3*, 8–18. [CrossRef]

29. Jin, W.; Choo, A.; Gromala, D.; Shaw, C.; Squire, P. A Virtual Reality Game for Chronic Pain Management: A Randomized, Controlled Clinical Study. *Stud. Health Technol. Inf.* **2016**, *220*, 154–160.

30. Fowler, C.A.; Ballistrea, L.M.; Mazzone, K.E.; Martin, A.M.; Kaplan, H.; Kip, K.E.; Murphy, J.L.; Winkler, S.L. A virtual reality intervention for fear of movement for Veterans with chronic pain: Protocol for a feasibility study. *Pilot Feasibility Stud.* **2019**, *5*, 146. [CrossRef]

31. Farrar, J.T.; Young, J.P., Jr.; LaMoreaux, L.; Werth, J.L.; Poole, M.R. Clinical importance of changes in chronic pain intensity measured on an 11-point numerical pain rating scale. *Pain* **2001**, *94*, 149–158. [CrossRef]

32. Delorme, A.; Makeig, S. EEGLAB: An open source toolbox for analysis of single-trial EEG dynamics including independent component analysis. *J. Neurosci. Methods* **2004**, *134*, 9–21. [CrossRef] [PubMed]

33. Gomez-Herrero, G.; Clercq, W.D.; Anwar, H.; Kara, O.; Egiazarian, K.; Huffel, S.V.; Paesschen, W.V. Automatic Removal of Ocular Artifacts in the EEG without an EOG Reference Channel. In Proceedings of the 7th Nordic Signal Processing Symposium— NORSIG 2006, Reykjavik, Iceland, 7–9 June 2006; pp. 130–133.

34. Melman, T.; Victor, J.D. Robust power spectral estimation for EEG data. *J. Neurosci. Methods* **2016**, *268*, 14–22. [CrossRef] [PubMed]

35. Higuchi, T. Approach to an irregular time series on the basis of the fractal theory. *Phys. D Nonlinear Phenom.* **1988**, *31*, 277–283. [CrossRef]

36. Loh, E.; Mirkowski, M.; Agudelo, A.R.; Allison, D.J.; Benton, B.; Bryce, T.N.; Guilcher, S.; Jeji, T.; Kras-Dupuis, A.; Kreutzwiser, D.; et al. The CanPain SCI clinical practice guidelines for rehabilitation management of neuropathic pain after spinal cord injury: 2021 update. *Spinal Cord* **2022**. [CrossRef]

37. Jensen, M.P.; Sherlin, L.H.; Gertz, K.J.; Braden, A.L.; Kupper, A.E.; Gianas, A.; Howe, J.D.; Hakimian, S. Brain EEG activity correlates of chronic pain in persons with spinal cord injury: Clinical implications. *Spinal Cord* **2013**, *51*, 55–58. [CrossRef]

38. Souza, R.H.C.E.; Naves, E.L.M. Attention Detection in Virtual Environments Using EEG Signals: A Scoping Review. *Front. Physiol.* **2021**, *12*, 727840. [CrossRef]

39. Lim, S.; Yeo, M.; Yoon, G. Comparison between Concentration and Immersion Based on EEG Analysis. *Sensors* **2019**, *19*, 1669. [CrossRef]

40. Fu, H.; Garrett, B.; Tao, G.; Cordingley, E.; Ofoghi, Z.; Taverner, T.; Sun, C.; Cheung, T. Virtual Reality–Guided Meditation for Chronic Pain in Patients With Cancer: Exploratory Analysis of Electroencephalograph Activity. *JMIR Biomed. Eng.* **2021**, *6*, e26332. [CrossRef]

41. Harmony, T. The functional significance of delta oscillations in cognitive processing. *Front. Integr. Neurosci.* **2013**, *7*, 83. [CrossRef]

42. Knyazev, G.G. Motivation, emotion, and their inhibitory control mirrored in brain oscillations. *Neurosci. Biobehav. Rev.* **2007**, *31*, 377–395. [CrossRef]

43. Herrmann, C.S.; Fründ, I.; Lenz, D. Human gamma-band activity: A review on cognitive and behavioral correlates and network models. *Neurosci. Biobehav. Rev.* **2010**, *34*, 981–992. [CrossRef]

44. Castelhano, J.; Rebola, J.; Leitão, B.; Rodriguez, E.; Castelo-Branco, M. To Perceive or Not Perceive: The Role of Gamma-band Activity in Signaling Object Percepts. *PLoS ONE* **2013**, *8*, e66363. [CrossRef]

45. Anderson, K.; Chirion, C.; Fraser, M.; Purcell, M.; Stein, S.; Vuckovic, A. Markers of Central Neuropathic Pain in Higuchi Fractal Analysis of EEG Signals From People With Spinal Cord Injury. *Front. Neurosci.* **2021**, *15*, 705652. [CrossRef]

46. Foss, J.M.; Apkarian, A.V.; Chialvo, D.R. Dynamics of Pain: Fractal Dimension of Temporal Variability of Spontaneous Pain Differentiates Between Pain States. *J. Neurophysiol.* **2006**, *95*, 730–736. [CrossRef]

47. Smits, F.M.; Porcaro, C.; Cottone, C.; Cancelli, A.; Rossini, P.M.; Tecchio, F. Electroencephalographic Fractal Dimension in Healthy Ageing and Alzheimer's Disease. *PLoS ONE* **2016**, *11*, e0149587. [CrossRef]

48. Sebastián, M.V.; Navascués, M.A.; Otal, A.; Ruiz, C.; Idiazábal, M.Á.; Stasi, L.L.D.; Díaz-Piedra, C. Fractal Dimension as Quantifier of EEG Activity in Driving Simulation. *Mathematics* **2021**, *9*, 1311. [CrossRef]

49. Austin, P.D.; Siddall, P.J.; Lovell, M.R. Feasibility and acceptability of virtual reality for cancer pain in people receiving palliative care: A randomised cross-over study. *Support. Care Cancer* **2022**, *30*, 3995–4005. [CrossRef]

50. Donati, A.R.C.; Shokur, S.; Morya, E.; Campos, D.S.F.; Moioli, R.C.; Gitti, C.M.; Augusto, P.B.; Tripodi, S.; Pires, C.G.; Pereira, G.A.; et al. Long-Term Training with a Brain-Machine Interface-Based Gait Protocol Induces Partial Neurological Recovery in Paraplegic Patients. *Sci. Rep.* **2016**, *6*, 30383. [CrossRef]
51. Ortiz-Catalan, M.; Guðmundsdóttir, R.A.; Kristoffersen, M.B.; Zepeda-Echavarria, A.; Caine-Winterberger, K.; Kulbacka-Ortiz, K.; Widehammar, C.; Eriksson, K.; Stockselius, A.; Ragnö, C.; et al. Phantom motor execution facilitated by machine learning and augmented reality as treatment for phantom limb pain: A single group, clinical trial in patients with chronic intractable phantom limb pain. *Lancet* **2016**, *388*, 2885–2894. [CrossRef]

Article

Low-Dimensional Dynamics of Brain Activity Associated with Manual Acupuncture in Healthy Subjects

Xinmeng Guo [1,2,*] **and Jiang Wang** [1]

1   School of Electrical and Information Engineering, Tianjin University, Tianjin 300072, China; jiangwang@tju.edu.cn
2   Academy of Medical Engineering and Translational Medicine, Tianjin University, Tianjin 300072, China
*   Correspondence: guoxm@tju.edu.cn

Abstract: Acupuncture is one of the oldest traditional medical treatments in Asian countries. However, the scientific explanation regarding the therapeutic effect of acupuncture is still unknown. The much-discussed hypothesis it that acupuncture's effects are mediated via autonomic neural networks; nevertheless, dynamic brain activity involved in the acupuncture response has still not been elicited. In this work, we hypothesized that there exists a lower-dimensional subspace of dynamic brain activity across subjects, underpinning the brain's response to manual acupuncture stimulation. To this end, we employed a variational auto-encoder to probe the latent variables from multichannel EEG signals associated with acupuncture stimulation at the ST36 acupoint. The experimental results demonstrate that manual acupuncture stimuli can reduce the dimensionality of brain activity, which results from the enhancement of oscillatory activity in the delta and alpha frequency bands induced by acupuncture. Moreover, it was found that large-scale brain activity could be constrained within a low-dimensional neural subspace, which is spanned by the "acupuncture mode". In each neural subspace, the steady dynamics of the brain in response to acupuncture stimuli converge to topologically similar elliptic-shaped attractors across different subjects. The attractor morphology is closely related to the frequency of the acupuncture stimulation. These results shed light on probing the large-scale brain response to manual acupuncture stimuli.

Keywords: acupuncture; EEG; dimensionality; neural subspace; latent variables; attractor

Citation: Guo, X.; Wang, J. Low-Dimensional Dynamics of Brain Activity Associated with Manual Acupuncture in Healthy Subjects. *Sensors* **2021**, *21*, 7432. https://doi.org/10.3390/s21227432

Academic Editor: Yvonne Tran

Received: 24 September 2021
Accepted: 6 November 2021
Published: 9 November 2021

Publisher's Note: MDPI stays neutral with regard to jurisdictional claims in published maps and institutional affiliations.

1. Introduction

Acupuncture, an ancient practice in traditional Chinese medicine (TCM), is gradually being recognized throughout the world as an important modality of alternative and complementary medicine [1,2]. The World Health Organization (WHO) and the National Institutes of Health (NIH) have reported that acupuncture is an efficient treatment for various conditions, such as addiction, headaches, myofascial pain, and lower back pain [3–6]. A number of available pieces of evidence have demonstrated that acupuncture may also help with stroke rehabilitation [7]. However, the scientific explanation of acupuncture's effects is still unknown. Clinical and experimental studies have indicated that acupuncture, as a complex somatosensory stimulation of the central nervous system, can mediate the electrical activity of autonomous neuronal networks [8,9]. Furthermore, neuroimaging data strongly suggest that widely distributed cortical and subcortical brain areas are recruited during acupuncture stimulation [10,11]. For example, Bai et al. demonstrated that acupuncture can increase activity in the amygdala, the perigenual anterior cingulate cortex (pACC), the periaqueductal gray (PAG), and the hypothalamus [12]. Therefore, more attention has been focused on probing brain activities during and after acupuncture stimulation.

In addition, an electroencephalogram (EEG) is an effective method for obtaining brain electrical signals, and is able to record spontaneous cerebral activity with a time resolution at the millisecond level. It has been widely used in clinical and experimental studies to analyze brain activity associated with acupuncture stimulation. Methods of characterizing brain activity based on EEG recordings can be divided into two categories. The first category is the statistical analysis of brain oscillatory activity, such as power spectral density, complexity, and coherence [13–15]. For example, Tanaka et al. investigated the variance of EEG power induced by acupuncture. They found that acupuncture could increase EEG power in all frequency bands, and this increment remained after acupuncture [16]. Furthermore, Qi et al. quantified the approximate entropy (ApEn) of EEG signals and confirmed the variance of ApEn in the prefrontal lobe, the posterior temporal lobe, and the occipital lobe before, during and after acupuncture stimulation [17]. The other category involves constructing a functional network based on various measurement of correlation or synchronization [18,19]. Yu et al. constructed the functional network of acupuncture EEG signals based on phase synchronization and found that acupuncture at ST36 can significantly improve the synchronization of alpha rhythms and enhance the small-world connection characteristics of the brain's functional network [20,21].

Brain activity is a high-dimensional dynamical process that evolves over time, and the data analysis methods above cannot be directly associated with brain dynamics, which poses a challenge in probing the dynamic response of the brain to acupuncture stimuli. As a usual feature of complex systems, the degrees of freedom traversed by its dynamics are much lower than the number of units comprising the system [22]. The human brain is such a complex system of numerous neurons coupled through synapses. Observations in electrophysiological experiments have demonstrated that the brain has low dimensionality at different levels, from macroscopic, to the mesoscopic and microscopic scales [23–25]. Based on this perspective, several neuroscientists have focused on investigating the low-dimensional dynamics of brain. They suggest that a low-dimensional representation of brain, known as "latent variables", can afford a deeper understanding of the core principles underpinning whole-brain patterns of neural activity [26–28]. For example, Cueva et al. found that low-dimensional dynamics provide a mechanism for the brain to solve the problem of storing information across time [29]. Abbaspourazad et al. extracted the low-dimensional dynamics in both spiking and LFP recordings within the motor cortex during reach-and-grasp tasks, and addressed that the multiscale, low-dimensional motor cortical state dynamics accounted for the neural control of motor behaviors [30].

Additionally, these latent variables are explanatory variables that are not directly observed but can be identified from the data using dimensionality reduction methods. These methods transform high-dimensional data into low-dimensional representations that retain important features of interest [31]. The variational auto-encoder (VAE) method is one of dimensionality reduction methods that consists of unsupervised neural networks, in which latent variables can be learned from the original high-dimensional datasets [32]. VAE is composed of an encoder and a decoder, the former is responsible for inferring the latent variables, and the latter is designed to generate a new dataset based on latent variables. This method shows good applicability in the study of brain activity. For example, Bi et al. put forward a semi-supervised VAE method to probe low-dimensional representations of ERPs, and found that the latent variables are of good applicability in brain-controlled vehicles [33]. Furthermore, the knowledge of low dimensional dynamics extracted from video-evoked cortical responses can predict its response with high accuracy, which has the potential to explain the cortical response scientifically [34]. Li et al. utilized VAE to learn the latent variables from the multichannel EEG signals and found that emotion recognition achieves excellent performance based on the learnt latent variables [35].

At present, study on brain activity under acupuncture stimuli mostly focus on the study of rhythm, complexity, synchronization, and functional networks. However, the brain is a high-dimensional, complex system composed of numerous neurons, and the response of the brain to acupuncture stimulation is associated with many distributed coupled cortical areas. To solve the problem of high dimensionality, we proposed to apply a dimensionality reduction method to probe the latent dynamics of brain activity associated with acupuncture stimulation. Latent variables can not only reflect the lower-dimensional features of brain activity, but can also yield clues about the underlying associated neural dynamics related to the intrinsic properties of external stimuli [36,37]. Specifically, we adopted the VAE method to extract latent variables from the experimental acupuncture signals, and further explored the brain activity associated with acupuncture stimuli.

Acupuncture is a complex interventional stimulation of the human body. Multiple stimulation parameters, including the needle sensation, acupoint specificity, acupuncture manipulation, and needle duration, have relevant influences on brain activity. Acupuncture manipulation is a key factor that determines the therapeutic effect of acupuncture. It is reported that acupuncture can reduce acute lower back pain for patients, and the improvement critically depends on the acupuncture manipulation. Therefore, this work focused on investigating the instant effect of acupuncture manipulation on brain activity. It was found that the characteristics of latent dynamics are associated with acupuncture manipulation. Overall, these results can provide a theoretical support for the selection of an appropriate acupuncture frequency for patients in clinical settings, and the proposed methods have potential in exploring the effects of acupuncture on brain activity.

This paper is organized as follows. In Section 2, the experimental acupuncture procedure and the corresponding method of analysis are introduced. In Section 3, the results are presented. Finally, the discussion and conclusion are provided in Sections 4 and 5, respectively.

2. Materials and Methods

2.1. Experiment Design and EEG Recording

Twelve right-handed healthy subjects (7 female, 5 male, mean age 23 years, range 22–25 years), who had never been treated with acupuncture, participated in the acupuncture experiment. They confirmed that they had not been taking any medication in the past 30 days and had no history of mental illness. Participants were informed about the needle stimulation in the acupuncture experiments and gave written informed consent to participate in the experiment. The Institutional Review Board of Armed Police Logistics College Affiliated Hospital approved our experimental protocol (LLKYPJ2010005).

In our experiment, acupuncture was administered manually at the ST36 (Zusanli) acupoint on the left leg (shown in Figure 1a) by a licensed acupuncturist using a single-use stainless steel needle of 0.2 mm in diameter and 40 mm in length. We adopt the twirling-twisting method with different frequencies as the acupuncture manipulation method. Specifically, the needle was twirled, mainly with the thumb forward, and the twisting was within a range of 90–180° and at a certain frequency. The subjects were randomly divided into three groups (four subjects in each group), which received manual acupuncture stimulation with different twirling and twisting frequencies of 50 times/min, 100 times/min, and 150 times/min, respectively.

The experiment was carried out in a dark, quiet room. The participants were asked to keep their eyes closed and stay awake to eliminate significant electromyoelectrical disturbance. For each subject, the entire experiment lasted about 59 min. The experimental procedure was carried out as follows (shown in Figure 1b): all subjects first rested for 10 min, then the acupuncture needle was inserted by the acupuncturist to a depth of 10 mm at the ST36 acupoint until deqi. The needle was kept inserted without operation for 10 min, referred to as the pre-acupuncture state (Pre-acu). Then, the twirling-twisting operation was conducted for 3 min (acupuncture, Acu). After the operation, it was necessary to keep the subject in a resting state for 10 min (post-acupuncture, Post-acu). This procedure was repeated 3 times. Finally, after removing the needle, the acupuncturist finished the experiment.

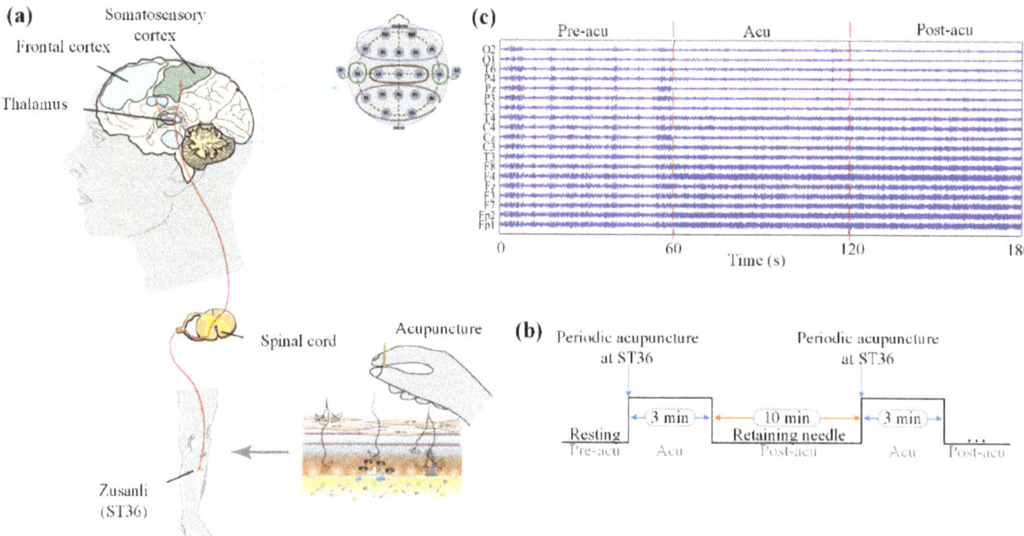

Figure 1. The schematic diagram of the experimental operation. (**a**) Schematic diagram of the acupuncture experiment. Electroencephalographic signals evoked by manual acupuncture at the ST36 acupoint of healthy subjects were directly recorded in three states: pre-acupuncture, acupuncture, and post-acupuncture. (**b**) A timeline of the detailed experimental procedure of manual acupuncture manipulation and (**c**) the EEG signals recorded.

EEG signals were recorded using a Neuroscan system with 19 Ag-AgCl electrodes, which were placed in accordance with the international standard 10–20 system. The reference electrode was located between electrodes A1 and A2, and the earlobe was used as the reference ground of the electrode. The data sampling frequency was 256 Hz, and the hardware filter passband was 0.5 Hz~100 Hz. Every subject selected a median of 1 min of EEG data of acupuncture for the elimination of the effect of the insertion or withdrawal of needle and other possible factors. For signal preprocessing, the noise in the EEG data was filtered out to extract effective data with a band-pass finite impulse digital filter with a band pass frequency ranging from 0.5 Hz to 30 Hz. Then, systematic effects which might be caused by referencing to a particular channel were removed by referencing the EEG data of each channel to the average of all channels. The EEG data after preprocessing are shown in Figure 1c.

2.2. Measurement of Dimensionality

Dimensionality, the minimal number of dimensions necessary to offer a precise representation of neural activity, is defined as [38]:

$$Dim(C) = \frac{(\mathrm{Tr}\,C)^2}{\mathrm{Tr}\,C^2} = \frac{(\sum_i \lambda_i)^2}{\sum_i \lambda_i^2},$$ (1)

where C is the covariance matrix of the activity vectors, and λ_i is the ith eigenvalue of the covariance matrix C. In this work, C is the covariance matrix of the electrical signals of 19 electrodes. $Dim(C) \in [1, 19]$, where $Dim(C) = 19$ indicates that the activity of the brain is independent and has equal variance, and $Dim(C) = 1$ demonstrates strongly correlated brain activity.

2.3. Method for Extracting Low-Dimensional Latent Variables

The variational auto-encoder (VAE) is a powerful deep learning method for extracting the latent variables from data, which occurs in a feedforward manner, consisting of symmetrical networks: the "encoder" and "decoder" (as shown in Figure 2). More specifically, the encoder is in charge of encoding the high-dimensional input into a low-dimensional representation, and the decoder is in charge of reestablishing the input data on the basis of the low-dimensional representation.

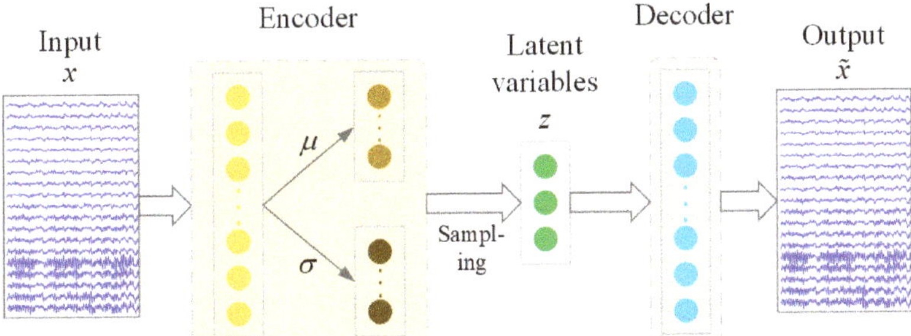

Figure 2. Neural network architectures of VAE. The encoder is in charge of encoding the high-dimensional input (x) into a low-dimensional representation (z), and the decoder is in charge of reestablishing the input data (x) on the basis of the low-dimensional representation (z).

Considering the dataset $\chi = \{x(t)\}_{t=1}^{N}$ of variable x, the VAE assumes that one random process involving an unobservable latent variable z generates all the data, which are produced from one prior distribution $p_\theta(z)$, thus x is determined by the conditional distribution $p_\theta(x|z)$ [35]. According to the Bayesian theory, the "decoder" network is in the form:

$$x \sim p_\theta(x|z)p_\theta(z), \tag{2}$$

and the "encoder" network is of the form:

$$p_\theta(z) \sim q_\phi(z|x)p(x). \tag{3}$$

The optimization function is defined based on minimizing the difference between the reconstructed data (output) and the original data (input), which is of the form:

$$\max E_{q_\phi(z|x)}[\log p_\theta(x|z)] - D_{KL}(q_\phi(z|x)\|p_\theta(z)). \tag{4}$$

According to the Monte Carlo estimation method, the first term in the equation above is calculated through sampling L times as follows:

$$E_{q_\phi(z|x)}[\log p_\theta(x|z)] = \frac{1}{L}\sum_{l=1}^{L}\log p_\theta(x(t)|z_l(t)) \tag{5}$$

The KL divergence of the approximate posterior $q_\phi(z|x)$ from the true prior $p_\theta(z)$ is computed through $-D_{KL}(q_\phi(z|x)\|p_\theta(z)) = \frac{1}{2}\sum_{j=1}^{J}\left(1 + \log\left(\sigma_j^2(t)\right) - \mu_j^2(t) - \sigma_j^2(t)\right)$, where J is the dimensionality of z.

We utilized stochastic gradient descent and a back-propagation method to optimize the unknown parameter θ and the latent variable z by minimizing the difference between the output data and the input data. In this work, the 3-min-long dataset under different states was cut into 18 10-s-long data segments; thus, the number of samples for one segment is 2560. Hence, the batch size for unsupervised VAE learning is set as 20 to balance the training speed. The VAE approach was realized through the Deep Learning Toolbox in Matlab (R2021b).

3. Results

3.1. The Oscillatory Properties of Brain Activity Evoked via Manual Acupuncture Stimulation

Brain activity is composed of high-dimensional complex oscillatory activity with rich rhythmic information. Therefore, the power spectrum density (PSD) of EEG signals was first investigated using the Welch method. Before acupuncture, the energy reaches two peaks near 1.2 Hz and 10 Hz, and the energy is mainly concentrated in the low-frequency band (1.2 Hz, the delta frequency band), as shown in Figure 3a. In the acupuncture state, the tendency of the energy distribution is similar to the pre-acupuncture state, but with a significant increment in energy in the delta and alpha frequency bands compared with the state before acupuncture. The results show that acupuncture at ST36 could affect the neural oscillatory activity, especially in the delta and alpha frequency bands.

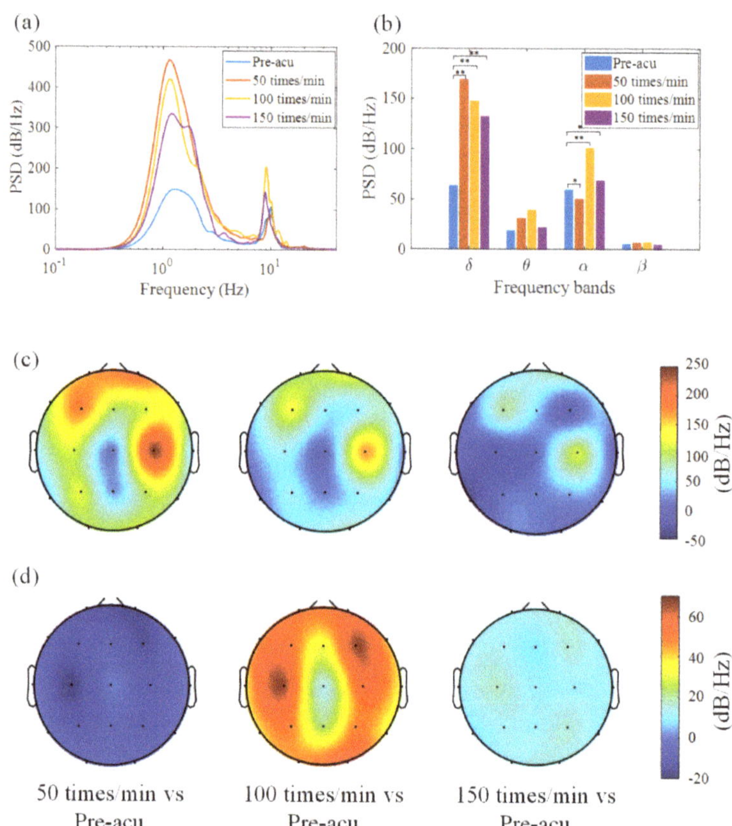

Figure 3. Brain activity associated with manual acupuncture stimulation. (**a**) Power spectrum characteristics of EEG data under different states. (**b**) PSD distribution in different frequency bands. $p < 0.05$ (*) and $p < 0.01$ (**) represent significant difference levels between pre-acupuncture and acu-puncture states. (**c**,**d**) Topographic map showing the variance of the PSD distribution between ac-tivity during and before different acupuncture manipulation states in (**c**) delta and (**d**) alpha fre-quency bands. Acupuncture can significantly affect the oscillatory activity in the delta and alpha frequency bands within EEGs. This variance is increased in the frontal and parietal lobes.

We further computed the average energy distribution across four sub-bands (delta, theta, alpha and beta), as shown in Figure 3b. Particularly, the energy in the delta frequency band was higher when the manipulation frequency was 50 times/min and 100 times/min. This phenomenon implies the emergence of resonance induced by acupuncture. As shown in Figure 3a,b, the neural activity oscillates at an inherent frequency (about 1.2 Hz). When the frequency of external stimulation comes close to this inherent frequency, the phenomenon of resonance occurs; thus, the oscillatory response in the delta band is amplified. The results indicate that the neural system may encode and transmit the acupuncture stimulus through resonance. Scientific studies have documented the experimental occurrence of resonance in electrical processes of the human brain, as recorded by EEG, elicited by mechanical tactile stimuli [39]. It can be inferred that resonance is one of the mechanisms by which the neural system encodes acupuncture stimulation.

In order to investigate the resonance effect of acupuncture on neural oscillations across brain regions, we calculated the PSD variance (the difference in the PSD value between the acupuncture state and the pre-acupuncture state). Figure 3c,d present the PSD variance in two typical frequency bands (delta and alpha). In the delta frequency band, energy in the frontal and parietal lobes is increased, especially in the left frontal lobe and the right parietal lobe. In the alpha frequency band, the energy is increased under acupuncture stimulation, except for the manipulation at 50 times/min. The findings obtained here are consistent with other experimental reports based on fMRI and PET data. Xiang et al. found that the brain regions that responded to acupuncture at ST36 only (specifically) were the inferior parietal lobe, the middle inferior gyrus, the posterior lobe of cerebellum, and the angular gyrus [40].

3.2. Dimensionality of Brain Activity

Recent research has investigated the dimensionality of neural ensembles from the sensory cortex of alert rats during periods of ongoing and stimulus-evoked activity, and found that stimuli could reduce the dimensionality of cortical activity [38]. Acupuncture is an external stimulation to the sensory system. It is of great importance to investigate whether the dimensionality of neural activity is affected by acupuncture. Figure 4a computes the dimensionality across all trials in the empirical dataset before and during acupuncture. The average dimensionality of brain activity in the pre-acupuncture state was larger than that in the acupuncture state. Moreover, the value of the dimensionality increased with an increase in the manipulation frequency. The dimensionality was minimal when the manipulation frequency was 50 times/min.

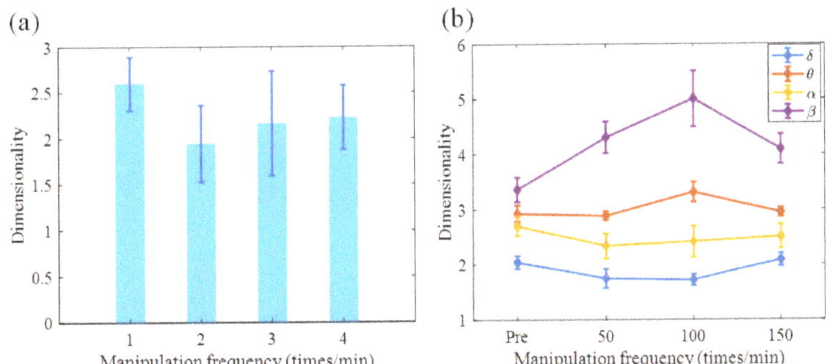

Figure 4. Dimensionality of brain activity. (**a**) Dependence of dimensionality of brain activity on acupuncture manipulation. (**b**) Dependence of dimensionality of brain activity in different sub-bands on acupuncture manipulation. Acupuncture can reduce the dimensionality of brain activity, especially with the manipulation at 50 times/min. The dimensionality in the delta and alpha frequency band was lower than that in the other two bands.

Furthermore, the dimensionality of neural activity in each sub-frequency band is explored in Figure 4b. The dimensionality in the delta and alpha frequency bands was smaller than that in the theta and beta frequency bands. In the delta and alpha frequency bands, the dimensionality was minimal when the manipulation frequency was 50 times/min, whereas in the theta and beta frequency bands, the dimensionality was maximized by acupuncture stimulation with a manipulation frequency of 100 times/min. Indeed, the oscillatory activity was more coherent in the delta and alpha frequency bands. It can be inferred that the enhancement of the correlated activity in the delta and alpha frequency bands induced by acupuncture could reduce the dimensionality of brain activity.

3.3. Low-Dimensional Dynamics of Brain Activity

Acupuncture's effects are higher-order processes that are produced by the collaborative involvement of various latent brain factors, including different brain areas and physical or functional brain networks [41]. For example, Dhond et al. have confirmed that acupuncture may exert its therapeutic effects on pain by modulating a distributed network of brain areas involved in sensory, autonomic, and cognitive/affect processing, including endogenous antinociceptive limbic networks, as well as cognitive and affective control centers within the prefrontal cortex and the medial temporal lobe [10]. Moreover, the relationships between acupuncture analgesia and attentional mechanisms have been gradually revealed [42]. As EEG results are an external manifestation of the latent brain factors' activities, it is of great importance to probe the low-dimensional dynamics of brain activity associated with acupuncture stimulation based on multichannel EEG signals.

We employed the VAE method to extract the low-dimensional latent variables from the EEGs recorded before and during acupuncture. First, the reconstruction performance of VAE under different assumed numbers of latent variables was investigated. The reconstruction performance was quantified as the mean correlation between the original and reconstructed EEG channel signals. As shown in Figure 5, the performance gradually improved with an increasing number of latent variables for all subjects. When the number of latent variables was greater than three, the model was able to obtain a reconstruction performance of more than 80% on the EEG dataset.

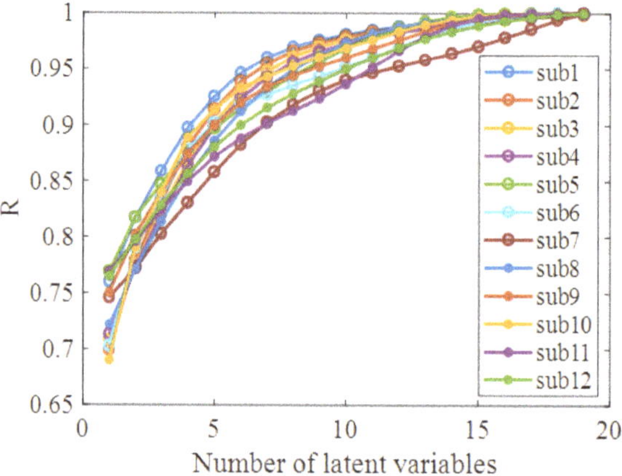

Figure 5. The reconstruction performance of VAE using different numbers of latent variables. The reconstruction performance increased with the enlargement of the latent variable number.

We further examined the dynamic properties of these latent variables extracted from the EEG dataset. For each acupuncture stimulation, we plotted the top 3 dimensions of latent variables in Figure 6. It was shown that all units in each acupuncture manipulation operation contributed to a span, which is known as a latent dynamic space. Each latent dynamic space captured a population-wide activity pattern. For different subjects, the latent factors of different states still formed a latent dynamic space, but they had different planes (Figure 6b). To test whether the neural latent dynamic spaces corresponded to different manipulation frequencies, we set the latent dynamic space formatted by the pre-acupuncture period as the reference plane (or null plane), and computed the angles between each plane (induced by each different acupuncture stimulation) and the reference plane. The measurement is depicted in Figure 6c. The statistical results shown in Figure 6d demonstrate that although the planes of different individuals varied, the angles between them and the reference plane remained unchanged with different subjects. Moreover, the angle (θ) linearly depends on the manipulation frequency with a high goodness of fit of 0.78.

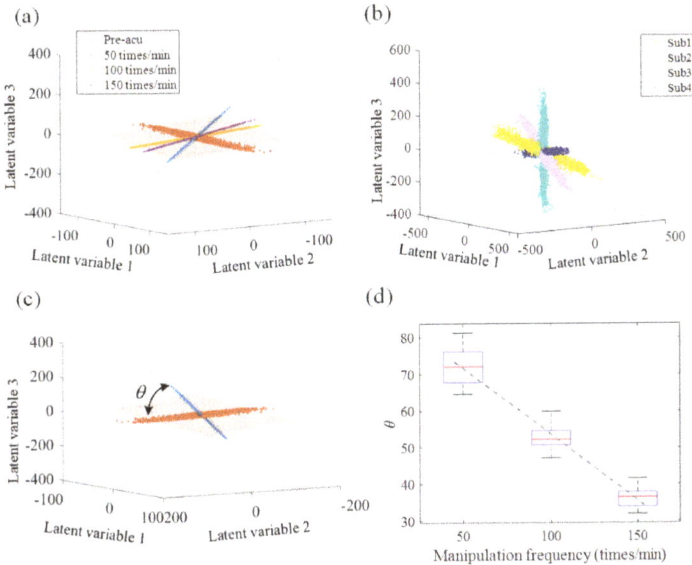

Figure 6. Latent variables in a 3-dimensional plane. (**a**) Different states of one subject. (**b**) Four ran-dom selected subjects in the pre-acupuncture period. Each color trace corresponds to a single trial. (**c**) Illustration of the variance of the latent dynamic space, where the angle between reference plane (pre-acupuncture state, orange) and acupuncture state (50 times/min, blue) was measured as θ. (**d**) Relationship between acupuncture manipulation and the plane included angle. Using VAE, the low-dimensional subspace of brain activity can be identified. The characteristics of the subspaces were determined by individuals and acupuncture stimulations.

In addition, we inspected the dynamics of the top three latent variables in each latent dynamic space, as shown in Figure 7. It was evident that the units representing time-varying activity in the neural space converged to an ellipse (defined as an attractor). The trajectory was mostly confined to the latent neural space, a plane shown in Figure 7 and spanned by the acupuncture modes p1 and p2. The arrow in each figure reflected the direction of the trajectory as it evolved over time. Intuitively, the long axes of the elliptic attractor increased. We computed the mean distance of the long and short axes across different trials and plotted them in Figure 8a. The quantitative results confirmed that the variance trends were influenced by different acupuncture manipulations. A one-way analysis of variance (ANOVA) was applied to determine whether there were any statistically significant differences in the attractors between acupuncture states. The index p was calculated based on the mean and variance of the length of the long and short axes of the elliptic attractors in each state. Table 1 indicates that the long and short axes of the attractor in each state had significant differences, where $p < 0.05$ (*) and $p < 0.01$ (**) stand for their significance levels in statistical analysis. Furthermore, the difference between p1 and p2 was calculated between any two states in Table 2. The maximum p-value was on the order of 10^{-3}, far less than 0.01. The obtained results confirmed the statistically significant differences of the attractors.

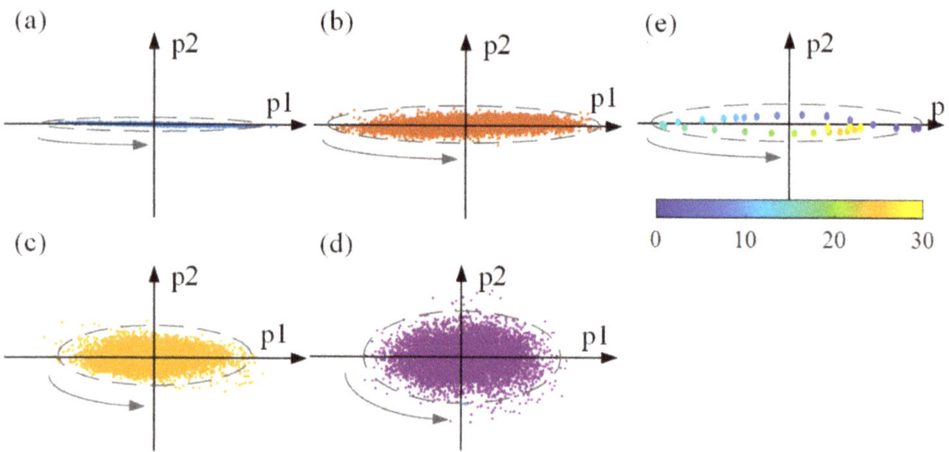

Figure 7. Schematic diagram of the trajectory under different acupuncture manipulation frequencies in latent dynamic space. (**a**) Pre-acupuncture, (**b**) 50 times/min, (**c**) 100 times/min, (**d**) 150 times/min. (**e**) An illustration exhibiting points' evolution over time in (**b**). The color labels present the time order of each point, and the time step between points is 1/256 s. It can be seen that the units representing time-varying activity in the neural space converge to an elliptic attractor.

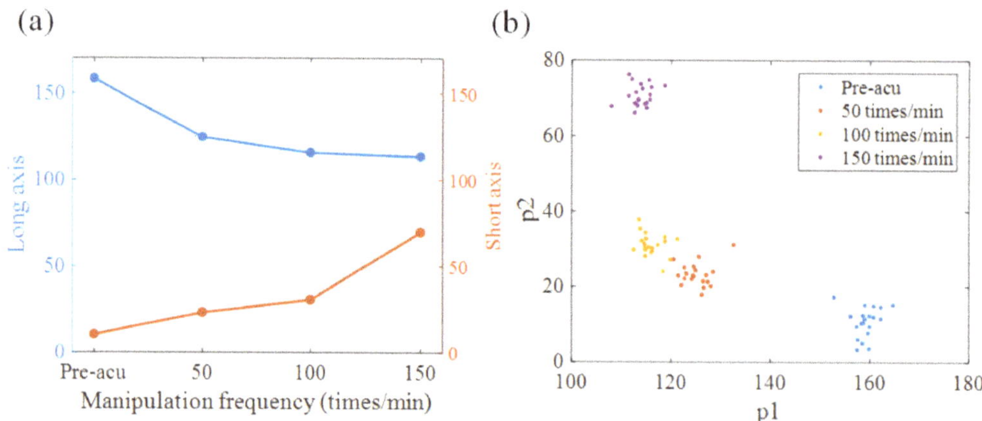

Figure 8. Statistical analysis of attractors of different states. (**a**) Dependence of long and short axes on manipulation frequency. (**b**) Cluster of manipulation operation based on attractors. The statistics of attractors can be discriminants for different brain states.

Table 1. ANOVA 1 analysis for comparison of the length of the long and short axes in different states.

Axis	Pre-Acu	50 Times/Min	100 Times/Min	150 Times/Min
p1 vs. p2	1.65×10^{-33}	5.08×10^{-29}	1.00×10^{-27}	3.07×10^{-17}

Table 2. ANOVA 1 analysis for comparison of the length of the long and short axes in different states, respectively.

Axis	Pre-Acu vs. 50 Times/Min	Pre-Acu vs. 100 Times/Min	Pre-Acu vs. 150 Times/Min	50 Times/Min vs. 100 Times/Min	50 Times/Min vs. 150 Times/Min	100 Times/Min vs. 150 Times/Min
p1	8.05×10^{-13}	1.89×10^{-28}	1.46×10^{-29}	1.28×10^{-3}	8.28×10^{-7}	2.61×10^{-3}
p2	1.34×10^{-3}	1.72×10^{-10}	7.20×10^{-28}	8.52×10^{-3}	2.03×10^{-26}	9.15×10^{-24}

Based on the different statistical characteristics of the attractors, the neural dynamics of different trials induced by different acupuncture manipulation conditions were clustered (as shown in Figure 8b). In order to automatically classify different states, four machine learning models—a support vector machine (SVM), the k-nearest neighbor (KNN) method, linear discriminant analysis (LDA), and decision trees (DTs)—were constructed. The length of the long and short axes extracted from the low-dimensional attractors were considered for the training of the classifier model. The average accuracy of the acupuncture classification was calculated by means of five-fold cross validation, conducted 10 times. Table 3 compares the mean classification accuracy obtained for these machine learning models. It indicates that all these four models were able to achieve more than 95% classification accuracy. This result suggests the universality of the proposed classification scheme based on the statistical characteristics of the attractors. Furthermore, the performance of LDA was better than that of the other three classifiers.

Table 3. Mean classification accuracy for various machine learning models.

Model	SVM	KNN	LDA	DT
Accuracy	97.5%	97.5%	98.8%	95.0%

4. Discussion

The present study was aimed at probing the low-dimensional dynamics of brain activity associated with acupuncture at the ST36 acupoint with different manipulation frequencies. Specifically, we studied the changes in the power spectrum of brain activity before and during acupuncture stimulation. We extracted the neural subspace and characterized the relationship between acupuncture stimuli and low-dimensional dynamics.

Using a manual acupuncture paradigm, in conjunction with brain electroencephalography (EEG) signal recording, we observed that acupuncture episodes were associated with increased spectral power in the delta and alpha frequency bands compared to episodes of resting, especially in the delta frequency band. This phenomenon suggests that stochastic resonance is a way in which the brain processes periodic acupuncture stimulation. Stochastic resonance is commonly understood to be the enhancement of the response of a nonlinear system in cases where the frequency of the external input is close to its intrinsic oscillatory frequency, with the help of noise [43,44]. Noise, which is ubiquitous in the brain, comes from synaptic transmission, channel gating, ion concentrations, and membrane conductance, and is possibly involved in stochastic resonance phenomena [45,46]. In the acupuncture experiment, when the stimulation frequency was close to the intrinsic frequency of the cerebral oscillations (the delta frequency band), the rhythmic activity of the cerebral oscillation was enhanced. This enhancement was mainly concentrated in the parietal lobe, which is associated with the somatosensory area. Resonance in the central nervous system of mammalians may account for their higher brain functions, such as human tactile sensations, visual perception, and animal feeding behavior [47,48]. In this study, we preliminarily found a resonant response of the brain to acupuncture stimulation. More experimental and analytical studies will be carried out to investigate the potential benefits of stochastic resonance in acupuncture information processing in the neural system.

Additionally, we found that acupuncture stimuli could reduce the dimensionality of the neural electrical response of the cerebral cortex. At present, the study of dimensionality in neural systems has attracted extensive attention [49–51]. Dimensionality analysis has been employed for various tasks and across neural systems [31,52]. For example, Rigotti et al. studied the relationship between the dimensionality of an evoked activity and task complexity, and suggested that the evoked dimensionality roughly amounted to the number of task conditions [53]. Acupuncture is a complex stimulation comprising multimodal sensory stimulations, including temperature, pressure, and noxious stimulations. Different manual acupuncture manipulations, such as lifting, thrusting, and twisting, contain different stimulating parameters, thus generating different responses to acupuncture [54]. The study of the dimensionality of brain activity under acupuncture stimuli will help to reveal the mechanisms underlying different acupuncture manipulations. Setting up an accurate experimental and theoretical connection between dimensionality and acupuncture manipulations, supported by an understanding of neural activity, is a significant question for further studies.

In this work, VAE was an efficient approach for reducing dimensionality and extracting latent variables from multichannel EEG signals. Essentially, the VAE adopted in this work was carried out in a feedforward manner, and this oversimplification of the network structure may result in lower effectiveness of VAE when the input becomes complex. One possible solution to this problem is to combine the recurrent network and VAE frameworks, which has been gradually applied in research on image recognition. In addition, the small world is a type of recurrent network with a smaller average transmission delay and more robust network connectivity. The combination of a small world network and the VAE framework may improve the processing performance for high-dimensional complex datasets and reduce the training time required.

Furthermore, using a dimensionality reduction method, we obtained a neural subspace of brain activity and found that the low-dimensional dynamics converged to topologically similar elliptic-shaped attractors. The brain state (pre-acupuncture or undergoing acupuncture with different manipulation frequencies) can be well classified based on the statistical characteristics of these attractors. The elliptical attractors implied characteristics of continuous fluctuation of the brain, which may result from internal variability (noise) and external stimuli. In a previous study [55], we observed fluctuations in the scaling of neural activity in a spontaneously active brain circuit. Olguin-Rodriguez et al. have investigated characteristic fluctuations around stable attractor dynamics extracted from highly nonstationary EEG recordings [56]. On the other hand, researchers have demonstrated that the dynamical regime of the sensory cortex converges to stable dynamics around a single stimulus-tuned attractor [57]. The attractor dynamics are not only associated with the properties of stimuli, but are associated with brain function. Finkelstein et al. showed that communication between brain regions can be gated via attractor dynamics, which control the degree of commitment to an action [58]. Therefore, it is of great importance to investigate the attractor dynamics of brain activity evoked by acupuncture stimuli, which will shed light on revealing the action mechanism of acupuncture.

Typical neural responses are shaped both by internal dynamics and various external stimuli. Even when exposed to the same external stimulation, different subjects responded differently, as their inherent internal dynamics are not quite the same. Consequently, the characteristics of low-dimensional dynamics extracted from multichannel EEG signals vary between individuals. Although differences between subjects and latent variables are informative for classification, there is still a key limitation of the proposed method, in that it cannot directly extract the stimulus-related variables from neural responses. Acupuncture can be regarded as a specific somatosensory stimulation on the acupoint, and can mediate the function of the human body via the nervous system. Furthermore, the VAE method neglects information about the relevant experimentally controlled variables. Therefore, in order to better probe the relationship between brain activity and acupuncture stimulation, we will decompose the acupuncture-evoked information from EEG signals, and further characterize the low-dimensional dynamics of acupuncture-evoked signals in the next step of our research. This further research will help to reveal the essential role of acupuncture.

As a complementary therapeutic treatment, acupuncture could improve symptoms in various neural diseases, such as depression, stroke rehabilitation, and Parkinson's disease [59–61]. Increasingly, clinical experiments have shown that the effectiveness of acupuncture is related to changes in brain activity. For example, Chae et al. documented a significant improvement in the motor function of PD patients after acupuncture treatment. The putamen and the primary motor cortex were activated when patients with PD received acupuncture treatment and these activations correlated with individual enhanced motor function [62]. Moreover, it was found that acupuncture can reduce drug addiction via direct activation of brain pathways [63]. In this work, we confirmed that acupuncture can affect the characteristics of the latent neural subspace. For different neural diseases, we proposed that abnormal brain activity may be reflected by the characteristics of this subspace as well. In future works, we will conduct further clinical experiments to validate the relationship between these latent neural dynamics and the therapeutic effects of acupuncture. These results can provide a theoretical support for the selection of appropriate acupuncture frequencies for patients in clinical settings, and the proposed methods have potential in relation to exploring the effects of acupuncture on brain activity.

5. Conclusions

In this work, the low-dimensional dynamics of brain activity associated with acupuncture stimuli was probed. We found that manual acupuncture stimuli can reduce the dimensionality of brain activity, which results from the enhancement of oscillatory activity in the delta and alpha frequency bands induced by acupuncture. Moreover, it was found that large-scale brain activity could be approximated through the dynamics of a relatively simple attractor contained within a low-dimensional neural space, and the attractor's morphology was closely related to the frequency of acupuncture stimulation. These results shed light on the large-scale brain response to manual acupuncture stimuli.

Author Contributions: Conceptualization, X.G. and J.W.; methodology, X.G. and J.W.; software, X.G. and J.W.; validation, X.G. and J.W.; formal analysis, X.G. and J.W.; investigation, X.G. and J.W.; resources, X.G. and J.W.; data curation, X.G. and J.W.; writing—original draft preparation, X.G. and J.W.; writing—review and editing, X.G. and J.W.; visualization, X.G. and J.W.; supervision, X.G. and J.W.; project administration, X.G. and J.W.; funding acquisition, X.G. and J.W. All authors have read and agreed to the published version of the manuscript.

Funding: This research was funded by the National Natural Science Foundation of China, grant numbers 62071324, 61871287, and 61471265.

Institutional Review Board Statement: The study was conducted and approved by Institutional Review Board of Armed Police Logistics College Affiliated Hospital, P. R. China (protocol code LLKYPJ2010005, date of approval: 5 May 2010).

Informed Consent Statement: Informed consent was obtained from all subjects involved in the study.

Data Availability Statement: Not applicable.

Acknowledgments: This work was supported by the National Natural Science Foundation of China (Grant Numbers: 62071324, 61871287, and 61471265).

Conflicts of Interest: The authors declare no conflict of interest.

References

1. Kaptchuk, T.J. Acupuncture: Theory, efficacy, and practice. *Ann. Intern. Med.* **2002**, *136*, 374–383. [CrossRef]
2. Han, J.S. Acupuncture and endorphins. *Neurosci. Lett.* **2004**, *361*, 258–261. [CrossRef]
3. Margolin, A.; Kleber, H.D.; Avants, S.K.; Konefal, J.; Gawin, F.; Stark, E.; Sorensen, J.; Midkiff, E.; Wells, E.; Jackson, T.R.; et al. Acupuncture for the treatment of cocaine addiction-A randomized controlled trial. *JAMA-J. Am. Med. Assoc.* **2002**, *287*, 55–63. [CrossRef]
4. Vickers, A.J.; Vertosick, E.A.; Lewith, G.; MacPherson, H.; Foster, N.E.; Sherman, K.J.; Irnich, D.; Witt, C.M.; Linde, K. Acupuncture for chronic pain: Update of an individual patient data meta analysis. *J. Pain* **2018**, *19*, 455–474. [CrossRef] [PubMed]

5. Liu, L.; Tian, T.; Li, X.; Wang, Y.; Xu, T.; Ni, X.; Li, X.; He, Z.; Gao, S.; Sun, M.; et al. Revealing the neural mechanism underlying the effects of acupuncture on migraine: A systematic review. *Front. Neurosci.* **2021**, *15*, 674852. [CrossRef]

6. Galasso, A.; Urits, I.; An, D.; Nguyen, D.; Borchart, M.; Yazdi, C.; Manchikanti, L.; Kaye, R.J.; Kaye, A.D.; Mancuso, K.F.; et al. A comprehensive review of the treatment and management of myofascial pain syndrome. *Curr. Pain Headache Rep.* **2020**, *24*, 8. [CrossRef]

7. Park, J.; Hopwood, V.; White, A.R.; Ernst, E. Effectiveness of acupuncture for stroke: A systematic review. *J. Neurol.* **2001**, *248*, 558–563. [CrossRef]

8. Wu, M.-T.; Hsieh, J.-C.; Xiong, J.; Yang, C.-F.; Pan, H.-B.; Chen, Y.-C.I.; Tsai, G.; Rosen, B.R.; Kwong, K.K. Central nervous pathway for acupuncture stimulation: Localization of processing with functional MR imaging of the brain-Preliminary experience. *Radiology* **1999**, *212*, 133–141. [CrossRef]

9. Lee, I.S.; Cheon, S.; Park, J.Y. Central and peripheral mechanism of acupuncture analgesia on visceral pain: A systematic review. *Evid.-Based Complement. Altern. Med.* **2019**, *2019*, 1304152. [CrossRef] [PubMed]

10. Dhond, R.P.; Kettner, N.; Napadow, V. Neuroimaging acupuncture effects in the human brain. *J. Altern. Complement. Med.* **2007**, *13*, 603–616. [CrossRef] [PubMed]

11. Zhang, J.; Wu, X.; Nie, D.; Zhuo, Y.; Li, J.; Hu, Q.; Xu, J.; Yu, H. Magnetic resonance imaging studies on acupuncture therapy in depression: A systematic review. *Front. Psychiatry* **2021**, *12*, 1336.

12. Bai, L.; Zhang, D.; Cui, T.-T.; Li, J.-F.; Gao, Y.-Y.; Wang, N.-Y.; Jia, P.-L.; Zhang, H.-Y.; Sun, Z.-R.; Zou, W.; et al. Mechanisms underlying the antidepressant effect of acupuncture via the CaMK signaling pathway. *Front. Behav. Neurosci.* **2020**, *12*, 563698. [CrossRef]

13. Song, Z.; Deng, B.; Wei, X.; Cai, L.; Yu, H.; Wang, J.; Wang, R.; Chen, Y. Scale-specific effects: A report on multiscale analysis of acupunctured EEG in entropy and power. *Phys. A* **2018**, *492*, 2260–2272. [CrossRef]

14. Hauck, M.; Schröder, S.; Meyer-Hamme, G.; Lorenz, J.; Friedrichs, S.; Nolte, G.; Gerloff, C.; Engel, A.K. Acupuncture analgesia involves modulation of pain-induced gamma oscillations and cortical network connectivity. *Sci. Rep.* **2017**, *7*, 16307. [CrossRef]

15. Yi, G.; Wang, J.; Tsang, K.-M.; Chan, W.-L.; Wei, X.; Deng, B.; Han, C. Ordinal pattern based complexity analysis for EEG activity evoked by manual acupuncture in healthy subjects. *Int. J. Bifurc. Chaos* **2014**, *24*, 1450018. [CrossRef]

16. Tanaka, Y.; Koyama, Y.; Joda, E. Effects of acupuncture to the sacral segment on the bladder activity and electroencephalogram. *Psychiatry Clin. Neurosci.* **2002**, *56*, 249–250. [CrossRef]

17. Qi, S.-Y.; Lin, D.; Lin, L.-L.; Huang, X.-Z.; Lin, S.; Yu, Y.-Y.; Cao, C.-H.; Wang, Z.-X. Using nonlinear dynamics and multivariate statistics to analyze EEG signals of insomniacs with the intervention of superficial acupuncture. *Evid.-Based Complement. Altern. Med.* **2020**, *2020*, 8817843. [CrossRef]

18. Si, X.; Han, S.; Zhang, K.; Zhang, L.; Sun, Y.; Yu, J.; Ming, D. The temporal dynamics of functional networks are modulated by acupuncture: An EEG microstates study. *Int. J. Psychophysiol.* **2021**, *168*, S196–S197. [CrossRef]

19. Wadhera, T. Brain network topology unraveling epilepsy and ASD Association: Automated EEG-based diagnostic model. *Expert Syst. Appl.* **2021**, *186*, 115762. [CrossRef]

20. Yu, H.; Liu, J.; Cai, L.; Wang, J.; Cao, Y.; Hao, C. Functional brain networks in healthy subjects under acupuncture stimulation: An EEG study based on nonlinear synchronization likelihood analysis. *Phys. A* **2017**, *468*, 566–577. [CrossRef]

21. Yu, H.; Wu, X.; Cai, L.; Deng, B.; Wang, J. Modulation of spectral power and functional connectivity in human brain by acupuncture stimulation. *IEEE Trans. Neural Syst. Rehabil. Eng.* **2018**, *26*, 977–986. [CrossRef]

22. Shine, J.M.; Hearne, L.; Breakspear, M.; Hwang, K.; Müller, E.J.; Sporns, O.; Poldrack, R.; Mattingley, J.B.; Cocchi, L. The low-dimensional neural architecture of cognitive complexity is related to activity in medial thalamic nuclei. *Neuron* **2019**, *104*, 849–855. [CrossRef] [PubMed]

23. Rossi-Pool, R.; Romo, R. Low dimensionality, high robustness in neural population dynamics. *Neuron* **2019**, *103*, 177–179. [CrossRef] [PubMed]

24. Xing, D.; Aghagolzadeh, M.; Truccolo, W. Low-dimensional motor cortex dynamics preserve kinematics information during unconstrained locomotion in nonhuman primates. *Front. Neurosci.* **2019**, *13*, 1046. [CrossRef]

25. Yoon, K.; Buice, M.A.; Barry, C.; Hayman, R.; Burgess, N.; Fiete, I.R. Specific evidence of low-dimensional continuous attractor dynamics in grid cells. *Nat. Neurosci.* **2013**, *16*, 1077–1084. [CrossRef]

26. Chialvo, D.R. Emergent complex neural dynamics. *Nat. Phys.* **2010**, *6*, 744–750. [CrossRef]

27. Shine, J.M.; Breakspear, M.; Bell, P.T.; Martens, K.A.E.; Shine, R.; Koyejo, O.; Sporns, O.; Poldrack, R.A. Human cognition involves the dynamics integration of neural activity and neuromodulatory systems. *Nat. Neurosci.* **2019**, *22*, 289–296. [CrossRef] [PubMed]

28. Kerkman, J.N.; Daffertshofer, A.; Gollo, L.L.; Breakspear, M.; Boonstra, T.W. Network structure of the human musculoskeletal system shapes neural interactions on multiple time scales. *Sci. Adv.* **2018**, *4*, eaat0497. [CrossRef]

29. Cueva, C.J.; Saez, A.; Marcos, E.; Genovesio, A.; Jazayeri, M.; Romo, R.; Salzman, C.D.; Shadlen, M.N.; Fusi, S. Low-dimensional dynamics for working memory and time encoding. *Proc. Natl. Acad. Sci. USA* **2020**, *117*, 23021–23032. [CrossRef]

30. Abbaspourazad, H.; Choudhury, M.; Wong, Y.T.; Pesaran, B.; Shanechi, M.M. Mulstiscale low-dimensional motor cortical state dynamics predict naturalistic reach-and-grasp behavior. *Nat. Commun.* **2021**, *12*, 607. [CrossRef]

31. Cunningham, J.P.; Yu, B.M. Dimensionality reduction for large-scale neural recordings. *Nat. Neurosci.* **2014**, *17*, 1500–1509. [CrossRef]

32. Kingma, D.P.; Welling, M. Auto-encoding variational Bayes. *arXiv* **2013**, arXiv:1312.6114.

33. Bi, L.Z.; Zhang, J.W.; Lian, J.L. EEG-based adaptive driver-vehicle interface using variational autoencoder and PI-TSVM. *IEEE Trans. Neural Syst. Rehabil. Eng.* **2019**, *27*, 2025–2033. [CrossRef] [PubMed]
34. Han, K.; Wen, H.; Shi, J.; Lu, K.-H.; Zhang, Y.; Fu, D.; Liu, Z. Variational autoencoder: An unsupervised model for encoding and decoding fMRI activity in visual cortex. *Neuroimage* **2019**, *198*, 125–136. [CrossRef]
35. Li, X.; Zhao, Z.; Song, D.; Zhang, Y.; Pan, J.; Wu, L.; Huo, J.; Niu, C.; Wang, D. Latent factor decoding of multi-channel EEG for emotion recognition through autoencoder-like neural networks. *Front. Neurosci.* **2020**, *14*, 87. [CrossRef]
36. Kobak, D.; Brendel, W.; Constantinidis, C.; Feierstein, C.E.; Kepecs, A.; Mainen, Z.; Qi, X.-L.; Romo, R.; Uchida, N.; Machens, C.K. Demixed principal component analysis of neural population data. *eLife* **2016**, *5*, e10989. [CrossRef] [PubMed]
37. Marton, C.D.; Schultz, S.R.; Averbeck, B.B. Learning to select actions shapes recurrent dynamics in the corticostriatal system. *Neural Netw.* **2020**, *132*, 375–393. [CrossRef]
38. Mazzucato, L.; Fontanini, A.; Camera, G.L. Stimuli reduce the dimensionality of cortical activity. *Front. Syst. Neurosci.* **2016**, *10*, 11. [CrossRef] [PubMed]
39. Manjarrez, E.; Diez-Martínez, O.; Méndez, I.; Flores, A. Stochastic resonance in human electroencephalographic activity elicited by mechanical tactile stimuli. *Neurosci. Lett.* **2002**, *324*, 213–216. [CrossRef]
40. Xiang, A.F.; Liu, H.; Liu, S.; Shen, X.Y. Brain regions responding to acupuncture stimulation of Zusanli (ST36) in healthy subjects analyzed on the basis of spontaneous brain activity. *Acupunct. Res.* **2019**, *44*, 66–70.
41. Lundeberg, T.; Lund, I.; Naslund, J. Acupuncture—Self-appraisal and the reward system. *Acupunct. Med.* **2007**, *25*, 87–99. [CrossRef]
42. Liu, T. Acupuncture: What underlies needle administration. *Evid.-Based Complement. Altern. Med.* **2009**, *6*, 185–193. [CrossRef]
43. Gammaitoni, L.; Hänggi, P.; Jung, P.; Marchesoni, F. Stochastic resonance. *Rev. Mod. Phys.* **1998**, *70*, 223–287. [CrossRef]
44. Moss, F.; Ward, L.M.; Sannita, W.G. Stochastic resonance and sensory information processing: A tutorial and review of application. *Clin. Neurophysiol.* **2004**, *115*, 267–281. [CrossRef] [PubMed]
45. Faisal, A.A.; Selen, L.P.J.; Wolpert, D.M. Noise in the nervous system. *Nat. Rev. Neurosci.* **2008**, *9*, 292–303. [CrossRef] [PubMed]
46. McDonnell, M.D.; Ward, L.M. The benefits of noise in neural systems: Bridging theory and experiment. *Nat. Rev. Neurosci.* **2011**, *12*, 415–425. [CrossRef]
47. Lugo, E.; Doti, R.; Faubert, J. Ubiquitous cross modal stochastic resonance in humans: Auditory noise facilitates tactile, visual and proprioceptive sensations. *PLoS ONE* **2008**, *3*, e2860. [CrossRef]
48. Richardson, K.A.; Imhoff, T.T.; Grigg, P.; Collins, J.J. Using electrical noise to enhance the ability of humans to detect subthreshold mechanical cutaneous stimuli. *Chaos* **1998**, *8*, 599–603. [CrossRef]
49. Recanatesi, S.; Ocker, G.K.; Buice, M.A.; Shea-Brown, E. Dimensionality in recurrent spiking networks: Global trends in activity and local origins in connectivity. *PLoS Comput. Biol.* **2019**, *15*, e1006446. [CrossRef]
50. Litwin-Kumar, A.; Harris, K.D.; Axel, R.; Sompolinsky, H.; Abbott, L. Optimal degrees of synaptic connectivity. *Neuron* **2017**, *93*, 1153–1164. [CrossRef]
51. Bar-Gad, I.; Morris, G.; Bergman, H. Information processing, dimensionality reduction and reinforcement learning in the basal ganglia. *Prog. Neurobiol.* **2003**, *71*, 439–473. [CrossRef]
52. Tang, Y.B.; Chen, D.; Li, X.L. Dimensionality reduction methods for brain imaging data analysis. *ACM Comput. Surv.* **2021**, *54*, 87. [CrossRef]
53. Rigotti, M.; Barak, O.; Warden, M.; Wang, X.-J.; Daw, N.D.; Miller, E.; Fusi, S. The importance of mixed selectivity in complex cognitive tasks. *Nature* **2013**, *497*, 585–590. [CrossRef]
54. Yu, Z.; Luo, L.; Li, Y.; Wu, Q.; Deng, S.; Lian, S.; Liang, F. Different manual manipulations and electrical parameters exert different therapeutic effects of acupuncture. *J. Tradit. Chin. Med.* **2014**, *34*, 754–758. [CrossRef]
55. Guo, X.; Yu, H.; Kodama, N.X.; Wang, J.; Galan, R.F. Fluctuation scaling of neuronal firing and bursting in spontaneously active brain circuits. *Int. J. Neural Syst.* **2020**, *30*, 1950017. [CrossRef]
56. Olguín-Rodríguez, P.V.; Arzate-Mena, J.D.; Corsi-Cabrera, M.; Gast, H.; Marín-García, A.; Mathis, J.; Ramos Loyo, J.; del Rio-Portilla, I.Y.; Rummel, C.; Schindler, K.; et al. Characteristic fluctuations around stable attractor dynamics extracted from highly nonstationary electroencephalographic recordings. *Brain Connect.* **2018**, *8*, 457–474. [CrossRef]
57. Hennequin, G.; Ahmadian, Y.; Rubin, D.B.; Lengyel, M.; Miller, K.D. The dynamics regime of sensory cortex: Stable dynamics around a single stimulus-tuned attractor account for patterns of noise variability. *Neuron* **2018**, *98*, 846–860. [CrossRef]
58. Finkelstein, A.; Fontolan, L.; Economo, M.N.; Li, N.; Romani, S.; Svoboda, K. Attractor dynamics gating cortical information flow during decision-making. *Nat. Neurosci.* **2021**, *24*, 843–850. [CrossRef]
59. Kawanokuchi, J.; Takagi, K.; Tanahashi, N.; Yamamoto, T.; Nagaoka, N.; Ishida, T.; Ma, N. Acupuncture treatment for social defeat stress. *Front. Behav. Neurosci.* **2021**, *15*, 685433. [CrossRef]
60. Yam, W.; Wilkinson, J.M. Is acupuncture an acceptable option in stroke rehabilitation? A survey of stroke patients. *Complement. Ther. Med.* **2010**, *18*, 143–149. [CrossRef] [PubMed]
61. Zeng, B.Y.; Zhao, K.C. Effect of acupuncture on the motor and nonmotor symptoms in Parkinson's disease: A review of clinical studies. *CNS Neurosci. Ther.* **2016**, *22*, 333–341. [CrossRef] [PubMed]
62. Chae, Y.; Lee, H.; Kim, H.; Kim, C.-H.; Chang, D.-I.; Kim, K.-M.; Park, H.-J. Parsing brain activity associated with acupuncture treatment in Parkinson's diseases. *Mov. Disord.* **2009**, *24*, 1704–1802. [CrossRef] [PubMed]
63. Lee, M.Y.; Lee, B.H.; Kim, H.Y.; Yang, C.H. Bidirectional role of acupuncture in the treatment of drug addiction. *Neurosci. Biobehav. Rev.* **2021**, *126*, 382–397. [CrossRef] [PubMed]

sensors

Article

Evaluation of Feature Selection Methods for Classification of Epileptic Seizure EEG Signals

Sergio E. Sánchez-Hernández, Ricardo A. Salido-Ruiz, Sulema Torres-Ramos and Israel Román-Godínez *

Division of Cyber-Human Interaction Technologies, University of Guadalajara (UdG),
Guadalajara 44100, Jalisco, Mexico; sergio.sanchez1153@alumnos.udg.mx (S.E.S.-H.);
ricardo.salido@academicos.udg.mx (R.A.S.-R.); sulema.torres@academicos.udg.mx (S.T.-R.)
* Correspondence: israel.roman@academicos.udg.mx

Abstract: Epilepsy is a disease that decreases the quality of life of patients; it is also among the most common neurological diseases. Several studies have approached the classification and prediction of seizures by using electroencephalographic data and machine learning techniques. A large diversity of features has been extracted from electroencephalograms to perform classification tasks; therefore, it is important to use feature selection methods to select those that leverage pattern recognition. In this study, the performance of a set of feature selection methods was compared across different classification models; the classification task consisted of the detection of ictal activity from the CHB-MIT and Siena Scalp EEG databases. The comparison was implemented for different feature sets and the number of features. Furthermore, the similarity between selected feature subsets across classification models was evaluated. The best F1-score (0.90) was reported by the K-nearest neighbor along with the CHB-MIT dataset. Results showed that none of the feature selection methods clearly outperformed the rest of the methods, as the performance was notably affected by the classifier, dataset, and feature set. Two of the combinations (classifier/feature selection method) reporting the best results were K-nearest neighbor/support vector machine and random forest/embedded random forest.

Keywords: EEG; epilepsy; seizure detection; machine learning; features; feature selection

Citation: Sánchez-Hernández, S.E.;
Salido-Ruiz, R.A.; Torres-Ramos, S.;
Román-Godínez, I. Evaluation of
Feature Selection Methods for
Classification of Epileptic Seizure
EEG Signals. *Sensors* **2022**, *22*, 3066.
https://doi.org/10.3390/s22083066

Academic Editors: Yvonne Tran and
Christian Baumgartner

Received: 14 January 2022
Accepted: 13 April 2022
Published: 16 April 2022

Publisher's Note: MDPI stays neutral
with regard to jurisdictional claims in
published maps and institutional affiliations.

1. Introduction

Epilepsy is one of the most common neurological diseases, affecting around 50 million people of all ages globally [1]. The Center for Surveillance, Epidemiology and Laboratory Services of the United States of America estimated that in 2010, the number of adults with active epilepsy in the United States was 2.3 million; by 2015, the estimate increased to 3 million adults [2]. An additional study calculated that the cumulative incidence of epilepsy, in Norwegian children at the age of ten was around 0.66%, with 0.62% having active epilepsy [3]. The authors in [4] reported six studies of epilepsy prevalence in Mexico, which found prevalence rates of 3.9 to 42.2 per 1000 inhabitants.

An important tool for the diagnosis and management of epilepsy is the electroencephalogram (EEG). As mentioned in [5] (p. ii2), EEG is a "convenient and relatively inexpensive way to demonstrate the physiological manifestations of abnormal cortical excitability that underlie epilepsy." Other diagnostic techniques used in conjunction with EEG include neuroimaging, metabolic tests, and genetic tests. EEG can be processed and classified by using machine learning methods. Several studies have applied machine learning to classify ictal EEG signals [6,7] and predict seizures [8,9]. Concerning signal description, several metrics have been used to characterize such problems, and to build machine learning models, some of those metrics are computed from an EEG. Some researchers have approached the classification problem by calculating statistical, entropy, and univariate linear metrics from EEG [6,8–11]; those metrics can be fed to a model in the form of a vector or a matrix. In addition, metrics can be estimated for the entire EEG bandwidth or for smaller sub-bands [12–14], the latter with the intention of obtaining a more detailed view of the signal. Furthermore, transformation of the EEG signal to different domains has been explored by applying Fourier [15], short-term Fourier [7,16,17], wavelet transform [15,16,18–20], and contourlet transform [21]. It has also been analyzed as a graph [8] and an image [16]. As a result of the increasing interest in this topic, Ref. [22] presented a complete summary of several descriptors for time-domain, frequency-domain, and time–frequency-domain, along with their interpretation, while applying to the epileptic seizure detection on EEG signals.

Regarding the classification problem, a diversity of machine learning models have been tested for epilepsy prediction; among these we have: decision trees [23–25], support vector machine [23,26,27], K-nearest neighbor [23], and random forest [27]. In recent years, diverse deep learning models have been tried for epilepsy-related tasks, e.g., convolutional neural networks [16,17,28] and long-short term memory [8,28].

An important step in classifier modeling is feature selection. When performing feature selection, it is important to take into account several facts: (a) there are numerous attributes that can be calculated from EEG signals—each one describing a particular aspect of EEG; (b) there is a strong relation between features and model accuracy; (c) the curse of dimensionality, which is related to the difficulty of optimizing a solution in high-dimensional spaces; (d) the complexity and interpretability of the resulting models, which are about reducing the time and costs by training simpler learning models or selecting the features that are more relevant and meaningful from the problem perspective [29]. In this sense, several feature selection methods have been used to overcome the aforementioned issues. Some of those techniques are statistical tests, information gain [30], principal component analysis [31], permutation importance [32], and recursive feature elimination (RFE) [33], among others [29].

In this regard, there are several works in the state-of-the-art focused on epileptic EEG signal classification; some of them use feature selection methods (FSMs) to improve their results and reduce the dimension of the feature vectors.

For example, principal component analysis was applied in [34] to obtain less correlated features; however, their main goal was to evaluate the effects of channel selection on epileptic analysis over adults and children, without considering the effect of the feature selection method in the classification's performance, and only one classification method was considered, in this case linear discriminant analysis.

In [24], RFE was used to rank features and a support vector machine (SVM) for classifying epilepsy, autism, and control groups in children. Even though they determined which features and combinations of features contributed the most to the classification accuracy, they did not analyze either several feature selection methods or other classification methods.

The authors of [25] evaluated one feature selection method (recursive feature elimination) and one feature set, in combination with seven classifiers to improve the classification accuracy of automatic seizure diagnosis. From the 11 features that were calculated from the EEG signal, they reduced them to 8 features. All the experiments were performed only on one adult dataset.

In [12], six FSMs along with nine classifiers were used for automatic seizure detection. The FSMs were evaluated to rank and reduce the number of features, ranking the important features using a *t*-test and selecting the top 20 or 25, without testing additional alternatives that may result in different rankings. Furthermore, the authors provided experimental results based on one dataset that belonged to children. Additionally, the hold-out cross-validation methodology was used, which is commonly used for bigger datasets.

FSMs have also been tested in signals other than EEG such as magnetic resonance images, as shown in [35]. The authors' objective was to compare three different FSMs (i.e., *t*-test filtering, the sparse-constrained dimensionality reduction model, and the support vector machine-recursive feature elimination) to determine which of them performed better when using an SVM as the classifier. However, the authors only tested the performance of the FSM on one classifier (SVM). Furthermore, they only tested on one dataset without considering if their result may change when either tested on different datasets or FSM sets.

In summary, it can be observed that, even though there is extensive research about the feature selection methods applied in seizure and seizure-free EEG signals' classification, such approximations do not allow having a general perspective of the advantages and disadvantages of using a particular combination of the feature selection method with a classification algorithm (C-FSM) to determine: (1) the effect of the dimensionality reduction on the performance of the classifiers; (2) the best C-FSM combinations; and (3) the amount of coincidence of the best-selected features among FSMs; all of that considering two different feature sets and two different databases (adults and children).

Hence, in this work, the CHB-MIT and Siena Scalp EEG databases were used along with two different feature sets to evaluate the combination of six FSMs and five classification models. The results of this work allow determining: the minimum number of features that can be chosen for each FSM without scarifying the classifiers' performances; the performance of several C-FSM combinations in order to discover if a relationship exists between the FSMs with a particular classification algorithm; and if there is a feature or feature set that remains across different C-FSM combinations.

2. Materials and Methods

In this section, the features, models, and training procedures are described. A general methodology overview is shown in Figure 1.

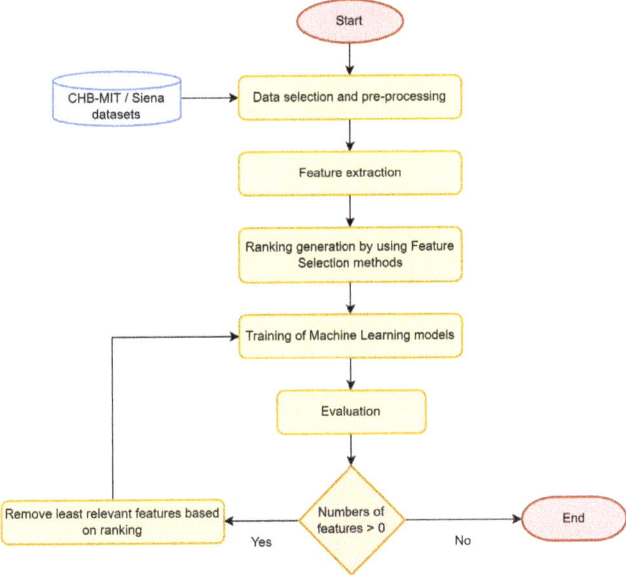

Figure 1. Flow diagram of the methodology.

2.1. Datasets

CHB-MIT was one of the two datasets used for this research. The dataset is available for download at Physionet [36] under Open Data Commons Attribution License v1.0.

The data were collected at the Boston Children's Hospital. The database contains scalp electroencephalograms of 23 pediatric patients having epilepsy [37]. The number of recordings per patient varies from 9 to 45; all of them contain a metadata file listing the channels' names and ictal activity intervals; most of the records have a duration of one hour. EEG signals were sampled at 256 Hz; electrodes were placed according to the 10–20 system [37]. Most of the recordings are provided following a bipolar longitudinal montage.

After exploring the data, recordings not containing ictal activity were discarded. Thus, an amount of 137 EEG recordings containing an overall amount of 181 ictal activity intervals from 23 patients were accomplished.

The second dataset used in this research was the Siena Scalp EEG database; it is also available on the Physionet site under a Creative Commons Attribution 4.0 license. The dataset was collected by the Unit of Neurology and Neurophysiology at the University of Siena. It contains EEG recordings of 14 patients, 9 males (ages 36–71) and 5 females (ages 20–58). There are a total of 41 EEG recordings, and these include 47 ictal activity intervals; the recordings' duration is variable, from 1 to 13 h. The start and end of each seizure are detailed in the metadata file provided by the original authors [38]. EEG signals were sampled at 512 Hz. Electrodes were placed according to the 10–20 system [38]. EEG channels provided in the dataset are monopolar.

Recordings were further processed to design a dataset for a bi-class classification problem: seizure or seizure-free.

2.2. Data Pre-Processing

The number of conserved bipolar channels for the CHB-MIT recordings was 21. A few of the channels do not follow the 10–20 positioning, but channels were considered because these are included in every recording of the dataset. The Siena Scalp recordings were converted to a longitudinal bipolar montage, to have a fair comparison between both datasets; 18 channels were conserved.

A second-order Butterworth high-pass filter was used for removing frequencies below 0.5 Hz; then, a notch filter was applied to remove power line frequency (60 Hz and 50 Hz for CHB-MIT and Siena datasets, respectively).

A window length of 2 s was chosen. Epoch length was based on [13], as they mentioned, when extracting spectral features, it is important to choose small epochs due to the non-stationarity of the EEG.

As the number of ictal segments was significantly lower than the non-ictal ones, a 50% overlap was applied to ictal windows; besides that, some non-ictal windows were removed in order to keep an approximate ratio of 9:1 between both classes.

Ten percent of the CHB-MIT epochs were kept apart for adjusting the parameters of the classification models.

Inspired by [12], windowed EEG signals were separated into the following 5 sub-bands: complete bandwidth (0.5–30 Hz), delta (0.5–4 Hz), theta (4–8 Hz), alpha (8–12 Hz), and beta (12–25 Hz). Second-order Butterworth band-pass filters were used for band separation.

2.3. Feature Extraction

Two different features sets were evaluated to observe the effect of varying the metrics and determine which one outperforms (see Table 1). Therefore, half of Feature Set 1 (FS1) was conformed by statistical metrics and was applied in the time-domain or frequency-domain. Specifically, the median frequency (i) was estimated to characterize the power spectrum of the EEG data. On the other hand, the variance (ii), skewness (iii), and kurtosis (iv) were estimated in time-domain to characterize the variability and the distribution of the EEG data. Other features such as the peak frequency (v) were used to describe the frequency of the highest peak in the power spectral density; the root mean square (vi), range (vii), and the number of zero crossings (viii) in the time-domain were used respectively to estimate the effective value of the signal, to measure the maximum wave amplitude, and to count the number of points where the EEG wave cuts the horizontal axis, changing its state from positive to negative and vice versa.

Table 1. Features' set.

Feature Set 1 (FS1)	Feature Set 2 (FS2)
Median Frequency (i) [39]	Minimum (ix) [34]
Variance (ii) [40]	Complexity (x) [34]
Skewness (iii) [40]	Mobility (xi) [34]
Kurtosis (iv) [40]	Interquartile Range (xii) [40]
Peak Frequency (v) [41]	Median Absolute Deviation (xiii) [42]
Root Mean Square (vi) [40]	Sample Entropy (xiv) [43]
Range (vii) [40]	Mean (xv) [40]
Number of Zero Crossings (viii) [34]	Standard Deviation (xvi) [40]

Those features were selected based on previous studies where the authors performed an extensive review on the state-of-the-art on epileptic seizure detection based on EEG [22] along with some other works [6,8,10,12,25,34] that have successfully applied such feature metrics. In particular, Features i–vi were selected based on [12]. The number of zero crossings (ZCs) was estimated instead of the ZC rate because all EEG epochs had the same length.

On the other hand, FS2 was formed by the minimum (ix), which describes the minimum value that a signal can take, complexity (x), used to describe the change in frequency, mobility (xi), which is a measure of the mean frequency, interquartile range (xii), which is a measure of statistical dispersion, the spread of the data or observations, median absolute deviation (xiii), which is a robust measure of how spread out a set of data is, sample entropy (xiv), used for assessing the complexity of physiological time series signals, mean (xv), and standard deviation (xvi), statistical parameters that describe the average value and the amount of variability, or dispersion, from the individual data values to this average value. These features have been utilized across different studies for the classification of ictal EEG data [6,10–12,34]. Parameters used for the sample entropy (SampEn) are r = 0.2, m = 2. When applied to seizure discrimination in EEG, it was previously observed [44] that there is no best parameter combination, but several optimal combinations, one of which is the one used in this work.

The features from FS1 and FS2 were computed per channel and frequency band. For both feature sets (FS1 and FS2), feature vectors of 840 and 720 in length were obtained for the CHB-MIT and Siena datasets, respectively.

After that, correlations between features in all bands and channels were identified in both feature sets. For each pair of features, the Pearson correlation coefficient was computed. If a combination had a coefficient over 0.95, one of the features of the pair would be discarded. As a result, from FS1, 105 and 113 features were removed for the CHB-MIT and Siena datasets, respectively; on the other hand, from FS2, 304 and 257 features were removed for the CHB-MIT and Siena datasets, respectively.

2.4. Classification Methods

The following five algorithms were selected for classifying ictal EEG signals:

1. Decision tree (DT): This is a hierarchical model composed of decision nodes and terminal leaves, each leaf have an output label. Decision nodes implement a test function $f(x)$, which is a discriminator dividing the input space into smaller regions. Among all possible splits, the DT looks for the one that minimizes impurity. There are several impurity measures, e.g., the Gini index and entropy. For a two-class problem, the Gini index is defined as [45]:

$$\phi(p, 1 - p) = 2p(1 - p) \tag{1}$$

where p is the probability of a sample reaching a node m, to belong to a class C. The classification and regression trees algorithm (CART) was applied in this research.

2. Support vector machine (SVM): This constructs a hyperplane or set of hyperplanes in a high-dimensional space that can be used for classification. Those hyperplanes have the largest distance to the nearest training data points (also known as the functional margin) [23]. The task of finding the optimal separating hyperplanes can be defined as [45]:

$$min \frac{1}{2}||w||^2 \text{ subject to } r^t(w^T x^t + w_0) \geq 1, \forall t \tag{2}$$

where w are the parameters defining the hyperplane, x^t are the instances of the training set, and r^t is the actual label. If the problem is not linearly separable, the problem can be mapped to a new space by using non-linear basis functions [45].

3. K-nearest neighbor (KNN): This is a classifier that learns by analogy. A target unknown instance is compared to all the instances in the training set, locating the k closest instances; the algorithm assigns the class that corresponds to the majority. "Closeness" is measured by using a distance metric; in this study, we used the Manhattan distance (selected through a parameter grid search). The Manhattan distance for a p-dimension space is defined as [46]:

$$d(i,j) = |x_{i1} - x_{j1}| + |x_{i2} - x_{j2}| + ... + |x_{ip} - x_{jp}| \tag{3}$$

4. Random forest (RF): This is an ensemble method, and each classifier of the ensemble is a DT. For each node of the DT, a random selection of features is used to determine the best split. RF also uses bagging (bootstrap aggregation), which means that the training set for each DT is sampled with replacement from the original training set. After model training, each DT votes, and the most voted class is assigned to the test instance [46].

5. Artificial neural network (ANN): This is a classifier that uses the idea of the perceptron, and it is referred to as a multi-layer perceptron. It can contain three or more layers, and these are: an input layer, one or more hidden layers, and an output layer [19]. Simply speaking, it is a set of input/output interconnected units, each connection having a weight associated with it. During the learning process, the ANN adjusts the weights of the connections in order to be able to predict the correct labels of new instances [46]. The neural network used in this study is described in Table 2.

Table 2. Structure of ANN layers.

Layer	Type	Size	Activation
1	Dense	100	ReLU
2	Dense	50	ReLU
3	Dense	1	Sigmoid

To select the best-suited parameters for the above machine learning models, a grid search was performed; accuracy was utilized as a comparison metric. For selecting the ANN structure, a different number of neurons and hidden layers (1 and 2 layers) were evaluated, resulting in our best architecture, the one depicted in Table 2. The configuration settings and parameters of the grid search are listed in Table 3 for FS1 and in Table 4 for FS2. Both FSs used the same parameter grid detailed in Table 3. It is worth mentioning that the selected classification algorithms have been previously implemented for epilepsy-related tasks; however, the model parameters were not inspired by any specific work; on the contrary, they were selected by performing a grid search.

Table 3. Configuration setting per classification method. It was adjusted from FS1.

Method	Parameter	Selected Value	Parameters Grid
DT	Min. samples per node Split criterion	8 Gini impurity	2, 4, 8, 16 Gini impurity
SVM	Regularization parameter Kernel Gamma coefficient	2 RBF $1/(n\_features \times var(x))$	0.1, 0.5, 1, 2, 10 Linear, RBF $1/(n\_features \times var(x)), 1/n\_features$
KNN	Num. of neighbors Distance metric Weight function	15 Manhattan Uniform	3, 5, 10, 15 Euclidean, Manhattan Uniform, distance-based
RF	Num. of trees Min. samples per node	150 4	30, 50, 100, 150 2, 4, 8, 16
ANN	Optimizer Loss measure Epochs	Adam Cross-entropy 200	

Table 4. Configuration setting per classification method. It was adjusted from FS2.

Method	Parameter	Selected Value
DT	Min. samples per node Split criterion	2 Gini impurity
SVM	Regularization parameter Kernel Gamma coefficient	2 RBF $1/(n\_features \times var(x))$
KNN	Num. of neighbors Distance metric Weight function	10 Euclidean Distance
RF	Num. of trees Min. samples per node	150 4
ANN	Optimizer Loss measure Epochs	Adam Cross-entropy 200

2.5. Feature Selection Methods

In this research, six FSMs were used. The metrics used to assign the importance of each feature are detailed below. The parameters selected for training each algorithm are mentioned in Table 5.

Table 5. Configuration setting per feature selection method.

Method	Parameter	Values
DT	Min. samples per node Split criterion	4 Gini impurity
SVM	Regularization parameter C Kernel	1 Linear kernel
LIME	Discretization of values	No
SHAP	Length of background dataset	50
ERF	Num. of trees Min. samples per node Split criterion	10 2 Gini impurity

1. Decision tree (DT): The measure used to assign the feature importance is the Gini importance. As described in [47], the importance of feature X_m in an RF can be measured by Equation (4)

$$Imp(X_m) = \frac{1}{N_T} \sum_{T} \sum_{t \in T : v(s_t) = X_m} p(t) \triangle i(s_t, t) \qquad (4)$$

where T is a set of DTs, $v(s_t)$ is the feature used to split node t, $\triangle i(s_t, t)$ is the impurity decrease in node t, and $p(t)$ is:

$$p(t) = N_t / N \tag{5}$$

N is the number of training samples, and N_t is the number of samples reaching the node t. As this study uses a single DT, the size of T is 1.

2. Support vector machine (SVM): Coefficients estimated by SVM can be utilized to produce features' ranking [33]. A linear SVM is trained on the input dataset, then features are ranked as per the absolute values of the hyperplane weights [48].

3. Local interpretable model-agnostic explanations (LIME): This is a technique for explaining the predictions of any classifier by learning an interpretable model locally around the prediction. For a classification model f and interpretable model class G, the explanation is obtained by optimizing Equation (6) [49].

$$\xi(x) = \underset{g \in G}{\operatorname{argmin}} L(f, g, \pi_x) + \Omega(g) \tag{6}$$

Function L measures the approximation of g to f in the locality defined by π_x. Function $\Omega(g)$ is a measure of the complexity of g. As per [49], the exponential kernel was used for π_x, the weighted square loss for L, and the linear model for G.

4. Shapley additive explanations (SHAP): This is a unified approach to interpreting model predictions. It assigns each feature an importance value or SHAP value. These are the Shapley values of a conditional expectation function of the original model [50]. The implementation applied in this study, kernel SHAP, optimizes (6), but it uses different forms of $\pi_{x'}$, L, and Ω:

$$\Omega(g) = 0 \tag{7}$$

$$\pi_{x'}(z') = \frac{M - 1}{(M \text{ choose } |z'|)|z'|(M - |z'|)} \tag{8}$$

$$L(f, g, \pi'_x) = \sum_{z' \in Z} [f(h_x(z')) - g(z')]^2 \pi_{x'}(z') \tag{9}$$

where g is the explanation model and follows a linear form, f is the classification model, M is the total number of features, and $|z'|$ is the number of used features. Parameter z' is the set of features represented as $\{0, 1\}^M$. As explained in [51], the function h_x maps the 1s of z' to the value from the instance to be explained x, and 0s are replaced by a random feature value of another instance sampled from the data.

5. Embedded random forest (ERF): "feature importance is measured by randomly permuting the feature in the out-of-bag samples and calculating the percent increase in misclassification rate as compared to the out-of-bag rate with all variables intact" [48] (p. 319). This technique was originally described in [52].

6. Reciprocal ranking (RR): Known also as inverse rank position [53], this is an ensemble method that merges ranked lists into a single one r based on Equation (10) [54]:

$$r(f) = \frac{1}{\sum_j \frac{1}{r_j(f)}} \tag{10}$$

where $r_j(f)$ is the ranking position assigned to a feature by the rest of the FSMs.

Instances of the SHAP background dataset (Table 5) were estimated by applying K-means ($k = 50$) to a subset of the training dataset. This was done in order to use a small, but representative set of instances for the estimation of the SHAP values. For feature evaluation in ERF, first, RF was trained, then permutation importance was estimated in a separate dataset.

The DT and SVM were chosen due to simplicity and training speed. SHAP, ERF, and RR were selected because, in [54], these methods returned good performance and consistency for a prediction task related to environmental datasets. LIME was considered because it is model-agnostic and allows model interpretation, similar to SHAP.

2.6. Model Training and Evaluation

Feature importance rankings were obtained for each combination of feature selection and classification method, and as a result, 30 feature rankings were estimated per dataset. For convenience, if any negative importance score was assigned to a feature, the absolute value was calculated. As LIME and SHAP compute feature importance per instance, the average was computed across all instances. When model training was required during ranking computation, 50% of the training dataset was passed to the model, and min–max normalization [46] was used for data scaling.

During the training and evaluation of the classification models, data were min–max-normalized and a 5×2 cross-validation (2-fold, 5 repetitions) was implemented for model evaluation [55]. On each database (i.e., CHB-MIT and Siena), patients' epochs were merged into a single dataset; then, each new dataset was randomized after each iteration of the validation methodology.

The 5×2 CV F-test procedure was originally proposed to compare supervised classification algorithms, even though it has been previously implemented for comparison of FSMs (not applied to EEG data) [56,57].

First, the classification models were trained by keeping all the features in the training set and assessed in order to compute their classification performance. Then, features having the smallest ranking criterion were removed, and the model was re-trained and re-evaluated. Feature removal was performed in steps of 50 features at a time. There were 25, 12, 6, and 1 features also considered during the evaluation. Rankings were computed at the beginning of the process.

It should be noted that the above pipeline was repeated 6 times per classification method, as there were 6 FSMs. In addition, models were trained per each dataset separately.

2.7. Computing and Software

The experiments were run on 2 different computing devices: a computer with Intel Core i7, 12 GB of RAM, and Ubuntu 18.04 and a server with Intel Xeon Gold and NVIDIA Tesla P100. Python 3.7 was used for coding all the experiments. Numpy (1.19.5) [58], pandas (1.2.2) [59], and scipy (1.6.1) [60] were used for data engineering, scikit-learn (0.24.1) [61] and tensorflow (2.4.1) [62] for building machine learning models and feature selection algorithms, and matplotlib (3.3.4) [63] for plotting. Other needed libraries were lime (0.2.0.1) [49] and shap (0.39.0) [50]. Some processing pipelines were run on a Jupyter Notebook in order to visualize the charts.

3. Results

3.1. Evaluation of Feature Dimensionality Reduction

This analysis was performed in order to visualize the effect of the reduction of the feature vector size on the classification models' performance. By doing this, it was possible to have a general overview of the robustness of the classification models regarding the reducing of the feature vectors that may result from feature selection methods.

Figures 2–5 present five plots depicting the average F1-score for the combination classifier (C), feature selection method (FSM), and a number of features (NF) (C-FSM-NF). Every subplot corresponds to a different classification model, namely (a) decision tree (DT), (b) support vector machine (SVM), (c) artificial neural network (ANN), (d) random forest (RF), and (e) K-nearest neighbor (KNN). Colored lines indicate the average performance for different NF values, and each color indicates a different FSM.

When the CHB-MIT dataset and the FS1 were used, the best performances were reported by the ANN model and the worsts by the DT and SVM models. The former reported the best F1-score with 0.86 corresponding to the combination ANN-ERF-250/200 (see Figure 2c). Additionally, it was observed that for every classifier, the RR feature selection method (brown line) decreased from the early stages (see Figure 2).

The second-best-performing classifier was RF (see Figure 2d); most of the F1-score values were between 0.80 and 0.85. Most of the curves showed similar tendencies, but the RR curve (brown line) went down faster, again.

KNN (see Figure 2e) returned few F1-scores that overcame the RF classifier, but it was less stable (i.e., its performance was more dependent on the FSM). A pair of the curves was over 0.80 (yellow and purple lines corresponding to the SVM and ERF feature selection methods, respectively). The rest of them presented a diminishing tendency that started from the early beginning.

The classifiers having the lowest performance were DT and SVM (see Figure 2a,b). The F1-score curves of the DT classifier had values slightly under 0.75. On the other hand, some of the SVM's F1-scores reached 0.75.

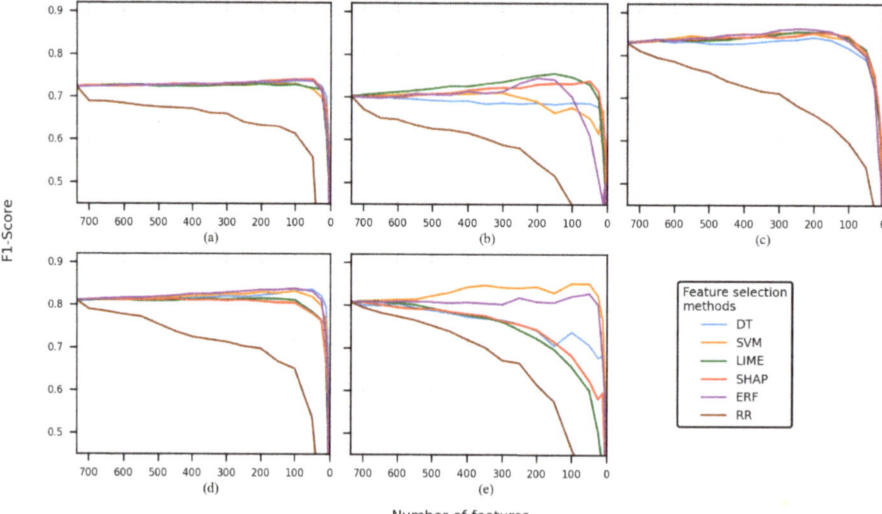

Figure 2. F1-score of the model was obtained by using a different number of features, using the CHB-MIT dataset and FS1. Classification model used: (**a**) decision tree, (**b**) support vector machine, (**c**) artificial neural network, (**d**) random forest, and (**e**) K-nearest neighbor.

When the Siena dataset and the FS1 were used, several classification models returned lower performance in comparison with the CHB-MIT experiments (see Figure 3a–c,e).

The best classification model was RF (see Figure 3d), and the combination RF-ERF reached an F1-score of 0.85. SVM and RR decreased faster than the rest of the FSMs.

The ANN model reached a performance of 0.80 for some combinations (ANN-LIME, ANN-SHAP). The RR curve decreased faster than the rest of the FSMs. SVM and the DT had a better performance than RR, but not as good as LIME and SHAP (see Figure 3c). The DT (see Figure 3a) showed a steadier behavior compared to all the classifiers, but its performance was around 0.70

Figure 3b depicts an interesting pattern. The F1-score was 0.5 at the beginning of the training, then several curves decreased almost from the beginning of the training (brown, blue, and purple lines). On the other hand, the LIME curve rose markedly as the number of features reduced.

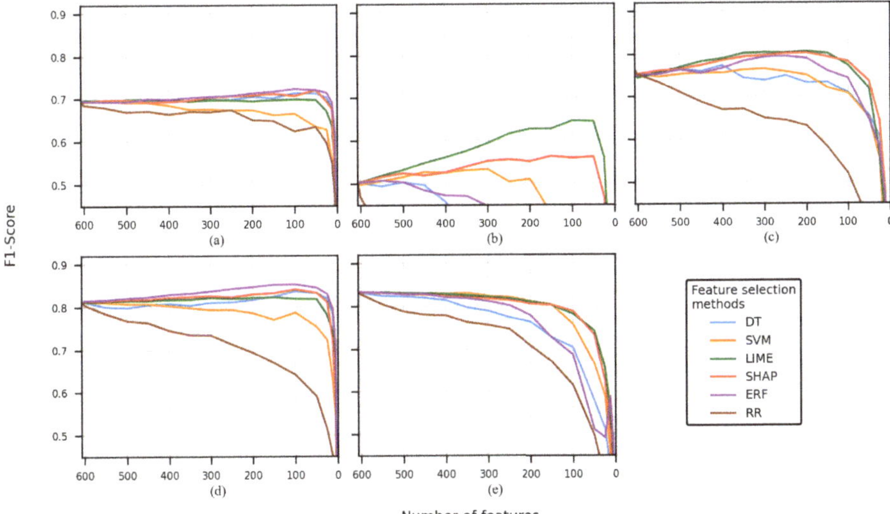

Figure 3. F1-score of the model is obtained by using a different number of features, using the Siena dataset and FS1. Classification model used: (**a**) decision tree, (**b**) support vector machine, (**c**) artificial neural network, (**d**) random forest, and (**e**) K-nearest neighbor.

When the CHB-MIT dataset and FS2 were used, the performance was better than the performance obtained during the FS1 experiments. Once again, the RR experiments tended to show poor performance and a faster decay in comparison to the rest of the FSMs.

The best performances were reported by the KNN model (see Figure 4e) and the worst by the DT and SVM models (see Figure 4a,b). KNN reported the best F1-score (0.90), and it corresponded to the experiments that removed a low number of features (e.g., 500 and 450). The combinations KNN-ERF-400/350/300/250/200 also reported an F1-score of 0.90.

The second-best-performing classifier was the ANN (see Figure 4c); most of the F1-score values were between 0.85 and 0.90. Most of the curves depicted a similar tendency, but the RR curve. RF (see Figure 4d) returned some F1-scores around 0.85, and these were slightly lower than the ANN scores.

When the Siena dataset and FS2 were used, the performances were a bit worse than the CHB-MIT experiments. The SVM classifier (see Figure 5b) depicted a similar pattern to the one observed in Figure 3b); when the number of features was reduced, the LIME and SHAP curves showed an increase in performance.

The best classification models were RF and KNN (see Figure 5d,e). These classifiers reported F1-scores around 0.85. For the KNN case, the RR curve (brown line) did not decay as fast as it did for the rest of the classifiers; however, RR was still the FSM with the worst performance.

Finally, to perform the comparison of the C-FSM combination and the feature selected, different cutoffs were selected. It is important to mention that based on the average of the F1-scores across every experiment, the decrease in classification performance between two consecutive cutoffs points was approximately equal, so based on the visual inspection of the Figures 2–5, four cutoffs were defined: 450, 150, 100, and 50 features. Notice that we discarded analyzing cutoff points under 50 features, because several models showed an F1-score lower than 0.6.

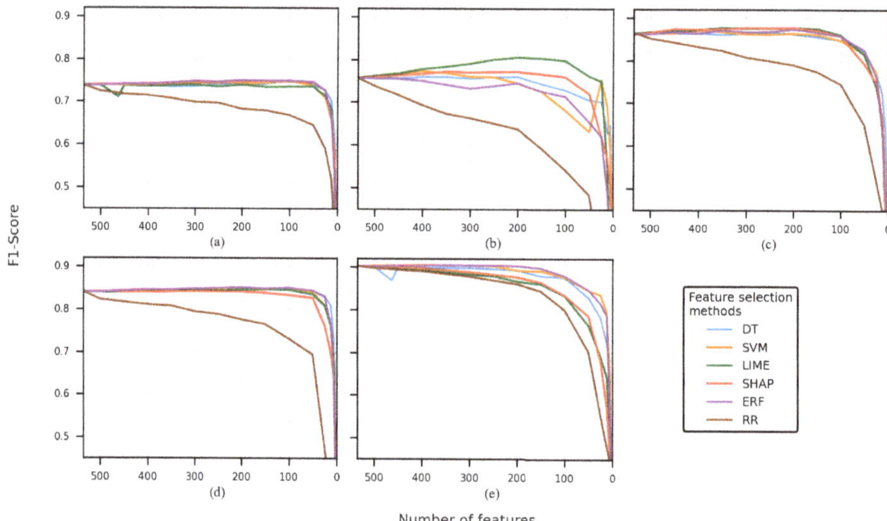

Figure 4. F1-score of the model is obtained by using a different number of features, using the CHB-MIT dataset and FS2. Classification model used: (**a**) decision tree, (**b**) support vector machine, (**c**) artificial neural network, (**d**) random forest, and (**e**) K-nearest neighbor.

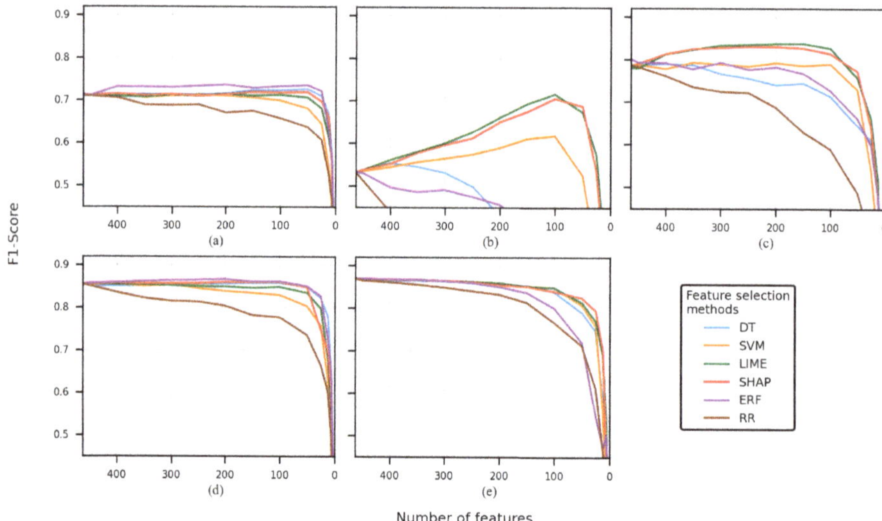

Figure 5. F1-score of the model is obtained by using a different number of features, using the Siena dataset and FS2. Classification model used: (**a**) decision tree, (**b**) support vector machine, (**c**) artificial neural network, (**d**) random forest, and (**e**) K-nearest neighbor.

3.2. Comparison of C-FSM Combinations

This analysis intended to observe, in detail, the performance of several C-FSM combinations. To do so, several cutoff points were chosen, then, to every cutoff point, the F1-score, sensitivity, and accuracy of all the combinations for a model classifier and feature selection method were computed. For the best-performing combinations of C-FSMs, the 5×2 CV F-test was applied to find statistically significant differences between the error rates.

Tables 6–9 depict the different C-FSM tuples and their respective F1-score, sensitivity, and accuracy for different sizes of the feature vector. In particular, these tables show the performances when the 450, 150, 100, and 50 best features were kept.

When 450 features were used, it is depicted in Table 6 that there was not an FSM that outperformed the rest. The best performances were obtained by the KNN and ANN models, the former using the FS2 and the combination KNN-SVM/SHAP/ERF (0.90) and the latter using FS1; the best combination was ANN-SVM/SHAP/ERF (0.84). On the other hand, the F1-scores of the Siena dataset experiments were lower, on average. The combinations having the best performance were KNN-SVM/LIME (0.83) and KNN-DT/SVM/LIME/SHAP/ERF/RR (0.86), for FS1 and FS2, respectively.

Observe that the experiments using FS2 tended to report better performances than those in the FS1 experiments, no matter the dataset used. Furthermore, it should be noted that there were large differences between the accuracy and the F1-score values, due to the large class imbalance. The performances of the DT and SVM classifiers were noticeably lower than ANN/RF/KNN, the SVM being the classifier with the worst performance values.

Table 6. Performance comparison using 450 features. The largest F1-scores are displayed in red bold and blue bold for FS1 and FS2, respectively. F1s = F1-score, Se = sensitivity, Acc = accuracy.

		DT			SVM			ANN			RF			KNN		
		CHB-MIT dataset														
FSM		F1s	Se	Acc	F1s	Se	Acc	F1s	Se	Acc	F1s	Se	Acc	F1s	Se	Acc
DT	FS1	0.72	0.73	0.94	0.69	0.53	0.95	0.82	0.76	0.96	0.81	0.69	0.96	0.78	0.65	0.96
	FS2	0.74	0.74	0.94	0.75	0.61	0.95	0.86	0.81	0.97	0.84	0.73	0.97	0.89	0.83	0.98
SVM	FS1	0.72	0.73	0.94	0.70	0.55	0.95	0.84	0.78	0.97	0.81	0.69	0.96	0.82	0.72	0.96
	FS2	0.73	0.74	0.94	0.76	0.63	0.96	0.86	0.80	0.97	0.84	0.73	0.97	0.90	0.84	0.98
LIME	FS1	0.72	0.72	0.94	0.72	0.58	0.95	0.83	0.77	0.96	0.80	0.68	0.96	0.78	0.65	0.96
	FS2	0.73	0.74	0.94	0.76	0.63	0.96	0.87	0.81	0.97	0.84	0.73	0.97	0.89	0.83	0.98
SHAP	FS1	0.72	0.73	0.94	0.70	0.56	0.95	0.84	0.77	0.97	0.81	0.69	0.96	0.78	0.65	0.96
	FS2	0.73	0.74	0.94	0.76	0.62	0.96	0.87	0.82	0.97	0.84	0.73	0.97	0.90	0.83	0.98
ERF	FS1	0.72	0.73	0.94	0.70	0.55	0.95	0.84	0.78	0.97	0.82	0.70	0.96	0.80	0.68	0.96
	FS2	0.74	0.74	0.94	0.75	0.61	0.95	0.86	0.81	0.97	0.84	0.73	0.97	0.90	0.84	0.98
RR	FS1	0.67	0.68	0.93	0.62	0.46	0.94	0.73	0.63	0.95	0.73	0.59	0.95	0.73	0.59	0.95
	FS2	0.71	0.72	0.94	0.71	0.56	0.95	0.84	0.78	0.97	0.81	0.69	0.96	0.89	0.82	0.97
		Siena Scalp EEG dataset														
FSM		F1s	Se	Acc	F1s	Se	Acc	F1s	Se	Acc	F1s	Se	Acc	F1s	Se	Acc
DT	FS1	0.69	0.69	0.91	0.48	0.33	0.90	0.75	0.69	0.93	0.80	0.68	0.95	0.82	0.77	0.95
	FS2	0.71	0.71	0.91	0.53	0.37	0.90	0.78	0.71	0.94	0.85	0.75	0.96	0.86	0.84	0.96
SVM	FS1	0.69	0.69	0.91	0.52	0.36	0.90	0.75	0.68	0.93	0.80	0.68	0.95	0.83	0.79	0.95
	FS2	0.71	0.72	0.91	0.53	0.37	0.90	0.78	0.74	0.94	0.85	0.75	0.96	0.86	0.84	0.96
LIME	FS1	0.69	0.70	0.91	0.54	0.38	0.91	0.78	0.72	0.94	0.81	0.69	0.95	0.83	0.79	0.95
	FS2	0.71	0.71	0.91	0.53	0.37	0.90	0.77	0.73	0.93	0.85	0.76	0.96	0.86	0.84	0.96
SHAP	FS1	0.69	0.70	0.91	0.51	0.35	0.90	0.77	0.71	0.94	0.81	0.70	0.95	0.82	0.79	0.95
	FS2	0.71	0.71	0.91	0.53	0.37	0.90	0.78	0.71	0.94	0.85	0.75	0.96	0.86	0.84	0.96
ERF	FS1	0.70	0.70	0.91	0.47	0.32	0.90	0.75	0.68	0.93	0.82	0.71	0.95	0.67	0.68	0.95
	FS2	0.71	0.72	0.91	0.53	0.36	0.90	0.79	0.75	0.94	0.85	0.76	0.96	0.86	0.84	0.96
RR	FS1	0.67	0.68	0.90	0.28	0.17	0.88	0.68	0.60	0.92	0.76	0.62	0.94	0.78	0.72	0.94
	FS2	0.70	0.71	0.91	0.51	0.35	0.90	0.78	0.73	0.94	0.85	0.75	0.96	0.86	0.83	0.96

Table 7 shows the classification metrics for the 150-feature experiments. The best performances, using the CHB-MIT dataset and FS1, were returned by the combination ANN-LIME/SHAP/ERF (0.85); for the FS2 case, this was KNN-SVM/ERF (0.89). On the other hand, the Siena experiments showed that the best-performing combinations included the ERF as an FSM; RF-ERF returned the largest F1-scores, and these were 0.85 and 0.86 for FS1 and FS2, respectively.

Table 7. Performance comparison using 150 features. The largest F1-scores are displayed in red bold and blue bold for FS1 and FS2, respectively. F1s = F1-score, Se = sensitivity, Acc = accuracy.

		DT			SVM			ANN			RF			KNN		
							CHB-MIT dataset									
FSM		F1s	Se	Acc	F1s	Se	Acc	F1s	Se	Acc	F1s	Se	Acc	F1s	Se	Acc
DT	FS1	0.73	0.74	0.94	0.68	0.53	0.94	0.83	0.77	0.96	0.82	0.71	0.96	0.70	0.55	0.95
	FS2	0.74	0.74	0.94	0.74	0.60	0.95	0.85	0.79	0.97	0.84	0.74	0.97	0.87	0.79	0.97
SVM	FS1	0.73	0.73	0.94	0.66	0.50	0.94	0.84	0.77	0.97	0.82	0.71	0.96	0.82	0.73	0.96
	FS2	0.74	0.75	0.94	0.72	0.58	0.95	0.86	0.79	0.97	0.84	0.74	0.97	0.89	0.82	0.97
LIME	FS1	0.72	0.73	0.94	0.75	0.62	0.95	0.85	0.80	0.97	0.81	0.69	0.96	0.69	0.54	0.95
	FS2	0.73	0.74	0.94	0.80	0.68	0.96	0.87	0.82	0.97	0.84	0.74	0.97	0.86	0.77	0.97
SHAP	FS1	0.73	0.74	0.94	0.73	0.59	0.95	0.85	0.80	0.97	0.80	0.68	0.96	0.71	0.56	0.95
	FS2	0.74	0.75	0.94	0.76	0.63	0.96	0.87	0.82	0.97	0.83	0.73	0.97	0.86	0.77	0.97
ERF	FS1	0.73	0.74	0.94	0.74	0.60	0.95	0.85	0.79	0.97	0.83	0.72	0.97	0.80	0.69	0.96
	FS2	0.74	0.75	0.94	0.72	0.58	0.95	0.86	0.80	0.97	0.84	0.74	0.97	0.89	0.83	0.98
RR	FS1	0.63	0.63	0.92	0.51	0.35	0.93	0.63	0.51	0.93	0.66	0.50	0.94	0.57	0.41	0.93
	FS2	0.67	0.68	0.93	0.59	0.42	0.93	0.77	0.68	0.95	0.76	0.62	0.96	0.84	0.74	0.97
							Siena Scalp EEG dataset									
FSM		F1s	Se	Acc	F1s	Se	Acc	F1s	Se	Acc	F1s	Se	Acc	F1s	Se	Acc
DT	FS1	0.70	0.70	0.91	0.30	0.18	0.88	0.73	0.66	0.93	0.82	0.71	0.95	0.72	0.64	0.93
	FS2	0.72	0.73	0.91	0.41	0.26	0.89	0.74	0.66	0.93	0.85	0.76	0.96	0.85	0.82	0.95
SVM	FS1	0.66	0.67	0.90	0.42	0.27	0.89	0.71	0.64	0.93	0.77	0.63	0.94	0.80	0.75	0.94
	FS2	0.70	0.71	0.91	0.61	0.45	0.91	0.78	0.72	0.94	0.83	0.73	0.95	0.85	0.81	0.95
LIME	FS1	0.69	0.70	0.91	0.62	0.47	0.92	0.80	0.76	0.94	0.82	0.71	0.95	0.80	0.74	0.94
	FS2	0.71	0.71	0.91	0.69	0.55	0.92	0.83	0.80	0.95	0.84	0.74	0.96	0.85	0.81	0.95
SHAP	FS1	0.71	0.71	0.91	0.56	0.40	0.91	0.79	0.75	0.94	0.83	0.73	0.95	0.80	0.75	0.94
	FS2	0.71	0.72	0.91	0.67	0.52	0.92	0.82	0.78	0.95	0.85	0.77	0.96	0.85	0.81	0.95
ERF	FS1	0.71	0.72	0.92	0.30	0.18	0.88	0.75	0.71	0.93	0.85	0.76	0.96	0.64	0.65	0.93
	FS2	0.72	0.73	0.92	0.40	0.25	0.89	0.76	0.71	0.93	0.86	0.77	0.96	0.83	0.78	0.95
RR	FS1	0.64	0.65	0.90	0.16	0.09	0.87	0.57	0.47	0.90	0.66	0.50	0.93	0.66	0.56	0.92
	FS2	0.67	0.68	0.90	0.29	0.17	0.87	0.62	0.52	0.91	0.78	0.65	0.94	0.81	0.76	0.94

In the 100-feature case (see Table 8), the behavior was similar to the 450-feature case, in the sense that the best-suited FSM depended on the feature set and dataset. For the CHB-MIT dataset, the combinations with the best performances were KNN-SVM (0.85) and KNN-DT/SVM/ERF (0.87) for FS1 and FS2, respectively. For the Siena dataset, the best performance was reported by RF-ERF (0.85 and 0.86).

Table 8. Performance comparison using 100 features. The largest F1-scores are displayed in red bold and blue bold for FS1 and FS2, respectively. F1s = F1-score, Se = sensitivity, Acc = accuracy.

		DT			SVM			ANN			RF			KNN		
							CHB-MIT dataset									
FSM		F1s	Se	Acc	F1s	Se	Acc	F1s	Se	Acc	F1s	Se	Acc	F1s	Se	Acc
DT	FS1	0.73	0.74	0.94	0.68	0.53	0.95	0.81	0.74	0.96	0.83	0.72	0.97	0.73	0.60	0.95
	FS2	0.74	0.75	0.94	0.72	0.58	0.95	0.84	0.78	0.97	0.84	0.74	0.97	0.87	0.79	0.97
SVM	FS1	0.73	0.73	0.94	0.67	0.52	0.94	0.83	0.76	0.96	0.83	0.72	0.97	0.85	0.76	0.97
	FS2	0.74	0.75	0.94	0.68	0.52	0.94	0.84	0.77	0.97	0.84	0.74	0.97	0.87	0.80	0.97
LIME	FS1	0.72	0.73	0.94	0.74	0.61	0.95	0.84	0.77	0.96	0.81	0.69	0.96	0.65	0.49	0.94
	FS2	0.73	0.74	0.94	0.79	0.68	0.96	0.86	0.80	0.97	0.84	0.74	0.97	0.83	0.73	0.96
SHAP	FS1	0.74	0.74	0.94	0.73	0.59	0.95	0.84	0.78	0.97	0.80	0.68	0.96	0.68	0.52	0.94
	FS2	0.74	0.75	0.94	0.75	0.63	0.95	0.85	0.80	0.97	0.83	0.72	0.97	0.83	0.73	0.97
ERF	FS1	0.74	0.74	0.94	0.70	0.55	0.95	0.83	0.75	0.96	0.83	0.73	0.97	0.82	0.72	0.96
	FS2	0.74	0.75	0.94	0.71	0.56	0.95	0.85	0.79	0.97	0.85	0.74	0.97	0.87	0.80	0.97
RR	FS1	0.61	0.62	0.91	0.44	0.29	0.92	0.59	0.46	0.93	0.65	0.49	0.94	0.46	0.30	0.92
	FS2	0.66	0.67	0.93	0.54	0.37	0.93	0.74	0.65	0.95	0.73	0.58	0.95	0.80	0.68	0.96

Table 8. *Cont.*

FSM		DT			SVM			ANN			RF			KNN		
		F1s	**Se**	**Acc**	**F1s**	**Se**	**Acc**	**F1s**	**Se**	**Acc**	**F1s**	**Se**	**Acc**	**F1s**	**Se**	**Acc**
								Siena Scalp EEG dataset								
DT	FS1	0.71	0.72	0.91	0.24	0.14	0.87	0.70	0.63	0.92	0.83	0.73	0.96	0.70	0.61	0.92
	FS2	0.72	0.72	0.91	0.36	0.22	0.88	0.71	0.63	0.92	0.85	0.76	0.96	0.83	0.80	0.95
SVM	FS1	0.66	0.67	0.90	0.35	0.22	0.89	0.70	0.63	0.92	0.78	0.66	0.95	0.75	0.69	0.93
	FS2	0.69	0.70	0.91	0.61	0.46	0.91	0.78	0.73	0.94	0.82	0.72	0.95	0.84	0.81	0.95
LIME	FS1	0.69	0.70	0.91	0.64	0.50	0.92	0.77	0.72	0.94	0.81	0.71	0.95	0.78	0.71	0.94
	FS2	0.71	0.72	0.91	0.71	0.58	0.93	0.82	0.78	0.95	0.84	0.75	0.96	0.84	0.80	0.95
SHAP	FS1	0.70	0.71	0.91	0.55	0.40	0.91	0.78	0.74	0.94	0.84	0.74	0.96	0.78	0.73	0.94
	FS2	0.71	0.72	0.91	0.70	0.57	0.93	0.81	0.77	0.94	0.85	0.77	0.96	0.83	0.80	0.95
ERF	FS1	0.72	0.72	0.92	0.32	0.20	0.88	0.74	0.67	0.93	**0.85**	0.76	0.96	0.62	0.63	0.92
	FS2	0.73	0.74	0.92	0.28	0.16	0.87	0.72	0.65	0.92	**0.86**	0.77	0.96	0.80	0.73	0.94
RR	FS1	0.62	0.63	0.89	0.14	0.07	0.87	0.51	0.39	0.89	0.64	0.47	0.92	0.61	0.49	0.91
	FS2	0.65	0.66	0.89	0.25	0.14	0.87	0.58	0.47	0.90	0.77	0.64	0.94	0.76	0.69	0.93

Table 9 shows that classifiers accounting for the best performances, RF and KNN, 0.84 and 0.85 being the largest values for both of the datasets. It is interesting to note that the best performances were similar for both datasets and the feature sets.

Table 9. Performance comparison using 50 features. The largest F1-scores are displayed in red bold and blue bold for FS1 and FS2, respectively. F1s = F1-score, Se = sensitivity, Acc = accuracy.

FSM		DT			SVM			ANN			RF			KNN		
		F1s	**Se**	**Acc**	**F1s**	**Se**	**Acc**	**F1s**	**Se**	**Acc**	**F1s**	**Se**	**Acc**	**F1s**	**Se**	**Acc**
								CHB-MIT dataset								
DT	FS1	0.73	0.74	0.94	0.68	0.53	0.94	0.79	0.70	0.96	0.83	0.72	0.97	0.70	0.56	0.95
	FS2	0.74	0.74	0.94	0.70	0.55	0.95	0.82	0.75	0.96	**0.84**	0.74	0.97	0.82	0.72	0.96
SVM	FS1	0.71	0.72	0.94	0.65	0.49	0.94	0.79	0.71	0.96	0.81	0.70	0.96	**0.85**	0.77	0.97
	FS2	0.73	0.74	0.94	0.63	0.47	0.94	0.81	0.73	0.96	0.83	0.73	0.97	**0.84**	0.75	0.97
LIME	FS1	0.72	0.73	0.94	0.73	0.59	0.95	0.81	0.73	0.96	0.78	0.65	0.96	0.60	0.44	0.94
	FS2	0.73	0.74	0.94	0.76	0.63	0.95	0.81	0.74	0.96	0.83	0.72	0.97	0.76	0.64	0.95
SHAP	FS1	0.74	0.74	0.94	0.73	0.59	0.95	0.80	0.73	0.96	0.78	0.65	0.96	0.62	0.46	0.94
	FS2	0.74	0.75	0.94	0.71	0.57	0.95	0.79	0.71	0.96	0.82	0.72	0.96	0.78	0.66	0.96
ERF	FS1	0.73	0.74	0.94	0.60	0.44	0.94	0.79	0.71	0.96	0.83	0.72	0.96	0.82	0.73	0.96
	FS2	0.74	0.75	0.94	0.65	0.49	0.94	0.82	0.74	0.96	**0.84**	0.74	0.97	**0.84**	0.75	0.97
RR	FS1	0.55	0.57	0.90	0.37	0.23	0.91	0.53	0.39	0.93	0.53	0.37	0.93	0.37	0.23	0.91
	FS2	0.64	0.65	0.92	0.48	0.32	0.92	0.64	0.52	0.94	0.69	0.54	0.95	0.70	0.55	0.95
								Siena Scalp EEG dataset								
FSM		**F1s**	**Se**	**Acc**	**F1s**	**Se**	**Acc**	**F1s**	**Se**	**Acc**	**F1s**	**Se**	**Acc**	**F1s**	**Se**	**Acc**
DT	FS1	0.71	0.71	0.91	0.20	0.11	0.87	0.65	0.56	0.91	0.83	0.73	0.95	0.57	0.47	0.90
	FS2	0.72	0.73	0.92	0.21	0.12	0.87	0.64	0.53	0.91	0.84	0.75	0.96	0.79	0.73	0.94
SVM	FS1	0.63	0.64	0.89	0.28	0.16	0.88	0.64	0.54	0.91	0.75	0.61	0.94	0.66	0.56	0.92
	FS2	0.68	0.68	0.90	0.52	0.36	0.90	0.72	0.66	0.92	0.80	0.69	0.95	0.80	0.74	0.94
LIME	FS1	0.69	0.70	0.91	0.64	0.50	0.92	0.71	0.64	0.92	0.81	0.71	0.95	0.74	0.66	0.93
	FS2	0.70	0.71	0.91	0.67	0.53	0.92	0.75	0.69	0.93	0.83	0.73	0.95	0.81	0.76	0.94
SHAP	FS1	0.71	0.73	0.92	0.56	0.40	0.91	0.73	0.68	0.93	0.83	0.73	0.95	0.73	0.64	0.93
	FS2	0.71	0.72	0.91	0.68	0.55	0.92	0.77	0.71	0.93	0.84	0.76	0.96	0.82	0.77	0.95
ERF	FS1	0.72	0.72	0.92	0.21	0.11	0.87	0.64	0.55	0.91	**0.84**	0.75	0.96	0.63	0.64	0.89
	FS2	0.73	0.73	0.92	0.19	0.10	0.87	0.66	0.55	0.91	**0.85**	0.76	0.96	0.71	0.64	0.92
RR	FS1	0.63	0.64	0.89	0.14	0.07	0.87	0.40	0.27	0.88	0.59	0.43	0.91	0.49	0.36	0.89
	FS2	0.63	0.64	0.89	0.19	0.10	0.87	0.48	0.35	0.89	0.73	0.59	0.93	0.71	0.62	0.92

It is worth noticing that the RF classifier presented the most steady performance (≈ 0.8) no matter the FSM, dataset, and feature reduction (see Tables 6–9).

In order to identify the significant differences between FSMs, Tables 10 and 11 show the results of the F-test; given a dataset, a number of features, and a feature set (FS1 or FS2), the F-test was applied to the results of the 5×2 CV experiments. During the testing, the best-performing C-FSM (as per the F1-score) was compared to the rest of the FSMs. It must be considered that the 5×2 CV F-test evaluates the error rates, not the F1-scores, so the accuracy is computed and depicted in Tables 6–9.

In the case of several combinations having the same F1-score, the sensitivity and accuracy were considered to choose the best-performing combinations.

Table 10. Comparison of the best C-FSM when FS1 is used. It is compared against the rest of the FSMs. • denotes $p < 0.05$.

Best FSM	Dataset	Classifier	DT	SVM	LIME	SHAP	ERF	RR
450 features								
SVM	CHB-MIT	ANN						•
ERF	CHB-MIT	ANN						•
SVM	Siena	KNN						•
LIME	Siena	KNN	•					•
150 features								
LIME	CHB-MIT	ANN	•					•
SHAP	CHB-MIT	ANN	•					•
ERF	Siena	RF	•	•	•	•		•
100 features								
SVM	CHB-MIT	KNN	•		•	•		
ERF	Siena	RF	•	•	•	•		•
50 features								
SVM	CHB-MIT	KNN	•		•	•	•	•
ERF	Siena	RF		•	•			•

Table 11. Comparison of the best C-FSM when FS2 is used. It is compared against the rest of the FSMs. • denotes $p < 0.05$.

Best FSM	Dataset	Classifier	DT	SVM	LIME	SHAP	ERF	RR
450 features								
SVM	CHB-MIT	KNN			•	•		•
ERF	CHB-MIT	KNN	•		•	•		•
DT	Siena	KNN						•
SVM	Siena	KNN					•	•
LIME	Siena	KNN						•
SHAP	Siena	KNN						
ERF	Siena	KNN		•				
150 features								
ERF	CHB-MIT	KNN	•		•	•		•
ERF	Siena	RF		•	•			
100 features								
SVM	CHB-MIT	KNN			•	•		•
ERF	CHB-MIT	KNN			•	•		•
ERF	Siena	RF		•	•			•
50 features								
SVM	CHB-MIT	KNN	•		•	•		•
ERF	CHB-MIT	KNN	•		•	•		•
ERF	Siena	RF		•	•			•

The 450 feature section of Tables 10 and 11 shows that most of the test results were not statistically significant ($p > 0.05$). In Table 10, there were three cases where the FSM error rates resulted in being statistically different from the rest of the FSMs, and these were RF-ERF-150 (Siena dataset), RF-ERF-100 (Siena dataset), and KNN-SVM-50 (CHB-MIT dataset). It must be noted that the differences in accuracy may be small (see Table 6–9), even if there are statistical differences.

In Table 11, there are no cases where an FSM was statistically different from the rest of the FSMs. It was observed that the RR experiments tended to show a statistical significance that did not depend on the number of features or the dataset.

Finally, Tables 10 and 11 show that the most common combinations were KNN-SVM and RF-ERF.

3.3. Comparison of Selected Features

To determine if there were coincidences in the features selected by the FSMs and if the FSMs assigned more importance to the same features, the Jaccard index [64] was used to calculate the similarity by pairs of FSMs. For this analysis, the FS2 experiments were chosen because those experiments produced higher performances in comparison with FS1.

When 450 features were used for training the KNN classifier (see Figure 6a,e), the feature sets practically overlapped; indices had values over 0.85 for the CHB-MIT dataset and the Siena dataset. This behavior was expected, as it is highly probable to select similar features when the number of features in a dataset is large, so the 450 case will not be further discussed.

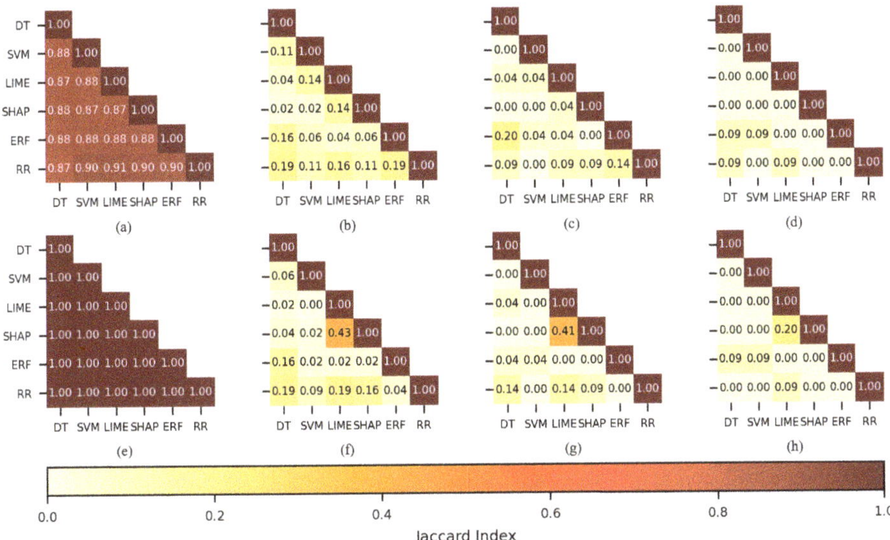

Figure 6. The similarity of the sets of features selected per each feature selection method when KNN was used as a classifier. CHB-MIT dataset: (**a**) best 450 features, (**b**) best 150, (**c**) best 100, and (**d**) best 50. Siena dataset: (**e**) best 450 features, (**f**) best 150, (**g**) best 100, and (**h**) best 50.

When 150 features were used during training (see Figure 6b,f), the Jaccard index for SHAP-LIME (0.43) was the largest of all the combinations when the Siena dataset was used; it was equivalent to 91 out of 150 features. For the CHB-MIT dataset, RR-ERF (0.19) and RR-DT (0.19) obtained the largest values. Figure 6c,d,g,h show a low coincidence for most of the FSMs when 100 or 50 features were used, and this applies to both datasets. The only notorious index was 0.41 belonging to SHAP-LIME (see Figure 6g).

In the case of the SVM classifier, by using 150 features (see Figure 7b,f), the largest values were obtained by SHAP-LIME (0.43 and 0.72); the FSMs coincided in selecting 91 features for the CHB-MIT dataset and 126 for the Siena dataset. The Jaccard indices computed for 100 features (see Figure 7c,g) showed a good similarity for SHAP-LIME (0.41), and the selected feature sets coincided in 59 features for both of the datasets. When 50 features were used for training (see Figure 7d,h), several combinations returned a similarity value of 0.2.

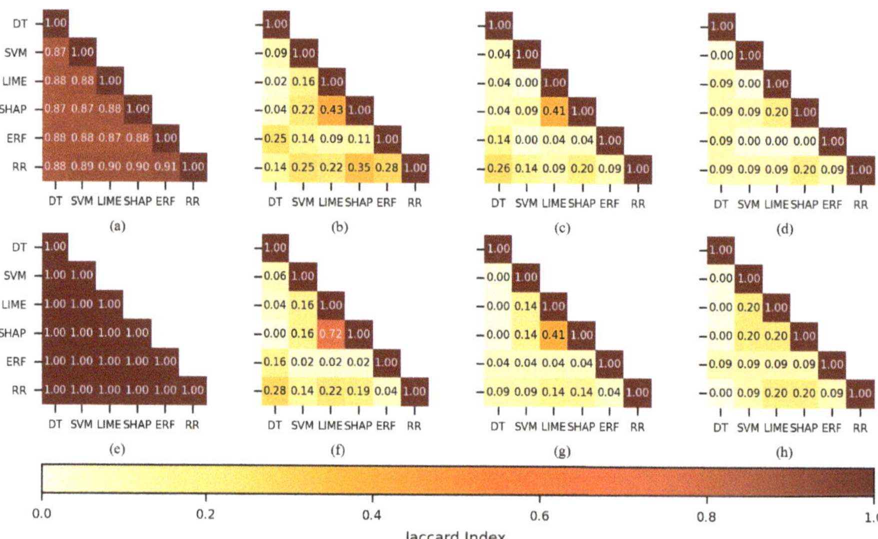

Figure 7. The similarity of the sets of features selected per each feature selection method when SVM was used as a classifier. CHB-MIT dataset: (**a**) best 450 features, (**b**) best 150, (**c**) best 100, and (**d**) best 50. Siena dataset: (**e**) best 450 features, (**f**) best 150, (**g**) best 100, and (**h**) best 50.

The ANN classifier followed a similar pattern as SVM; SHAP-LIME obtained larger indices than most of the combinations, and these were 0.72 and 0.52 for 150 features, 0.71 and 0.50 for 100, and 1.0 and 0.33 for 50 (see Figure 8b–d,f–h), respectively. It should be noted that a Jaccard index of 1 means that both sets totally overlapped.

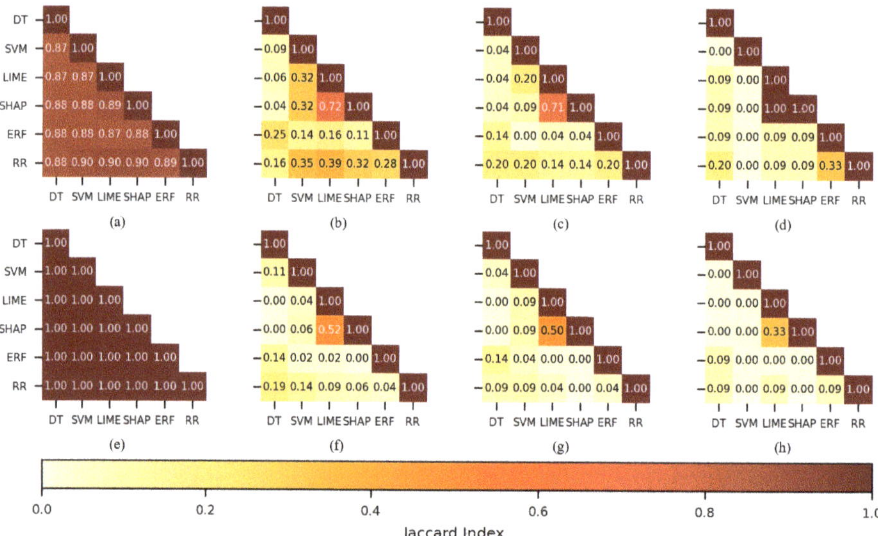

Figure 8. The similarity of the sets of features selected per each feature selection method when ANN was used as a classifier. CHB-MIT dataset: (**a**) best 450 features, (**b**) best 150, (**c**) best 100, and (**d**) best 50. Siena dataset: (**e**) best 450 features, (**f**) best 150, (**g**) best 100, and (**h**) best 50.

Interestingly, when the DT classifier was used, the SHAP-LIME similarity was equal to or lower than other combinations. When using 150 features, the largest indices were returned by the combination RR-SHAP (see Figure 9b,f). Figure 9c,g,h show that the largest similarity values were returned by combinations including LIME or SHAP.

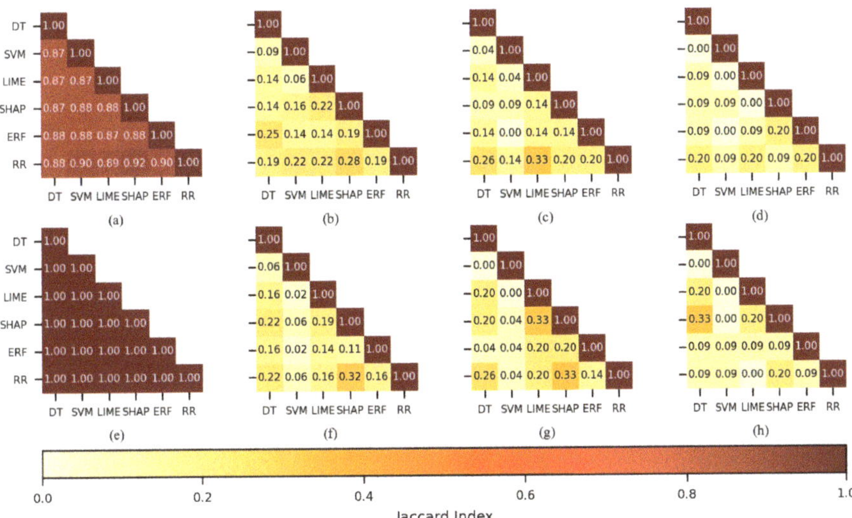

Figure 9. The similarity of the sets of features selected per each feature selection method when DT was used as a classifier. CHB-MIT dataset: (**a**) best 450 features, (**b**) best 150, (**c**) best 100, and (**d**) best 50. Siena dataset: (**e**) best 450 features, (**f**) best 150, (**g**) best 100, and (**h**) best 50.

Comparable to the DT, the RF case showed the largest values when a combination included LIME or SHAP. A remarkable fact is that there were two combinations having an index value of 0.5 (see Figure 10c,d).

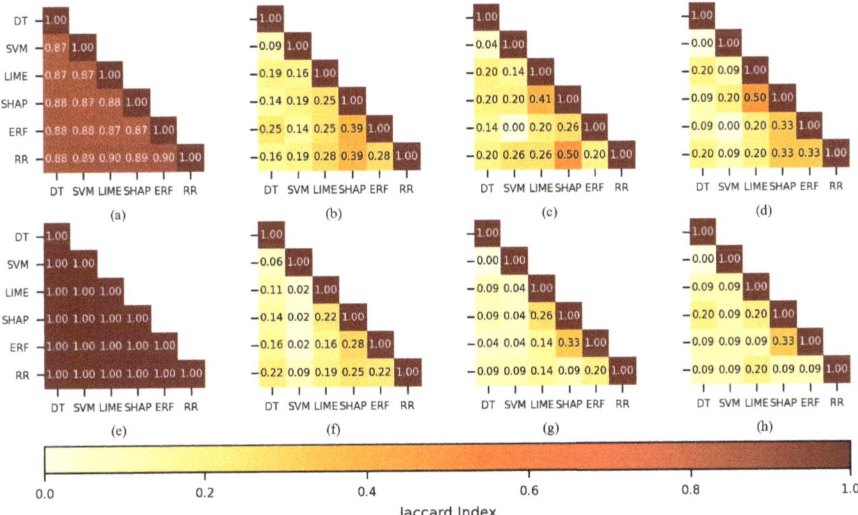

Figure 10. The similarity of the sets of features selected per each feature selection method when RF was used as a classifier. CHB-MIT dataset: (**a**) best 450 features, (**b**) best 150, (**c**) best 100, and (**d**) best 50. Siena dataset: (**e**) best 450 features, (**f**) best 150, (**g**) best 100, and (**h**) best 50.

In order to compare the selected features for the best C-FSM combinations in Table 9, Figures 11 and 12 show the top-10 features for KNN-SVM and RF-ERF, respectively. The former figure corresponds to the CHB-MIT dataset, while the latter to the Siena dataset. Each feature is defined as follows: EEG band/bipolar channel/metric. For example, in Figure 11, the feature with the greatest importance value is "alpha_FP1-F3_skew", which indicates that the most important feature was the skewness measured in the bipolar channel FP1-F3 on the alpha band. It is important to note that the selected features may vary due to several factors, including seizure type, epileptogenic region, and EEG montage, among others.

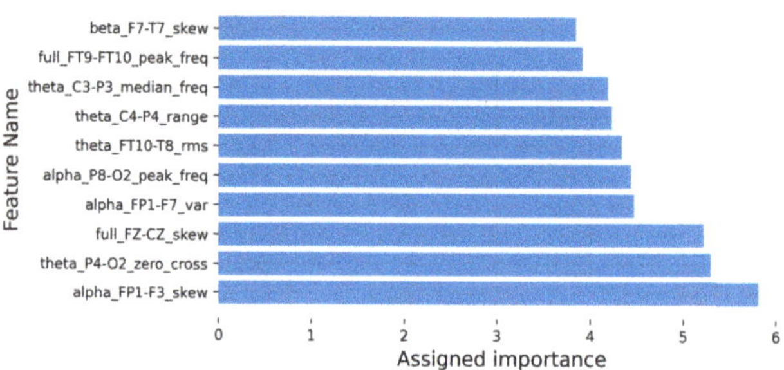

Figure 11. Features having the largest assigned importance. The results correspond to the combination KNN-SVM and the CHB-MIT dataset. Importance values are not normalized.

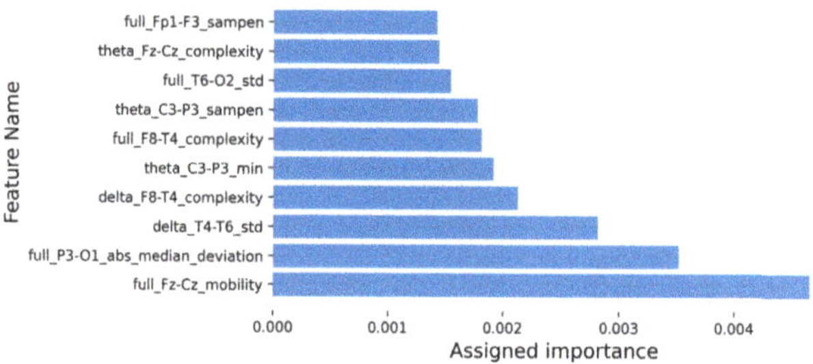

Figure 12. Features having the largest assigned importance. The results correspond to the combination RF-ERF and the Siena dataset. Importance values are not normalized.

4. Discussion

In the evaluation of feature dimensionality reduction analysis, it was observed that the best machine learning classifiers were ANN, RF, and KNN, taking into account neither the database nor the FS. This is evident by looking at the performance tendency of every combination of C-FSMs (colored lines). It is important to note that there was a recurrent behavior in all the combinations, that is almost all of them started showing a performance decrease when the feature vectors were around a length of 50. Lower than that, the scores began to be around 0.6 or less. Therefore, a feature vector of a length of 50 is the minimum suggested for having good classification performance using some of the C-FSM combinations, for FS in epilepsy databases. Moreover, if FS2 is used, the scores achieved by most of the C-FSM combinations were improved and steadier. In this sense, the DT and RF classifiers were less affected by the dimensionality reduction no matter the database, the FS, nor the FSM. This could be explained by the nature of those algorithms, that is both algorithms use a ranking metric to determine the importance of each feature vector in a particular classification task.

Regarding to the comparison of C-FSM combinations, it can be noticed that the models that were trained with CHB-MIT data had better F1-scores than the ones trained with

Siena data. We considered that the performance difference was due to the CHB-MIT database being larger than the Siena Scalp EEG database. Actually, the number of CHB-MIT epochs used during experimentation was more than twice the number of Siena epochs. Another aspect to consider is the number of channels, 21 and 18 for the CHB-MIT and Siena datasets, respectively.

In addition, we can observe in Tables 6–9 that the greater the number of features, the ANN and KNN showed better F1-scores; on the other hand, the lower the number of features, KNN kept showing the best F1-scores, and RF emerged with better F1-scores than the ANN. Note that, even though KNN presented the highest F1-scores, the difference between KNN and RF reduced as the number of features decreased.

Furthermore, in Tables 10 and 11, it can be seen that the smaller the number of features, the greater the number of significant differences was between the best C-FSM and the other FSMs. In this sense, these significant differences in the classifiers' performance indicate a relationship between the classifier and the feature selection method; in this case, for a lower number of features, KNN was better when using SVM and ERF, while RF was better when using ERF.

The main purpose of this study was not to train a classifier for seizure classification from EEG data; however, it can be considered that the performances obtained, in particular for KNN and RF, were similar to other studies when experiments were conducted under similar conditions, specifically the same classifier and EEG database.

In this sense, comparing our work with the state-of-the-art, in [12], the authors used the CHB-MIT database, seven feature selection methods, nine classifiers, and selecting the top 20 or 25 features, obtaining the best mean classification error of 0.12 by using the KNN classifier. Nonetheless, the authors tested their methodology neither using a wider spectrum of feature vector sizes, nor using different databases. In this work, we found that having the top 50 features, on the same database, the combinations KNN-SVM-50 (FS1), KNN-SVM-50 (FS2), and KNN-ERF-50 (FS2) (see Table 9) returned a mean classification error of 0.03. Moreover, the results in [12] correspond to a balanced dataset, while in this work, the proportion between seizure and no-seizure was 1:9, respectively, our methodology being more appropriate for seizure and no-seizure detection considering that epilepsy datasets are naturally unbalanced.

Additionally, Kathi and Ingle [25] presented an accuracy of 0.97 and an F1-score of 0.97 by using KNN and 11 feature metrics, testing their methodology on the Bonn University database [65] and using an equal proportion of healthy and seizure instances for the training and test sets. Furthermore, they reported a reduction in the feature set used to compute the feature vectors instead of directly reducing the size of the feature vectors. Hence, the authors did not evaluate either different sizes of the feature vectors or the different feature selection methods. On the contrary, in the present work, several experiments reported an accuracy of 0.97, but had a lower F1-score (0.84 and 0.85); however, these lower scores were expected considering that, for unbalanced datasets, the F1-score is more reliable.

On the other hand, RR presented the worst results no matter the classifier used, contrasting with [54], who reported RR as one of the FSMs that performed better and also provided good stability across datasets. However, the nature of the datasets used in [54] was different from EEG, which indicates that RR is not appropriate to be used for seizure and no-seizure detection.

Finally, concerning the comparison of selected features, by using the Jaccard index, it was observed that most of the time, SHAP-LIME returned the largest levels of similarity, meaning that both FMSs usually selected the same features, even though they did not present the best performance on the classification; on the contrary, SVM and ERF, which were the best FSMs for KNN and RF, respectively, did not present a higher Jaccard index.

Large values in SHAP-LIME were partially expected because, as explained in [49,50], SHAP and LIME are designed to be model-agnostic and to explain the classification model. In addition, the SHAP implementation applied in this study (kernel SHAP) followed a similar approach to LIME, but different functions were applied for the estimation of the locality and the similarity between the classification f and interpretable g model. In the case of SVM and ERF, the low levels of the similarity of features selected were partially explained because the nature and training procedure of SVM are quite different from tree-based models.

Therefore, this experiment showed that two different feature spaces might result in a good classification performance, such as the features selected for KNN-SVM and RF-ERF (Figures 11 and 12). On the contrary, the RR method presented good similarity values compared to other FSMs; nonetheless, it presented the worst classification performances.

5. Conclusions and Future Work

In the present study, six FSMs were compared to define, first, the minimum number of features that can be chosen for each FSM without sacrificing the classifiers' performances, second, the performance of several C-FSM combinations to discover if a relationship exists between the FSMs with a particular classification algorithm, and third, if there is a feature or feature set that remains across different C-FSM combinations.

We can conclude that when the number of selected features was drastically reduced (100 and 50 features), the differences between classifiers' performance increased, but none of the FSMs showed a predominance over the rest. Furthermore, it was observed that it was possible to perform a large reduction of the number of features while having a low impact on the model performance until having a 50-feature vector length.

The results indicated that the classifiers' performance might be affected by diverse factors such as the EEG database, the features, and the number of features. However, the combinations KNN-SVM and RF-ERF are advised. Furthermore, the use of RR is not appropriate for seizure EEG data, as it yielded lower performances than the rest of the FSMs, and it was more time-consuming.

Regarding the proposed feature sets, FS2 is suggested to be used on seizure and no-seizure classification problems, given that the performance of several C-FSMs was improved while using it.

In future work, some opportunity areas could be explored: First, a more extended analysis is required to evaluate the FSMs in combination with deep learning models, including a more extensive parameter tuning process and the use of more complex features. Furthermore, considering that our results showed that RF-ERF obtained one of the best performances, it would be interesting to perform an evaluation of tree-based models for feature selection and classification of seizure and no-seizure EEG data.

Second, FSM stability was not evaluated in this study; the evaluation of stability with instance perturbation, as proposed by [48], would help to evaluate the robustness of an FSM against small variations in the EEG dataset, because variations can be caused by inter-subject variability or noise.

Finally, assessing a subject cross-validation methodology would be interesting for future work to test inter-subject performances.

Author Contributions: Conceptualization, S.E.S.-H.; methodology, S.E.S.-H., R.A.S.-R., S.T.-R., and I.R.-G.; software, S.E.S.-H.; validation, S.E.S.-H., R.A.S.-R., S.T.-R., and I.R.-G.; formal analysis, S.E.S.-H., R.A.S.-R., S.T.-R., and I.R.-G.; investigation, S.E.S.-H., R.A.S.-R., S.T.-R., and I.R.-G.; data curation, S.E.S.-H.; writing—original draft preparation, S.E.S.-H.; writing—review and editing, R.A.S.-R., S.T.-R., and I.R.-G.; visualization, S.E.S.-H., R.A.S.-R., S.T.-R., and I.R.-G.; supervision, R.A.S.-R., S.T.-R., and I.R.-G. All authors have read and agreed to the published version of the manuscript.

Funding: This research received no external funding.

Institutional Review Board Statement: Not applicable.

Informed Consent Statement: Not applicable.

Data Availability Statement: Publicly available datasets were analyzed in this study. The CHB-MIT Scalp EEG Database is available at https://physionet.org/content/chbmit/1.0.0/ (accessed on

18 October 2021) and the Siena Scalp EEG Database is available at https://physionet.org/content/siena-scalp-eeg/1.0.0/ (accessed on 18 October 2021).

Acknowledgments: We appreciate the facilities provided to the project "Identificación de patrones en registros EEG de crisis epilépticas utilizando métodos de inteligencia artificial explicable" by the Data Analysis and Supercomputing Center (CADS, for its acronym in Spanish) of the University of Guadalajara through the use of the Leo Átrox Supercomputer.

Conflicts of Interest: The authors declare no conflict of interest.

References

1. World Health Organization Epilepsy. 2019. Available online: https://www.who.int/health-topics/epilepsy#tab=tab_1 (accessed on 31 October 2021).
2. Zack, M.M.; Kobau, R. National and State Estimates of the Numbers of Adults and Children with Active Epilepsy—United States, 2015. *MMWR Morb. Mortal. Wkly. Rep.* **2017**, *66*, 821. [CrossRef]
3. Aaberg, K.M.; Gunnes, N.; Bakken, I.J.; Lund Søraas, C.; Berntsen, A.; Magnus, P.; Lossius, M.I.; Stoltenberg, C.; Chin, R.; Surén, P. Incidence and Prevalence of Childhood Epilepsy: A Nationwide Cohort Study. *Pediatrics* **2017**, *139*, e20163908. [CrossRef] [PubMed]
4. Noriega-Morales, G.; Shkurovich-Bialik, P. Situación de la epilepsia en México y América Latina. *Anales Médicos Asoc. Médica Cent. Médico ABC* **2020**, *65*, 224–232. [CrossRef]
5. Smith, S.J.M. EEG in the diagnosis, classification, and management of patients with epilepsy. *J. Neurol. Neurosurg. Psychiatry* **2005**, *76*, ii2–ii7. [CrossRef] [PubMed]
6. Gill, A.F.; Fatima, S.A.; Usman Akram, M.; Khawaja, S.G.; Awan, S.E. Analysis of EEG Signals for Detection of Epileptic Seizure Using Hybrid Feature Set. In *Theory and Applications of Applied Electromagnetics*; Springer: Cham, Switzerland, 2015; pp. 49–57. [CrossRef]
7. Pinto-Orellana, M.A.; Cerqueira, F.R. Patient-Specific Epilepsy Seizure Detection Using Random Forest Classification over One-Dimension Transformed EEG Data. In *Intelligent Systems Design and Applications*; Springer: Cham, Switzerland, 2017; pp. 519–528. [CrossRef]
8. Tsiouris, K.M.; Pezoulas, V.C.; Zervakis, M.; Konitsiotis, S.; Koutsouris, D.D.; Fotiadis, D.I. A Long Short-Term Memory deep learning network for the prediction of epileptic seizures using EEG signals. *Comput. Biol. Med.* **2018**, *99*, 24–37. [CrossRef] [PubMed]
9. Yang, Y.; Zhou, M.; Niu, Y.; Li, C.; Cao, R.; Wang, B.; Yan, P.; Ma, Y.; Xiang, J. Epileptic Seizure Prediction Based on Permutation Entropy. *Front. Comput. Neurosci.* **2018**, *12*, 55. [CrossRef] [PubMed]
10. Usman, S.M.; Hassan, A. Efficient Prediction and Classification of Epileptic Seizures Using EEG Data Based on Univariate Linear Features. *J. Comput.* **2018**, *13*, 616–621. [CrossRef]
11. Siddiqui, M.K.; Islam, M.Z.; Kabir, M.A. Analyzing performance of classification techniques in detecting epileptic seizure. In *Advanced Data Mining and Applications*; Springer: Cham, Switzerland, 2017; pp. 386–398. [CrossRef]
12. Fergus, P.; Hussain, A.; Hignett, D.; Al-Jumeily, D.; Abdel-Aziz, K.; Hamdan, H. A machine learning system for automated whole-brain seizure detection. *Appl. Comput. Inform.* **2016**, *12*, 70–89. [CrossRef]
13. Shoeb, A.; Guttag, J. Application of Machine Learning to Epileptic Seizure Detection. In Proceedings of the 27th International Conference on International Conference on Machine Learning, Omnipress, Haifa, Israel, 21–24 June 2010; pp. 975–982. [CrossRef]
14. Birjandtalab, J.; Baran Pouyan, M.; Cogan, D.; Nourani, M.; Harvey, J. Automated Seizure Detection Using Limited-Channel EEG and Non-Linear Dimension Reduction. *Comput. Biol. Med.* **2017**, *82*, 49–58. [CrossRef]
15. Zhan, Q.; Hu, W. An Epilepsy Detection Method Using Multiview Clustering Algorithm and Deep Features. *Comput. Math. Methods Med.* **2020**, *2020*, 5128729. [CrossRef] [PubMed]
16. Rashed-Al-Mahfuz, M.; Moni, M.A.; Uddin, S.; Alyami, S.A.; Summers, M.A.; Eapen, V. A Deep Convolutional Neural Network Method to Detect Seizures and Characteristic Frequencies Using Epileptic Electroencephalogram (EEG) Data. *IEEE J. Transl. Eng. Health Med.* **2021**, *9*, 1–12. [CrossRef] [PubMed]
17. Usman, S.M.; Khalid, S.; Aslam, M.H. Epileptic Seizures Prediction Using Deep Learning Techniques. *IEEE Access* **2020**, *8*, 39998–40007. [CrossRef]
18. Deivasigamani, S.; Senthilpari, C.; Yong, W.H. Machine learning method based detection and diagnosis for epilepsy in EEG signal. *J. Ambient. Intell. Humaniz. Comput.* **2021**, *12*, 4215–4221. [CrossRef]
19. Glory, H.A.; Vigneswaran, C.; Jagtap, S.S.; Shruthi, R.; Hariharan, G.; Sriram, V.S.S. AHW-BGOA-DNN: A novel deep learning model for epileptic seizure detection. *Neural Comput. Appl.* **2021**, *33*, 6065–6093. [CrossRef]
20. Kitano, L.A.S.; Sousa, M.A.A.; Santos, S.D.; Pires, R.; Thome-Souza, S.; Campo, A.B. Epileptic Seizure Prediction from EEG Signals Using Unsupervised Learning and a Polling-Based Decision Process. In *Artificial Neural Networks and Machine Learning—ICANN 2018*; Springer: Cham, Switzerland, 2018; pp. 117–126. [CrossRef]
21. Srinath, R.; Gayathri, R. Detection and classification of electroencephalogram signals for epilepsy disease using machine learning methods. *Int. J. Imaging Syst. Technol.* **2021**, *31*, 729–740. [CrossRef]

22. Boonyakitanont, P.; Lek-Uthai, A.; Chomtho, K.; Songsiri, J. A review of feature extraction and performance evaluation in epileptic seizure detection using EEG. *Biomed. Signal Process. Control* **2020**, *57*, 101702. [CrossRef]

23. Hussain, L. Detecting epileptic seizure with different feature extracting strategies using robust machine learning classification techniques by applying advance parameter optimization approach. *Cogn. Neurodyn.* **2018**, *12*, 271–294. [CrossRef] [PubMed]

24. Bosl, W.J.; Loddenkemper, T.; Nelson, C.A. Nonlinear EEG biomarker profiles for autism and absence epilepsy. *Neuropsychiatr. Electrophysiol.* **2017**, *3*, 1. [CrossRef]

25. Khati, R.M.; Ingle, R. Feature extraction for epileptic seizure detection using machine learning. *Curr. Med. Res. Pract.* **2020**, *10*, 266–271. [CrossRef]

26. Sharif, B.; Jafari, A.H. Prediction of epileptic seizures from EEG using analysis of ictal rules on Poincaré plane. *Comput. Methods Programs Biomed.* **2017**, *145*, 11–22. [CrossRef]

27. Manzouri, F.; Heller, S.; Dümpelmann, M.; Woias, P.; Schulze-Bonhage, A. A Comparison of Machine Learning Classifiers for Energy-Efficient Implementation of Seizure Detection. *Front. Syst. Neurosci.* **2018**, *12*, 43. [CrossRef]

28. Liu, Y.; Huang, Y.X.; Zhang, X.; Qi, W.; Guo, J.; Hu, Y.; Zhang, L.; Su, H. Deep C-LSTM Neural Network for Epileptic Seizure and Tumor Detection Using High-Dimension EEG Signals. *IEEE Access* **2020**, *8*, 37495–37504. [CrossRef]

29. Remeseiro, B.; Bolon-Canedo, V. A review of feature selection methods in medical applications. *Comput. Biol. Med.* **2019**, *112*, 103375. [CrossRef] [PubMed]

30. Quinlan, J.R. Induction of decision trees. *Mach. Learn.* **1986**, *1*, 81–106. [CrossRef]

31. Hotelling, H. Analysis of a complex of statistical variables into principal components. *J. Educ. Psychol.* **1933**, *24*, 417–441. [CrossRef]

32. Altmann, A.; Toloşi, L.; Sander, O.; Lengauer, T. Permutation importance: A corrected feature importance measure. *Bioinformatics* **2010**, *26*, 1340–1347. [CrossRef]

33. Guyon, I.; Weston, J.; Barnhill, S.; Vapnik, V. Gene Selection for Cancer Classification using Support Vector Machines. *Mach. Learn.* **2002**, *46*, 389–422. [CrossRef]

34. Bergil, E.; Bozkurt, M.R.; Oral, C. An Evaluation of the Channel Effect on Detecting the Preictal Stage in Patients with Epilepsy. *Clin. EEG Neurosci.* **2020**, *52*, 376–385. [CrossRef] [PubMed]

35. Lai, C.; Guo, S.; Cheng, L.; Wang, W. A Comparative Study of Feature Selection Methods for the Discriminative Analysis of Temporal Lobe Epilepsy. *Front. Neurol.* **2017**, *8*, 633. [CrossRef] [PubMed]

36. Goldberger, A.; Amaral, L.; Glass, L.; Hausdorff, J.; Ivanov, P.; Mark, R.; Mietus, J.; Moody, G.; Peng, C.; Stanley, H. PhysioBank, PhysioToolkit, and PhysioNet: Components of a new research resource for complex physiologic signals. *Circulation* **2000**, *101*, E215–E220. [CrossRef] [PubMed]

37. Shoeb, A.H. Application of Machine Learning to Epileptic Seizure onset Detection and Treatment. Ph.D. Thesis, Massachusetts Institute of Technology, Cambridge, MA, USA, 2009.

38. Detti, P.; Vatti, G.; Zabalo Manrique de Lara, G. EEG Synchronization Analysis for Seizure Prediction: A Study on Data of Noninvasive Recordings. *Processes* **2020**, *8*, 846. [CrossRef]

39. Phinyomark, A.; Thongpanja, S.; Hu, H.; Phukpattaranont, P.; Limsakul, C. The Usefulness of Mean and Median Frequencies in Electromyography Analysis. In *Computational Intelligence in Electromyography Analysis—A Perspective on Current Applications and Future Challenges*; IntechOpen: London, UK, 2012; pp. 195–220. [CrossRef]

40. Kokoska, S.; Zwillinger, D. *CRC Standard Probability and Statistics Tables and Formulae*; CRC Press: Boca Raton, FL, USA, 2000.

41. Aarabi, A.; Fazel-Rezai, R.; Aghakhani, Y. A fuzzy rule-based system for epileptic seizure detection in intracranial EEG. *Clin. Neurophysiol.* **2009**, *120*, 1648–1657. [CrossRef] [PubMed]

42. Bruce, P.; Bruce, A. *Practical Statistics for Data Scientists: 50+ Essential Concepts Using R and Python*, 1st ed.; O'Reilly Media: Sebastopol, CA, USA, 2017.

43. Richman, J.S.; Moorman, J.R. Physiological time-series analysis using approximate entropy and sample entropy. *Am. J. Physiol. Heart Circ. Physiol.* **2000**, *278*, H2039–H2049. [CrossRef]

44. Song, Y.; Crowcroft, J.; Zhang, J. Automatic epileptic seizure detection in EEGs based on optimized sample entropy and extreme learning machine. *J. Neurosci. Methods* **2012**, *210*, 132–146. [CrossRef]

45. Alpaydin, E. *Introduction to Machine Learning*, 2nd ed.; MIT Press: Cambridge, MA, USA, 2010.

46. Han, J.; Pei, J.; Kamber, M. *Data Mining: Concepts and Techniques*, 3rd ed.; Elsevier: Waltham, MA, USA, 2012.

47. Louppe, G.; Wehenkel, L.; Sutera, A.; Geurts, P. Understanding variable importances in forests of randomized trees. *Advances in Neural Information Processing Systems*; Curran Associates, Inc.: Red Hook, NY, USA, 2013; Volume 26.

48. Saeys, Y.; Abeel, T.; Van de Peer, Y. Robust Feature Selection Using Ensemble Feature Selection Techniques. In *Machine Learning and Knowledge Discovery in Databases*; Springer: Berlin/Heidelberg, Germany, 2008; pp. 313–325. [CrossRef]

49. Ribeiro, M.; Singh, S.; Guestrin, C. "Why Should I Trust You?": Explaining the Predictions of Any Classifier. In Proceedings of the 2016 Conference of the North American Chapter of the Association for Computational Linguistics: Demonstrations, Association for Computational Linguistics, San Francisco, CA, USA, 13–17 August 2016; pp. 97–101. [CrossRef]

50. Lundberg, S.M.; Lee, S.I. A Unified Approach to Interpreting Model Predictions. In Proceedings of the 31st International Conference on Neural Information Processing Systems, Long Beach, CA, USA, 4–9 December 2017; pp. 4768–4777. [CrossRef]

51. Molnar, C. A Guide for Making Black Box Models Explainable. 2018. Available online: https://christophm.github.io/interpretable-ml-book (accessed on 31 October 2021).

52. Breiman, L. Random forests. *Mach. Learn.* **2001**, *45*, 5–32. [CrossRef]
53. Chatzichristofis, S.A.; Arampatzis, A.; Boutalis, Y.S. Investigating the behavior of compact composite descriptors in early fusion, late fusion, and distributed image retrieval. *Radioengineering* **2010**, *19*, 725–733.
54. Effrosynidis, D.; Arampatzis, A. An evaluation of feature selection methods for environmental data. *Ecol. Inform.* **2021**, *61*, 101224. [CrossRef]
55. Alpaydm, E. Combined 5×2 cv F Test for Comparing Supervised Classification Learning Algorithms. *Neural Comput.* **1999**, *11*, 1885–1892. [CrossRef]
56. Han, X.; Chang, X.; Quan, L.; Xiong, X.; Li, J.; Zhang, Z.; Liu, Y. Feature subset selection by gravitational search algorithm optimization. *Inf. Sci.* **2014**, *281*, 128–146. [CrossRef]
57. Cantu-Paz, E. Feature subset selection, class separability, and genetic algorithms. In *Genetic and Evolutionary Computation Conference*; Springer: Berlin/Heidelberg, Germany, 2004; pp. 959–970.
58. Harris, C.R.; Millman, K.J.; van der Walt, S.J.; Gommers, R.; Virtanen, P.; Cournapeau, D.; Wieser, E.; Taylor, J.; Berg, S.; Smith, N.J.; et al. Array programming with NumPy. *Nature* **2020**, *585*, 357–362. [CrossRef]
59. McKinney, W. Data Structures for Statistical Computing in Python. In Proceedings of the 9th Python in Science Conference, Austin, TX, USA, 28 June–3 July 2010; pp. 56–61. [CrossRef]
60. Virtanen, P.; Gommers, R.; Oliphant, T.E.; Haberland, M.; Reddy, T.; Cournapeau, D.; Burovski, E.; Peterson, P.; Weckesser, W.; Bright, J.; et al. SciPy 1.0: Fundamental Algorithms for Scientific Computing in Python. *Nat. Methods* **2020**, *17*, 261–272. [CrossRef] [PubMed]
61. Pedregosa, F.; Varoquaux, G.; Gramfort, A.; Michel, V.; Thirion, B.; Grisel, O.; Blondel, M.; Prettenhofer, P.; Weiss, R.; Dubourg, V.; et al. Scikit-learn: Machine Learning in Python. *J. Mach. Learn. Res.* **2011**, *12*, 2825–2830. [CrossRef]
62. Abadi, M.; Agarwal, A.; Barham, P.; Brevdo, E.; Chen, Z.; Citro, C.; Corrado, G.S.; Davis, A.; Dean, J.; Devin, M.; et al. TensorFlow: Large-Scale Machine Learning on Heterogeneous Systems. Software. 2015. Available online: tensorflow.org (accessed on 3 September 2021).
63. Hunter, J.D. Matplotlib: A 2D graphics environment. *Comput. Sci. Eng.* **2007**, *9*, 90–95. [CrossRef]
64. Jaccard, P. Étude comparative de la distribution florale dans une portion des Alpes et des Jura. *Bull. Soc. Vaudoise Sci. Nat.* **1901**, *37*, 547–579.
65. Andrzejak, R.G.; Lehnertz, K.; Mormann, F.; Rieke, C.; David, P.; Elger, C.E. Indications of nonlinear deterministic and finite-dimensional structures in time series of brain electrical activity: Dependence on recording region and brain state. *Phys. Rev. E* **2001**, *64*, 061907. [CrossRef]

sensors

Article

Wavelet Ridges in EEG Diagnostic Features Extraction: Epilepsy Long-Time Monitoring and Rehabilitation after Traumatic Brain Injury [†]

Yury Vladimirovich Obukhov [1], Ivan Andreevich Kershner [1,*], Renata Alekseevna Tolmacheva [1], Mikhail Vladimirovich Sinkin [2,3] and Ludmila Alekseevna Zhavoronkova [4]

[1] Kotelnikov Institute of Radio Engineering and Electronics of RAS, Mokhovaya St. 11-7, 125009 Moscow, Russia; yuvobukhov@mail.ru (Y.V.O.); tolmatcheva@yandex.ru (R.A.T.)

[2] Department of Neurosurgery of the Sklifosovsky Research Institute for Emergency Medicine of Moscow Healthcare Department, Bolshaya Sukharevskaya Square 3, 129090 Moscow, Russia; sinkinmv@sklif.mos.ru or mvsinkin@gmail.com

[3] Laboratory of Invasive Neurointerfaces of the Research Institute TechnoBioMed, A.I. Yevdokimov Moscow State University of Medicine and Dentistry, Delegatskaya St. 20 p.1, 127473 Moscow, Russia

[4] Laboratory of General and Clinical Neurophysiology of the Institute of Higher Nervous Activity and Neurophysiology of RAS, Butlerova St. 5a, 117485 Moscow, Russia; lzhavoronkova@hotmail.com or Lzhavor@nsi.ru

* Correspondence: ivan.kershner@gmail.com

† This paper is an extended version of our paper published in Proceedings of the 2020 International Conference on Information Technology and Nanotechnology (ITNT), Samara, Russian, 26–29 May 2020; doi:10.1109/ITNT49337.2020.9253182.

Citation: Obukhov, Y.V.; Kershner, I.A.; Tolmacheva, R.A.; Sinkin, M.V.; Zhavoronkova, L.A. Wavelet Ridges in EEG Diagnostic Features Extraction: Epilepsy Long-Time Monitoring and Rehabilitation after Traumatic Brain Injury. *Sensors* **2021**, *21*, 5989. https://doi.org/10.3390/s21185989

Academic Editor: Yvonne Tran

Received: 13 August 2021
Accepted: 3 September 2021
Published: 7 September 2021

Publisher's Note: MDPI stays neutral with regard to jurisdictional claims in published maps and institutional affiliations.

Abstract: Interchannel EEG synchronization, as well as its violation, is an important diagnostic sign of a number of diseases. In particular, during an epileptic seizure, such synchronization occurs starting from some pairs of channels up to many pairs in a generalized seizure. Additionally, for example, after traumatic brain injury, the destruction of interneuronal connections occurs, which leads to a violation of interchannel synchronization when performing motor or cognitive tests. Within the framework of a unified approach to the analysis of interchannel EEG synchronization using the ridges of wavelet spectra, two problems were solved. First, the segmentation of the initial data of long-term monitoring of scalp EEG with various artifacts into fragments suspicious of epileptic seizures in order to reduce the total duration of the fragments analyzed by the doctor. Second, assessments of recovery after rehabilitation of cognitive functions in patients with moderate traumatic brain injury. In the first task, the initial EEG was segmented into fragments in which at least two channels were synchronized, and by the adaptive threshold method into fragments with a high value of the EEG power spectral density. Overlapping in time synchronized fragments with fragments of high spectral power density was determined. As a result, the total duration of the fragments for analysis by the doctor was reduced by more than 60 times. In the second task, the network of phase-related EEG channels was determined during the cognitive test before and after rehabilitation. Calculation-logical and spatial-pattern cognitive tests were used. The positive dynamics of rehabilitation was determined during the initialization of interhemispheric connections and connections in the frontal cortex of the brain.

Keywords: electroencephalogram; wavelet spectrum; ridge; segmentation; phase connectivity; epilepsy; traumatic brain injury

1. Introduction

Wavelet transform (WT) is widely used in the processing and analysis of non-stationary signals [1–5]. Since the 1990s, in various fields of biology and medicine [6], in neurophysiology [7], discrete and continuous wavelet transforms have been used to extract diagnostic information from signals and images of various types.

Considering EEG as a simultaneously amplitude and phase modulated analytical signal, and if the scanning wavelet width is narrower than the changes in signal phase, then it is possible to use the property of the wavelet spectrum ridge, namely those that the amplitude and phase of the signal are equal to the amplitude and phase of the wavelet spectrum ridge [8–11]. Thus, defining the ridge as the absolute maximum of the wavelet spectrum at each moment (reference point) of time, we obtain instantaneous values of the amplitude, frequency and phase of the signal. This very useful property of WT ridges makes it easy to find interchannel synchronized EEG fragments during epileptic seizures (ESs) in long-term clinical monitoring data, restoring the cognitive functions of patients after moderate traumatic brain injury using interchannel phase coupling link analysis, and other tasks of EEG diagnostics.

WT is used for EEG decomposition into time-frequency fragments for the subsequent detection of epileptic seizures (ESs). Currently, there are many publications on the use of various classifiers for the detection and prediction of an epileptic seizure in EEG signals using various classifiers [12–19]. Initial data on epilepsy monitoring should be preliminarily processed, including removal of artifacts and filtering noise to get a clean epilepsy EEG signal for the next step, feature extraction and classification [18,19].

In decision support systems, methods based on the analysis of EEG patterns are most often used and one of them is the "Persyst" system by Persyst Development Corpo ation (https://www.persyst.com, accessed on 13 August 2021). To detect ES in the time domain, discrete-time sequences are analyzed into which the original EEG signal is divided. One of such methods is based on tracking successive extrema in the selected time interval of the signal and evaluating the histogram of the amplitude difference and time separation between the maximum and minimum values of the histogram [20]. The different approaches for detecting ESs which were proposed in the time domain are the calculation of signal energy [21]; the frequency characteristics of the signal were studied: the index of the phase slope of multichannel EEG [22]; frequency-moment signatures [23]; entropy features [24]; Bayesian linear discriminant analyses of lacunarity and fluctuation index [25], four-level Daubechies wavelet transform [26] and five-level wavelet decomposition method [27]. The most promising method of EEG analysis is the study of the parameters of the ridges of wavelet spectrograms. In the EEG signals for the detection of epileptic seizures, the dynamics of synchronization and changes in the phase ratio before, during and after the seizures are monitored.

At present, attempts are being made to improve methods for detecting ESs in EEG. Paper [28] describes a way to improve the support vector machine method by adding an adaptive median feature baseline correction method. A combination of methods is also used to search for ESs, for example, complementary ensemble empirical mode decomposition with extreme gradient boosting [15]. A method has been proposed that combines time-domain feature analysis and entropy calculation [16]. A similar combination was also presented in work [14], but the study of parameters in the time domain was used to segment the signal sections, in order to then carry out analysis using machine learning methods. To differentiate ESs from non-seizure events, neural networks [13] and similar methods are used, such as the method of binarization of frequency and temporal features of signal fragments [29].

It should be noted that the estimation of the accuracy and specificity of the classification was carried out on EEG fragments previously selected and annotated by EEG neurosciensists as ictal and interictal events. The most representative databases are the EPILEPSIAE database [30], the Temple University Hospital EEG Data Corpus [31], Bonn epilepsy dataset etc.

Clinical EEG investigations of epilepsy consist in long-term (several days) monitoring of multichannel EEG using scalp or intracrinial electrodes in the presence of various artifacts: the electrical activity that is not recorded in the cerebral zone, such as that due to the equipment, patient behavior or the environment; eye movement and chewing are common events that can often be confused with a spike; signals instrument fluctuations and artifacts of vital activity [32]. It can be seen from the review that the methods for removing artifacts described in the literature are mainly focused on removing one type of possible artifacts or ocular and muscular ones present in the real initial data of long-term EEG monitoring. In general, we can conclude that the problem of automated removal of artifacts of various types from the initial data of long-term EEG monitoring has not been fully resolved. Additionally, this article describes canonical correlation analysis as a successful method for removing muscle artifacts.

One of the most important characteristics of ES is abnormal inter-channel synchronization or so-called coherency. To assess interchannel EEG synchronization, canonical correlation analysis [33], normalized cross-correlation and imaginary part of coherency or phase synchronization are used [34]. The main disadvantage of estimation coherence is the necessity to average it over time epochs and frequency ranges [35]. The study of short-term frequency synchronization of signals in two EEG channels by comparing their WT ridges frequencies during a previously selected by physicians is presented in ES [36].

We did not find in the literature any information on taking into account one of the most important feature of ESs—EEG interchannel synchronization for detecting ESs. So this article describes a new approach to the segmentation of the initial long-term clinical multichannel EEG monitoring data of patients with epilepsy into temporal fragments suspicious of an ES, to reduce the quantity of EEG fragments. This approach is based at first on the EEG WT ridges segmentation of the into frequency-synchronized fragments, and secondly with a thresholding of the ridge spectral power density.

Another part of this paper is devoted to a new approach to the diagnosis and assessment of rehabilitation of patients after traumatic brain injury (TBI). TBI is an insult to the brain from an external mechanical force, which can lead to permanent or temporary impairment of cognitive, physical, and psychosocial functions. The most used EEG methods of investigation TBI are spectral analysis, absolute and relative amplitude and power, coherence, and symmetry between homologous pairs of electrodes (see review [36]). A multivariate support system has been developed to quantify and classify by Random Forest classifier TBI stage based on analysis of EEG power in various frequency ranges [37]. A study [38] investigated the possibility of detecting moderate TBI according to the Glasgow Coma Scale [39] by EEG amplitude analysis and convolutional neural network classification. Recently, a single channel system was developed for real-time mild TBI detection with Convolutional Neural Networks classifier of EEG power in different frequency ranges [40]. The proposed method can be applied for screening of the moderate TBI and for selection of the patients for further diagnostics and treatment. In [41] the analysis of the EEG data applying the energy, sample entropy, approximate entropy, Lempel–Ziv complexity features demonstrated the increase in sample entropy was related with the functional recovery, i.e., the rehabilitation dynamics of the injured brain region. EEG-based neurofeedback is used for cognitive rehabilitation of patients with TBI [42].

Our approach is based on the analysis of the network of phase sinchronized EEG channels WT ridges in patients with moderate TBI. The interchannel phase difference of the EEG is determined during cognitive tests at the points of the frequency-modulated wavelet spectra ridges. We investigate the neurons connectivity disruption of the brain after TBI and consider the inter-channel phase connectivity between EEG channels during cognitive tests. It does not depend on the EEG signal amplitude. In this paper Section 2 contains the basic formulae and conditions for their application. Section 3 describes a new approach to segmentation of long-term EEGs into temporal fragments suspicious of an ES by interchannel WT ridges frequency sincronization and power spectral density thresholding. Section 4 describes a new approach to determine the evaluation of rehabilitation positive dynamics of patients with moderate TBI.

2. Materials and Methods

We studied long-term (from several hours to several days) initial EEG records of preoperative patients with epilepsy, obtained in laboratory of invasive neurointerfaces of the Research Institute TechnoBioMed. A.I. Yevdokimov Moscow State University of Medicine and Dentistry. The segmentation method was used for several days' 19-channel EEG. The records were carried out according to the 10–20 system [43] in reference montage with a sampling rate of 256 Hz. Power supply artifacts were removed from all EEG channels using a notch filter at frequencies multiples of 50 Hz. The use of Morlet WT in the frequency range from 0.5 to 22 Hz, so myographic artifacts were rejected.

The records of 19-channel EEG were considered, therefore the quantity of pairs of channels is 171 for the group of control volunteers (18 subjects) and for the group of patients with moderate TBI (12 subjects), where three patients had repeated EEG records after rehabilitation in two cognitive tests. Cognitive tests were calculation-logical (CT1) and spatial-pattern (CT2). During the CT1 test, the doctor randomly spoke words from the category of "clothing" or "food" to the subject. During the test, the subject counted in their mind the number of items belonging to one of these categories. At the end of the test, the subject announced the result of the number of items. On the CT2 test, the doctor named an arbitrary time. The subject had to represent the position of the hands-on-dial in accordance with the indicated time. If both clock hands were in the same half of the dial, he said "yes", and if they were in different halves, he kept silent. Investigations of control volunteers and patients with moderate TBI were carried out at the National Medical Research Center for Neurosurgery named after Academician N.N. Burdenko. All subjects were right-handed and signed written consent to participate in the research in accordance with the provisions of the Helsinki Agreement. The rehabilitation was performed for 1–2 months. The time of the rehabilitation was 40–45 min two times a week. The criteria for the inclusion of patients in the investigation were the ability to stand on their own and the ability to follow the doctor's instructions, and also the absence of hemiparesis and other neurological disorders. The international 10–20 system of the position of scalp electrodes was used for EEG record. The recording time for every test was 60 s. EEG recording was carried out both during the tests and without them. The sampling rate of the EEG was 250 Hz in the processing of EEG signals. The original signals were recorded with a high-pass filter with a cut-off frequency of 0.5 Hz, a low pass filter with a cut-off frequency of 70 Hz. Then, a notch filter at frequencies multiples of 50 Hz and a Butterworth filter were used. The signals were filtered by a fourth-order Butterworth bandpass filter with a bandwidth from 2 to 10 Hz. The EEG records were analyzed without selecting individual fragments of the signal. However, the removal of outliers in the EEG signals was done with the Huber's X84 method [44].

We considered EEG as an analytical signal with time-varying amplitude and frequency. The analytic signal was first locally represented as a modulated oscillation, demodulated by its own instantaneous frequency, and then Taylor-expanded at each point in time. We represent this signal as the following function:

$$S(t) = A_S(t)\exp(i\Phi_S(t)), \tag{1}$$

where $A_S(t)$ is the amplitude and $\Phi_S(t)$ is the phase of the signal. Continuous wavelet transform of signal $S(t)$ is represented as:

$$W(a,b) = \int_{-\infty}^{\infty} S(t)\psi_{a,b}^{*}(t)\,dt \tag{2}$$

$$\psi_{a,b}(t) = \frac{1}{\sqrt{|a|}}\psi\left(\frac{t-b}{a}\right) \tag{3}$$

where $a, b, a \neq 0$ are the real numbers defining the scale and the shift. We used the following function (Morlet mother wavelet) that was employed in the Matlab software:

$$\psi(t) = \frac{1}{\sqrt{\pi f_b}}\exp(-t^2/f_b)\exp(2\pi i f_c t) \tag{4}$$

where f_b is a positive and related with the variance of Gaussian function and f_c is a positive value that corresponds with central frequency. The Morlet wavelet transform can be represented as follows:

$$W(a,b) = M(a,b)\exp(i\Phi(a,b)), \tag{5}$$

where $M(a,b)$ is the absolute value of wavelet transform and $\Phi(a,b)$ is the phase of wavelet transform (2).

Substituting expressions (3) and (4) in formula (2), we obtain:

$$W(a,b) = \frac{1}{\sqrt{a\pi f_b}} \int_{-\infty}^{\infty} A_S(t)\exp\left(-\frac{(t-b)^2}{a^2 f_b}\right)\exp\left(i\left(\Phi_S(t) - 2\pi f_c\frac{t-b}{a}\right)\right)dt, \tag{6}$$

usually (in Matlab) $f_b = f_c = 1$.

Integral (6) is approximately calculated by the method of stationary phase [45]. Under certain conditions, the main contribution to the integral is made by the imaginary part of the exponential function, since the contributions of rapidly changing phases cancel each other out, and the contribution is made by the values located at the point of the stationary phase. The stationary phase method is applicable when the amplitude $A(t)$ of the signal exhibits relatively slow changes compared to fast changes in the total signal associated with fast changes in phase, for example, and asymptotic properties are satisfied concerning the window $\psi(t)$ under the following assumptions; so that the following conditions are satisfied [11]:

$$\left|\frac{d\Phi_S(t)}{dt}\right| \gg \left|\frac{1}{A_S(t)}\frac{dA_S(t)}{dt}\right|, \left|\frac{1}{A_S(t)}\frac{dA_S(t)}{dt}\right| \ll \left|\frac{1}{\psi(t)}\frac{d\psi(t)}{dt}\right| \tag{7}$$

The relationship between the phase from expression (5) and the phase from expression (6) is given as follows:

$$\Phi(t) = \Phi_S(t) - 2\pi f_c\left(\frac{t-b}{a}\right) \tag{8}$$

For the stationary phase $\Phi(t)$, we have

$$\frac{d\Phi(t)}{dt} = \Phi'_S(t) - \frac{2\pi f_c}{a} = 0 \tag{9}$$

Such a condition is satisfied at $t = t(a)$. To estimate the integral from formula (6), we expanded the phase $\Phi(t)$ in a Taylor series up to a polynomial of the second degree in the neighborhood of point $t = t(a)$ till the order $(t - t(a))^2$:

$$\Phi(t) \approx \Phi_S(t(a)) - 2\pi f_c\left(\frac{t(a)-b}{a}\right) + \frac{1}{2}\Phi''_S(t(a))(t-t(a))^2 \tag{10}$$

Below, we use the notation $\Phi_S \equiv \Phi_S(t(a))$ and $\Phi''_S \equiv \Phi''_S(t(a))$.

After substituting formula (10) into formula (6), we obtained an approximate value for the phase and absolute value of the wavelet transform:

$$\Phi(a,b) \approx \Phi_S - 2\pi f_c\left(\frac{t(a)-b}{a}\right) + \frac{1}{2}\arctan\left(\frac{a^2 f_b}{2}\Phi''_S\right) + \frac{2(t(a)-b)^2\Phi''_S}{4 + a^4 f_b^2(\Phi''_S)^2} \tag{11}$$

$$M(a,b) \approx A_S(t(a))\left(1 + \frac{a^4 f_b^2}{4}(\Phi''_S t(a))^2\right)^{-\frac{1}{4}}\exp\left(-\frac{(t(a)-b)^2 a^2 f_b(\Phi''_S)^2}{4 + a^4 f_b^2(\Phi''_S)^2}\right) \tag{12}$$

Expression (12) shows that the maximum of wavelet transform absolute value was reached at $b = t(a)$. The instantaneous frequency at the ridge point f_r at time moment $t = t(a)$ was calculated using expression (9):

$$f_r(t(a)) = 2\pi \frac{f_c}{a} \tag{13}$$

In this case, the maximal value of the wavelet transform (ridge) of the signal is given by

$$\max_a |W(a,b)| \approx A_S(t(a)) \left(1 + \frac{a^4 f_b^2}{4} (\Phi_S'' t(a))^2 \right)^{-\frac{1}{4}} \tag{14}$$

and the phase is approximated of a ridge point as

$$\Phi(a,b) \approx \Phi_S + \frac{1}{2} \arctan \left(\frac{a^2 f_b}{2} \Phi_S'' \right) \tag{15}$$

As in [46], the relationship between the frequency of Fourier spectrum of the wavelet transform and the scales a of the wavelet transform (2) is given as follows:

$$f = \frac{f_0}{2a} + \frac{\sqrt{2 + 4(\pi f_0)^2}}{4\pi a} \cong \frac{f_0}{a} = \frac{1}{a} \tag{16}$$

where f_0 is a wavelet central frequency and it is considered that $4(\pi f_0)^2 \gg 2$. So, for the ridge points (f_r, t) we have:

$$W_r(t) = \max_f |W(f,t)|, \ f_r(t) = \operatorname{argmax}_f |W(f,t)|,$$

$$\Phi_S(t) \cong \Phi_r(f,t) = \arctan \left(\frac{Im(W(t,f_r))}{Re(W(t,f_r))} \right), \tag{17}$$

when the condition

$$\frac{\Phi_S''}{2f_r^2} = \frac{f_r'}{2f_r^2} \ll 1, \tag{18}$$

is satisfied.

Summarizing, it should be noted that, in contrast to other works, the obtained simple method for determining the ridge points as the maximum of the modulus of the wavelet spectrum at each time point was undoubtedly easy to calculate.

3. EEG Segmentation

This chapter describes an EEG segmentation method based on the study of Morlet wavelet transform ridges, which allows finding time intervals of interest in ES detection, which is used to analyze continuous long-term EEG monitoring data during post-processing. The long-term EEG segmentation method consists of the following stages, as shown in Figure 1: 1. signal filtration at frequencies multiples 50 Hz, 2. wavelet Morlet transform of signals, 3. determination of wavelet spectrogram ridges, 4. marking time intervals with interchannel synchronization, 5. marking time intervals with power spectral density (PSD) values above the threshold, 6. intersections of time intervals, 7. visualization of a segmented signal with marked time intervals.

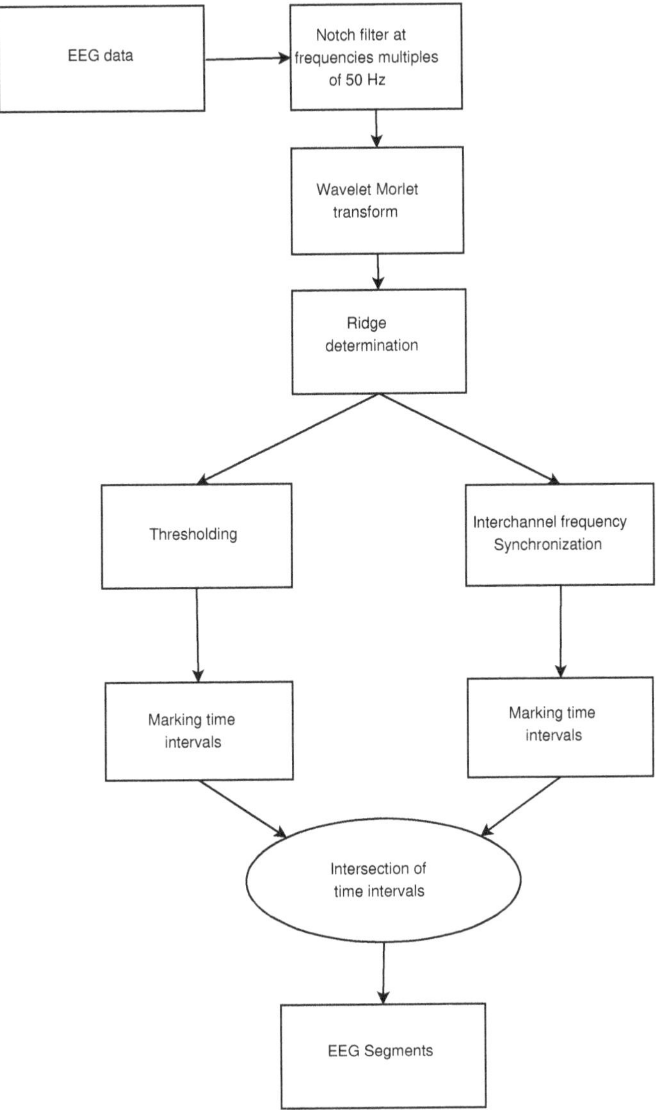

Figure 1. Block diagram of a long-term EEG segmentation.

Let us show how segmentation was carried out using the example of an EEG recording fragment containing an ES. For each EEG channel, we calculated the wavelet spectrogram (2) and the ridges of the wavelet spectrogram (17) in frequency range [0.5; 22] Hz.

Generalized ESs were characterized by changes in power in several EEG channels and the synchronization of different channels pairs. In order to estimate the inter-channel synchronization, the modulus of the frequency difference at the points of the ridges was calculated for each pair of channels. If the modulus of the difference was less than ε, then there was synchronization $Sync_{i,j}$, otherwise, it was not:

$$Sync_{i,j}(k) = \begin{cases} 1, |f_{ri}(k) - f_{rj}(k)| \leq \varepsilon \\ 0, |f_{ri}(k) - f_{rj}(k)| > \varepsilon \end{cases} \tag{19}$$

where f_{ri}, f_{rj} are the frequencies of the ridges of the wavelet spectrograms on the i and j EEG channels, k is the point of the ridge.

Figure 2 shows the projection of the wavelet spectrogram onto the PSD-frequency plane of the sinusoidal signals with a frequency of 2 and 2.5 Hz. At $\varepsilon = 0.5$ the peaks were distinguishable. On smaller epsilons, the peaks could merge.

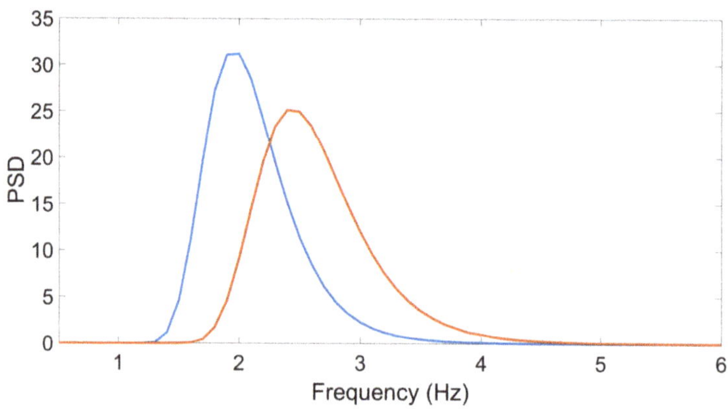

Figure 2. PSD-Frequency projection of wavelet spectrums of two sinusoidal signals: blue is 2 Hz, orange is 2.5 Hz.

Nearby points at which condition (19) was satisfied were combined into fragments. Fragments between which the time interval was less than 10 s were combined into one. For each fragment, the beginning and end times of synchronization in pairs of channels were calculated. Table 1 shows a histogram of the number of synchronized fragments depending on the duration in 19 pairs of EEG channels. For neurophysiological considerations, this work considered fragments with a duration of 10 s or more.

Table 1. Histogram of the number of synchronized fragments depending on the duration in 19 pairs of EEG derivations.

Pairs of Channels	Fragment Duration, s				
	>2 ≤ 5	>5 ≤ 10	>10 ≤ 15	>15 ≤ 30	>30
FP1-F7	3040	621	131	44	4
F7-T3	3683	430	84	29	9
T3-T5	4265	336	68	27	6
T5-O1	4210	375	77	40	4
FP1-F3	3413	534	90	20	1
F3-C3	3784	350	43	17	2
C3-P3	4140	397	41	15	1
P3-O1	4047	401	63	31	1
FZ-CZ	4250	362	35	18	1
CZ-PZ	4223	409	54	24	2
FZ-Pz	4044	329	31	16	0
FP2-F4	3352	535	99	46	7
F4-C4	3310	405	78	25	3
C4-P4	3869	421	75	15	1
P4-O2	4028	426	76	32	6
FP2-F8	3024	678	130	48	3
F8-T4	3385	518	88	46	5
T4-T6	3946	454	67	28	3
T6-O2	4190	470	76	28	3

Figure 3 shows an example of ES fragment from observed EEG recording with visualization of the presence of synchronization in channels pairs.

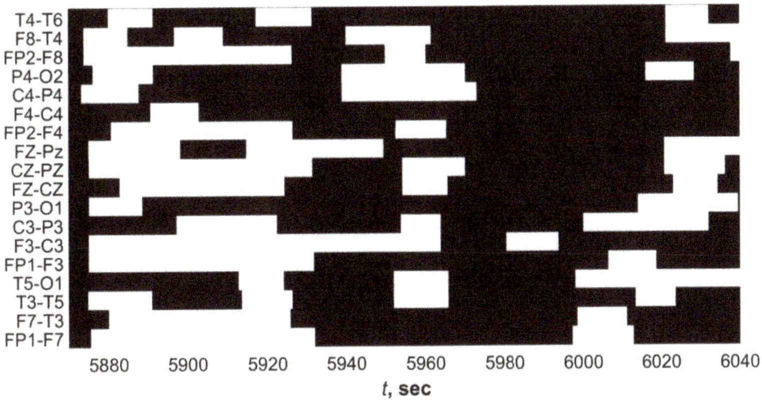

Figure 3. Fragment of daily EEG signal with ES illustrating the frequency of the wavelet spectrogram ridge synchronization in different pairs of EEG channels. Black shows the presence of synchronization. The ordinate shows the labels of channel pairs.

As can be seen, EEG synchronization could be observed in not all channels simultaneously (Figure 3). In this example, synchronization was observed in most channel pairs from about 5970 s to 6000 s, but synchronization began to appear earlier in a smaller number of pairs.

The time intervals in which the inter-channel synchronization in the frequency of the ridges was recorded could correspond to both ES and artifacts of chewing, sleep, and random physical influences on the electrodes, which generated artifacts of a non-epileptic nature.

A characteristic feature of the ES was a sharp change in the amplitude over a short period of time. Therefore, in addition to searching for time intervals in which there was synchronization on several pairs of EEG derivations, the detection of areas with high values of the power spectral density (PSD) was carried out. In order to understand the idea of the method, let us consider the histogram of the ridge points of the wavelet spectrogram, calculated for one of the leads. The peak of the histogram contained about 1.2×10^6 points, the maximum PSD value at which the number of ridge points tended to 0, about $2.8 \times 10^7 \mu \, V^2/Hz$. Such a histogram gave a large peak in the region of low PSD values and did not allow us to estimate the distribution of the ridge points, therefore, Figure 4 shows the "window" of the PSD histogram. As can be seen from the figure, the number of ridge points with low PSD values was large and could be interpreted as noise. It was necessary to separate the informative points of the ridge from the noise.

In order to separate the points of the ridge of the wavelet spectrogram related to high-amplitude electrical activity from noise, it was required to find the threshold value of PSD Tr. The ridge PSD_r values of the wavelet spectrogram were determined as follows:

$$PSD_r(t) = \begin{cases} PSD_r(t), PSD_r(t) \geq Tr \\ 0, PSD_r(t) < Tr \end{cases} \qquad (20)$$

The points of the ridge $PSD_r(t)$ lying between the nearest points $PSD_r(t) = 0$ was called the ridge segment. Figure 5 shows a histogram of the number of ridge segments from the PSD threshold Tr to $Tr = 5 \times 10^5 \mu \, V^2/Hz$. At large values of PSD_r, the number of segments tended to be 0. The PSD_r values could differ greatly not only from patient to patient but also by channels; therefore, it was required to determine the threshold value adaptively. Figure 5 shows a sharp decrease in the number of fragments with an increase in the threshold value. To identify the threshold value, the second derivative of segments quantity from the threshold changes was analyzed. After reaching a local maximum (circle mark in Figure 5), it became negligible. This means that the segments quantity linearly decreased with growing threshold. In Figure 5 threshold PSD value $Tr = 1.5 \times 10^5 \mu\,V^2/Hz$. With this choice of the threshold, most ESs detected by the expert and a small number of artifacts like ES were observed. Figure 6 shows an example of a segmented ridge of a wavelet spectrum containing an ES.

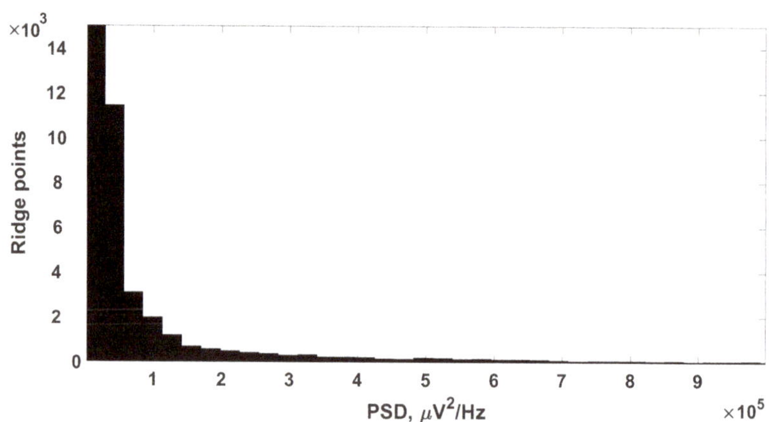

Figure 4. The window of the histogram of the PSD of the wavelet spectrogram ridge of the long-term EEG signal.

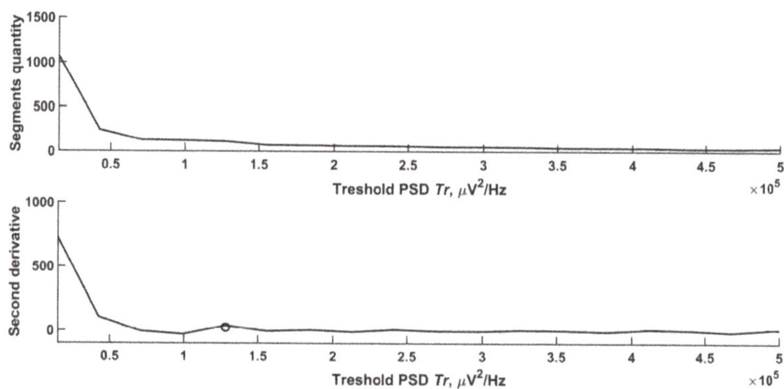

Figure 5. Histogram of the number of ridge segments from the threshold PSD Tr. The circle marks the local maxima at threshold value $Tr = 1.27 \times 10^5 \mu\,V^2/Hz$.

Figure 7 shows a fragment of a daily EEG signal showing an ES. Marks of the expert neurophysiologist are green vertical lines; the blue line, repeating the waveform, marks the fragment on which the synchronization was recorded on several pairs of EEG derivations; the dotted rectangles mark the areas found by the threshold method. Thus, the method of application for the search for ES is shown.

For the 5-hour EEG 2017 the overall synchronized fragments duty more than 10 were found (see Table 1), 112 segments were detected by thresholding, and finally we obtained nine intersected segments. with total duration total duration about 4 min. Earlier [47], we showed that by processing synchronous video of these nine fragments four fragments were recognized as moving artefacts. As a result, there were five segments left with a total duration of 4 min.

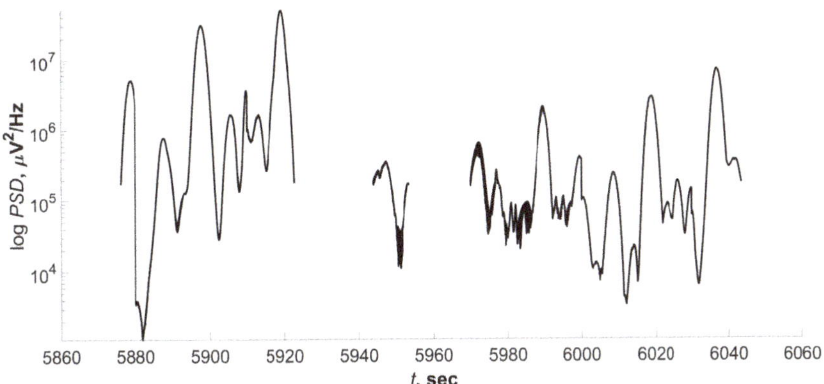

Figure 6. The segmented wavelet spectrogram ridge of the EEG signal fragment typical for ES. The upper figure shows a segmented ridge PSD_r, the bottom one shows f_r.

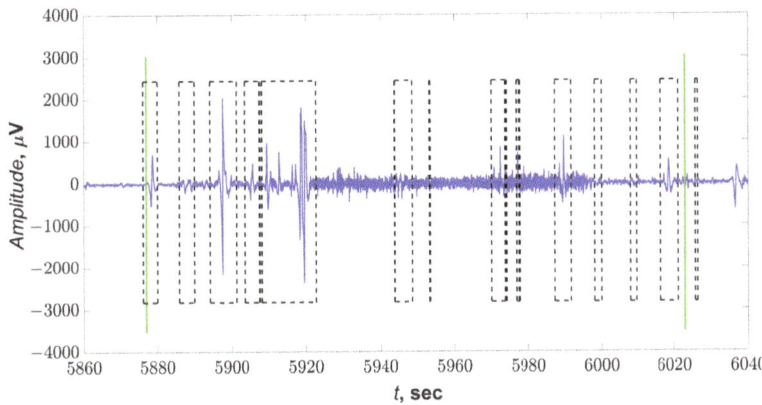

Figure 7. Fragment of an EEG with ES. Green vertical lines are expert marks. The blue line repeating the signal waveform is a mark obtained by searching for synchronized channel pairs. Dotted squares are marks obtained by the threshold method for detecting ES.

The detection of ES on EEG is complicated by the presence of many non-epileptic artifacts in signals received from scalp electrodes: electromyographic, motor, instrumental human actions, etc. An overview of various types of artifacts is given in [48]. Therefore, there is a need to develop methods to differentiate ES from artifacts of a non-epileptic nature.

To solve this problem, an algorithm was proposed, which consisted of studying the broadband peaks of the wavelet spectrograms, which were characteristic of an ES and a chewing artifact. Let us make a comparison using the example of wavelet spectra of an epileptic seizure (Figure 8) and chewing (Figure 9). We analyzed slices of wavelet spectrograms frequency $f_{cur}(t)$ higher ridge frequency $f_r(t)$, for example, at Figures 8 and 9 $f_{cur}(t) = 4$ Hz (green line).

For each slice, we calculated Fourier spectra. Figure 10 Fourier spectra of ES and chewing artifact at $f_{cur} = 4$ Hz. The frequency of the main peak and full width at half maximum (FWHM) of the Fourier spectrum were calculated.

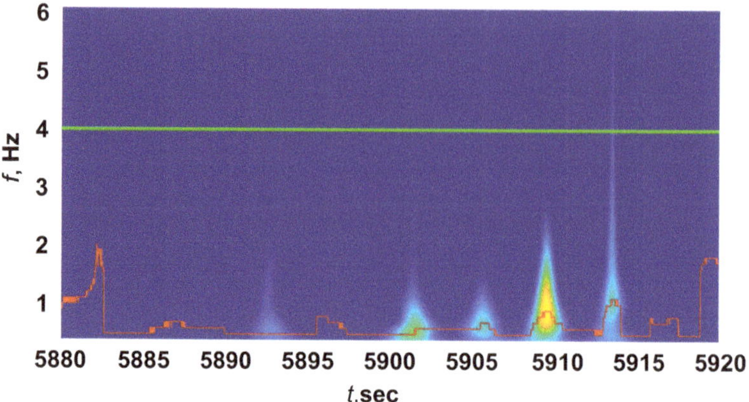

Figure 8. Wavelet spectrogram of an EEG with ES. Green line corresponds to slice of the wavelet spectrogram at the frequency $f_{cur} = 4$ Hz.

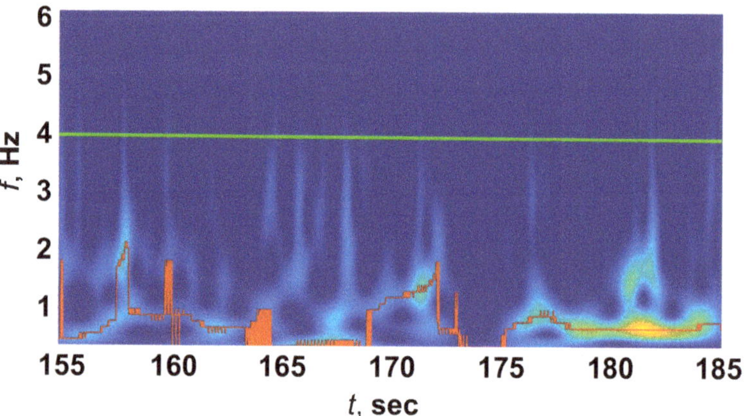

Figure 9. Wavelet spectrogram of an EEG with chewing artifact. Green line corresponds to slice of the wavelet spectrogram at the frequency $f_{cur} = 4$ Hz.

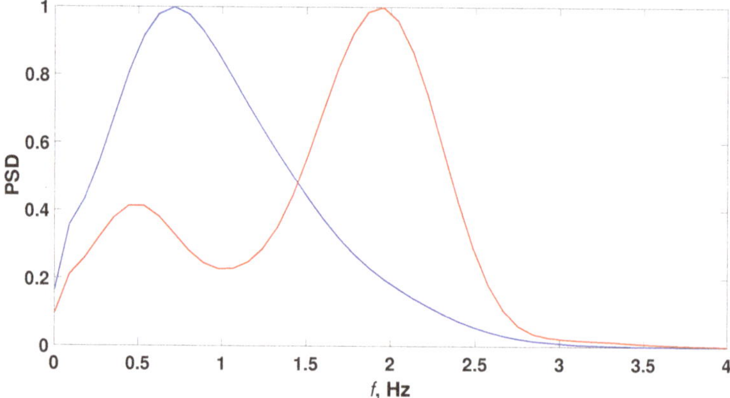

Figure 10. Fourier spectra of wavelet spectrograms slices $f_{cur} = 4$ Hz. The red line is the Fourier Spectrum of the ES slice; the blue line is the Fourier spectrum of the chewing artifact slice.

Figure 10 shows differences between ES and chewing artifacts. The main peak frequency of the Fourier spectrum was at 0.71 Hz for the chewing artifact, and at 1.86 Hz for the ES. Two peaks could be observed in ES slice spectrum, this can be interpreted as the presence of spike-wave activity. There was a difference in FWHM of the peaks of the Fourier spectrum of the slices: for a chewing artifact, it was almost 2 times more than for an epileptic seizure. This may mean that the seizure period was more stable than chewing.

4. The Estimation of Inter-Channel EEG Phase Connectivity in Patients with TBI

Various methods of EEG phase coherence are used to estimate the connectivity of brain regions. Usually, the phase coherency of signals is used for the estimation of the inter-channel connectivity of EEG [33,49,50]. Coherency $Coh_{xy}(f)$ is defined by the normalized complex cross-correlation C_{xy} of signals $x(t)$ and $y(t)$:

$$Coh_{xy}(f) = |<C_{xy}>|, C_{xy} = \frac{S_{xy}(f)}{(|S_{xx}(f)||S_{yy}(f)|)^{1/2}}, \tag{21}$$

and a phase coherency is defined as $|<exp(i\Delta\Phi)>|$, where $|<\bullet>|$ is an averaging [33].

In coherency analysis of non-stationary EEG, it is necessary to average $exp(i\Delta\Phi)$ over different time intervals (epochs), and it is the first problem. The presence of the peak in the histogram of the phase difference in different epochs determines the presence or the absence of the phase synchronization in the absence of a peak. In addition to this, $Coh_{xy}(f)$ is averaged in preliminary selected frequency bands that are specified using neurophysiological data. Usually, these bands correspond to the delta (2–4 Hz), theta (4–8 Hz), alpha (8–12 Hz), beta (12–25 Hz) EEG, and other rhythms, and this is the second problem. These disadvantages of the coherency analysis that leads to instability in the definition of the inter-channel EEG connectivity. The validity of the coherent analysis of non-stationary EEG signals is questioned [34].

Another method for the estimation of the phase connectivity is to determinate the analytical signal $x^*(t) = x(t) + iH(x(t))$, where $H(x(t))$ is the Hilbert transform [51]. Then, the phase of signal $x^*(t)$ is calculated as the arccosine (arcsine) of the ratio of the real (imaginary) part $x^*(t)$ to its modulus [52]. The phase synchronization of two signals takes place when:

$$|\Delta\Phi_{x,y}(t)| \leq const, \tag{22}$$

where $\Delta\Phi_{x,y}(t) = n\Phi_x(t) - m\Phi_y(t)$, Φ is a phase of the signal; n, m are integers. Then, the angular frequency of the signal can be found by the phase differentiating with respect to time. Numerical differentiation in the presence of phase fluctuations is an unstable procedure. Additionally, the disadvantage of the approach associated with the calculation of analytical signals is that it is well applicable for narrowband signals and not good enough for broadband signals [53].

The paper describes the methods and results of determining the phase-connected pairs of EEG channels of patients with moderate TBI before and after rehabilitation, which can be used to estimate the dynamics of treatment and rehabilitation of patients. The method of the estimation of the inter-channel EEG phase synchronization at the points of the ridges $f_r(t_i)$ of their wavelet-spectrograms (6) is considered as an inverse task for the task of modeling ridges:

$$f_r(t_i) = \arg\left\{ \max_{f(t_i)\in[1:25\text{ Hz}]} (|W(t_i, f(t_i))|) \right\}, \tag{23}$$

on the condition (18). $\Phi_x(t) \cong \Phi_r(f,t) = \arctan\left(\frac{Im(W(t,f_r))}{Re(W(t,f_r))}\right)$ according to (17), when the condition (18) is satisfied.

However, the ridge $W_r(t)$ can be considered as the frequency-modulated signal. It is necessary to take an unmodulated oscillation [54]:

$$x = A_0 \sin(\omega_0 t + \Phi_0), \tag{24}$$

and enter a variable frequency $\omega = \omega_0 + \Delta\omega\xi(t) = \omega(t)$, where $\xi(t)$ is some unknown function, and $\omega(t)$ is known.

Then if $\Phi_0 = 0$:

$$x = A_0 \sin\left(\omega_0 t + \Delta\omega \int_0^t \xi(t)\, dt\right) = A_0 \sin(\omega(t)t), \tag{25}$$

Then it is possible to estimate the phase of the ridge as [55]:

$$\Phi(t, f_r) = 2\pi f_r(t)t, \tag{26}$$

Figure 11 represents two ridge frequencies of the Morlet wavelet transform for two EEG channels. Ridge points are points of the maximum power spectral density. Fp1 EEG channel is indicated by the blue line. Fp2 EEG channel is indicated by the red line. The abscissa is the time in seconds; the ordinate is the frequency in Hz.

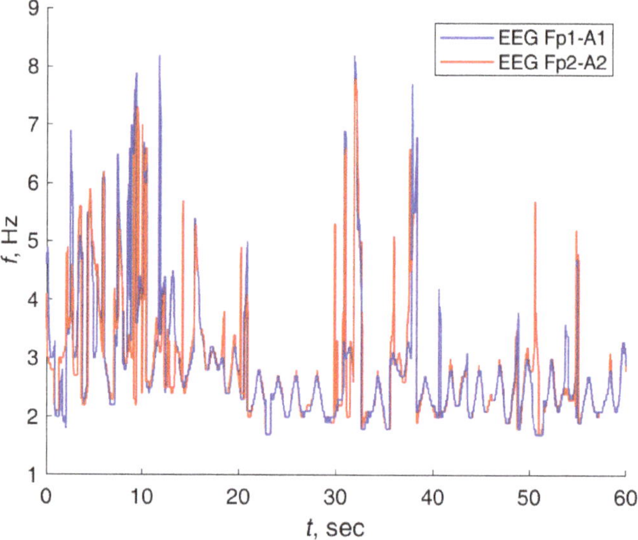

Figure 11. Ridge frequencies of the Morlet wavelet transform for two EEG channels. Fp1 EEG channel is indicated by the blue line. Fp2 EEG channel is indicated by the red line.

The EEG frequencies coincided in some time fragments. The phase of the ridge could be estimated with the formula (26) if the ridge frequency was known.

The phases of the EEG signals were calculated and compared at the points of the ridges (t_i, f_r) of their wavelet spectrograms in EEG records both with cognitive tests and without tests. Then, the phase difference of two signals $x(t)$ and $y(t)$ in two EEG channels was calculated. Next, the normalized histogram of portions $\rho_{x,y} = n_{x,y}/N$ in different pairs of EEG channels was calculated, where $n_{x,y}$ is the quantity of reference points of ridges with $|\Delta\Phi_{x,y}(t)| < 0.01\pi$, N is a total quantity of EEG signal reference points in the test.

Figure 12 represents the normalized histograms of portions of the phase difference at the ridge points of the wavelet spectrograms of two EEG channels for the case of a phase coupled pair of EEG channels Fp1-Fp2, which were obtained by two methods.

Figure 12a demonstrates the first way based on the calculation of the phase according to the formula (26). Figure 12b demonstrates the second way based on the calculation of the phase according to the formula (17).

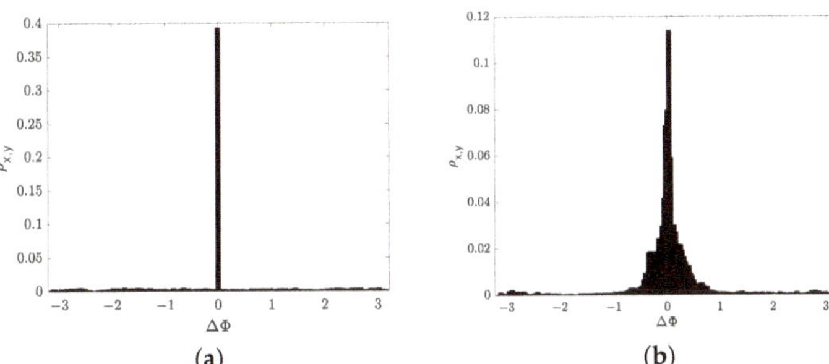

Figure 12. Histograms of portions $\rho_{x,y}$ for the phase difference at ridge points for two EEG channels. This example shows a phase-connected pair of Fp1-Fp2 EEG channels. (**a**) The histogram is obtained from the phase calculation by (26). (**b**) The histogram is obtained from the phase calculation by (17).

Figure 12a demonstrates that the histogram of portions of the phase difference at the points of the ridge of the wavelet spectrograms calculated by the first way (26) had a higher and sharper peak versus the second way (17) (Figure 12b). Below, we will calculate the phase by formula (26).

Let $A = \max(\rho_{x,y})$ be the maximum values of the histogram in the cognitive test and let $B = \max(\rho_{x,y})$ are the maximum values of the histogram in the EEG record without a test. It is convenient to consider the difference $D = A - B$, which was sorted in order to increase $\max(\rho_{x,y})$. Figure 13 demonstrates the dependence of D sorted in increasing order versus the numbers of a pair of EEG channels and its derivative for a healthy subject in the CT1 test.

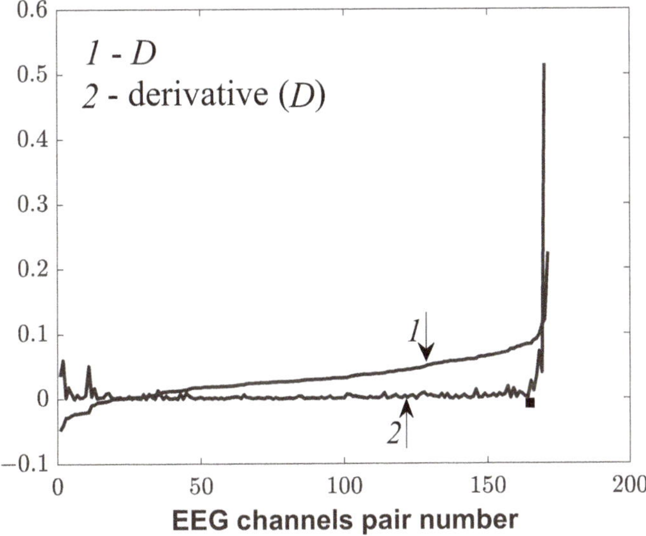

Figure 13. The dependence of D sorted in increasing order (line1) versus the numbers of a pair of EEG channels and its derivative (line 2) for a control subject in the CT1 test.

Figure 13 shows that the curve of the graph appears at some point D. It is advisable to consider pairs of channels with numbers greater than at point sharp of increasing of derivative D (black point) as phase-connected pairs. Thus, phase-connected pairs of EEG channels were identified before and after rehabilitation of patients with moderate TBI.

Figure 14 shows a block diagram of the developed algorithm for the determination of phase-connected EEG channels. The developed algorithm for the determination of phase-connected EEG channels consisted of the following stages, as shown in Figure 14: 1. Preprocessing of signals. It was outlier removing; notch filter at frequencies multiples of 50 Hz; filtering of signals with a Butterworth filter; 2. Calculation of wavelet spectra and ridges; 3. Calculation of the ridges phase at each point of the wavelet spectra ridges. Calculation histograms of the phase difference portions $(\rho_{x,y})$ in two channels for 171 channels pairs. The determination $\max(\rho_{x,y})$ for each channel pairs; 4. Calculation of the difference between $\max(\rho_{x,y})$ with cognitive test and without a test (D), sorting in in-creasing order of D. Calculation derivative (D); 5. The determination of phase connected EEG channels. If the derivative (D) sharply increased with the growing pair, the pairs with numbers greater than at the sharp point of the increasing derivative (D) were considered as a phase connected pairs. If the derivative (D) did not sharply increase, it was impossible to identify phase connected pairs.

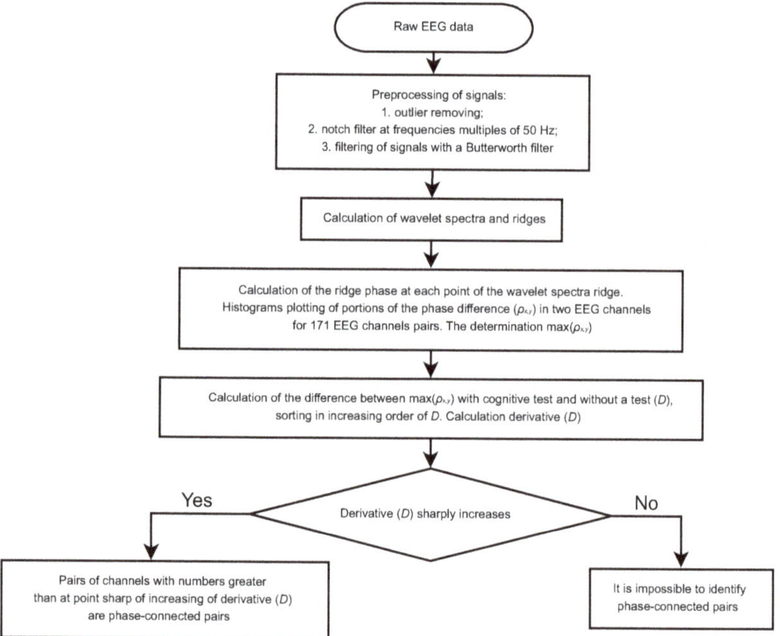

Figure 14. The block diagram of the developed algorithm for the determination of phase-connected EEG channels.

Figure 15 demonstrates the phase-coupled pairs of EEG channels for seven healthy subjects during the EEG recording in the CT1 test.

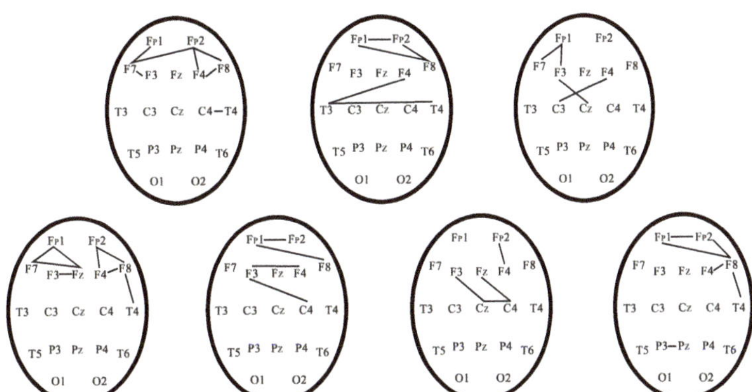

Figure 15. Phase-connected pairs of EEG channels of control subjects during the EEG recording in the CT1 test.

Figure 15 represents that the frontal regions and interhemispheric connections are activated in cognitive tests (CT1). Interhemispheric connections and connections in the frontal cortex in control subjects are activated during CT1 test in accordance with published work [56]. However, each control subject and patient with TBI are characterized by different phase-connected pairs due to the individuality of each person during the CT1 test. Therefore, we considered phase-connected pairs individually for each subject.

Figure 16 demonstrates the phase-connected pairs of EEG channels for the seven control subjects during the EEG recording in the CT2 test.

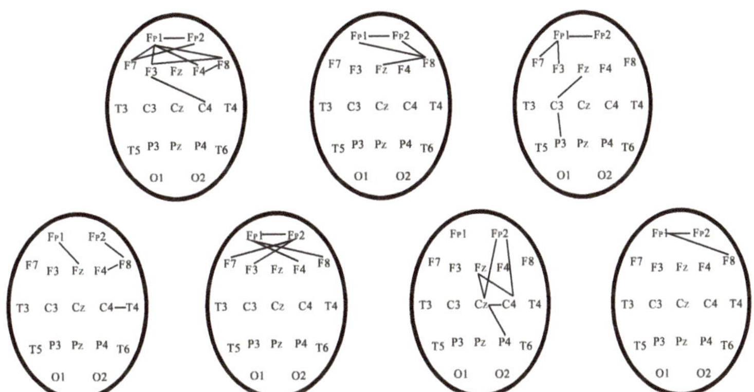

Figure 16. Phase-connected pairs of EEG channels of the control subjects during the EEG recording in the CT2 test.

Figure 16 represents that the frontal regions and interhemispheric connections were activated in cognitive tests (CT2). Interhemispheric connections and connections in the frontal cortex in control subjects were activated during CT2 test in accordance with published work [56]. However, each control subject and patient with TBI were characterized by different phase-connected pairs due to the individuality of each person during the CT2 test. Therefore, we considered phase-connected pairs individually for each subject.

Figure 17 demonstrates phase-connected pairs of EEG channels for three patients with TBI during the EEG recording in CT1 and CT2 tests.

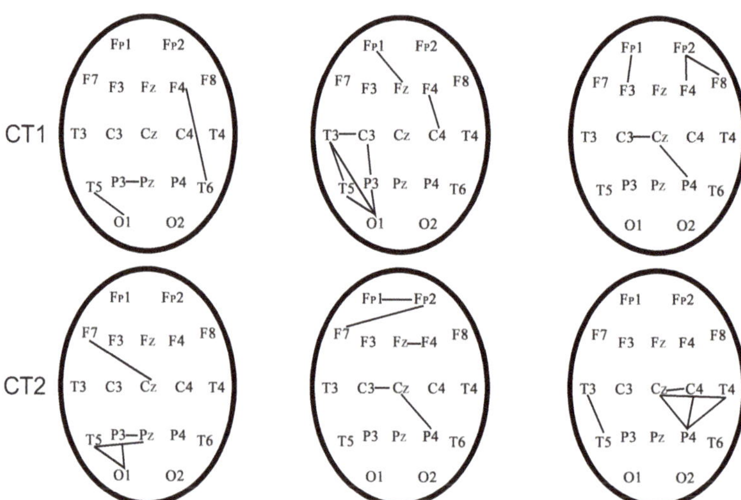

Figure 17. Phase-connected pairs of EEG channels of patients with TBI during the EEG recording in CT1 and CT2 tests.

Figure 17 shows that phase-connected pairs appeared more in the parietal and occipital regions, than in the interhemispheric and frontal cortex in patients with TBI during CT1 and CT2 tests.

Additionally, the dynamics of inter-channel EEG synchronization of three patients with TBI before and after the rehabilitation was also investigated. The phase-connected EEG pairs in patients before and after rehabilitation were compared with the phase-connected pairs of the control group for each test. If interhemispheric connections or connections in the frontal cortex were activated in patients, as in control subjects in cognitive tests (CT1 and CT2), it could be concluded that the cognitive function had positive dynamics.

Figure 18 demonstrates that the positive dynamics could be seen of the rehabilitation of a patient with TBI in the CT1 test. If interhemispheric connections or connections in the frontal cortex in the CT1 test appeared after rehabilitation, as in the control subjects, the positive dynamics of rehabilitation could be concluded.

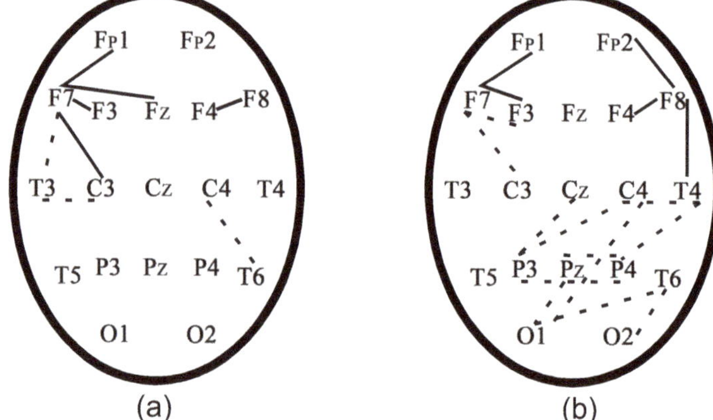

Figure 18. Phase-connected pairs of EEG channels of patients with TBI before (dotted lines) and after the rehabilitation (solid lines) in the CT1 test. (**a**) Patient 1. (**b**) Patient 2.

Figure 19 demonstrates that the positive dynamics could be seen of the rehabilitation of patients with TBI in the CT2 test.

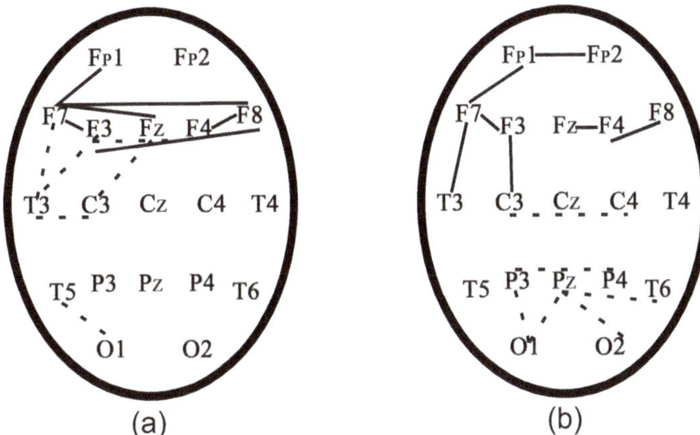

Figure 19. Phase-connected pairs of EEG channels of a patients with TBI before (dotted lines) and after the rehabilitation (solid lines) in the CT2 test. (**a**) Patient 1. (**b**) Patient 2.

Figure 19 demonstrates that the positive dynamics could be seen of the rehabilitation of a patient with TBI in the CT2 test because interhemispheric connections or connections in the frontal cortex were activated in patients, as in control subjects. If interhemispheric connections or connections in the frontal cortex in the CT2 test appeared after rehabilitation, as in the control subjects, it the positive dynamics of rehabilitation could be concluded.

Let us consider an example of the lack of progress of the rehabilitation of a patient with TBI. Figure 20 demonstrates the dependence of D sorted in increasing order versus the numbers of pairs of EEG channels and its derivative for a patient with TBI.

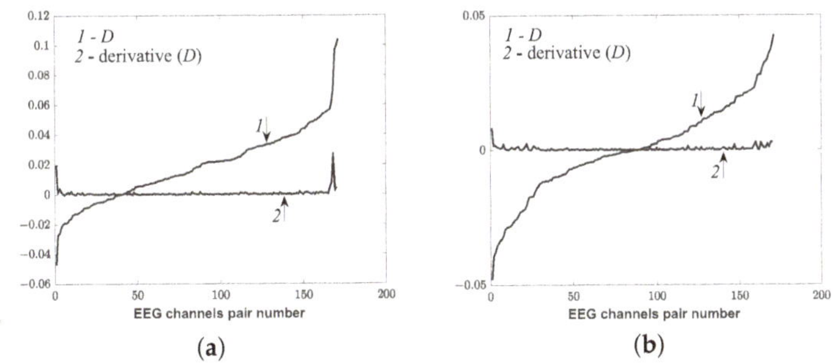

Figure 20. The dependence of D sorted in increasing order (line 1) versus the numbers of a pair of EEG channels and its derivative (line 2) for a patient with TBI in the CT1 test. (**a**) before the rehabilitation; (**b**) after the rehabilitation.

Figure 20b shows that there was no sharp increase in the D derivative after the rehabilitation during the cognitive calculate-logical test, in contrast to Figure 20a. Thus, it was impossible to clearly determine the quantitative signs and identify phase-connected pairs by the suggested method. It could be concluded that there was no progress in rehabilitation during the cognitive calculation-logical test.

5. Conclusions

The paper presents an approach for segmenting long-term 19-channel EEG monitoring data. For the signals, the ridges of Morlet wavelet transform were calculated. Interchannel synchronization was used as a new feature of epileptic seizure. We also used the adaptive thresholding of the wavelet spectrogram ridges for signal segmentation. The intersection of the synchronized and the power spectral density intervals were obtained. As a result, the total duration of the fragments for analysis by the doctor was reduced by more than 60 times. It was shown that the frequency of the peak of the Fourier spectrum of the cutoff of the wavelet spectrogram at a frequency higher than the frequency of the ridge during an epileptic discharge was 2.5 times higher than the frequency of the Fourier peak corresponding to chewing. The Fourier peak full width at half maximum of the chewing artifact was 2 times larger than that of ES.

A comparison of the phases of EEG at the points of the Morlet wavelet spectrogram ridges were used for evaluation the EEG interchannel phase synchronization during cognitive tests in control subjects and patients with moderate TBI. Calculation-logical and spatial-pattern cognitive tests were used. Interhemispheric connections and connections in the frontal cortex in control subjects are initiated during the cognitive tests. The possibility of determining the positive dynamics of rehabilitation during the initialization of interhemispheric connections and connections in the frontal cortex of the brain or the absence of progress in rehabilitation has been shown.

Author Contributions: Conceptualization, Y.V.O., M.V.S. and L.A.Z.; methodology, Y.V.O.; software, I.A.K. and R.A.T.; validation, Y.V.O., M.V.S., L.A.Z., I.A.K. and R.A.T.; formal analysis, Y.V.O., I.A.K. and R.A.T.; investigation, Y.V.O., I.A.K. and R.A.T.; resources, M.V.S. and L.A.Z.; data acquisition, M.V.S., L.A.Z. and R.A.T.; data curation, M.V.S. and L.A.Z.; writing—original draft preparation, Y.V.O., I.A.K. and R.A.T.; writing—review and editing, Y.V.O., I.A.K. and R.A.T.; visualization, I.A.K. and R.A.T.; supervision, Y.V.O.; project administration, Y.V.O.; funding acquisition, Y.V.O. All authors have read and agreed to the published version of the manuscript.

Funding: The work was carried out within the framework of the state task.

Institutional Review Board Statement: The study was conducted according to the guidelines of the Declaration of Helsinki, and approved by the Ethics Committee of National Medical Research Center for Neurosurgery named after Academician N.N. Burdenko and by the Interuniversity Ethics Committee.

Informed Consent Statement: Informed consent was obtained from all subjects involved in the study.

Data Availability Statement: The data presented in this study are available on request from the corresponding author. The clinical data are not publicly available due to the ethical policy of the institute.

Conflicts of Interest: The authors declare no conflict of interest. The funders had no role in the design of the study; in the collection, analysis, or interpretation of data; in the writing of the manuscript; or in the decision to publish the results.

References

1. Mallat, S.G. Multifrequency channel decompositions of images and wavelet models. *IEEE Trans. Acoust. Speech Signal Process.* **1989**, *37*, 2091–2110. [CrossRef]
2. Mallat, S. *A Wavelet Tour of Signal Processing*; Elsevier: Amsterdam, The Netherlands, 1999.
3. Euskai, M.; Beylkin, G.; Coifman, E. *Wavelets and Their Applications and Data Analysis*; Jones and Bartlett Publishers: Boston, MA, USA, 1992.
4. Daubechies, I. *Ten Lectures on Wavelets*; Society for Industrial and Applied Mathematics: Philadelphia, PA, USA, 1992.
5. Meyer, Y. *Wavelets: Algorithms and Applications*; Society for Industrial and Applied Mathematics: Philadelphia, PA, USA, 1993; Volume 142.
6. Unser, M.; Aldroubi, A. A review of wavelets in biomedical applications. *Proc. IEEE* **1996**, *84*, 626–638. [CrossRef]
7. Hramov, A.E.; Koronovskii, A.A.; Makarov, V.A.; Pavlov, A.N.; Sitnikova, E. *Wavelets in Neuroscience*; Springer: Berlin/Heidelberg, Germany, 2015.
8. Bowyer, S.M. Coherence a measure of the brain networks: Past and present. *Neuropsychiatr. Electrophysiol.* **2016**, *2*, 1. [CrossRef]

9. Delprat, N.; Escudié, B.; Guillemain, P.; Kronland-Martinet, R.; Tchamitchian, P.; Torresani, B. Asymptotic wavelet and Gabor analysis: Extraction of instantaneous frequencies. *IEEE Trans. Inf. Theory* **1992**, *38*, 644–664. [CrossRef]
10. Lilly, J.M.; Olhede, S.C. On the analytic wavelet transform. *IEEE Trans. Inf. Theory* **2010**, *56*, 4135–4156. [CrossRef]
11. Guillemain, P.; Kronland-Martinet, R. Characterization of acoustic signals through continuous linear time-frequency representations. *Proc. IEEE* **1996**, *84*, 561–585. [CrossRef]
12. Alotaiby, T.N.; Alshebeili, S.A.; Alshawi, T.; Ahmad, I.; Abd El-Samie, F.E. EEG seizure detection and prediction algorithms: A survey. *EURASIP J. Adv. Signal Process.* **2014**, *2014*, 183. [CrossRef]
13. Xiong, Z.; Wang, H.; Zhang, L.; Fan, T.; Shen, J.; Zhao, Y.; Liu, Y.; Wu, Q. A Study on Seizure Detection of EEG Signals Represented in 2D. *Sensors* **2021**, *21*, 5145. [CrossRef]
14. Ahmad, Z.K.; Singh, V.; Khan, Y.U. Sequential Segmentation of EEG Signals for Epileptic Seizure Detection using Machine Learning. In Proceedings of the 2019 2nd International Conference on Signal Processing and Communication (ICSPC), Coimbatore, India, 29–30 March 2019; pp. 258–262.
15. Wu, J.; Zhou, T.; Li, T. Detecting epileptic seizures in EEG signals with complementary ensemble empirical mode decomposition and extreme gradient boosting. *Entropy* **2020**, *22*, 140. [CrossRef]
16. Rani, T.P.; Chellam, G.H. A novel peak signal feature segmentation process for epileptic seizure detection. *Int. J. Inf. Technol.* **2021**, *13*, 423–431. [CrossRef]
17. Hussein, R.; Palangi, H.; Ward, R.K.; Wang, Z.J. Optimized deep neural network architecture for robust detection of epileptic seizures using EEG signals. *Clin. Neurophysiol.* **2019**, *130*, 25–37. [CrossRef]
18. Liu, G.; Xiao, R.; Xu, L.; Cai, J. Minireview of Epilepsy Detection Techniques Based on Electroencephalogram Signals. *Front. Syst. Neurosci.* **2021**, *15*, 44. [CrossRef]
19. Saminu, S.; Xu, G.; Shuai, Z.; Abd El Kader, I.; Jabire, A.H.; Ahmed, Y.K.; Karaye, I.A.; Ahmad, I.S. A Recent Investigation on Detection and Classification of Epileptic Seizure Techniques Using EEG Signal. *Brain Sci.* **2021**, *11*, 668. [CrossRef]
20. Runarsson, T.P.; Sigurdsson, S. On-line detection of patient specific neonatal seizures using support vector machines and half-wave attribute histograms. In Proceedings of the International Conference on Computational Intelligence for Modelling, Control and Automation and International Conference on Intelligent Agents, Web Technologies and Internet Commerce (CIMCA-IAWTIC'06), Vienna, Austria, 28–30 November 2005; Volume 2, pp. 673–677.
21. Yoo, J.; Yan, L.; El-Damak, D.; Altaf, M.A.B.; Shoeb, A.H.; Chandrakasan, A.P. An 8-channel scalable EEG acquisition SoC with patient-specific seizure classification and recording processor. *IEEE J.-Solid-State Circuits* **2012**, *48*, 214–228. [CrossRef]
22. Rana, P.; Lipor, J.; Lee, H.; Van Drongelen, W.; Kohrman, M.H.; Van Veen, B. Seizure detection using the phase-slope index and multichannel ECoG. *IEEE Trans. Biomed. Eng.* **2012**, *59*, 1125–1134. [CrossRef]
23. Khamis, H.; Mohamed, A.; Simpson, S. Frequency–moment signatures: A method for automated seizure detection from scalp EEG. *Clin. Neurophysiol.* **2013**, *124*, 2317–2327. [CrossRef]
24. Acharya, U.R.; Molinari, F.; Sree, S.V.; Chattopadhyay, S.; Ng, K.H.; Suri, J.S. Automated diagnosis of epileptic EEG using entropies. *Biomed. Signal Process. Control* **2012**, *7*, 401–408. [CrossRef]
25. Zhou, W.; Liu, Y.; Yuan, Q.; Li, X. Epileptic seizure detection using lacunarity and Bayesian linear discriminant analysis in intracranial EEG. *IEEE Trans. Biomed. Eng.* **2013**, *60*, 3375–3381. [CrossRef]
26. Niknazar, M.; Mousavi, S.; Vahdat, B.V.; Sayyah, M. A new framework based on recurrence quantification analysis for epileptic seizure detection. *IEEE J. Biomed. Health Inform.* **2013**, *17*, 572–578. [CrossRef]
27. Liu, Y.; Zhou, W.; Yuan, Q.; Chen, S. Automatic seizure detection using wavelet transform and SVM in long-term intracranial EEG. *IEEE Trans. Neural Syst. Rehabil. Eng.* **2012**, *20*, 749–755. [CrossRef] [PubMed]
28. Raghu, S.; Sriraam, N.; Gommer, E.D.; Hilkman, D.M.; Temel, Y.; Rao, S.V.; Kubben, P.L. Adaptive median feature baseline correction for improving recognition of epileptic seizures in ICU EEG. *Neurocomputing* **2020**, *407*, 385–398. [CrossRef]
29. Lee, J.; Shin, H.C. Burst Suppression Segmentation of EEG Using Adaptive Binarization in Time and Frequency Domains. *IEEE Access* **2019**, *7*, 54550–54561. [CrossRef]
30. Klatt, J.; Feldwisch-Drentrup, H.; Ihle, M.; Navarro, V.; Neufang, M.; Teixeira, C.; Adam, C.; Valderrama, M.; Alvarado-Rojas, C.; Witon, A.; et al. The EPILEPSIAE database: An extensive electroencephalography database of epilepsy patients. *Epilepsia* **2012**, *53*, 1669–1676. [CrossRef]
31. Obeid, I.; Picone, J. The temple university hospital EEG data corpus. *Front. Neurosci.* **2016**, *10*, 196. [CrossRef]
32. Mannan, M.M.N.; Kamran, M.A.; Jeong, M.Y. Identification and removal of physiological artifacts from electroencephalogram signals: A review. *IEEE Access* **2018**, *6*, 30630–30652. [CrossRef]
33. Nolte, G.; Bai, O.; Wheaton, L.; Mari, Z.; Vorbach, S.; Hallett, M. Identifying true brain interaction from EEG data using the imaginary part of coherency. *Clin. Neurophysiol.* **2004**, *115*, 2292–2307. [CrossRef]
34. Kulaichev, A. The informativeness of coherence analysis in EEG studies. *Neurosci. Behav. Physiol.* **2011**, *41*, 321. [CrossRef]
35. Rudrauf, D.; Douiri, A.; Kovach, C.; Lachaux, J.P.; Cosmelli, D.; Chavez, M.; Adam, C.; Renault, B.; Martinerie, J.; Le Van Quyen, M. Frequency flows and the time-frequency dynamics of multivariate phase synchronization in brain signals. *Neuroimage* **2006**, *31*, 209–227. [CrossRef]
36. Ianof, J.N.; Anghinah, R. Traumatic brain injury: An EEG point of view. *Dement. Neuropsychol.* **2017**, *11*, 3–5. [CrossRef]

37. Haveman, M.E.; Van Putten, M.J.; Hom, H.W.; Eertman-Meyer, C.J.; Beishuizen, A.; Tjepkema-Cloostermans, M.C. Predicting outcome in patients with moderate to severe traumatic brain injury using electroencephalography. *Crit. Care* **2019**, *23*, 1–9. [CrossRef]

38. Lai, C.Q.; Ibrahim, H.; Abdullah, M.Z.; Azman, A.; Abdullah, J.M. Detection of moderate traumatic brain injury from resting-state eye-closed electroencephalography. *Comput. Intell. Neurosci.* **2020**, *2020*, 8923906. [CrossRef]

39. Udekwu, P.; Kromhout-Schiro, S.; Vaslef, S.; Baker, C.; Oller, D. Glasgow Coma Scale score, mortality, and functional outcome in head-injured patients. *J. Trauma Acute Care Surg.* **2004**, *56*, 1084–1089. [CrossRef] [PubMed]

40. Dhillon, N.S.; Sutandi, A.; Vishwanath, M.; Lim, M.M.; Cao, H.; Si, D. A Raspberry Pi-based traumatic brain injury detection system for single-channel electroencephalogram. *Sensors* **2021**, *21*, 2779. [CrossRef] [PubMed]

41. Cheng, Q.; Yang, W.; Liu, K.; Zhao, W.; Wu, L.; Lei, L.; Dong, T.; Hou, N.; Yang, F.; Qu, Y.; et al. Increased sample entropy in EEGs during the functional rehabilitation of an injured brain. *Entropy* **2019**, *21*, 698. [CrossRef]

42. Arroyo-Ferrer, A.; de Noreña, D.; Serrano, J.I.; Ríos-Lago, M.; Romero, J.P. Cognitive rehabilitation in a case of traumatic brain injury using EEG-based neurofeedback in comparison to conventional methods. *J. Integr. Neurosci.* **2021**, *20*, 449–457. [CrossRef] [PubMed]

43. Homan, R.W.; Herman, J.; Purdy, P. Cerebral location of international 10–20 system electrode placement. *Electroencephalogr. Clin. Neurophysiol.* **1987**, *66*, 376–382. [CrossRef]

44. Hampel, F.R.; Ronchetti, E.M.; Rousseeuw, P.J.; Stahel, W.A. *Robust Statistics: The Approach Based on Influence Functions*; John Wiley & Sons: Hoboken, NJ, USA, 2011; Volume196.

45. Wong, R. *Asymptotic Approximations of Integrals*; Society for Industrial and Applied Mathematics: Philadelphia, PA, USA, 2001.

46. Pavlov, A.N.; Hramov, A.E.; Koronovskii, A.A.; Sitnikova, E.Y.; Makarov, V.A.; Ovchinnikov, A.A. Wavelet analysis in neurodynamics. *Phys.-Uspekhi* **2012**, *55*, 845. [CrossRef]

47. Murashov, D.M.; Obukhov, Y.V.; Kershner, I.A.; Sinkin, M.V. An algorithm for detecting events in video EEG monitoring data of patients with craniocerebral injuries. *Comput. Opt.* **2021**, *45*, 301–305. [CrossRef]

48. Islam, M.K.; Rastegarnia, A.; Yang, Z. Methods for artifact detection and removal from scalp EEG: A review. *Neurophysiol. Clinique/Clin. Neurophysiol.* **2016**, *46*, 287–305. [CrossRef]

49. Wendling, F.; Ansari-Asl, K.; Bartolomei, F.; Senhadji, L. From EEG signals to brain connectivity: A model-based evaluation of interdependence measures. *J. Neurosci. Methods* **2009**, *183*, 9–18. [CrossRef]

50. Zhan, Y.; Halliday, D.; Jiang, P.; Liu, X.; Feng, J. Detecting time-dependent coherence between non-stationary electrophysiological signals—A combined statistical and time–frequency approach. *J. Neurosci. Methods* **2006**, *156*, 322–332. [CrossRef] [PubMed]

51. Rosenblum, M.; Pikovsky, A.; Kurths, J.; Schäfer, C.; Tass, P.A. Phase synchronization: From theory to data analysis. In *Handbook of Biological Physics*; Elsevier: Amsterdam, The Netherlands, 2001; Volume 4, pp. 279–321.

52. Vakman, D.; Vaĭnshteĭn, L. Amplitude, phase, frequency-fundamental concepts of oscillation theory. *Sov. Phys. Uspekhi* **1977**, *20*, 1002. [CrossRef]

53. Le Van Quyen, M.; Foucher, J.; Lachaux, J.P.; Rodriguez, E.; Lutz, A.; Martinerie, J.; Varela, F.J. Comparison of Hilbert transform and wavelet methods for the analysis of neuronal synchrony. *J. Neurosci. Methods* **2001**, *111*, 83–98. [CrossRef]

54. Kharkevich, A.; Weiss, G. Spectra and analysis. *Phys. Today* **1961**, *14*, 61. [CrossRef]

55. Tolmacheva, R.; Obukhov, Y.V.; Polupanov, A.; Zhavoronkova, L. New approach to estimation of interchannel phase coupling of electroencephalograms. *J. Commun. Technol. Electron.* **2018**, *63*, 1070–1075. [CrossRef]

56. Zhavoronkova, L.; Shevtsova, T.; Maksakova, O. *How Does Human Brain Simultaneously Solve Two Problems*; LAP LAMBERT Academic: Saarbrucken, Germany, 2017; Volume 59.

Article

Motion Artifacts Correction from Single-Channel EEG and fNIRS Signals Using Novel Wavelet Packet Decomposition in Combination with Canonical Correlation Analysis

Md Shafayet Hossain [1], Muhammad E. H. Chowdhury [2,*], Mamun Bin Ibne Reaz [1,*], Sawal Hamid Md Ali [1], Ahmad Ashrif A. Bakar [1], Serkan Kiranyaz [2], Amith Khandakar [2], Mohammed Alhatou [3], Rumana Habib [4] and Muhammad Maqsud Hossain [5]

1 Department of Electrical, Electronic and Systems Engineering, Universiti Kebangsaan Malaysia, Bangi 43600, Malaysia; p108100@siswa.ukm.edu.my (M.S.H.); sawal@ukm.edu.my (S.H.M.A.); ashrif@ukm.edu.my (A.A.A.B.)
2 Department of Electrical Engineering, Qatar University, Doha 2713, Qatar; mkiranyaz@qu.edu.qa (S.K.); amitk@qu.edu.qa (A.K.)
3 Neuromuscular Division, Department of Neurology, Al-Khor Branch, Hamad General Hospital, Doha 3050, Qatar; malhatou@hamad.qa
4 Department of Neurology, BIRDEM General Hospital, Dhaka 1000, Bangladesh; drhrumana4@gmail.com
5 NSU Genome Research Institute (NGRI), North South University, Dhaka 1229, Bangladesh; muhammad.maqsud@northsouth.edu
* Correspondence: mchowdhury@qu.edu.qa (M.E.H.C.); mamun@ukm.edu.my (M.B.I.R.)

Citation: Hossain, M.S.; Chowdhury, M.E.H.; Reaz, M.B.I.; Ali, S.H.M.; Bakar, A.A.A.; Kiranyaz, S.; Khandakar, A.; Alhatou, M.; Habib, R.; Hossain, M.M. Motion Artifacts Correction from Single-Channel EEG and fNIRS Signals Using Novel Wavelet Packet Decomposition in Combination with Canonical Correlation Analysis. *Sensors* **2022**, *22*, 3169. https://doi.org/10.3390/s22093169

Academic Editor: Yvonne Tran

Received: 24 February 2022
Accepted: 25 March 2022
Published: 21 April 2022

Publisher's Note: MDPI stays neutral with regard to jurisdictional claims in published maps and institutional affiliations.

Abstract: The electroencephalogram (EEG) and functional near-infrared spectroscopy (fNIRS) signals, highly non-stationary in nature, greatly suffers from motion artifacts while recorded using wearable sensors. Since successful detection of various neurological and neuromuscular disorders is greatly dependent upon clean EEG and fNIRS signals, it is a matter of utmost importance to remove/reduce motion artifacts from EEG and fNIRS signals using reliable and robust methods. In this regard, this paper proposes two robust methods: (i) Wavelet packet decomposition (WPD) and (ii) WPD in combination with canonical correlation analysis (WPD-CCA), for motion artifact correction from single-channel EEG and fNIRS signals. The efficacy of these proposed techniques is tested using a benchmark dataset and the performance of the proposed methods is measured using two well-established performance matrices: (i) difference in the signal to noise ratio (ΔSNR) and (ii) percentage reduction in motion artifacts (η). The proposed WPD-based single-stage motion artifacts correction technique produces the highest average ΔSNR (29.44 dB) when db2 wavelet packet is incorporated whereas the greatest average η (53.48%) is obtained using db1 wavelet packet for all the available 23 EEG recordings. Our proposed two-stage motion artifacts correction technique, i.e., the WPD-CCA method utilizing db1 wavelet packet has shown the best denoising performance producing an average ΔSNR and η values of 30.76 dB and 59.51%, respectively, for all the EEG recordings. On the other hand, for the available 16 fNIRS recordings, the two-stage motion artifacts removal technique, i.e., WPD-CCA has produced the best average ΔSNR (16.55 dB, utilizing db1 wavelet packet) and largest average η (41.40%, using fk8 wavelet packet). The highest average ΔSNR and η using single-stage artifacts removal techniques (WPD) are found as 16.11 dB and 26.40%, respectively, for all the fNIRS signals using fk4 wavelet packet. In both EEG and fNIRS modalities, the percentage reduction in motion artifacts increases by 11.28% and 56.82%, respectively when two-stage WPD-CCA techniques are employed in comparison with the single-stage WPD method. In addition, the average ΔSNR also increases when WPD-CCA techniques are used instead of single-stage WPD for both EEG and fNIRS signals. The increment in both ΔSNR and η values is a clear indication that two-stage WPD-CCA performs relatively better compared to single-stage WPD. The results reported using the proposed methods outperform most of the existing state-of-the-art techniques.

Keywords: motion artifact; electroencephalogram (EEG); functional near-infrared spectroscopy (fNIRS); wavelet packet decomposition (WPD); canonical correlation analysis (CCA)

1. Introduction

Due to the paradigm shift of hospital-based treatment in the direction of wearable and ubiquitous monitoring, nowadays, the acquisition and processing of vital physiological signals have become prevalent in the ambulatory setting. Since the acquisition of physiological signals is inclined to movement artifacts that happen due to the deliberate and/or voluntary movement of the patient during signal procurement utilizing wearable devices, restricting patients totally from physical movements, intentional and/or unintentional, is exceptionally troublesome. As a result, the physiological signals may get corrupted to some degree by motion artifacts. In some instances, this defilement may end up so conspicuous that the recorded signals may lose their usability unless the movement artifacts are diminished significantly.

Electroencephalogram (EEG) measures the electrical activity of the human brain quantitatively which took place due to the firing of neurons [1] and such brain activity is recorded utilizing a good number of cathodes which are located at different regions of the scalp [2]. EEG is one of the key diagnostic tests for epileptic seizure detection [3,4]. Other decisive utilization of EEG includes the estimation of drowsiness levels [5–8], emotion detection [9], cognitive workload [6,10], and brain-computer interfaces (BCIs) [11–16]. All of which have potential applications in the personal healthcare domain. Lately, the implementation of EEG-based biometric systems utilizing the inborn anti-spoofing capability of EEG signals was studied and appeared to be promising [17].

The functional near-infrared spectroscopy (fNIRS), a non-invasive optical brain imaging technique, measures changes in hemoglobin (Hb) concentrations inside the human brain [18] by employing light of various wavelengths in the infrared band and estimating the difference in the optical absorption [19]. Medical applications of fNIRS mainly focus on the noninvasive measurement of brain functions [20,21], cognitive tasks identification [22,23], and BCI [24–26].

Apart from movement artifacts, physiological signals undergo other types of artifacts as well. Gradient artifacts (GA) and pulse artifacts (PA) are the two most frequent artifacts observed in EEG during the simultaneous EEG-fMRI tests [27–29]. On the other hand, event-related fNIRS signals are regularly sullied by heartbeat, breath, Mayer waves, etc., as well as extra-cortical physiological clamors from the superficial layers [30].

Numerous attempts were made to reduce motion artifacts from EEG previously, which were summarized in [31,32]. In [33], the performance of motion artifacts correction techniques utilizing discrete wavelet transform (DWT) [34], empirical mode decomposition (EMD) [35], ensemble empirical mode decomposition (EEMD) [36], EMD along with canonical correlation analysis (EMD-CCA), EMD with independent component analysis (EMD-ICA), EEMD with ICA (EEMD-ICA), and EEMD with CCA (EEMD-CCA) were reported. Maddirala and Shaik [37] used singular spectrum analysis (SSA) [38], whereas DWT along with the thresholding technique was utilized in [39]. Gajbhiye et al. [40] employed wavelet-based transform along with the total variation (TV) and weighted TV (WTV) denoising techniques, whereas in [41], wavelet domain optimized Savitzky–Golay filter was proposed for the removal of motion artifacts from EEG. Recently, Hossain et al. [42] utilized variational mode decomposition (VMD) [43] for the correction of motion artifacts from EEG signals.

In the last few decades, multiple motion artifacts removal techniques were proposed [44–46] for the removal of motion artifacts from the fNIRS signal. Sweeney et al. [47] used adaptive filter, Kalman Filter, and EEMD-ICA. Scholkmann et al. [48] utilized the moving standard deviation and spline interpolation method, whereas in [49], a wavelet-based method was proposed. The authors of [33] used DWT, EMD, EEMD, EMD-ICA,

EEMD-ICA, EMD-CCA, and EEMD-CCA. In [50], Barker et al. used an autoregressive model-based algorithm, while kurtosis-based wavelet transform was proposed in [51], and Siddiquee et al. [52] utilized nine-degree of freedom inertia measurement unit (IMU) data to mathematically estimate the movement artifacts in the fNIRS signal using autoregressive exogenous (ARX) input model. A hybrid algorithm was proposed in [53] to filter out the movement artifacts from fNIRS signals where both the spline interpolation method and Savitzky–Golay filtering were employed. Very recently, the two-stage VMD-CCA technique was employed in [42].

The development of robust algorithms that can successfully reduce motion artifacts significantly from EEG and fNIRS data is critical; otherwise, the signals' interpretation could be erroneous by medical doctors and/or machine-learning-based applications. As mentioned earlier, DWT, EMD, EEMD, VMD, DWT-ICA, EMD-ICA, EEMD-ICA, EMD-CCA, EEMD-CCA, VMD-CCA, etc. were the most commonly used methods for the correction of motion artifacts from EEG and fNIRS signals. ICA and CCA cannot be used independently for single-channel EEG/fNIRS motion artifacts correction as the input of ICA/CCA algorithms require at least two (or more) channels data, whereas DWT, EMD, EEMD, VMD, etc. algorithms suffer from several limitations which are discussed in the discussion section of this paper. Additionally, there is still room for improvement for ΔSNR and η values which can be achieved using other effective novel methods. Therefore, in this paper, two novel motion artifacts removal techniques have been proposed which can eliminate motion artifacts from single-channel EEG and fNIRS signals to a great extent. The first is a single-stage motion artifacts correction technique using the wavelet packet decomposition (WPD), whereas the other novel method is WPD in combination with CCA (WPD-CCA), a two-stage motion artifacts removal technique, as the name suggests.

In this extensive study, for the correction of motion artifact from EEG and fNIRS signals using the WPD method, four different wavelet packet families (Daubechies (dbN), Symlets (symN), Coiflets (coifN), Fejer-Korovkin (fkN)) have been used with three different vanishing moments (for each of the wavelet packets) that resulted in a total of 12 different investigations. The wavelet packets used in the WPD method are db1, db2, db3, sym4, sym5, sym6, coif1, coif2, coif3, fk4, fk6, and fk8. To the best of our knowledge, the WPD algorithm has not been used for the removal of motion artifacts from single-channel EEG and fNIRS signals to date. WPD-CCA method is another novel contribution of this research work where Daubechies and Fejer-Korovkin wavelet packet families are utilized. In the WPD-CCA technique, db1, db2, db3, fk4, fk6, and fk8 have been used separately, resulting in six different investigations to reduce motion artifacts from EEG and fNIRS signals more efficiently.

The rest of this paper is organized as follows: Section 2 discusses the theoretical background of the different algorithms (WPD, CCA, WPD-CCA) investigated here, while Section 3 provides brief information about the EEG and fNIRS benchmark dataset and experimental methodology. Section 4 provides the results of the artifact removal techniques proposed in this work and Section 5 covers the discussion. Finally, the paper is concluded in Section 6.

2. Theoretical Background

2.1. Wavelet Packet Decomposition (WPD)

Using the WPD technique, signals can be decomposed into a wavelet packet basis at diverse scales [54,55]. For j-level decomposition, a wavelet packet basis is represented by multiple signals $\left[(n - 2^j k)\right]_{k \in \mathbb{Z}}$, where $i \in \mathbb{Z}^+$, $0 \leq i \leq 2j - 1$. The wavelet packet bases $\psi_j^i(n)$ are produced recursively from the scaling and wavelet functions, $\psi_1^0(n) = \phi(n)$ and $\psi_1^1(n) = \psi(n)$, respectively, as follows:

$$\psi_j^{2i}(n) = \sum_k h(k)\psi_{j-1}^i\left(n - 2^{j-1}k\right) \tag{1}$$

$$\psi_j^{2i+1}(n) = \sum_k g(k)\psi_{j-1}^i\left(n - 2^{j-1}k\right) \tag{2}$$

where $h(n)$ represents lowpass filter and $g(n)$ is the highpass filter defined as [54,56]:

$$h(k) = \left\langle \psi_j^{2i}(u), \psi_{j-1}^i\left(u - 2^{j-1}k\right) \right\rangle \tag{3}$$

$$g(k) = \left\langle \psi_j^{2i+1}(u), \psi_{j-1}^i\left(u - 2^{j-1}k\right) \right\rangle \tag{4}$$

The decomposition of a signal $x(n)$ onto the wavelet basis $j(n)$ at level j can be expressed as:

$$x(n) = \sum_{i,k} X_j^i \psi_j^i\left(n - 2^j k\right) \tag{5}$$

where $X_j^i(k)$ signifies the kth wavelet coefficient of the packet i, at level j. Here, $X_j^i(k)$ represents the intensity of the localized wavelet $\psi_j^i(n - 2^j k)$, defined by:

$$X_j^i(k) = \left\langle x(n),\ \psi_j^i\left(n - 2^j k\right) \right\rangle \tag{6}$$

Let $x(n)$ represent a recorded EEG/fNIRS signal which can be expressed as the sum of a source signal $s(n)$ and a motion artifact signal $v(n)$ as follows:

$$x(n) = s(n) + v(n) \tag{7}$$

In general, the source signal $s(n)$ is assumed to be normally distributed having a mean value equals to zero, $s(n) \sim N(0, \sigma)$, where σ^2 characterizes the variance of $s(n)$ [57]. On the other hand, general assumptions regarding the artifact signal $v(n)$ includes temporal localization, not normally distributed with high local variance.

According to [58], $X_j^i(k)$, can be represented as the sum of $S_j^i(k)$ and $V_j^i(k)$, where $X_j^i(k)$, $S_j^i(k)$, and $V_j^i(k)$ are the wavelet coefficients of $x(n)$, $s(n)$, and $v(n)$, respectively:

$$X_j^i(k) = S_j^i(k) + V_j^i(k) \tag{8}$$

It is noteworthy to mention that the wavelet coefficients $V_j^i(k)$ will be sparse as well as the non-zero coefficients will have a relatively higher magnitude as the variance of $v(n)$ is locally high, which would cause an increase in the local variance of the recorded EEG/fNIRS signal $x(n)$.

2.2. Canonical Correlation Analysis (CCA)

CCA [59] is one of the most popular blind source separation methods which has the capability of dissociating multiple mixed or noisy signals. Assuming linear mixing, square mixing, and stationary mixing [60], the CCA technique computes an un-mixing matrix $\mathbf{W}$, which helps identify the unknown independent components $\hat{\mathbf{S}}$ from a matrix $\mathbf{X}$, which is a recorded multi-channel signal as follows:

$$\hat{\mathbf{S}} = \mathbf{WX} \tag{9}$$

CCA also estimates the unknown independent components $\hat{\mathbf{S}}$ using Equation (9) utilizing second-order statistics (SOS). CCA forcefully makes the sources to be auto-correlated maximally as well as makes the sources mutually uncorrelated [61]. Let us assume $\mathbf{y}$ as a linear combination of neighboring samples for an input signal $\mathbf{x}$ (i.e., $y(t) = x(t-1) + x(t+1)$) [62]. Consider the linear combinations of the components in $\mathbf{x}$ and $\mathbf{y}$, known as the the canonical variates:

$$x = \mathbf{w}_\mathbf{x}^T\left(\mathbf{x} - \bar{\mathbf{x}}\right) \tag{10}$$

$$y = \mathbf{w}_\mathbf{y}^T\left(\mathbf{y} - \bar{\mathbf{y}}\right) \tag{11}$$

where $\mathbf{w}_\mathbf{x}$ and $\mathbf{w}_\mathbf{y}$ represents the weight matrices. CCA computes $\mathbf{w}_\mathbf{x}$ and $\mathbf{w}_\mathbf{y}$ in such a way so that the correlation ρ between x and y will be maximized [62]:

$$\rho = \frac{\mathbf{w}_x^T \mathbf{C}_{xy} \mathbf{w}_y^T}{\sqrt{\mathbf{w}_x^T \mathbf{C}_{xx} \mathbf{w}_x \mathbf{w}_y^T \mathbf{C}_{yy} \mathbf{w}_y}} \tag{12}$$

where $\mathbf{C}_{xx}$ and $\mathbf{C}_{yy}$ signify the nonsingular within-set covariance matrices and $\mathbf{C}_{xy}$ represent the between-sets covariance matrix. The maximized ρ is calculated by setting the derivatives of Equation (12) (with respect to $\mathbf{w}_x$ and $\mathbf{w}_y$) equal to zero:

$$\begin{aligned}
\mathbf{C}_{xx}^{-1} \mathbf{C}_{xy} \mathbf{C}_{yy}^{-1} \mathbf{C}_{yx}^T \hat{\mathbf{w}}_x &= \rho^2 \hat{\mathbf{w}}_x \\
\mathbf{C}_{yy}^{-1} \mathbf{C}_{yx} \mathbf{C}_{xx}^{-1} \mathbf{C}_{xy}^T \hat{\mathbf{w}}_y &= \rho^2 \hat{\mathbf{w}}_y
\end{aligned} \tag{13}$$

$\mathbf{w}_x$ and $\mathbf{w}_y$ can then be found out as the eigenvectors of the matrices $\mathbf{C}_{xx}^{-1} \mathbf{C}_{xy} \mathbf{C}_{yy}^{-1} \mathbf{C}_{yx}^T$ and $\mathbf{C}_{yy}^{-1} \mathbf{C}_{yx} \mathbf{C}_{xx}^{-1} \mathbf{C}_{xy}^T$, respectively, and the corresponding eigenvalues ρ^2 are the squared canonical correlations. It is sufficient to solve only one of the eigenvalue equations to obtain the un-mixing matrix $\mathbf{W}$ as the solutions are related. Furthermore, the underlying source signals $\hat{\mathbf{S}}$ can be estimated.

The components that seem to be artifacts can then be discarded by simply setting the corresponding columns of the $\hat{\mathbf{S}}$ matrix to zero before the signal reconstruction.

2.3. WPD-CCA

The WPD algorithm can be utilized to decompose a single-channel signal into multi-channel signal $\mathbf{X}$ where each column of matrix $\mathbf{X}$ represents the detailed and approximated sub-band signals. The total number of generated sub-band signals would be equal to 2^j, where j denotes the level, a priori. To estimate the underlying true sources $\hat{\mathbf{S}}$ (Equation (9)), these generated sub-band signals can then be used as the multi-channel input signals to the CCA algorithm. After that, the component/s of $\hat{\mathbf{S}}$ which seem to be artifacts can be discarded by making the corresponding columns of the matrix $\hat{\mathbf{S}}$ equal to zero. Bypassing this newly obtained source matrix through the inverse of the un-mixing matrix $\mathbf{W}^{-1}$, the multi-channel signals $\hat{\mathbf{X}}$ can be obtained. Finally, the cleaner signal $\hat{x}$ can be produced by simply summing all the columns of the matrix $\hat{\mathbf{X}}$.

3. Methods

This section describes the benchmark dataset used, pre-processing, study design, motion component identification, and evaluation metrics.

3.1. Dataset Description

A publicly available PhysioNet dataset [32,33,63] is used in this study that contains "reference ground truth" and motion corrupted signals for both EEG and fNIRS modalities. The details of the data recording procedure for EEG and fNIRS modalities were mentioned in [47]. During the data acquisition, two channels having the same hardware properties were placed on the test subject's scalp at very close proximity (20 mm for EEG modality and 30 mm for fNIRS modality), where the first channel was impacted with motion artifacts for 10–25 s at regular 2 min interval and the second channel was left untouched and undisturbed for the entire recording period. From the unimpacted channel (2nd channel), the EEG/fNIRS signal was extracted, which was free from motion artifacts and referred to as "reference ground truth" signal, whereas the impacted channel (1st channel) provided EEG/fNIRS signal corrupted with motion artifacts. It is worthwhile to mention that both the motion corrupted and "reference ground truth" signals were extracted simultaneously from channels 1 and 2, respectively, for approximately 9 min for each of the trial/test subjects. Additionally, the same channels were used to extract EEG/fNIRS data from all of the test subjects.

the single-stage motion artifact correction technique, i.e., WPD. Among these 12 wavelet packets, 6 wavelet packets (db1, db2, db3, fk4, fk6, and fk8) were used in the WPD-CCA method due to the relatively better performance shown by Daubechies and Fejer-Korovkin wavelet packet families incorporated in the WPD technique. As several wavelet packets were used in this study, in the rest of the manuscript, a subscript is added with WPD to denote the corresponding wavelet packet used. As an example, $WPD_{(db1)}$ would refer to that the db1 wavelet packet is used.

With the availability of sub-band signals decomposed using the WPD technique, the artifact components can then be selected and removed. All the remaining sub-band signals can then either be added up to reconstruct a cleaner signal or all the sub-band signals can be fed as inputs to the CCA algorithm to determine the motion corrupted components to enhance the signal quality further.

CCA technique needs the number of input channels to be at least two or greater. In this work, single-channel EEG and fNIRS signals have been evaluated for the correction of motion artifacts. Hence, it is required to generate several sub-band signals which would be used as the inputs for the CCA algorithm. Six different WPD-CCA-based ($WPD_{(db1)}$-CCA, $WPD_{(db2)}$-CCA, $WPD_{(db3)}$-CCA, $WPD_{(fk4)}$-CCA, $WPD_{(fk6)}$-CCA, and $WPD_{(fk8)}$-CCA) two-stage artifacts removal technique has been realized for both single-channel EEG and fNIRS signals.

3.4. Removal of Motion Artifact Components Using "Reference Ground Truth" Method

A common challenge in eliminating motion artifacts utilizing the aforementioned artifact removal approaches is consistently identifying and removing the motion corrupted components from the signal of interest and reconstructing a cleaner signal. The available reference "ground truth" signal of EEG and fNIRS modalities were used to identify the motion corrupted components as well as test the efficacy of the proposed algorithms. If a component of the decomposed signal is removed and the signal is rebuilt using the other components, the correlation coefficient between the newly reconstructed signal and the ground truth signal will only rise if the removed component has motion artifacts. Using this basic yet efficient notion, motion artifact-affected components of the decomposed signal were discovered and discarded to reconstruct a cleaner signal, ensuring the best performance of each suggested technique during evaluation.

Figure 3a shows an example motion corrupted EEG signal and below Figure 3b represents the corresponding 16 sub-band components generated from that corresponding EEG signal using $WPD_{(sym4)}$ algorithm. Figure 4a depicts an example motion corrupted EEG signal and Figure 4b represents the resultant 16 CCA components where the input of the CCA method was 16 sub-band signals generated from the motion corrupted EEG signal using $WPD_{(coif1)}$.

Similarly, Figures 5a and 6a show two different motion corrupted fNIRS signals, whereas Figures 5b and 6b represent the sub-band signals generated from $WPD_{(db1)}$, and 16 output CCA components where the input of the CCA algorithm consisted of 16 sub-band signals generated from the motion corrupted EEG signal using $WPD_{(fk8)}$, respectively.

From visual inspection of the components generated from the single-stage (WPD) and two-stage (WPD-CCA) motion artifacts removal techniques, it can be stated that in most of the cases, motion artifacts components are usually found in one or two approximation sub-band/CCA components. Although this was the case for most of the EEG and fNIRS recordings, rather than blindly discarding these one or two sub-band/CCA components as motion artifact components, only those components were discarded that, when removed, improved the correlation coefficient of the reconstructed signal in comparison with the available reference "ground truth" signal.

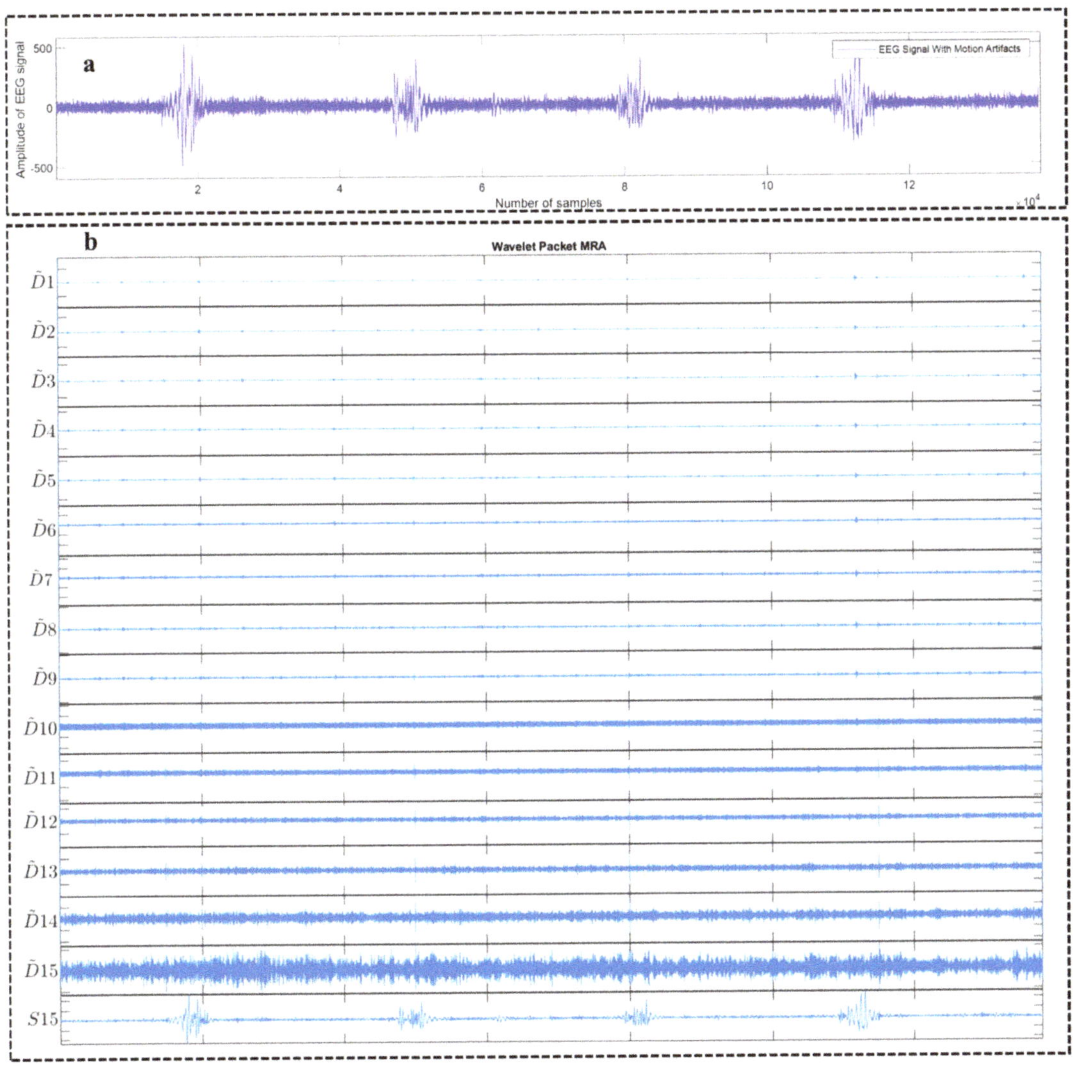

Figure 3. An example motion-corrupted single-channel EEG signal (**a**) and the corresponding 16 sub-band components generated using $\text{WPD}_{(sym4)}$ algorithm (**b**). S15 denotes the Approximation sub-band signal having the lowest center frequency compared to the other sub-band signals, i.e., D1–D15.

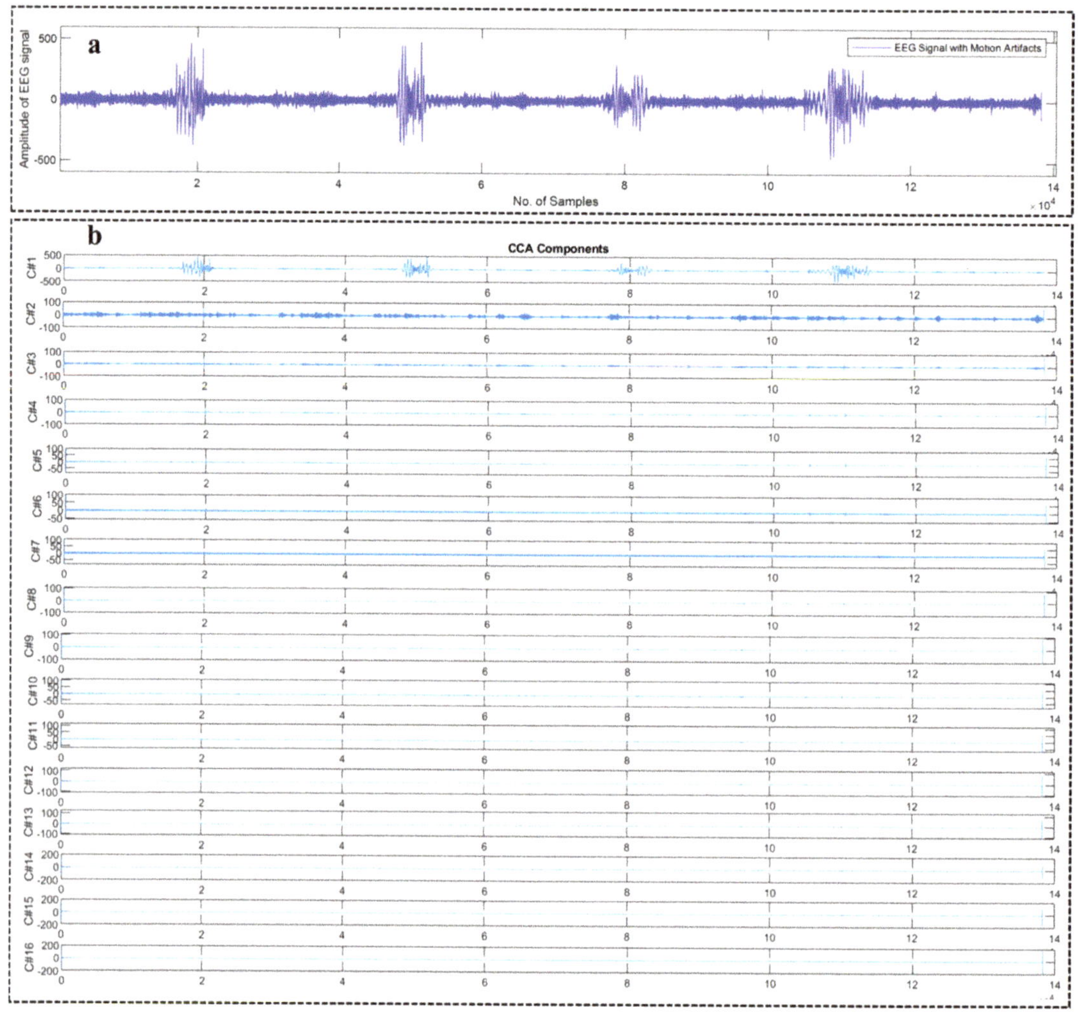

Figure 4. An example motion-corrupted single-channel EEG signal (**a**) and the corresponding 16 CCA components generated from the CCA algorithm (**b**).

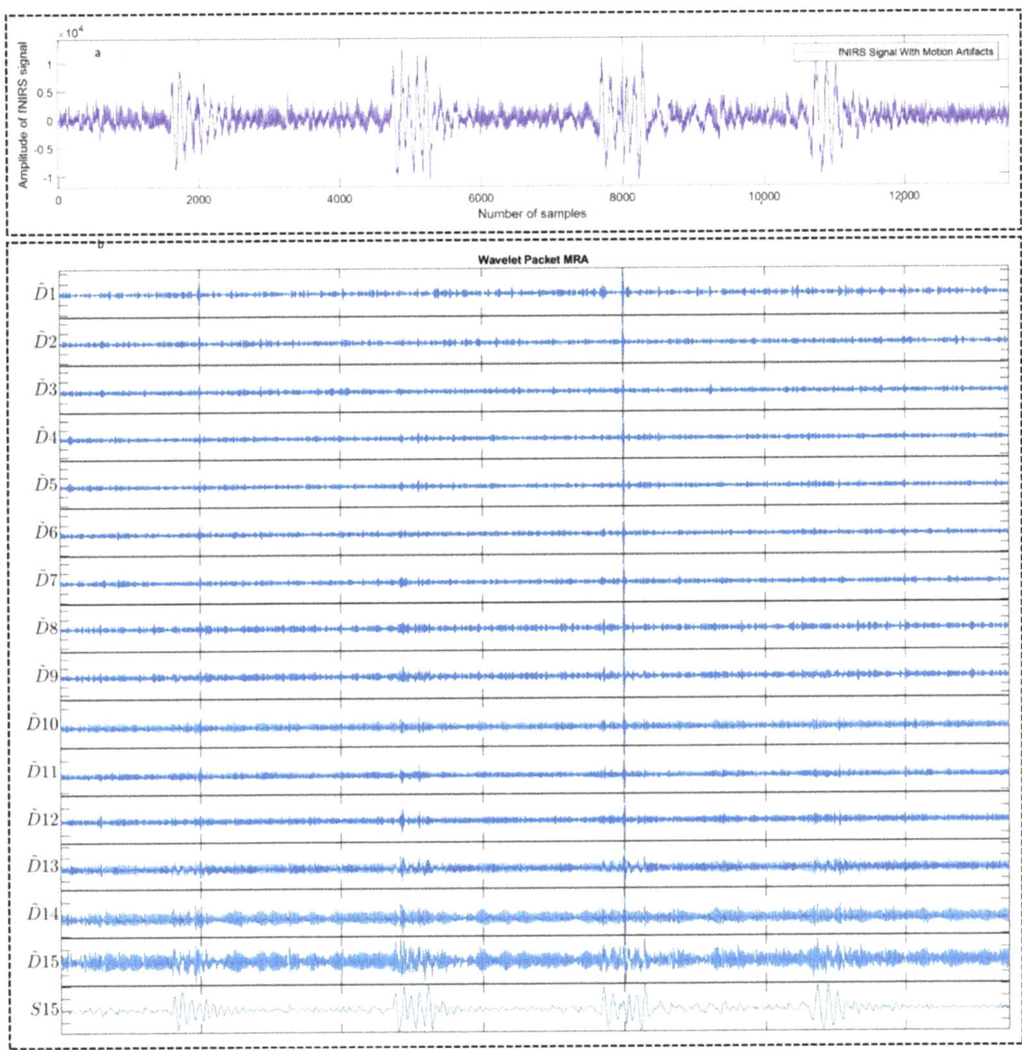

Figure 5. An example motion-corrupted single-channel fNIRS signal (**a**) and the corresponding 16 sub-band components generated using WPD$_{(db1)}$ algorithm (**b**). S15 denotes the Approximation sub-band signal having the lowest center frequency compared to the other sub-band signals, i.e., D1–D15.

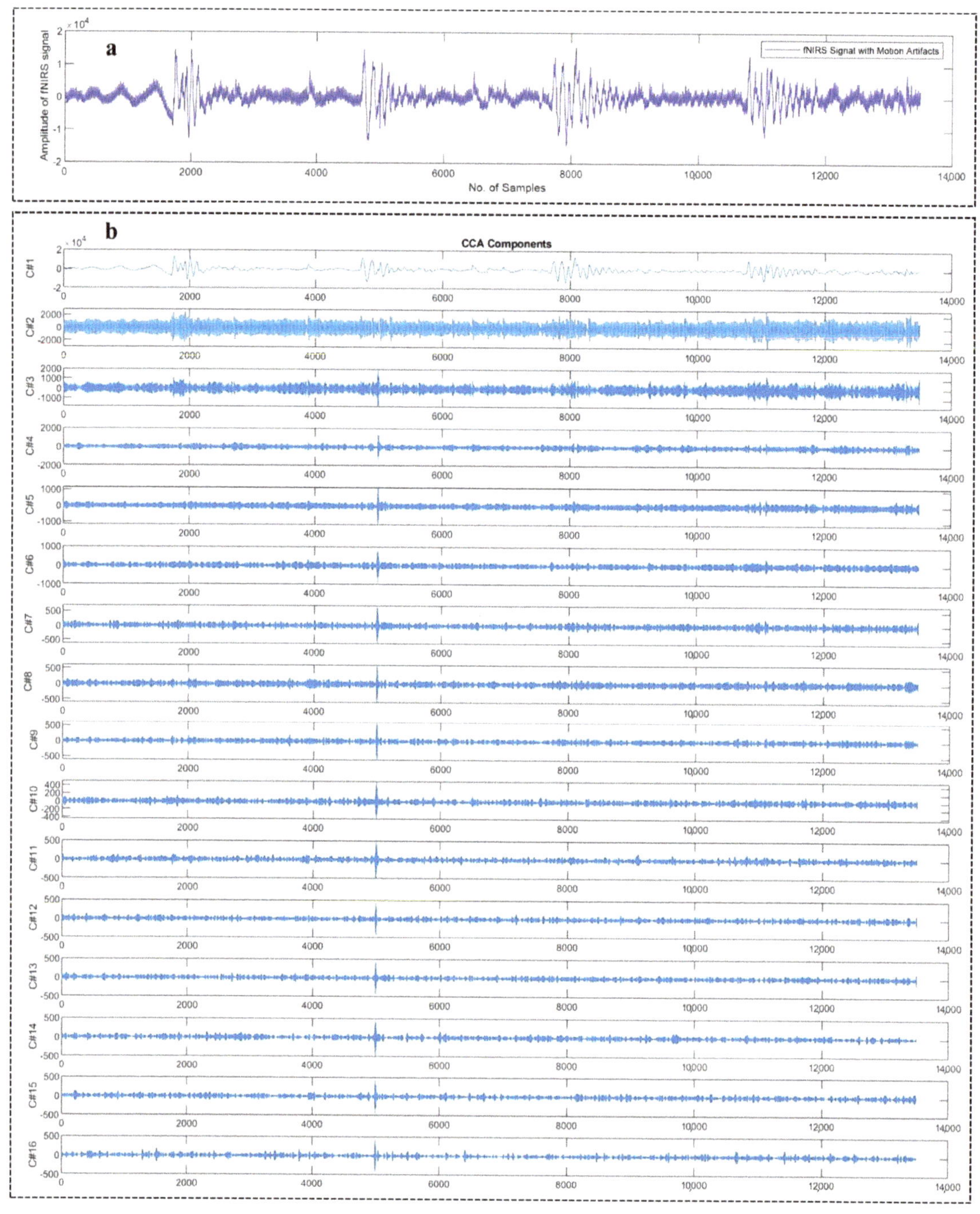

Figure 6. An example motion-corrupted single-channel EEG signal (**a**) and the corresponding 16 CCA components generated from the CCA algorithm (**b**).

3.5. Performance Metrics

The efficacy and performance of each proposed artifact removal approach can be computed using the provided reference "ground truth" signal for each modality, as detailed before. Since the objective of each proposed technique is to reduce artifacts from the motion-artifact contaminated signal, calculating ΔSNR and percentage reduction in motion artifacts can assess the efficacy of that corresponding technique's capacity to remove artifacts. Hence, the difference in SNR before and after artifact removal (ΔSNR), and the improvement in correlation between motion corrupted and reference "ground truth" signals, expressed by the percentage reduction in motion artifact η [33], are utilized as performance metrics.

For the calculation of ΔSNR, the following formula is used which was given in [33]:

$$\Delta SNR = 10\, log_{10}\left(\frac{\sigma_x^2}{\sigma_{e_{after}}^2}\right) - 10\, log_{10}\left(\frac{\sigma_x^2}{\sigma_{e_{before}}^2}\right) \tag{14}$$

where σ_x^2, $\sigma_{e_{before}}^2$, and $\sigma_{e_{after}}^2$ represent the variance of the reference "ground truth", motion corrupted signal, and cleaned signal, respectively.

To calculate the percentage reduction in motion artifact η, the following formula is used [33]:

$$\eta = 100\left(1 - \frac{\rho_{clean} - \rho_{after}}{\rho_{clean} - \rho_{before}}\right) \tag{15}$$

where ρ_{before} is the correlation coefficient between the reference "ground truth" and motion-corrupted signals. The correlation coefficient between the reference "ground truth" and the cleaned signals is denoted by ρ_{after}, whereas ρ_{clean} is the correlation between the reference "ground truth" and motion corrupted signals over the epochs where motion artifact is absent.

In this study, we considered $\rho_{clean} = 1$, as in an ideal situation, the "reference ground truth" and the motion corrupted signal over the artifacts-free epochs would always be completely correlated. Hence, the following equation was used to estimate η:

$$\eta = 100\left(1 - \frac{1 - \rho_{after}}{1 - \rho_{before}}\right) \tag{16}$$

4. Results

The results obtained in this work, using the various novel artifact removal techniques are mentioned below where the performance metrics were calculated using Equations (14) and (16).

4.1. Motion Artifact Correction from EEG Data

All the algorithms (18 in total) were applied on all the 23 recordings of EEG. Figure 7a–d depicts four different examples of EEG recordings after the correction of the motion artifact using $WPD_{(db2)}$, $WPD_{(db3)}$, $WPD_{(fk6)}$, and $WPD_{(fk8)}$ methods, respectively, whereas Figure 8a,b illustrates example EEG signals after the motion artifact correction using $WPD_{(db1)}$-CCA and $WPD_{(fk4)}$-CCA techniques, respectively.

WPD: Among all the 12 different approaches ($WPD_{(db1)}$, $WPD_{(db2)}$, $WPD_{(db3)}$, $WPD_{(sym4)}$, $WPD_{(sym5)}$, $WPD_{(sym6)}$, $WPD_{(coif1)}$, $WPD_{(coif2)}$, $WPD_{(coif3)}$, $WPD_{(fk6)}$, $WPD_{(fk6)}$, and $WPD_{(fk8)}$), the highest average ΔSNR of 29.44 dB with a standard deviation of 9.93 was found when $WPD_{(db2)}$ algorithm was employed over all (23) EEG recordings. The best average percentage reduction in artifact was provided by $WPD_{(db1)}$ algorithm (53.48%) among these 12 single-channel motion artifact correction techniques.

WPD-CCA: Six different approaches namely $WPD_{(db1)}$-CCA, $WPD_{(db2)}$-CCA, $WPD_{(db3)}$-CCA, $WPD_{(fk4)}$-CCA, $WPD_{(fk6)}$-CCA, and $WPD_{(fk8)}$-CCA were investigated, all of which are two-stage motion artifacts correction techniques. The best average ΔSNR was found to be 30.76 dB when $WPD_{(db1)}$-CCA technique was applied over all the EEG records. The highest average percentage reduction in artifact was also provided by the same algorithm, which is 59.51% among these six single-channel motion artifact correction techniques for EEG modality.

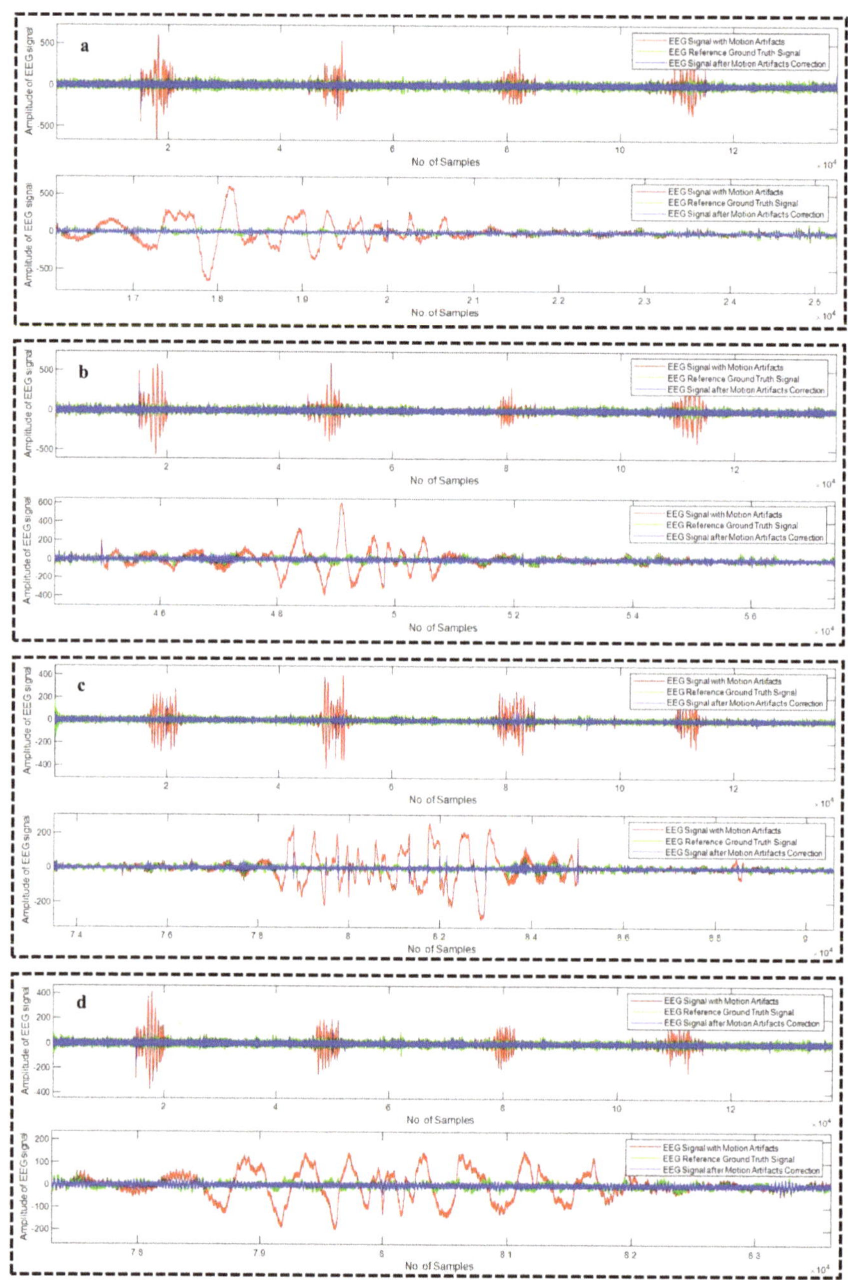

Figure 7. Motion artifact correction from different example EEG signals using WPD$_{(db2)}$ (**a**), WPD$_{(db3)}$ (**b**), WPD$_{(fk6)}$ (**c**), and WPD$_{(fk8)}$ (**d**) techniques.

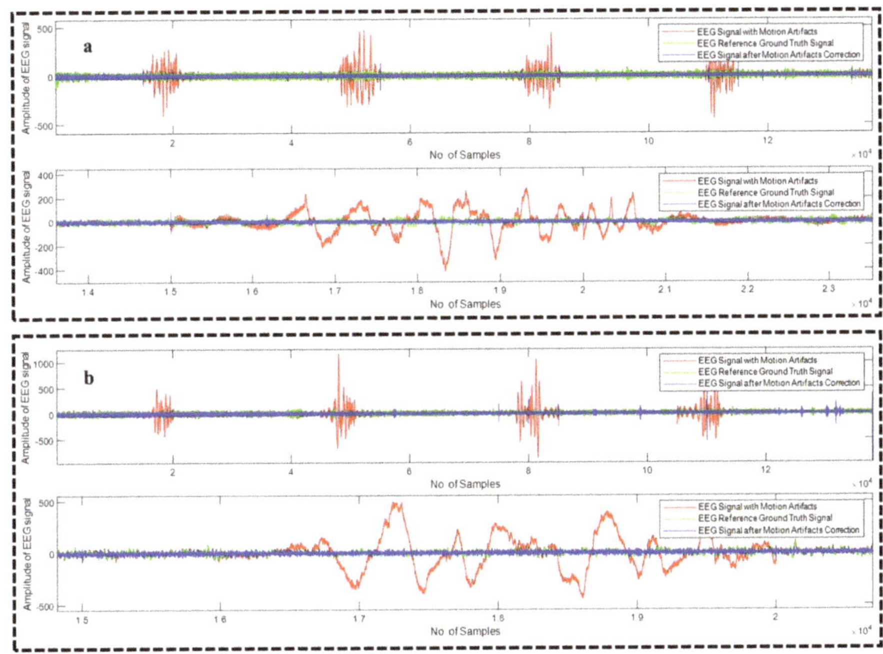

Figure 8. Motion artifact from example EEG signals using $WPD_{(db1)}$-CCA (**a**) and $WPD_{(fk4)}$-CCA (**b**) techniques.

4.2. Motion Artifact Correction from fNIRS Data

All the algorithms (18 in total) were applied on all the 16 recordings of the fNIRS modality. Figure 9a–d depicts four different example fNIRS signals after the correction of the motion artifact using $WPD_{(sym5)}$, $WPD_{(sym6)}$, $WPD_{(coif2)}$, and $WPD_{(coif1)}$ techniques, respectively, whereas Figure 10a,b illustrate example fNIRS signals after the motion artifact correction using $WPD_{(db1)}$-CCA and $WPD_{(fk4)}$-CCA techniques, respectively.

WPD: Among all the 12 different approaches ($WPD_{(db1)}$, $WPD_{(db2)}$, $WPD_{(db3)}$, $WPD_{(sym4)}$, $WPD_{(sym5)}$, $WPD_{(sym6)}$, $WPD_{(coif1)}$, $WPD_{(coif2)}$, $WPD_{(coif3)}$, $WPD_{(fk6)}$, $WPD_{(fk6)}$, and $WPD_{(fk8)}$), the highest average ΔSNR of 16.03 dB with a standard deviation of 4.31 was found when $WPD_{(db1)}$ algorithm was employed over all (16) fNIRS recordings. The best average percentage reduction in artifact was provided by $WPD_{(fk4)}$ algorithm among these 12 single-channel motion artifact correction techniques.

WPD-CCA: Finally, the six different approaches namely $WPD_{(db1)}$-CCA, $WPD_{(db2)}$-CCA, $WPD_{(db3)}$-CCA, $WPD_{(fk4)}$-CCA, $WPD_{(fk6)}$-CCA, and $WPD_{(fk8)}$-CCA, all of which are two-stage motion artifacts correction techniques, were investigated for fNIRS modality. The best average ΔSNR was found to be 16.55 dB when $WPD_{(db1)}$-CCA technique was applied over all the 16 fNIRS records. The highest average percentage reduction in artifact (41.40%) was provided by $WPD_{(fk8)}$-CCA technique among these six single-channel motion artifact correction techniques for fNIRS modality.

Table 1 summarizes the results obtained (average ΔSNR and average percentage reduction in motion artifacts η) using the artifact removal techniques proposed in this paper, i.e., $WPD_{(db1)}$, $WPD_{(db2)}$, $WPD_{(db3)}$, $WPD_{(sym4)}$, $WPD_{(sym5)}$, $WPD_{(sym6)}$, $WPD_{(coif1)}$, $WPD_{(coif2)}$, $WPD_{(coif3)}$, $WPD_{(fk6)}$, $WPD_{(fk6)}$, $WPD_{(fk8)}$, $WPD_{(db1)}$-CCA, $WPD_{(db2)}$-CCA, $WPD_{(db3)}$-CCA, $WPD_{(fk4)}$-CCA, $WPD_{(fk6)}$-CCA, and $WPD_{(fk8)}$-CCA for all the EEG (23) and fNIRS (16) recordings. The values inside first brackets in Table 1 denote the corresponding standard deviations.

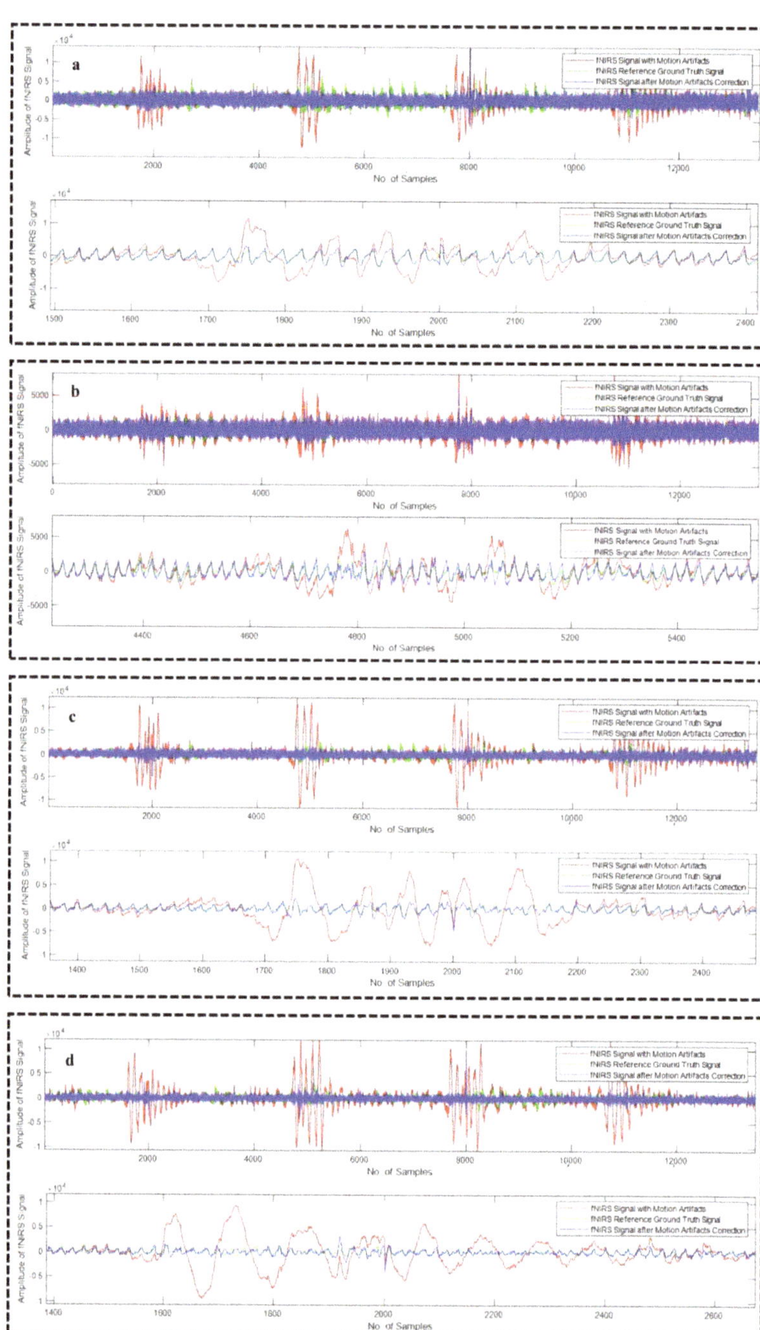

Figure 9. Motion artifact correction from example fNIRS signals using WPD$_{(sym5)}$ (**a**), WPD$_{(sym6)}$ (**b**), WPD$_{(coif2)}$ (**c**), and WPD$_{(coif1)}$ (**d**) techniques.

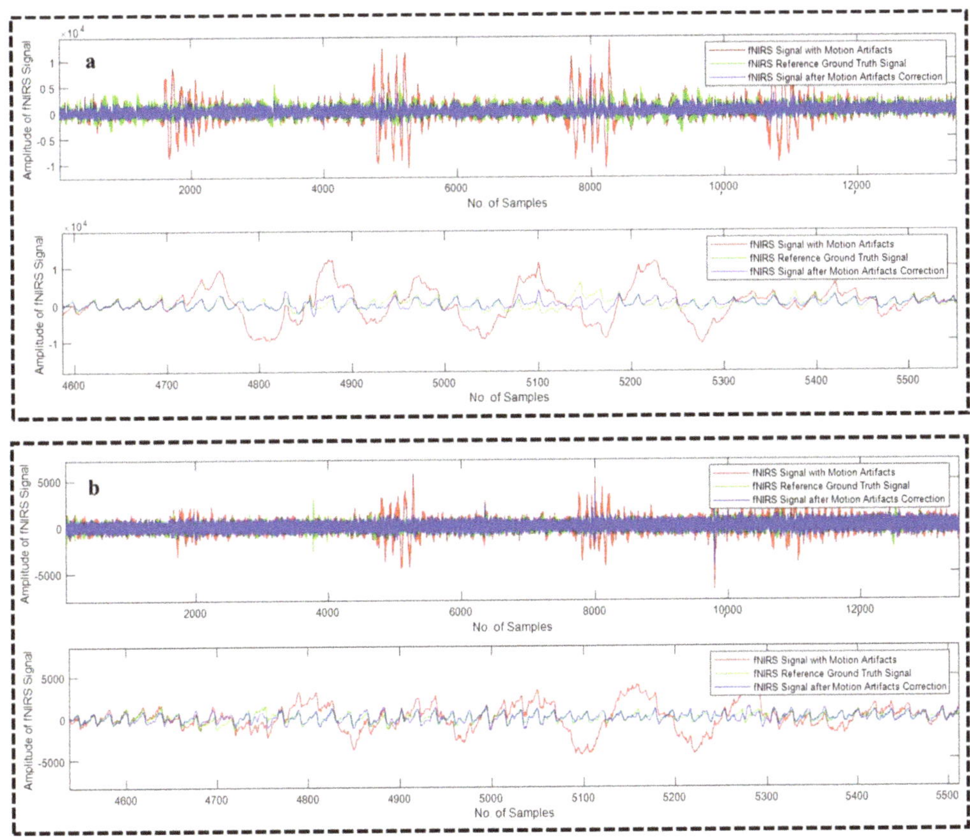

Figure 10. Motion artifact correction from example fNIRS signals using WPD$_{(db1)}$-CCA (**a**) and WPD$_{(fk4)}$-CCA (**b**) techniques.

It is evident from the results of Table 1 that the cleaner EEG signals reconstructed using the WPD$_{(db1)}$ technique provided the highest average η value (53.48%, with corresponding ΔSNR value of 29.26 dB) compared to the other 11 types of single-stage motion artifact correction approaches, whereas the greatest average ΔSNR value (29.44 dB) was provided by WPD$_{(db2)}$ with corresponding average η value of 51.40%. Among these 12 different single-stage artifact removal approaches, the lowest average η (50.00%) and smallest ΔSNR (29.08 dB) was produced by the WPD$_{(coif3)}$ method. When two-stage motion artifacts removal techniques were employed (WPD-CCA) using six different wavelet packets separately, the best average correlation improvement (59.51%) and best average ΔSNR value (30.76 dB) was produced by the WPD$_{(db1)}$-CCA approach, whereas the lowest performance was recorded utilizing the WPD$_{(fk8)}$-CCA technique (average ΔSNR and η values of 28.86 dB and 55.88%, respectively). Overall, an increase of 11.28% in the average percentage reduction in motion artifacts was found, while the best-performing two-stage WPD$_{(db1)}$-CCA was incorporated compared to the best-performing single-stage motion artifact correction technique, namely WPD$_{(db1)}$. Additionally, the average ΔSNR value improved by 4.48% (from 29.44 dB to 30.76 dB), while the best performing two-stage WPD$_{(db1)}$-CCA technique was utilized instead of the best-performing single-stage WPD$_{(db2)}$ method for the correction of motion artifacts from single-channel EEG recordings.

Table 1. Average ΔSNR and average percentage reduction in artifacts (η) for all the EEG and fNIRS recordings. Corresponding standard deviations are shown inside the bracket. (*) represents the best-performing metrics.

Type	Technique	EEG (23 Records)		fNIRS (16 Records)	
		Average ΔSNR (in dB)	Average η (in %)	Average ΔSNR (in dB)	Average η (in %)
Single-stage motion artifact correction techniques	WPD$_{(db1)}$	29.26 (10.29)	53.48 (33.35) *	16.03 (4.31)	26.21 (26.38)
	WPD$_{(db2)}$	29.44 (9.93) *	51.40 (33.59)	15.99 (4.49)	25.92 (28.86)
	WPD$_{(db3)}$	29.37 (10.01)	50.74 (33.55)	15.71 (4.52)	26.05 (29.11)
	WPD$_{(sym4)}$	29.27 (10.05)	50.40 (33.50)	15.54 (4.55)	26.14 (29.18)
	WPD$_{(sym5)}$	29.19 (10.09)	50.20 (33.47)	15.43 (4.57)	26.17 (29.22)
	WPD$_{(sym6)}$	29.11 (10.12)	50.05 (33.43)	15.35 (4.59)	26.16 (29.24)
	WPD$_{(coif1)}$	29.43 (9.94)	51.34 (33.59)	15.97 (4.49)	25.94 (28.88)
	WPD$_{(coif2)}$	29.25 (10.06)	50.35 (33.49)	15.51 (4.56)	26.15 (29.19)
	WPD$_{(coif3)}$	29.08 (10.13)	50.00 (33.42)	15.33 (4.60)	26.15 (29.25)
	WPD$_{(fk4)}$	29.21 (9.87)	52.58 (33.48)	16.11 (4.42) *	26.40 (27.53) *
	WPD$_{(fk6)}$	29.32 (10.03)	50.55 (33.51)	15.59 (4.54)	26.20 (29.08)
	WPD$_{(fk8)}$	29.15 (10.10)	50.15 (33.45)	15.38 (4.58)	26.25 (29.18)
Two-stage motion artifact correction techniques	WPD$_{(db1)}$-CCA	30.76 (12.29) *	59.51 (25.99) *	16.55 (6.29) *	36.58 (11.22)
	WPD$_{(db2)}$-CCA	30.35 (12.50)	57.57 (25.89)	14.50 (5.85)	39.62 (10.59)
	WPD$_{(db3)}$-CCA	29.42 (12.57)	56.52 (25.71)	13.72 (5.82)	40.39 (10.60)
	WPD$_{(fk4)}$-CCA	30.36 (12.65)	58.83 (25.93)	14.97 (6.25)	38.32 (10.90)
	WPD$_{(fk6)}$-CCA	29.12 (13.00)	56.81 (25.16)	13.81 (5.70)	40.48 (10.43)
	WPD$_{(fk8)}$-CCA	28.86 (12.77)	55.88 (25.10)	12.41 (5.51)	41.40 (10.08) *

From Table 1, the cleaner fNIRS signals reconstructed using WPD$_{(fk4)}$ technique provided the highest average η value (26.40%) compared to the other 11 types of single-stage motion artifact correction approaches. The greatest average ΔSNR value (16.11 dB) was also provided by the same approach. Among these 12 different single-stage artifact removal approaches, the lowest average η (25.92%) was produced by WPD$_{(db2)}$, whereas the smallest ΔSNR value (15.33 dB) was produced by WPD$_{(coif3)}$. When two-stage motion artifacts removal techniques were employed (WPD-CCA) using six different wavelet packets for all the fNIRS signals, the best average correlation improvement (41.40%) was produced by the WPD$_{(fk8)}$-CCA technique and the lowest average percentage reduction in artifacts (36.58%) was generated from WPD$_{(db1)}$-CCA. On the other hand, the best average ΔSNR value (16.55 dB) was obtained from the WPD$_{(db1)}$-CCA technique, and the WPD$_{(fk8)}$-CCA technique produced the lowest ΔSNR value of 12.41 dB. Overall, an increase of 56.82% in percentage reduction in motion artifacts was found while the best performing two-stage motion artifacts technique, i.e., WPD$_{(fk8)}$-CCA was incorporated compared to the best performing single-stage motion artifact correction technique namely WPD$_{(fk4)}$. Additionally, an increase of 2.73% in ΔSNR value was found when best performing two-stage WPD$_{(db1)}$-CCA was employed instead of the best-performing single-stage WPD$_{(fk4)}$ technique.

From Table 1, it is clear that two-stage artifacts correction techniques performed relatively better compared to the single-stage artifacts correction approaches for both EEG and fNIRS modalities.

The authors of [37] found that no brain activity was registered in trials 12 and 15. Moreover, they found a poor correlation coefficient over the clean epochs of the recordings of 12 and 15, and hence, they carried out their investigation on the remaining 21 recordings of EEG. We have also observed a similar situation in this work. Trials 12 and 15 consistently produced very bad performance metrics (ΔSNR and η values), while both single-stage and two-stage artifact reduction techniques were applied proposed in this paper.

Table 2 illustrates the average ΔSNR and average percent reduction in motion artifacts using $WPD_{(db1)}$, $WPD_{(sym4)}$, $WPD_{(coif1)}$, and $WPD_{(fk4)}$. This time, the faulty trials (trials 12 and 15) were excluded and the experiments were conducted on the remaining 21 sets of EEG recordings. The motion corrupted signal was decomposed into 16 sub-band components using WPD and then the cleaner signals were generated by simply discarding the lowest-frequency approximation sub-band component (for example, Figure 3, S15 component) and adding the remaining 15 sub-band components (D1 to D15) directly. During this process, the reference ground truth signal was only used to compute the performance metrics.

Table 2. Average ΔSNR and average percentage reduction in artifacts (η) for 21 recordings of EEG modality. Corresponding standard deviations are shown inside the first bracket. (*) denotes the best-performing metrics.

Type	Method	EEG (21 Records)	
		Average ΔSNR (in dB)	Average η (in %)
Single-stage motion artifact correction techniques	$WPD_{(db1)}$	26.20 (6.35)	60.22 (21.79) *
	$WPD_{(sym4)}$	26.46 (6.56)	57.23 (22.11)
	$WPD_{(coif1)}$	26.70 (6.54) *	58.19 (22.04)
	$WPD_{(fk4)}$	26.36 (6.36)	59.37 (21.90)

From Table 2, it is clear that the cleaner EEG signals reconstructed using the $WPD_{(db1)}$ technique provided the highest average η value (60.22%, corresponding ΔSNR value of 26.20 dB) compared to the other three types of single-stage motion artifact correction approaches, whereas the greatest average ΔSNR value (26.70 dB) was produced by $WPD_{(coif1)}$ with an average η value of 58.19%. The values obtained following this process is a clear indication that without the availability of "reference ground truth signal", correction of motion artifacts from EEG signal is still possible. The similar approach can also be used for motion artifacts correction from fNIRS signals, but will be considered in a future work.

5. Discussion

In this paper, we have proposed two novel methods (WPD and WPD-CCA) using four different wavelet packet families with three different vanishing moments, resulting in 18 different techniques ($WPD_{(db1)}$, $WPD_{(db2)}$, $WPD_{(db3)}$, $WPD_{(sym4)}$, $WPD_{(sym5)}$, $WPD_{(sym6)}$, $WPD_{(coif1)}$, $WPD_{(coif2)}$, $WPD_{(coif3)}$, $WPD_{(fk6)}$, $WPD_{(fk6)}$, $WPD_{(fk8)}$, $WPD_{(db1)}$-CCA, $WPD_{(db2)}$-CCA, $WPD_{(db3)}$-CCA, $WPD_{(fk4)}$-CCA, $WPD_{(fk6)}$-CCA, and $WPD_{(fk8)}$-CCA) for the correction of motion artifacts from single-channel EEG and fNIRS recordings. The performance metrics (ΔSNR and η) calculated and reported in the "Results" section utilizing these 18 approaches are a clear indication of the efficacy of our proposed techniques. Both the Daubechies and Fejer-Korovkin wavelet packet families relatively performed better compared to the Symlet and Coiflet wavelet packet families in removing motion artifacts from EEG and fNIRS recordings. For this reason, while implementing the two-stage artifacts correction technique, we have used only the Daubechies and Fejer-Korovkin wavelet packet families.

As previously stated, DWT, EMD, EEMD, VMD, EMD-ICA, EMD-CCA, EEMD-ICA, EEMD-CCA, VMD-CCA, SSA, and DWT, along with approximation sub-band filtering, adaptive filtering (ARX model with exogenous input), etc., were commonly employed for the correction of movement artifacts from motion corrupted EEG and fNIRS signals. Each of these methods suffers from some limitations.

Using DWT-based approaches, to improve signal quality from motion-corrupted physiological data, selecting the suitable wavelet is critical and rather complex. To date, there is no hard and fast rule for selecting the appropriate wavelet for the specific physiological signal of interest; instead, wavelets are often selected depending on the morphology of the signal. As a result, improper wavelet selection would result in inefficient denoising.

The EMD-based motion artifact reduction approach suffers heavily from the "mode mixing" issue [33], which may result in an incorrect outcome. To fix this problem, the EEMD approaches are employed [33,36]. Although EEMD is not affected by the mode mixing problem, it still requires a prior declaration of the number of ensembles to be employed, which is determined through trial and error basis [33].

To make use of the SSA algorithm, for the correction of movement artifacts from physiological signals, a prior declaration of the window length and the required number of reconstruction components is necessary, which makes SSA inefficient as well [37].

The authors of [40] employed DWT along with approximation sub-band filtering using total variation (TV) and weighted TV. While reconstructing the cleaner signal, the first three high-frequency detailed sub-band signals were rejected, since they included no important information from the EEG signal. However, detecting non-useful sub-band signals when utilizing DWT-based algorithms is very challenging for removing motion artifacts from EEG and fNIRS signals. Furthermore, the value of the regularization factor used to address the optimization problem of TV and MTV approaches was picked without explanation.

Siddiquee et al. [52] studied the autoregressive exogenous input model (adaptive technique) to model motion corrupted segments as output and IMU data as exogenous input. Only four test participants' fNIRS data were used by the authors to demonstrate the efficacy of their prescribed approach. One of the most important aspects of adopting this technique is the precise synchronization of fNIRS and IMU data. Furthermore, if the epoch duration of the motion artifacts is sufficiently long (specifically, the sample size), modeling the artifacts mathematically using the least square method would necessitate higher-order models, which would eventually cause instability. Hence, incorporating this method to remove motion artifacts would be extremely difficult in a real-world scenario.

ICA and CCA algorithms are multi-channel signal processing algorithms, meaning there must be two (or more) channel data values as input. Therefore, ICA and CCA algorithms cannot be incorporated independently for the processing of single-channel data. Additionally, since ICA uses higher-order statistics (HOS) and CCA uses second-order statistics (SOS) [33], the CCA algorithm is computationally efficient in comparison with ICA. That is why previous studies as well as this study used the CCA algorithm as a second-stage signal processing method.

WPD is the more generalized version of DWT, but the former provides better signal decomposition which enhances the signal quality for further processing. Additionally, WPD is better in denoising in the sense that there is no necessity of identifying and discarding any sub-band signals other than the motion corrupted sub-band component. Additionally, the results obtained in this work utilizing the WPD method for 12 different wavelet packets, show a little variation while computing ΔSNR and η. This is a clear indication that applying WPD compared to the DWT is much more robust and efficient in terms of performance metrics improvement.

Although the two-stage motion artifacts removal approaches (WPD-CCA) proposed in this paper performed better compared to the single-stage artifacts correction techniques using WPD, the WPD-CCA technique will not be able to identify the motion corrupted CCA components in the absence of a ground truth signal, which is a limitation of two-stage artifacts removal technique. Hassan et al. provided an alternate technique in [65], in which the authors employed the autocorrelation function to detect the motion corrupted components. The automated artifact component selection approach introduced in [65] employing the autocorrelation function has not been experimented within this study and will be considered in a future study.

However, even in the absence of the "reference ground truth" signal, our proposed single-stage motion artifact reduction approach (WPD) would produce optimal results. While decomposing the signal of interest (EEG/fNIRS) using WPD, it was visually seen that the approximation sub-band component (having the lowest frequency band compared to the rest of the sub-band components) included the highest percentage of motion artifacts. Hence, discarding this noisy sub-band component and reconstruction of the signal using the remaining sub-band signals would reduce the motion artifacts to a great extent. The validation of this statement is supported by Table 2, where the performance metrics (ΔSNR and η) were reported and produced reasonable noise reduction.

Throughout this work, while estimating the percentage reduction in motion artifacts η, we have considered Equation (16), instead of Equation (15), where we have assumed that $\rho_{clean} = 1$ as in an ideal situation, the "reference ground truth" and the motion corrupted signal over the artifacts-free epochs would always be completely correlated. However, in practice, the value of ρ_{clean} would always be less than 1, because it is impossible to extract a "reference ground truth" signal which would completely be similar compared with a motion-corrupted signal during the artifacts-free epochs. It is counter-intuitive that a lower value of ρ_{clean} would produce a lower value of η; it is just the opposite. For example, let $\rho_{before} = 0.6$; $\rho_{after} = 0.8$; $\rho_{clean} = 0.95$, from Equation (15), we would get η equals 57.14% and Equation (16) would give 50%. That is why choosing $\rho_{clean} = 1$ would give a worst-case scenario result. Additionally, this same formula is used in [40–42] assuming the ideal "reference ground truth signal".

6. Conclusions

In this extensive study, two novel motion artifact removal techniques have been proposed, namely wavelet packet decomposition (WPD), and WPD in combination with canonical correlation analysis (WPD-CCA) for EEG and fNIRS modalities. Furthermore, the proposed algorithms were investigated by 18 different approaches where four different wavelet packet families namely Daubechies, Symlet, Coiflet, and Fejer-Korovkin wavelet packet families were utilized. WPD-CCA techniques can be used on single-channel recordings as the WPD algorithm can decompose a single-channel signal into a predefined number of sub-band components which can be fed as the input channels for the CCA algorithm. The performance parameters obtained from all these approaches are a clear indication of the efficacy of these algorithms. The novel $WPD_{(db1)}$-CCA and $WPD_{(fk8)}$-CCA technique provided the best performance in terms of the percentage reduction in motion artifacts (59.51% and 41.40%) when analyzing the EEG and fNIRS data, respectively. On the other hand, the $WPD_{(db1)}$-CCA technique generated the highest average ΔSNR (30.76 dB and 16.55 dB) for both EEG and fNIRS signals. An alternative approach for removing motion artifacts from EEG signals using the WPD method has also been proposed where the lowest-frequency approximation sub-band component was discarded and a clean EEG signal was reconstructed by adding up the remaining sub-band components. By computing the performance metrics, it has been shown that this single-stage motion artifacts correction technique is also capable of removing motion artifacts to a great extent. In the future, deep learning-based models will be investigated for the automated detection and removal of artifacts in physiological signals (EEG, ECG, EMG, PPG, fNIRS, etc.). New methods based on the use of different multivariate signal processing approaches will be developed for the elimination of other artifacts from the EEG and fNIRS signals that are recorded using multiple electrodes.

Author Contributions: Conceptualization, M.S.H., M.E.H.C., M.B.I.R., S.H.M.A., A.A.A.B., S.K., A.K., M.A., R.H. and M.M.H.; data curation, M.S.H. and M.B.I.R.; formal analysis, M.S.H.; funding acquisition, M.E.H.C. and M.B.I.R.; methodology, M.S.H., M.E.H.C., M.B.I.R., S.H.M.A., A.K. and R.H.; project administration, M.E.H.C. and M.B.I.R.; resources, M.E.H.C.; software, M.E.H.C.; supervision, M.E.H.C. and M.B.I.R.; validation, M.S.H.; visualization, M.S.H.; writing—original draft, M.S.H., M.E.H.C., M.B.I.R., S.H.M.A., A.A.A.B., S.K., A.K., M.A., R.H. and M.M.H.; writing—review and editing, M.S.H., M.E.H.C., M.B.I.R. and M.M.H. All authors have read and agreed to the published version of the manuscript.

Funding: This work was made possible by the Qatar National Research Fund (QNRF) NPRP12S-0227-190164 and an International Research Collaboration Co-Fund (IRCC) grant: IRCC-2021-001, as well as Universiti Kebangsaan Malaysia (UKM) under Grant GUP-2021-019, and Grant DIP-2020-004. The statements made herein are solely the responsibility of the authors.

Institutional Review Board Statement: Not applicable.

Informed Consent Statement: Not applicable.

Data Availability Statement: The dataset used in this study is publicly available in the PhysioNet database and the authors of this study did not collect the dataset. Sweeney et al. [32,33,63] collected this dataset with ethical approval.

Acknowledgments: The dataset used in this experiment is kindly shared in the PhysioNet database by Sweeney et al. [32,33,63].

Conflicts of Interest: The authors declare no conflict of interest.

References

1. Henry, J.C. Electroencephalography: Basic principles, clinical applications, and related fields. *Neurology* **2006**, *67*, 2092–2092-a. [CrossRef]
2. Nuwer, M. Assessment of digital EEG, quantitative EEG, and EEG brain mapping: Report of the American Academy of Neurology and the American Clinical Neurophysiology Society. *Neurology* **1997**, *49*, 277–292. [CrossRef] [PubMed]
3. Shoeb, A.; Guttag, J.; Schachter, S.; Schomer, D.; Bourgeois, B.; Treves, S.T. Detecting seizure onset in the ambulatory setting: Demonstrating feasibility. In Proceedings of the 2005 IEEE Engineering in Medicine and Biology 27th Annual Conference, Shanghai, China, 17–18 January 2006; pp. 3546–3550.
4. Sharma, R.R.; Varshney, P.; Pachori, R.B.; Vishvakarma, S.K. Automated system for epileptic EEG detection using iterative filtering. *IEEE Sens. Lett.* **2018**, *2*, 1–4. [CrossRef]
5. Berka, C.; Levendowski, D.J.; Westbrook, P.; Davis, G.; Lumicao, M.N.; Olmstead, R.E.; Popovic, M.; Zivkovic, V.T.; Ramsey, C.K. EEG quantification of alertness: Methods for early identification of individuals most susceptible to sleep deprivation. In Proceedings of the Biomonitoring for Physiological and Cognitive Performance during Military Operations, Orlando, FL, USA, 23 May 2005; pp. 78–89.
6. Berka, C.; Levendowski, D.J.; Cvetinovic, M.M.; Petrovic, M.M.; Davis, G.; Lumicao, M.N.; Zivkovic, V.T.; Popovic, M.V.; Olmstead, R. Real-time analysis of EEG indexes of alertness, cognition, and memory acquired with a wireless EEG headset. *Int. J. Hum. Comput. Interact.* **2004**, *17*, 151–170. [CrossRef]
7. Papadelis, C.; Kourtidou-Papadeli, C.; Bamidis, P.D.; Chouvarda, I.; Koufogiannis, D.; Bekiaris, E.; Maglaveras, N. Indicators of sleepiness in an ambulatory EEG study of night driving. In Proceedings of the 2006 the International Conference of the IEEE Engineering in Medicine and Biology Society, New York, NY, USA, 30 August–3 September 2006; pp. 6201–6204.
8. Tripathy, R.; Acharya, U.R. Use of features from RR-time series and EEG signals for automated classification of sleep stages in deep neural network framework. *Biocybern. Biomed. Eng.* **2018**, *38*, 890–902. [CrossRef]
9. Gupta, V.; Chopda, M.D.; Pachori, R.B. Cross-subject emotion recognition using flexible analytic wavelet transform from EEG signals. *IEEE Sens. J.* **2018**, *19*, 2266–2274. [CrossRef]
10. Stevens, R.; Galloway, T.; Berka, C. Integrating EEG models of cognitive load with machine learning models of scientific problem solving. *Augment. Cogn. Past Present Future* **2006**, *2*, 55–65.
11. Bell, C.J.; Shenoy, P.; Chalodhorn, R.; Rao, R.P. Control of a humanoid robot by a noninvasive brain–computer interface in humans. *J. Neural Eng.* **2008**, *5*, 214. [CrossRef]
12. Lee, J.C.; Tan, D.S. Using a low-cost electroencephalograph for task classification in HCI research. In Proceedings of the 19th Annual ACM Symposium on User Interface Software and Technology, Montreux, Switzerland, 15–18 October 2006; pp. 81–90.
13. Sullivan, T.J.; Deiss, S.R.; Jung, T.-P.; Cauwenberghs, G. A brain-machine interface using dry-contact, low-noise EEG sensors. In Proceedings of the 2008 IEEE International Symposium on Circuits and Systems, Seattle, WA, USA, 18–21 May 2008; pp. 1986–1989.
14. Wolpaw, J.R.; McFarland, D.J. Control of a two-dimensional movement signal by a noninvasive brain-computer interface in humans. *Proc. Natl. Acad. Sci. USA* **2004**, *101*, 17849–17854. [CrossRef]
15. Gaur, P.; Pachori, R.B.; Wang, H.; Prasad, G. A multi-class EEG-based BCI classification using multivariate empirical mode decomposition based filtering and Riemannian geometry. *Expert Syst. Appl.* **2018**, *95*, 201–211. [CrossRef]
16. Gaur, P.; Pachori, R.B.; Wang, H.; Prasad, G. An automatic subject specific intrinsic mode function selection for enhancing two-class EEG-based motor imagery-brain computer interface. *IEEE Sens. J.* **2019**, *19*, 6938–6947. [CrossRef]
17. Rahman, A.; Chowdhury, M.E.; Khandakar, A.; Kiranyaz, S.; Zaman, K.S.; Reaz, M.B.I.; Islam, M.T.; Ezeddin, M.; Kadir, M.A. Multimodal EEG and Keystroke Dynamics Based Biometric System Using Machine Learning Algorithms. *IEEE Access* **2021**, *9*, 94625–94643. [CrossRef]
18. Sangani, S.; Lamontagne, A.; Fung, J. Cortical mechanisms underlying sensorimotor enhancement promoted by walking with haptic inputs in a virtual environment. *Prog. Brain Res.* **2015**, *218*, 313–330. [PubMed]
19. Bunce, S.C.; Izzetoglu, M.; Izzetoglu, K.; Onaral, B.; Pourrezaei, K. Functional near-infrared spectroscopy. *IEEE Eng. Med. Biol. Mag.* **2006**, *25*, 54–62. [CrossRef]
20. Huppert, T.J.; Diamond, S.G.; Franceschini, M.A.; Boas, D.A. HomER: A review of time-series analysis methods for near-infrared spectroscopy of the brain. *Appl. Opt.* **2009**, *48*, D280–D298. [CrossRef]

21. Holper, L.; Muehlemann, T.; Scholkmann, F.; Eng, K.; Kiper, D.; Wolf, M. Testing the potential of a virtual reality neurorehabilitation system during performance of observation, imagery and imitation of motor actions recorded by wireless functional near-infrared spectroscopy (fNIRS). *J. Neuroeng. Rehabil.* **2010**, *7*, 57. [CrossRef]

22. Izzetoglu, K.; Bunce, S.; Onaral, B.; Pourrezaei, K.; Chance, B. Functional optical brain imaging using near-infrared during cognitive tasks. *Int. J. Hum. Comput. Interact.* **2004**, *17*, 211–227. [CrossRef]

23. Cui, X.; Bray, S.; Bryant, D.M.; Glover, G.H.; Reiss, A.L. A quantitative comparison of NIRS and fMRI across multiple cognitive tasks. *Neuroimage* **2011**, *54*, 2808–2821. [CrossRef]

24. Matthews, F.; Pearlmutter, B.A.; Wards, T.E.; Soraghan, C.; Markham, C. Hemodynamics for brain-computer interfaces. *IEEE Signal Process. Mag.* **2007**, *25*, 87–94. [CrossRef]

25. Khan, M.J.; Hong, K.-S. Passive BCI based on drowsiness detection: An fNIRS study. *Biomed. Opt. Express* **2015**, *6*, 4063–4078. [CrossRef]

26. Hong, K.-S.; Naseer, N.; Kim, Y.-H. Classification of prefrontal and motor cortex signals for three-class fNIRS–BCI. *Neurosci. Lett.* **2015**, *587*, 87–92. [CrossRef] [PubMed]

27. Chowdhury, M.E.; Khandakar, A.; Hossain, B.; Alzoubi, K. Effects of the phantom shape on the gradient artefact of electroencephalography (EEG) data in simultaneous EEG–fMRI. *Appl. Sci.* **2018**, *8*, 1969. [CrossRef]

28. Chowdhury, M.E.; Khandakar, A.; Mullinger, K.; Hossain, B.; Al-Emadi, N.; Antunes, A.; Bowtell, R. Reference layer artefact subtraction (RLAS): Electromagnetic simulations. *IEEE Access* **2019**, *7*, 17882–17895. [CrossRef]

29. Chowdhury, M.E.; Khandakar, A.; Mullinger, K.J.; Al-Emadi, N.; Bowtell, R. Simultaneous EEG-fMRI: Evaluating the effect of the EEG cap-cabling configuration on the gradient artifact. *Front. Neurosci.* **2019**, *13*, 690. [CrossRef]

30. Nguyen, H.-D.; Yoo, S.-H.; Bhutta, M.R.; Hong, K.-S. Adaptive filtering of physiological noises in fNIRS data. *Biomed. Eng. Online* **2018**, *17*, 1–23. [CrossRef] [PubMed]

31. Islam, M.K.; Rastegarnia, A.; Yang, Z. Methods for artifact detection and removal from scalp EEG: A review. *Neurophysiol. Clin. Clin. Neurophysiol.* **2016**, *46*, 287–305. [CrossRef] [PubMed]

32. Sweeney, K.T.; Ward, T.E.; McLoone, S.F. Artifact removal in physiological signals—Practices and possibilities. *IEEE Trans. Inf. Technol. Biomed.* **2012**, *16*, 488–500. [CrossRef] [PubMed]

33. Sweeney, K.T.; McLoone, S.F.; Ward, T.E. The use of ensemble empirical mode decomposition with canonical correlation analysis as a novel artifact removal technique. *IEEE Trans. Biomed. Eng.* **2012**, *60*, 97–105. [CrossRef]

34. Akansu, A.N.; Haddad, R.A.; Haddad, P.A.; Haddad, P.R. *Multiresolution Signal Decomposition: Transforms, Subbands, and Wavelets*; Academic Press: Cambridge, MA, USA, 2001.

35. Huang, N.E.; Shen, Z.; Long, S.R.; Wu, M.C.; Shih, H.H.; Zheng, Q.; Yen, N.-C.; Tung, C.C.; Liu, H.H. The empirical mode decomposition and the Hilbert spectrum for nonlinear and non-stationary time series analysis. *Proc. R. Soc. Lond. Ser. A Math. Phys. Eng. Sci.* **1998**, *454*, 903–995. [CrossRef]

36. Wu, Z.; Huang, N.E. Ensemble empirical mode decomposition: A noise-assisted data analysis method. *Adv. Adapt. Data Anal.* **2009**, *1*, 1–41. [CrossRef]

37. Maddirala, A.K.; Shaik, R.A. Motion artifact removal from single channel electroencephalogram signals using singular spectrum analysis. *Biomed. Signal Process. Control* **2016**, *30*, 79–85. [CrossRef]

38. Vautard, R.; Yiou, P.; Ghil, M. Singular-spectrum analysis: A toolkit for short, noisy chaotic signals. *Phys. D Nonlinear Phenom.* **1992**, *58*, 95–126. [CrossRef]

39. Kumar, P.S.; Arumuganathan, R.; Sivakumar, K.; Vimal, C. Removal of ocular artifacts in the EEG through wavelet transform without using an EOG reference channel. *Int. J. Open Probl. Compt. Math* **2008**, *1*, 188–200.

40. Gajbhiye, P.; Tripathy, R.K.; Bhattacharyya, A.; Pachori, R.B. Novel approaches for the removal of motion artifact from EEG recordings. *IEEE Sens. J.* **2019**, *19*, 10600–10608. [CrossRef]

41. Gajbhiye, P.; Mingchinda, N.; Chen, W.; Mukhopadhyay, S.C.; Wilaiprasitporn, T.; Tripathy, R.K. Wavelet Domain Optimized Savitzky–Golay Filter for the Removal of Motion Artifacts From EEG Recordings. *IEEE Trans. Instrum. Meas.* **2020**, *70*, 1–11. [CrossRef]

42. Hossain, M.S.; Reaz, M.B.; Chowdhury, M.E.; Ali, S.H.; Bakar, A.A.A.; Kiranyaz, S.; Khandakar, A.; Alhatou, M.; Habib, R. Motion Artifacts Correction from EEG and fNIRS Signals using Novel Multiresolution Analysis. *IEEE Access* **2022**, *10*, 29760–29777. [CrossRef]

43. Dragomiretskiy, K.; Zosso, D. Variational mode decomposition. *IEEE Trans. Signal Process.* **2013**, *62*, 531–544. [CrossRef]

44. Robertson, F.C.; Douglas, T.S.; Meintjes, E.M. Motion artifact removal for functional near infrared spectroscopy: A comparison of methods. *IEEE Trans. Biomed. Eng.* **2010**, *57*, 1377–1387. [CrossRef]

45. Cooper, R.; Selb, J.; Gagnon, L.; Phillip, D.; Schytz, H.W.; Iversen, H.K.; Ashina, M.; Boas, D.A. A systematic comparison of motion artifact correction techniques for functional near-infrared spectroscopy. *Front. Neurosci.* **2012**, *6*, 147. [CrossRef]

46. Brigadoi, S.; Ceccherini, L.; Cutini, S.; Scarpa, F.; Scatturin, P.; Selb, J.; Gagnon, L.; Boas, D.A.; Cooper, R.J. Motion artifacts in functional near-infrared spectroscopy: A comparison of motion correction techniques applied to real cognitive data. *Neuroimage* **2014**, *85*, 181–191. [CrossRef]

47. Sweeney, K.T.; Ayaz, H.; Ward, T.E.; Izzetoglu, M.; McLoone, S.F.; Onaral, B. A methodology for validating artifact removal techniques for physiological signals. *IEEE Trans. Inf. Technol. Biomed.* **2012**, *16*, 918–926. [CrossRef]

48. Scholkmann, F.; Spichtig, S.; Muehlemann, T.; Wolf, M. How to detect and reduce movement artifacts in near-infrared imaging using moving standard deviation and spline interpolation. *Physiol. Meas.* **2010**, *31*, 649. [CrossRef] [PubMed]

49. Molavi, B.; Dumont, G.A. Wavelet-based motion artifact removal for functional near-infrared spectroscopy. *Physiol. Meas.* **2012**, *33*, 259. [CrossRef] [PubMed]

50. Barker, J.W.; Aarabi, A.; Huppert, T.J. Autoregressive model based algorithm for correcting motion and serially correlated errors in fNIRS. *Biomed. Opt. Express* **2013**, *4*, 1366–1379. [CrossRef] [PubMed]

51. Chiarelli, A.M.; Maclin, E.L.; Fabiani, M.; Gratton, G. A kurtosis-based wavelet algorithm for motion artifact correction of fNIRS data. *NeuroImage* **2015**, *112*, 128–137. [CrossRef]

52. Siddiquee, M.R.; Marquez, J.S.; Atri, R.; Ramon, R.; Perry Mayrand, R.; Bai, O. Movement artefact removal from NIRS signal using multi-channel IMU data. *Biomed. Eng. Online* **2018**, *17*, 120. [CrossRef]

53. Jahani, S.; Setarehdan, S.K.; Boas, D.A.; Yücel, M.A. Motion artifact detection and correction in functional near-infrared spectroscopy: A new hybrid method based on spline interpolation method and Savitzky–Golay filtering. *Neurophotonics* **2018**, *5*, 015003. [CrossRef]

54. Mallat, S. *A Wavelet Tour of Signal Processing*; Mallat, S., Ed.; Academic Press: Cambridge, MA, USA, 2009.

55. Jaffard, S.; Meyer, Y.; Ryan, R.D. *Wavelets: Tools for Science and Technology*; Society For Industrial & Applied (SIAM): Philadelphia, PA, USA, 2001.

56. Farooq, O.; Datta, S. Mel filter-like admissible wavelet packet structure for speech recognition. *IEEE Signal Process. Lett.* **2001**, *8*, 196–198. [CrossRef]

57. Sanei, S.; Chambers, J.A. *EEG Signal Processing*; John Wiley & Sons: Hoboken, NJ, USA, 2013.

58. Gwin, J.T.; Gramann, K.; Makeig, S.; Ferris, D.P. Removal of movement artifact from high-density EEG recorded during walking and running. *J. Neurophysiol.* **2010**, *103*, 3526–3534. [CrossRef]

59. Hotelling, H. Relations between two sets of variates. In *Breakthroughs in Statistics*; Springer: Berlin/Heidelberg, Germany, 1992; pp. 162–190.

60. James, C.J.; Hesse, C.W. Independent component analysis for biomedical signals. *Physiol. Meas.* **2004**, *26*, R15. [CrossRef]

61. De Clercq, W.; Vergult, A.; Vanrumste, B.; Van Paesschen, W.; Van Huffel, S. Canonical correlation analysis applied to remove muscle artifacts from the electroencephalogram. *IEEE Trans. Biomed. Eng.* **2006**, *53*, 2583–2587. [CrossRef] [PubMed]

62. Borga, M.; Knutsson, H. A canonical correlation approach to blind source separation. In *Report LiU-IMT-EX-0062*; Department of Biomedical Engineering, Linkoping University: Linköping, Sweden, 2001.

63. Goldberger, A.L.; Amaral, L.A.; Glass, L.; Hausdorff, J.M.; Ivanov, P.C.; Mark, R.G.; Mietus, J.E.; Moody, G.B.; Peng, C.-K.; Stanley, H.E. PhysioBank, PhysioToolkit, and PhysioNet: Components of a new research resource for complex physiologic signals. *Circulation* **2000**, *101*, e215–e220. [CrossRef] [PubMed]

64. Tsipouras, M.G. Spectral information of EEG signals with respect to epilepsy classification. *EURASIP J. Adv. Signal Process.* **2019**, *2019*, 10. [CrossRef]

65. Hassan, M.; Boudaoud, S.; Terrien, J.; Karlsson, B.; Marque, C. Combination of canonical correlation analysis and empirical mode decomposition applied to denoising the labor electrohysterogram. *IEEE Trans. Biomed. Eng.* **2011**, *58*, 2441–2447. [CrossRef] [PubMed]

Article

Generalized Deep Learning EEG Models for Cross-Participant and Cross-Task Detection of the Vigilance Decrement in Sustained Attention Tasks

Alexander Kamrud *, Brett Borghetti , Christine Schubert Kabban and Michael Miller

Department of Electrical and Computer Engineering, Air Force Institute of Technology, Wright-Patterson Air Force Base, OH 45433, USA; brett.borghetti@afit.edu (B.B.); christine.schubert@afit.edu (C.S.K.); michael.miller@afit.edu (M.M.)
* Correspondence: alexander.kamrud.1@us.af.mil

Abstract: Tasks which require sustained attention over a lengthy period of time have been a focal point of cognitive fatigue research for decades, with these tasks including air traffic control, watchkeeping, baggage inspection, and many others. Recent research into physiological markers of mental fatigue indicate that markers exist which extend across all individuals and all types of vigilance tasks. This suggests that it would be possible to build an EEG model which detects these markers and the subsequent vigilance decrement in any task (i.e., a task-generic model) and in any person (i.e., a cross-participant model). However, thus far, no task-generic EEG cross-participant model has been built or tested. In this research, we explored creation and application of a task-generic EEG cross-participant model for detection of the vigilance decrement in an unseen task and unseen individuals. We utilized three different models to investigate this capability: a multi-layer perceptron neural network (MLPNN) which employed spectral features extracted from the five traditional EEG frequency bands, a temporal convolutional network (TCN), and a TCN autoencoder (TCN-AE), with these two TCN models being time-domain based, i.e., using raw EEG time-series voltage values. The MLPNN and TCN models both achieved accuracy greater than random chance (50%), with the MLPNN performing best with a 7-fold CV balanced accuracy of 64% (95% CI: 0.59, 0.69) and validation accuracies greater than random chance for 9 of the 14 participants. This finding demonstrates that it is possible to classify a vigilance decrement using EEG, even with EEG from an unseen individual and unseen task.

Keywords: EEG; deep learning; vigilance decrement; sustained attention; mental fatigue; cross-participant; cross-task; task-generic

Citation: Kamrud, A.; Borghetti, B.; Schubert Kabban, C.; Miller, M. Generalized Deep Learning EEG Models for Cross-Participant and Cross-Task Detection of the Vigilance Decrement in Sustained Attention Tasks. *Sensors* **2021**, *21*, 5617. https://doi.org/10.3390/s21165617

Academic Editor: Yvonne Tran

Received: 30 July 2021
Accepted: 17 August 2021
Published: 20 August 2021

Publisher's Note: MDPI stays neutral with regard to jurisdictional claims in published maps and institutional affiliations.

1. Introduction

Mental fatigue is a significant contributor to a decline in performance for sustained attention type tasks [1,2], also known as vigilance tasks. Vigilance tasks require operators to remain focused and alert to stimulus during a task [3], and in the control and surveillance of today's automated systems, vigilance typically suffers either due to the low level of workload and stimulus associated with the task [4] or due to the mental demands vigilance requires over a lengthy task [5].

A decline in performance during these vigilance tasks is called a vigilance decrement, and it is defined as a decrease in probability of detecting rare but significant events within vigilance tasks [6]. Some form of mental fatigue is typically associated with a vigilance decrement, and this mental fatigue has been linked to increased human error rate [7–9]. If this mental fatigue could be detected using artificial intelligence (AI), then systems could be developed to regulate mental fatigue by varying levels of stimulus to aid in sustained attention [10,11] or by providing recovery time [12].

Mental fatigue has also been linked to specific changes in physiological measures, such as specific increases and decreases in magnitude for the average spectral power of different frequency bands for electroencephalography (EEG) signals [13,14]. Recent machine learning research has utilized EEG signals to classify mental fatigue in specific tasks such as driving [15]; however, a task-generic model which can accurately classify either mental fatigue or a vigilance decrement through EEG signals has not yet been generated. The EEG markers of mental fatigue during vigilance tasks are consistent across both participants and different types of tasks, and mental fatigue is typically always associated with a vigilance decrement in vigilance tasks [16], thus, a model could be built which is capable of performing classification of a vigilance decrement in any vigilance task, through detection of mental fatigue in EEG signals, in any individual's EEG (i.e., a cross-participant model).

Recently, Yin et al., pursued the goal of building a task-generic cross-participant mental fatigue detector using extreme learning machines (ELMs) [17]. Two tasks were used which had the participants replicate the role of an aircraft's automated cabin air management system. Eight participants performed Task 1, and six different participants performed Task 2. Each task varied parameters within the task to create "low" and "high" mental fatigue conditions, with these conditions then corresponding to labeled trials of their respective condition. Models were then built from the EEG data for each task and each condition using entropy features and spectral features (average power of the theta, alpha, beta, and gamma bands) as input features. The models were then tested upon the participant data of the opposite task, with classification accuracies ranging from 65% to 69%. An issue with relating this to vigilance decrement detection is that the tasks simply varied parameters within the task to create "low" and "high" mental fatigue conditions. These conditions were then used as the labels to train the classifier. This means the classifier was trained to identify EEG signals which correspond to these "low" and "high" mental fatigue conditions and not actual vigilance decrements. For proper identification of a vigilance decrement, instead an objective measure of the participant's performance which is associated with the vigilance decrement (such as accuracy and/or response time) would need to be recorded and used to generate the labels of vigilance decrement vs. no decrement for the machine learning classifier. Another issue is that it is unclear if the two stated tasks are analogous to two separate tasks in the real-world, such as the difference between driving and monitoring closed-circuit security cameras, as both tasks used in the experiment had participants performing the same role of the aircraft automated cabin air management system, with only certain parameters and conditions being varied between the two tasks. This suggests that their results are applicable to a varied version of the same type of task that is not truly task-generic.

In this research, we build three different cross-participant models which use EEG signals to perform task-generic classification of the vigilance decrement on any individual. Two of the models are time-domain based, meaning they use the raw EEG time-series voltage values as their data, and the third model is frequency-domain based, using spectral features extracted from the average power of the five clinical EEG frequency bands. The data is comprised of two EEG datasets, with each dataset containing different participants and each dataset containing different vigilance tasks (three different tasks in total). These datasets were collected by the 711th Human Performance Wing (HPW) in partnership with the University of Dayton through two different experiments for the purpose of studying event related potentials (ERPs) during a vigilance decrement across various vigilance tasks [18,19]. Models are trained on data from two of the vigilance tasks and only a subset of the participants and then tested using data from a separate vigilance task that the model has not seen, as well as participants that the model has not seen, which is crucial in order to avoid overestimated test accuracies in cross-participant EEG models [20].

The significant contribution of this research is a model which is capable of detecting a vigilance decrement in unseen participants in an unseen task, as evidenced by the best performing model with 7-fold CV accuracy significantly greater than random chance at 64% (95% CI: 0.59, 0.69). This finding is novel as the cross-participant model was tested with a separate task that the model had not seen, meaning the vigilance decrement was classified in unseen participants in an unseen task, and thus far, a task-generic model which is capable of vigilance decrement classification in an unseen task has not yet been created. Previous work by Yin et al. in building a task-generic model did not utilize a different type of task in order to validate their model and instead only varied parameters within a single task of operating an aircraft's cabin air management system in order to create two tasks. Our research in contrast utilized three different types of tasks (the air traffic controller task, the line task, and the 3-stimulus oddball task), each of which are well established in the literature as different kinds of vigilance-type tasks.

This paper has the following structure. First, in Section 2, background is provided for the vigilance decrement and how it is linked to EEG. Next, in Section 3, we provide our methodology, first providing details on the datasets collected and the tasks used within those datasets, followed by details for the training and testing of all three models. Then, in Section 4, results are presented for all three models. Finally, in Section 5, results are compared and discussed, with conclusions and future work following in Section 6.

2. Related Work

Decision making and how it deteriorates in stressful work environments has been extensively studied since the late 1800s [6]. One of the main phenomena studied has been the concept of vigilance, which is the quality or state of being wakeful or alert. Tasks which require vigilance fall under a taxonomy developed by Parasuraman and Davies [21], with the taxonomy classifying tasks into different categories based on specific information-processing transactions within the tasks themselves, such as signal discrimination (successive or simultaneous), task complexity, event rate, and sensory modality. For signal discrimination, simultaneous tasks are ones in which the critical stimulus and non-critical stimulus are both present at the same time for participants to use for comparison. Successive tasks, however, do not provide these stimulus to the participant at the same time, and therefore, it requires the participant to hold the non-critical stimulus in memory.

2.1. Vigilance Decrement

Extensive research over the decades on vigilance and the vigilance decrement has found that the behavioral cause of the decrement is due to performing attention-demanding tasks over an extended period of time, ranging from tens of minutes to hours, depending on the task and its cognitive demand [22]. Performing these attention-demanding tasks for extended periods of time results in mental fatigue and/or a decrease in sustained attention [23], with mental fatigue being defined as a gradual and cumulative phenomenon that is increased in magnitude by time spent on a tedious but mentally demanding task [24].

Numerous factors have also been found to affect the magnitude and timing of the vigilance decrement [25]. For magnitude, simultaneous stimulus, shorter signals [26,27], task type/source complexity [28], and stimulus event rate [29,30], all result in a greater vigilance decrement. For timing, the vigilance decrement varies depending on the task demands, with the vigilance decrement occurring earlier in more difficult tasks [22], and typically occurring within the first 20–35 min of a task, with half of the decrement occurring in the first 15 min [31].

Performance Measurement

To identify in data whether a vigilance decrement has occurred, some measure of task performance through either accuracy, shown in Equation (1), response time (RT), shown in Equation (2), or both, is needed. Accuracy and RT are frequently correlated, such that slower responses are more accurate and vice versa, and this is referred to as the speed-accuracy trade-off [32,33].

$$\text{Accuracy} = \frac{\text{hits} + \text{correct rejections}}{\text{hits} + \text{false alarms} + \text{misses} + \text{correct rejections}}. \tag{1}$$

$$\text{Response Time} = T_{response} - T_{stimulus}. \tag{2}$$

Due to this correlation, it is best to use both accuracy and RT to assess performance, and many different measures have been developed to combine both speed and accuracy into a single measure of performance. For example, there is the Inverse Efficiency Score (IES), which is the ratio of the mean RT and the proportion of correct responses (PC) [34], the Rate-Correct Score (RCS) which is the inverse of the correct RT-based IES [35], the Balanced Integration Score (BIS) which is a combined z-score of RT and accuracy [36], and many others. Recently, research by Mueller et al. examined 12 different measures of accuracy and RT on a vigilance task to determine their sensitivity to the vigilance decrement and found that most single measures which combined accuracy and RT were slight improvements over just accuracy or RT alone [37]. While they found that the Linear Ballistic Accumulator model was the most sensitive and representative measure of the vigilance decrement, they also noted that it was difficult and cumbersome to use and recommended the BIS measure overall.

The BIS measure is designed to give equal weights to both PCs and RTs, hence the name Balanced Integration Score, and is shown below in Equation (3). First, the PCs and RTs are standardized as shown in Equations (4) and (5), with participants j and standard deviations s, and then once standardized, the standardized RT is subtracted from the standardized PC. This gives the difference in standardized mean correct RTs and PCs. z_{pc} and z_{rt} can be calculated individually for each participant j, giving the BIS measurement for only that participant, or across all participants, giving the BIS measurement for the group.

$$BIS_j = z_{PC_j} - z_{RT_j}. \tag{3}$$

$$z_{pc_j} = \frac{PC_j - \overline{PC}}{s_{pc_j}}. \tag{4}$$

$$z_{rt_j} = \frac{RT_j - \overline{RT}}{s_{rt_j}}. \tag{5}$$

When calculating measures such as BIS from data collected during a vigilance task, trials must be binned in some manner for the standardized measures of z_{pc} and z_{rt} to be calculated. A common method is to divide the trials over the duration of the experiment into four time segments (bins) [19,37,38]. Once the number of bins is selected, BIS can then be calculated and compared for each bin to determine whether a vigilance decrement has occurred for the participant; a decreasing BIS indicates a decrement in vigilance. A typical method is to plot the bins on a graph to view the participant's performance over the course of the task as well as to plot a line of best fit (least squares) to see how their performance trended over the course of the task, with a negative slope indicating a vigilance decrement over the course of the entire task.

2.2. EEG

Physiological measurements such as EEG, electrocardiography (ECG), and electrooculography (EOG) have been progressively utilized to better understand the underlying mechanisms of mental fatigue and the vigilance decrement over the past two decades, with EEG receiving significant attention in research for its insight into the status of the brain [16]. EEG signals are a measure of the electrical activity in the brain using electrodes distributed over the scalp, and EEG is often referred to by its different clinical frequency bands, namely delta (2–4 Hz), theta (4–7 Hz), alpha (8–12 Hz), beta (13–29 Hz), and gamma (33–80 Hz). A physiological measurement such as EEG has the advantage of providing a more objective measurement of fatigue than a behavioral measure, as behavioral measures are subjective in nature and left to the experimenter's or participant's judgment. EEG studies investigating neural correlates of fatigue have found differing results based on the type of fatigue that the participant is experiencing, with the primary difference being fatigue from sleepiness (sleep fatigue) versus accumulating fatigue from cognitive processes and mental workload (mental fatigue). For example, neural correlates of sleep fatigue have been found to differ based on the task that is being performed. Driver fatigue research found that symptoms associated with sleepiness (e.g., prolonged eye closure) correlated to increases in spectral power for the alpha and beta bands [13], while in pilot fatigue studies, sleepiness was more associated with the opposite effect, with decreases in spectral power for the alpha band [39,40]. Mental fatigue, however, has shown consistent neural correlates of increased spectral power for the alpha band across tasks [16]. This allows for the detection of mental fatigue across tasks and across participants. However, given that both types of fatigue can contribute to changes in performance, such as the vigilance decrement, yet have differing neural correlates, it is important to distinguish sleep fatigue from mental fatigue to reduce confounding variables.

Utilizing these neural correlates of EEG has been useful for both within-participant and cross-participant detection of the vigilance decrement, with all previous research being within-task detection of the vigilance decrement. EEG spectral features have been common features used to detect drowsiness, mental fatigue, and alertness [41,42]. Power spectral density (PSD) in combination with independent component analysis [42], the mean power of the frequency bands and their ratios [41,43,44], power spectral indices of wavelet transforms [45], and full spectrum log power are all spectral features that have been used [46]. Directed connectivity has also been utilized using relative wavelet transform entropy and partial directed coherence to estimate the strength and directionality of information flow between EEG nodes [47,48].

3. Methods

3.1. Datasets

In this study, two existing EEG datasets are utilized, each collected through experiments conducted previously by the United States Air Force Research Laboratory, 711th Human Performance Wing (HPW), in partnership with the University of Dayton. These experiments were each conducted for the purpose of studying ERPs during a vigilance decrement within various vigilance tasks [18,19]; however, the experiments were conducted separately and did not coincide. All data was de-identified before it was shared with us for our experiments, and because it was de-identified existing data, a Human Research Protection Plan review determined this research to be not involving "human subjects" under US Common Rule (32 CFR 219) on 6 June 2020.

In one experiment, 32 participants (10 men and 22 women, ages ranging from 18 to 36 with a mean of 22.7, with 27 being right-handed) completed three different tasks across a two hour session in the following order: the Hitchcock Air Traffic Controller (ATC) Task [49], the Psychomotor Vigilance Test (PVT) [50], and the 3-Stimulus Oddball Task [51]. The PVT was omitted from our research as the task length was short in duration (<10 min) along with a few amount of trials (<100), making it difficult to segment into bins and quantify with the BIS measure. The Hitchcock ATC task and 3-Stimulus Oddball Task were performed as described in Sections 3.2.1 and 3.2.2, and trials for each task occurred as follows. The ATC Task included 200 practice trials with feedback provided every 50 trials, then a short break followed by 1600 trials without feedback or breaks. The 3-Stimulus Oddball Task included 20 practice trials, a short break, and 4 blocks of 90 trials each, with performance feedback after each block. Practice trials across both tasks are not utilized in analysis or model training. Some participants had incomplete data, and only the data from the 14 participants with complete datasets were analyzed.

The second experiment consisted of two sessions for each participant, conducted over two separate days, and utilized the line task described in Section 3.2.3. Each day, participants performed 200 practice trials and 4 blocks of 400 experimental trials each, with a short few minute break offered between each block. There were 29 participants; however, only 26 of the participants returned the second day. The data from all 29 participants was utilized in the current study. Participant demographics were not available for this.

Experiment details and EEG/ERP analysis can be found in references [18,19], with summary information provided here. For both datasets, the tasks were presented on an

LCD 60 Hz monitor using Psychophysics Toolbox [52] within MATLAB. EEG was recorded using a BioSemi Active II 64 + 2 electrode cap (10–20 system) with the 2 reference electrodes placed over the mastoids (with no additional detail provided), with a sampling rate of 512 Hz. Vertical EOG (VEOG) and Horizontal EOG (HEOG) were also recorded [18,19]. Baseline resting EEG was recorded before starting the experiment and checked for artifacts. Voltage offsets were reduced to less than 40 mV to ensure low impedance, and any high impedance electrodes were re-gelled and re-applied.

3.2. Vigilance Decrement Tasks
3.2.1. Hitchcock Air Traffic Controller Task

The Hitchcock ATC Task was designed to test theories surrounding sustained attention, workload, and performance, within a standardized controllable task that is relatively more representative of the real world [53]. Stimulus of a filled red circle and three concentric white circles are continually displayed to the participant. Two white line segments are then displayed over these stimuli in different configurations, as seen in Figure 1. The red circle represents a city, and the white line segments represent aircraft. Participants are instructed to respond (through press of a key on a keyboard) only if the two jets are on a collision course with one another, i.e., the white lines are colinear. If they are, this is a critical event, and a small minority of trials are critical events (3.3%), the rest being non-critical as seen in Figure 1. The stimulus appear every 2 s and only remain on screen for 300 ms.

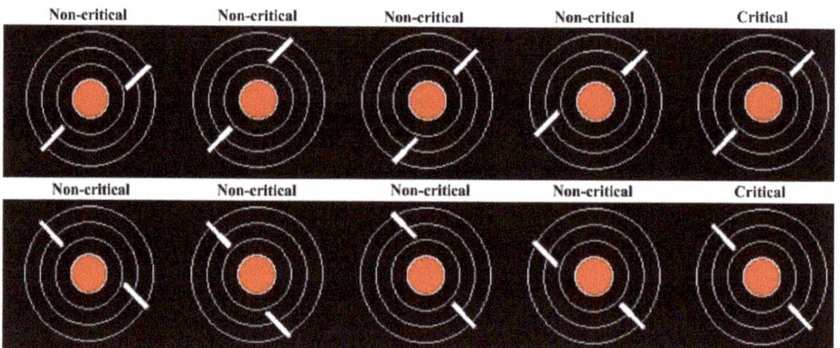

Figure 1. Examples of the different Air Traffic Controller Task stimuli [19].

3.2.2. 3-Stimulus Oddball Task

The 3-Stimulus Oddball Task was designed to assess how individuals discriminate targets, non-target distractors, and standard distractors, in various challenging scenarios [51]. In this task, three different visual stimuli can appear: targets, non-target distractors, and standard distractors. Targets and non-target distractors each appear separately in 10% of trials, and standard distractors appear in the remaining 80% of trials. As seen in Figure 2, the target is a large circle, the standard distractor a small circle, and the non-target distractor a large square. Stimuli are every 2 s with a 75 ms duration. Participants are instructed to respond only to targets by pressing a response key on a keyboard, ignoring non-target distractors and standard distractors.

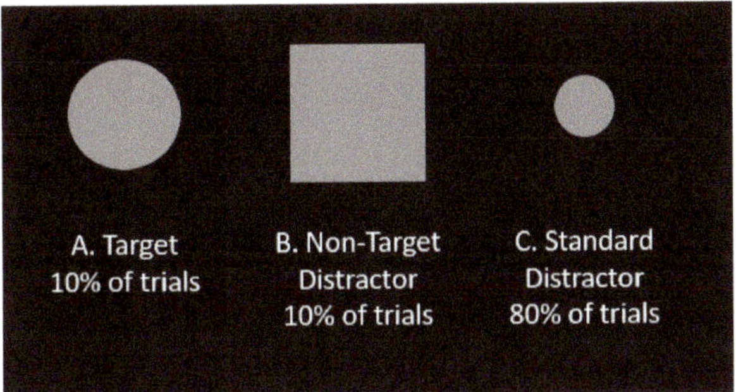

Figure 2. Shapes for the 3-Stimulus Task. The target is a large circle (**A**), the non-target distractor a large square (**B**), and the standard distractor a small circle (**C**) [18].

3.2.3. Line Task

In the line task, participants observe a series of pairs of parallel lines and select whether or not each stimulus is critical. The critical stimuli vary among four conditions for the task, and with critical stimuli comprising 10% of the stimuli. The parallel lines are 0.75 mm in width and variable in length based on trial condition [29]. The first and second conditions are successive-discrimination tasks, meaning the participant has to hold the critical stimulus in memory. In the first condition, the set of lines both being 1.46 cm (short) is the critical stimulus, with both lines being 1.8 cm (long) as the non-critical stimulus. In the second condition, these are reversed. The third and fourth conditions are simultaneous-discrimination tasks, meaning the participant is provided both the critical and non-critical stimulus at the same time for comparison. In the third condition, the critical condition occurs when the lines are different in length while in the fourth condition, these are reversed. Critical stimuli are sequenced such that there are at least four non-critical stimuli in between each pair of critical stimuli. Each participant completed both simultaneous and successive discrimination conditions (counterbalanced across sessions). Stimulus appeared on screen for 150 ms and total trial duration was randomized to be between 1.3 s and 1.7 s. Figure 3 shows an example of the line stimulus.

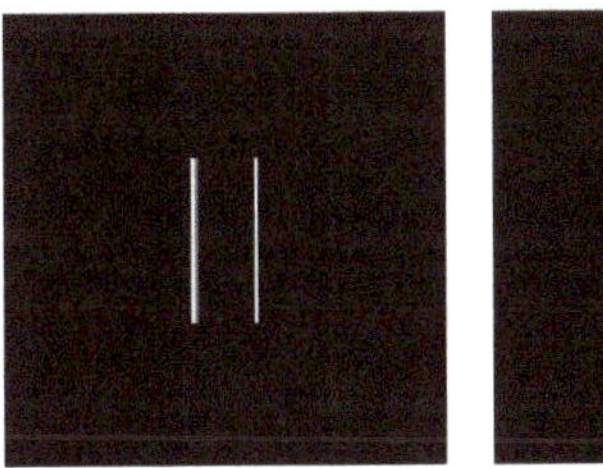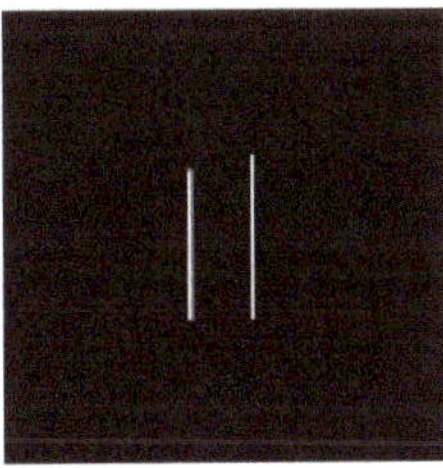

Figure 3. Examples of the different line task stimuli, with lines being the same length on the left and different lengths on the right.

3.3. Preprocessing and Epoching of EEG Signals

Preprocessing of EEG data was performed through script batch processing using EEGLAB [54] and consisted of a combination of best practice steps from both Makoto's preprocessing pipeline [55] and the PREP pipeline [56]. Details for these steps can be found in Appendix A but worth noting is that the data is downsampled to 250 Hz, and that EOG is used for Independent Component Analysis (ICA) to remove eyeblink artifacts from the EEG. All tasks were relatively similar in trial duration, ranging from 1 s to 1.7 s, with inter-trial duration ranging from 1.2 s to 2 s. To avoid an epoching window which extends into the following trial for some tasks but not others, a 1 s epoching window was selected based on both trial duration and inter-trial duration. Additionally, analysis performed by the 711 HPW demonstrated that a 1 s window following stimulus-onset contained the majority of EEG activity for each task [18,19]. This resulted in a sequence length of 250 for observations across all three tasks.

For labeling of the EEG signals, trials are divided over the duration of the task into four time segments (bins) for each task, and the BIS measure (described in Section 2.1) is used to determine participant performance for each bin, with BIS values and the corresponding z-scores calculated separately for each individual. Using this method resulted in a BIS measure of a participant's performance for the first, second, third, and fourth quarters of the task, allowing analysis of a participant's performance as the task progressed in time. Performance for each task and each participant are plotted in Figure 4, including the best-fit line for each task and each participant. From the best fit lines in Figure 4, it can be seen that every participant, for every task, was at their highest performing state in the 1st bin, meaning every bin following the 1st bin was a vigilance decrement in comparison to the 1st bin. However, across the tasks, participants had varying performance following the 1st bin as can be seen in Figure 4, with some experiencing their largest decrement in the 2nd, 3rd, or 4th bins. This makes labelling across all four bins difficult while trying to also maintain a balanced dataset. Given this challenge, we opted to use the 1st and 4th bins for our model creation, labelling the 1st bin as attentive, and the 4th bin as a decrement, resulting in a perfectly balanced dataset.

Proper labeling of the data is crucial for a machine learning model, and utilizing only the 1st bin as attentive maximizes tying the most attentive trials to their respective neural correlates. Additionally, the underlying mechanism that allows success in building a task-generic model is that mental fatigue is consistent in producing a vigilance decrement in these tasks and that it is consistent in its neural correlates across different types of vigilance tasks [16]. As mental fatigue has been shown to accumulate over the duration of a vigilance task, the EEG data for the last bin is most likely to have the neural correlates of mental fatigue. As the last bin is a vigilance decrement for all participants across all tasks, using the 1st and 4th bins should maximize the likelihood that the data is labeled properly and will contain the underlying neural correlates to best ensure its success.

3.4. Model Creation

To be effective in detection across participants, a model must be highly generalizable and resistant to the effects of non-stationarity and individual differences. For training and testing of a cross-participant model, this requires that data from participants used for model training must not be used for model validation or testing [20]. This is due to the individual differences and non-stationarity that are inherent within EEG data. If this rule is not followed, the model will likely have overestimated test accuracies, and additionally, the model will not train to be generalizable to a more general population, as the model will learn parameters which are likely only accurate for those participants. Additionally, as this is a task-generic model, the model should be tested with a vigilance task that is unseen by the model. To follow these guidelines, we adopted a leave-two-participants-out cross-validation (L2PO-CV) training method for all three models, resulting in 7-folds. The ATC and line tasks were used to train the model, with the 3-Stimulus Oddball Task used for validation. This L2PO-CV method was used for training and validation of all three models. Both the ATC task and the line tasks have the greatest amount of trials, with each participant having performed four times more trials in each of those tasks than the 3-Stimulus Oddball task, resulting in a more desirable ratio of training to validation data than if the ATC or line tasks were used for validation. Additionally, this ensures there is training data from both experiments to allow additional generalization for the model, as the line task was performed in a separate experiment, with an independently selected pool of participants. Ideally, CV would be performed across all three tasks; however, this was infeasible due to the immense amount of training time it would require. All together this results in training folds with 41 participants and 53,600 observations total and validation folds with 2 participants and 360 observations total.

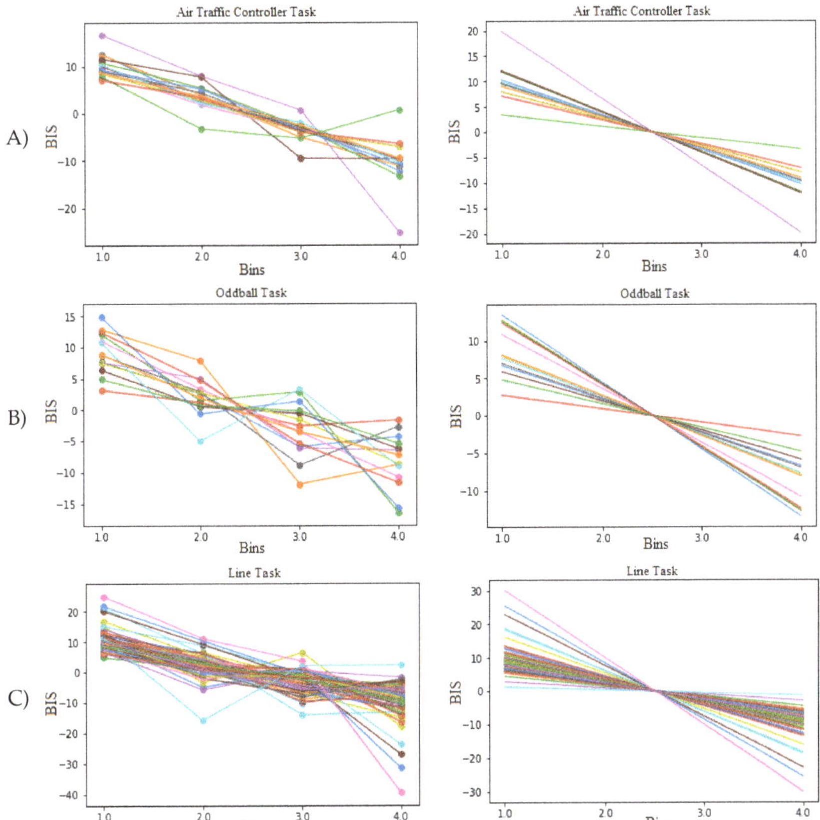

Figure 4. BIS measures and the corresponding best-fit lines for: (**A**) Air Traffic Controller task (top), (**B**) Oddball task (middle), and (**C**) Line task (bottom). The lines represent a participant's BIS measures over the duration of the task, with lines on the right being best-fit lines. BIS measures vary from bin to bin for each participant, with some participants decreasing steadily throughout the entire task, some decreasing initially and then recovering, or some alternating between decreasing and increasing BIS. Note that every participant's best-fit line has a negative slope, indicating that every participant's first bin is their most attentive bin with their largest BIS measure.

As these cross-participant models are also task-generic, features must be invariant for not only the participants but also the task. For the frequency-domain model, the average power of the five traditional EEG frequency bands for all 64 scalp electrodes were selected as features, resulting in 320 spectral features for each observation, as literature demonstrated that the average power correlates with mental fatigue and is invariant across task, time, and participant [16]. However, an alternative to performing feature extraction by hand is to have the model extract salient features itself. Recently, autoencoders (AEs) have been shown to be more effective than handcrafted features in their ability to compose meaningful latent features from EEG across various classification tasks [57–59]. Another recent deep learning innovation is Temporal Convolutional Networks (TCNs), which are a new type of architecture for time-series data. TCNs have the advantage of processing a sequence of any length without having a lengthy memory history, leading to much faster training and convergence when compared to Long Short-Term Memory (LSTM) models [60]. For the time-domain models, a TCN-AE is used for one of the models, and a TCN for the other. In the next two sections, general information on TCNs and AEs is provided, followed by the proposed architectures, hyperparameters, and training and testing parameters for all three models.

3.5. Temporal Convolutional Networks

A TCN is a type of convolutional neural network (CNN) for 1D sequence data and was recently developed by Bai et al. [60]. A TCN utilizes dilated convolutions to process a sequence of any length, without having a lengthy memory history. TCNs are typically causal, meaning there is no information leakage from the future to the past; however, they can be non-causal as well. The primary elements of a TCN consist of the dilation factor d, the number of filters n, and the kernel size k. The dilation factor controls how deep the network is, with dilations typically consisting of a list of multiples of two. Figure 5 provides a visual example of a causal TCN and aids in understanding the dilated convolutions on a sequence, with the dilation list in the figure being (1,2,4,8). The kernel size controls the volume of the sequence to be considered within the convolutions, with Figure 5 showing a kernel size of 2. Finally, the filters are similar as they are in a standard CNN and can be thought of as the number of features to extract from the sequence.

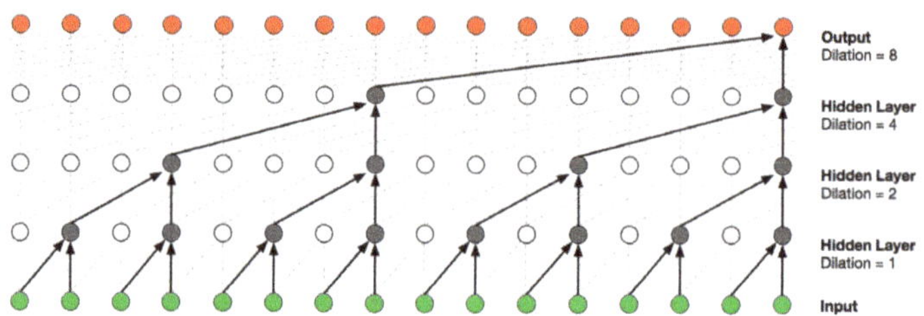

Figure 5. Visual illustration of a causal TCN [61]. This TCN has a block size of 1, a dilation list (1,2,4,8) (i.e., dilation factor 8), and a kernel size of 2. This results in a receptive field of $2 \cdot 1 \cdot 8 = 16$.

These combined elements form a block as in Figure 5, and blocks can be stacked as they are in Figure 6. This increases the receptive field, which is the total length the TCN captures in processing and is a function of the number of TCN blocks, the kernel size, and the final dilation, as shown in Equation (6). It is common to have a receptive field which matches the input sequence length; however, the receptive field is flexible and can be designed to process any length, which is a primary advantage of TCNs. Other advantages include their ability to be trained faster than LSTMs/Gated Recurrent Unit (GRU) models of similar length, having a longer memory than LSTMs/GRUs when capacity of the networks is equivalent and having similar or better performance than LSTMs/GRUs on a number of sequence related datasets [60,62]

$$R_{field} = K_{size} \cdot N_{blocks} \cdot d_{final}.$$ (6)

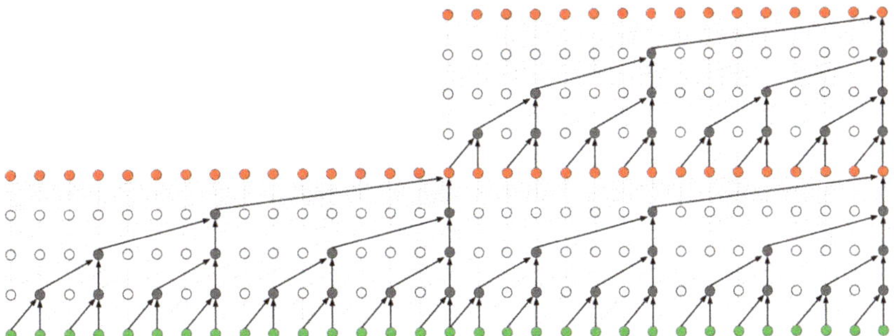

Figure 6. Visual illustration of a causal TCN with stacked blocks [62]. This TCN has a block size of 2, a dilation list (1,2,4,8) (i.e., dilation factor 8), and a kernel size of 2. This results in a receptive field of $2 \cdot 2 \cdot 8 = 32$.

3.6. Autoencoders

An autoencoder (AE) is a type of neural network architecture for unsupervised learning that is primarily used for reproduction of what is input into the network [63]. This is done through the use of two separate networks. One network named the *encoder* $f(\mathbf{x})$ compresses the input into a lower-dimensional representation called the *code* or the *latent-space* $\mathbf{h} = f(\mathbf{x})$ and another network named the *decoder* reconstructs the input from the code $\mathbf{r} = \mathbf{g}(\mathbf{h})$. An example of a standard AE architecture can be seen in Figure 7. Because of the nature of the encoder, AEs are useful for dimensionality reduction, are powerful feature detectors, and can also be used for unsupervised pretraining of deep neural networks [64].

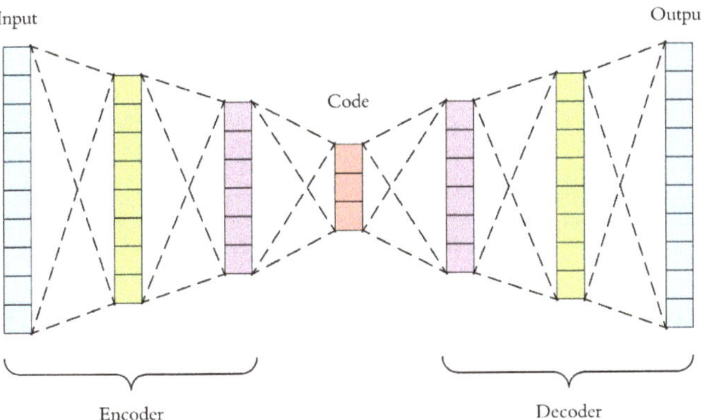

Figure 7. Visual representation of a standard AE architecture [65].

In Figure 7, the code $\mathbf{h}$ is constrained to have a smaller dimension than the input $\mathbf{x}$. This is called being *undercomplete* and is typical of an AE, as it forces the AE to capture the most salient features of the training data, and thus, the AE does not overfit the training data and copy it perfectly [63].

3.7. Frequency-Domain Model

The frequency-domain model was a fully connected MLPNN as can be seen in Figure 8 and utilized spectral features extracted from the 1s epoched EEG signal using complex Morlet wavelet transforms in MATLAB to determine the mean power of the five traditional frequency bands: delta (2–4 Hz), theta (4–7 Hz), alpha (8–12 Hz), beta (13–29 Hz), and gamma (33–80 Hz) (details of this process are out of scope for this paper, and we refer the reader to Chapters 12 and 13 in Mike Cohen's book, *Analyzing Neural Time Series*

Data [66]). With 64 channels from the 64 electrode cap, this resulted in 320 spectral features for each observation ($5 \times 64 = 320$). To improve model training, the spectral features were standardized and also log transformed.

The model consisted of three hidden layers with hidden units *hu*, each followed by a dropout layer with dropout rate *dr*, with the ReLU activation function used for each hidden layer. As specified at the beginning of Section 3.4, L2PO-CV was used for training and validation of the MLPNN model. The Adam optimizer [67] was used to train the models for 300 epochs by minimizing the binary cross-entropy loss, and a hyperparameter sweep was performed over the hidden units *hu*, the dropout rate *dr*, and the learning rate *lr*.

3.8. Time-Domain Models

3.8.1. TCN-AE

The TCN-AE architecture was modeled after work done by Thill et al., who recently developed one of the first published TCN-AE architectures for unsupervised anomaly detection in time series data for health monitoring of machines [68]. They credit the success of this model architecture to the architecture's ability to compose and encode salient latent features from the data, doing so unsupervised. This architecture involves first training the AE to have the ability to reconstruct the EEG signal with minimal loss. Then the encoder of the trained AE encodes the EEG signal to its latent representation, and those latent features are used for training of a classification model. Their architecture was used as a basis for the TCN-AE model of this research, as the goal for this TCN-AE was to encode the most salient features of the EEG data, and then use those features as input to a fully connected neural network (FCN) classifier to perform classification.

Figure 8. Visual representation of MLPNN classifier. The MLPNN architecture consists of three fully-connected hidden layers with hidden units *hu* and the ReLU activation, each followed by a dropout layer with a dropout rate *dr*.

The architecture of the TCN-AE is included below in Figure 9, with the encoder on the left, the decoder on the right, and the latent space in the bottom center. The encoder takes as input the EEG signal with dimensions of 250×64, with the 250 representing the sequence length of the 1s epoch downsampled to 250 Hz and the 64 representing the different features from the 64 electrodes. The first layer is a TCN with hyperparameters as specified in Section 3.5, with *d* representing the dilation factor, *k* the kernel size, *b* the number of blocks, and *n* the number of filters. The TCN also used batch normalization, dropout, and recurrent dropout, with the dropout rate *dr* set as a hyperparameter. This is followed by a 1D convolution (Conv1D) with a kernel size of 1 for further dimensionality reduction and additional non-linearity [68], with *L* representing the number of filters for this convolution layer, which also represents the number of latent features, as there is no further dimensionality reduction after this layer. The ReLU activation function is used for both the TCN and Conv1D layers. Temporal average pooling is then performed with a size of 5 to reduce the sequence length by a factor of 5. This results in the latent space having a sequence length of $50 \times L$ number of features.

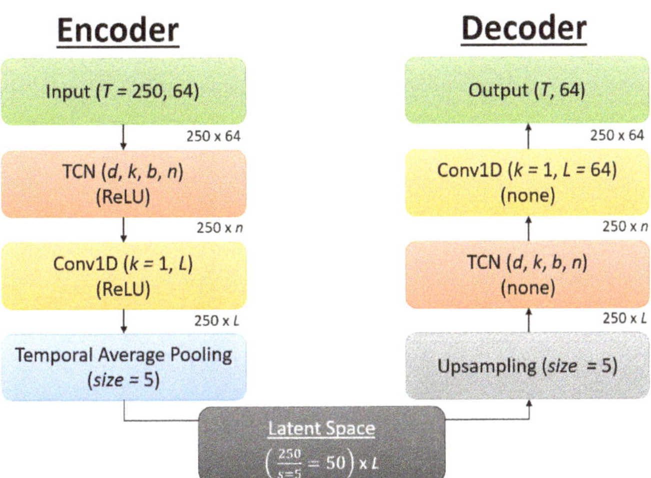

Figure 9. Visual representation of the TCN-AE architecture. Each block corresponds to a layer, with hyperparameters for that layer *italicized*. The activation function for the TCN and Conv1D layers is in parentheses, using ReLU for the encoder and no activation function for the decoder. The dimensions for the input are also provided in the upper-right of each layer as it passes throughout the architecture, with the dimensions starting at $T = 250$ for the sequence length and 64 representing the features (corresponding to the 64 electrodes). The latent space dimensions are $50 \times L$, with L being a hyperparameter.

The decoder is similar to the encoder in its architecture, albeit in reverse. The sequence is first upsampled back to its original length of 250 using nearest neighbor interpolation. The sequence is then passed into a TCN which again has hyperparameters $d, k, b,$ and n, followed by a Conv1D layer which increases the dimensionality of the sequence back to its original size of 64. There is no activation function for the TCN and Conv1D layers in the decoder, as this allows the values of the sequence length to take on any value to recreate the original signal.

L2PO-CV was used for training and validation of the reconstruction phase of the AE, with EEG signals standardized by channel for faster model convergence. The Adam optimizer [67] was used to train the autoencoder for 50 epochs for reconstruction of the EEG signal by minimizing the MSE loss, and hyperparameters were grid-searched using Ray Tune version 1.3.0, with the hyperparameters consisting of the dilation factor d, the kernel size k, the number of blocks b, the number of filters n, the number of latent features L, the dropout rate dr, and the learning rate lr.

Once the autoencoder was trained for reconstruction, the weights of the encoder were locked and the encoder was then used to encode input sequences into latent features. The latent features were then flattened and used as input features into a FCN classifier. The TCN-AE architecture in its entirety can be seen in Figure 10. The FCN classifier had two hidden layers, each with the ReLU activation function, followed by a dropout layer, and a output layer using the sigmoid function. L2PO-CV was used for training and validation of the FCN for classification. The Adam optimizer [67] was used to train the models by minimizing the binary cross-entropy loss, and a hyperparameter sweep was performed over the number of hidden units for each layer, the dropout rate, and the learning rate.

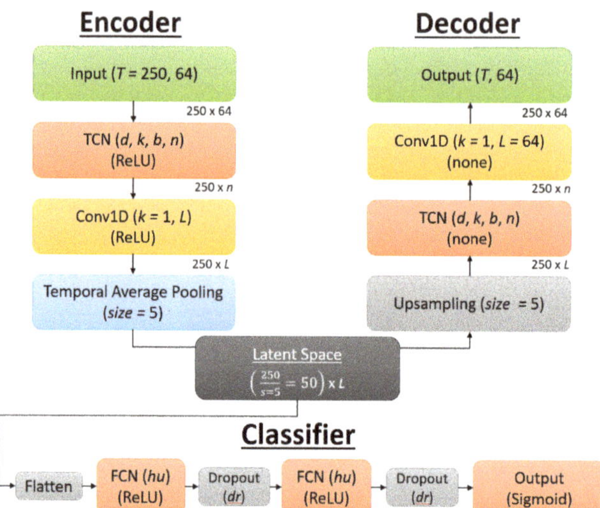

Figure 10. Visual representation of the TCN-AE classifier. The Encoder and Decoder comprise the AE architecture, with the latent space then used as input to the FCN classifier shown at the bottom. The FCN classifier architecture consists of two fully-connected hidden layers with hidden units hu, each followed by a dropout layer with a dropout rate dr.

3.8.2. TCN

The TCN model can be seen in Figure 11 and was similar to the encoder portion of the TCN-AE architecture in that it consists of a TCN layer and a Conv1D layer; however, this model differs in that prediction is performed after the Conv1D layer, using an output layer with a sigmoid activation function. The TCN layer has hyperparameters as specified in Section 3.5, with d representing the dilation factor, k the kernel size, b the number of blocks, and n the number of filters. The TCN also used batch normalization, dropout, and recurrent dropout, with the dropout rate dr set as a hyperparameter. The Conv1D has a kernel size of 1 and a filter size of 4, providing dimensionality reduction before the output layer. The ReLU activation function is used for both the TCN layer and the Conv1D layer. L2PO-CV was used for training and validation of the TCN for classification, with EEG signals standardized by channel for faster model convergence. The Adam optimizer [67] was used to train the models for 100 epochs by minimizing the binary cross-entropy loss, and a hyperparameter sweep was performed using Ray Tune and grid search over the dilation factor d, the kernel size k, the number of blocks b, the number of filters n, the dropout rate dr, and the learning rate lr.

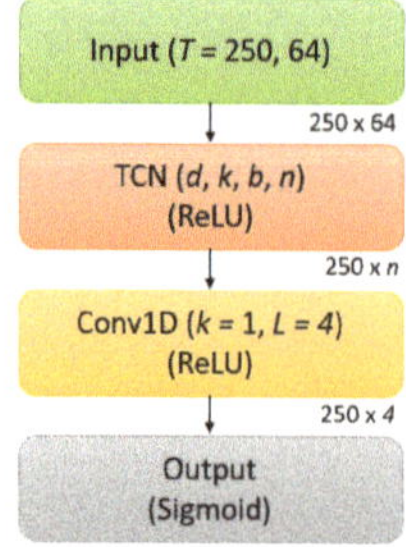

Figure 11. Visual representation of the TCN classifier. Each block corresponds to a layer, with hyperparameters for that layer *italicized*, and the activation function in parentheses.

4. Results

Below are the results for both the frequency-domain model and the time-domain models. For each model, the best hyperparameter configuration is presented along with its CV balanced accuracy and confidence interval (CI). As accuracy is a binomial distribution, approximate binomial confidence intervals are used. Specifically we utilize Agresti Coull confidence intervals, as they typically maintain α while not being overly conservative [69]. Each model's CV balanced accuracy and its 95% Agresti Coull confidence interval are compared to random chance, i.e., a naïve classifier with accuracy of 50% (accuracy is 50% as this is a binary classification task). Validation accuracies are also provided for each participant by the participant's ID, along with their 95% confidence interval. At the end of this section, a table is provided with the participant validation accuracies for each model and the 7-fold CV accuracy for each model.

4.1. Frequency-Domain Model

Hyperparameter sweeps for the MLPNN model resulted in the best network achieving a 7-fold CV balanced accuracy of 64% (95% CI: 0.59, 0.69) and 7-fold CV area under the receiver operating characteristic (AUROC) of 0.71 with the following hyperparameters: hidden units of (250, 200, 150) (by layer), learning rate of 0.00001, and dropout rate of 0.5. This results in the model having CV accuracy statistically greater than random chance as evidenced by the confidence interval. Figure 12 depicts the validation accuracies for each participant for the MLPNN model, with nine participants having validation accuracies statistically greater than random chance. Participants 2, 3, 7, 8, and 11 did not have validation accuracies greater than random chance.

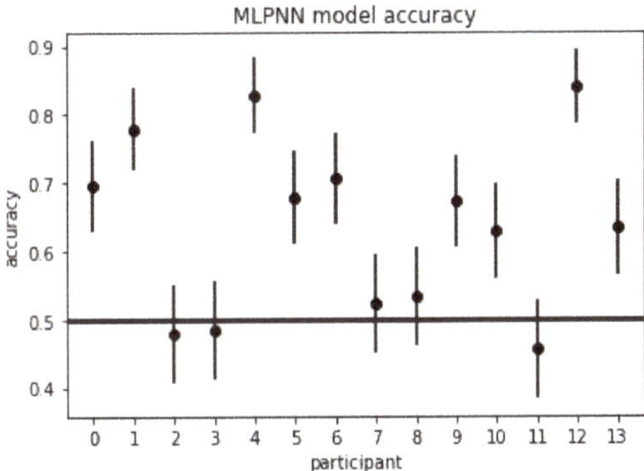

Figure 12. Participant validation accuracies for the MLPNN model, with 9 participants having validation accuracies statistically greater than random chance. Participants 2, 3, 7, 8, and 11 did not have validation accuracies greater than random chance. This model achieved a 7-fold CV accuracy of 64% (95% CI: 0.59, 0.69).

4.2. Time-Domain Model—TCN-AE

The best hyperparameters found for the TCN-AE signal reconstruction had the following configuration: dilations (1, 2, 4, 8, 16, 32), kernel size of 2, number of filters 36, number of blocks 2, learning rate of 0.0001, and dropout rate of 0.0; and resulted in a receptive field of $2 \cdot 2 \cdot 32 = 128$. For the classifier portion of the TCN-AE, all hyperparameter sweeps resulted in similar performance, with accuracies ranging between 48% and 52% for 7-fold CV balanced accuracy, with no set of hyperparameters resulting in a model which performed statistically better than chance. Individual participant accuracies were also investigated for each hyperparameter sweep, with two or less participants having significant performance for the hyperparameter sweeps. No participants had validation accuracies statistically greater than random chance.

4.3. Time-Domain Model—TCN

The best hyperparameter sweep for the TCN model yielded a 7-fold CV balanced accuracy of 56% (95% CI: 0.51, 0.61) and 7-fold CV AUROC of 0.57 with the following hyperparameters: dilations (1, 2, 4, 8, 16, 32), kernel size of 4, number of filters 10, number of blocks 2, learning rate of 0.0001, and dropout rate of 0.5; and resulted in a receptive field of $4 \cdot 2 \cdot 32 = 256$. This results in the model having CV accuracy statistically greater than random chance as evidenced by the confidence interval. Figure 13 depicts the validation accuracies for each participant for the TCN model, with 3 participants (1, 7, and 12) having validation accuracies statistically greater than random chance.

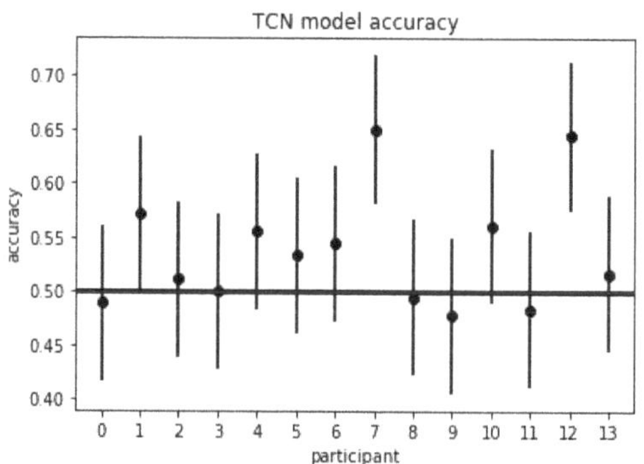

Figure 13. Participant validation accuracies for the TCN model, with 3 participants (1, 7, and 12) having validation accuracies statistically greater than random chance. This model achieved a 7-fold CV accuracy of 56% (95% CI: 0.51, 0.61).

Table 1 provides the participant validation accuracies and the 7-fold CV accuracy for all three models.

Table 1. Vigilance decrement classification model performance results for each model type. Participant validation accuracies and the 7-fold CV accuracy are provided for each model, with 95% confidence intervals provided in parentheses. For both participant accuracies and the 7-fold CV results across all participants, **Bold** signifies statistical significance of accuracy over random chance (defined as 50% for this binary classification task) as evidenced by the 95% confidence interval.

Participant #	MLPNN Val Acc	TCN-AE Val Acc	TCN Val Acc
0	**0.69 (0.62, 0.76)**	0.51 (0.44, 0.58)	0.49 (0.42, 0.56)
1	**0.78 (0.71, 0.83)**	0.49 (0.42, 0.56)	**0.57 (0.50, 0.64)**
2	0.48 (0.41, 0.55)	0.49 (0.42, 0.57)	0.51 (0.44, 0.58)
3	0.48 (0.41, 0.56)	0.52 (0.45, 0.59)	0.50 (0.43, 0.57)
4	**0.83 (0.77, 0.88)**	0.49 (0.42, 0.57)	0.56 (0.48, 0.63)
5	**0.68 (0.61, 0.74)**	0.55 (0.48, 0.62)	0.53 (0.46, 0.60)
6	**0.71 (0.63, 0.77)**	0.47 (0.40, 0.54)	0.54 (0.47, 0.62)
7	0.52 (0.45, 0.59)	0.56 (0.48, 0.63)	**0.65 (0.58, 0.72)**
8	0.53 (0.46, 0.60)	0.56 (0.49, 0.63)	0.49 (0.42, 0.57)
9	**0.67 (0.60, 0.74)**	0.54 (0.47, 0.62)	0.48 (0.41, 0.55)
10	**0.63 (0.56, 0.69)**	0.49 (0.42, 0.57)	0.56 (0.49, 0.63)
11	0.46 (0.38, 0.53)	0.46 (0.38, 0.53)	0.48 (0.41, 0.56)
12	**0.84 (0.78, 0.89)**	0.48 (0.41, 0.55)	**0.64 (0.57, 0.71)**
13	**0.63 (0.56, 0.70)**	0.53 (0.46, 0.60)	0.52 (0.44, 0.59)
7-fold CV	**0.64 (0.59, 0.69)**	0.52 (0.47, 0.57)	**0.56 (0.51, 0.61)**

5. Discussion

The frequency-domain model (MLPNN) had the highest level of performance of the three model types, with 7-fold CV accuracy significantly greater than random chance at 64% (95% CI: 0.59, 0.69), and nine of the fourteen participants having validation accuracies significantly greater than random chance, as evidenced by their respective 95% confidence intervals. The best time-series domain model (TCN) also had 7-fold CV accuracy significantly greater than random chance at 56% (95% CI: 0.51, 0.61); however, only three of the fourteen participants had validation accuracies significantly greater than random chance. Additionally, the MLPNN had significantly greater CV model accuracy than the TCN model, as evidenced by the 95% confidence interval for the difference between the two classifiers not containing 0, i.e., model accuracy difference of 8% (95% CI: 0.01, 0.15), and the MLPNN also had significantly more participants with validation accuracies greater than random chance than the TCN model, as evidenced by the McNemar's test statistic of $4.5 \geq 3.84$ ($p < 0.034$, $\alpha = 0.05$). Two of the participants in the MLPNN model, 4 and 12, had validation accuracies greater than 80%. Participant IDs of this model that did not have validation accuracies significantly greater than random chance were participants 2, 3, 7, 8, and 11, with Participant 7 having the worst validation accuracy of 46%. Participant 7, however, was the participant with the highest validation accuracy for the TCN model, with Participant 12 being the second highest. Participants having such differing levels of performance across all three model types suggests that low model performance for the TCN and the TCN-AE was not due to certain individual participants having poor quality of data.

One reason for the significant difference between the MLPNN and TCN models could lie in their difference of domains, i.e., frequency vs. time. The literature suggests that changes in the average power of specific bands correlates to mental fatigue in sustained attention tasks [16], which also correlates to a vigilance decrement, and if these spectral features are the most salient information for mental fatigue, then there is no additional information gained by the network utilizing raw time-series signals versus spectral features. Furthermore, TCN performance is contingent on being able to learn that these spectral features are important given only the time-series signals, whereas these spectral features *are* the input for the MLPNN, so the MLPNN does not have to learn them. Thus, the MLPNN may have an advantage over the time-series domain models in that it could already have the most salient features to perform classification.

BIS measures of the 3-stimulus oddball task were investigated to determine if they correlated to model performance of the MLPNN model. If BIS measures are correlated to model performance, this would suggest that the magnitude of decline in a participant's task performance is correlated to how well the model can classify the EEG; i.e., the worse a decline in a participant's performance, the better the model can classify the EEG. Additionally, if the MLPNN model uses neural correlates of mental fatigue to perform classification, this would also suggest that as a participant becomes more mentally fatigued, they suffer a larger decline in task performance. To investigate if there was a correlation, BIS slopes of each participant, as well as the difference between the BIS measure of the first and last bins of each participant, were compared to the MLPNN model performance for that participant. These values are provided in Table 2. The BIS slopes and MLPNN validation accuracies were not found to be correlated ($\rho = 0.07$, $p = 0.82$) nor were the BIS difference values and MLPNN validation accuracies ($\rho = -0.09$, $p = 0.76$).

For the MLPNN model, as this is an artificial neural network, there is no way to know for certain if the model is utilizing neural correlates of mental fatigue to determine if there is a vigilance decrement. However, if the model is utilizing neural correlates of mental fatigue, the lack of correlation between BIS measures and model performance suggests that the magnitude of the mental fatigue does not correlate to the magnitude of the vigilance decrement or that the correlation is participant specific, i.e., some participants could be heavily fatigued and only suffer a slight decrease in performance, while some participants may have a significant decrease in performance when even moderately fatigued. In addition, the vigilance decrement is a measure of task performance, and thus, in general, factors other than fatigue can affect a person's performance, such as outside distractions, lack of motivation to perform well, etc. It is possible that, even in a lab environment, factors such as this affected participant performance, resulting in a large BIS slope or BIS difference for certain participants, yet with only minimal mental fatigue accumulation.

Table 2. This table provides the BIS slope and difference between the BIS measures of first and last bin for the oddball task for each participant. Validation accuracy for the frequency-domain MLPNN model is also provided for each participant. **Bold** signifies statistical significance of accuracy over random chance (defined as 50% for this binary classification task) as evidenced by the 95% confidence interval.

Participant #	BIS Slope	BIS Difference (1st Bin–4th Bin)	MLPNN Val Acc
0	−4.46	12.02	**0.69 (0.62, 0.76)**
1	−8.48	21.67	**0.78 (0.71, 0.83)**
2	−8.42	28.53	0.48 (0.41, 0.55)
3	−1.84	4.85	0.48 (0.41, 0.56)
4	−5.34	14.07	**0.83 (0.77, 0.88)**
5	−3.92	12.68	**0.68 (0.61, 0.74)**
6	−7.22	21.75	**0.71 (0.63, 0.77)**
7	−4.64	11.56	0.52 (0.45, 0.59)
8	−5.30	16.19	0.53 (0.46, 0.60)
9	−5.11	19.81	**0.67 (0.60, 0.74)**
10	−8.96	30.53	**0.63 (0.56, 0.69)**
11	−5.40	16.16	0.46 (0.38, 0.53)
12	−3.21	10.41	**0.84 (0.78, 0.89)**
13	−8.24	24.08	**0.63 (0.56, 0.70)**

The literature also notes that the neural correlates of mental fatigue and sleep fatigue manifest differently depending on the task and that they can be opposites of one another, yet both types of fatigue affect task performance in a similar manner. Given this, it could be that some of the participants accumulated sleep fatigue, as opposed to mental fatigue, as the task continued on, resulting in a decrease in performance but with neural correlates which differ from mental fatigue. As these neural correlates can be opposites of one another (e.g., an increase in spectral power for the alpha band as opposed to a decrease), it would be difficult for the model to generalize both of these types of fatigue.

Additional challenges associated with this work include that each of the vigilance tasks used in the datasets were visual tasks as opposed to other types of tasks (e.g., auditory). In order to further validate the model, data from vigilance tasks outside of the visual domain should be used to test the model. Additionally, in order to build a truly task-generic model, training data will likely be needed from these different task domains. To continue to properly validate the model with an unseen task, this would require at least two tasks worth of data from each of the different task domains (one for training and one for validation), requiring a diverse amount of data from many different experiments.

6. Conclusions and Future Work

In conclusion, the model type that was most capable of classifying the vigilance decrement in an unseen task and unseen participant out of the models examined was the MLPNN frequency-domain model, utilizing spectral features extracted from the EEG, namely the average power of the five traditional EEG frequency bands. This finding is significant as thus far, a task-generic EEG cross-participant model of the vigilance decrement, i.e., a model capable of classifying the vigilance decrement in an unseen task and unseen participants, has not been built or validated. Previous work by Yin et al. in building a task-generic model did not utilize a different type of task in order to validate their model and instead only varied parameters within a single task of operating an aircraft's cabin air management system in order to create two tasks. In contrast, the advantage with our research is that it utilized three different types of tasks (the air traffic controller task, the line task, and the 3-stimulus oddball task), all of which are well established in the literature as vigilance type tasks. Additionally, having utilized two tasks for training from two separate experiments as opposed to only one task is likely to provide additional generalization of the model.

To improve model performance, future work should incorporate more vigilance tasks for both training and testing as more EEG vigilance type datasets become available. CV should also be performed across all tasks to investigate if certain tasks provide more or less generalization and task invariance to the model.

Selection of specific spectral features, such as certain frequency bands, should also be explored. By selecting only certain frequency bands, and/or certain regions of the head, model performance could be improved, as currently, the model utilizes a large number of features (320), but only certain features or regions of the brain may be needed in order for the model to accurately classify the vigilance decrement, and removing these unnecessary features could reduce overfitting of the model. This feature importance of the neural network model could be determined through visualization techniques which allow for visual inspection of the model features which result in maximum discrimination between the two classes (vigilance decrement vs. not). Further investigation into mental fatigue vs. sleep fatigue could also be useful. Experiments which note the sleepiness of participants throughout the experiment, either through objective measurements such as prolonged eye closure, or through subjective measurements such as observation and surveys, could result in separate data for neural correlates of mental fatigue vs. sleep fatigue. These experiments could then be used for separate training and testing of the model, and this could reveal if incorporating both types of fatigue either aids the model or hinders it.

To further validate the model, future experiments should investigate devising tasks which result in an increase in vigilance. Currently, every participant experiences a vigilance decrement over the duration of each task, as the tasks are designed to do so. However, this presents a concern for model validation as the data is homogeneous across every participant. Ideally, for model training and validation, there would be data for both a vigilance decrement and a vigilance increase to ensure the model could differentiate between the two and to ensure the model is not classifying based solely on task duration. These tasks could perhaps be achieved through planned breaks throughout the task; however, these experiments would require further validation themselves to ensure they reliably produce an increase in vigilance.

Separate but related work which should stem from this research would be to use EEG to determine when an individual is dropping below a standard level of performance. The vigilance decrement is useful as it informs when someone is experiencing a decrease in performance; however, this decrease in performance is relative to the person's own baseline level of performance. In certain tasks, it would be valuable to predict when an individual's expected performance would be too low for successful task completion. This research has demonstrated that EEG can be utilized to determine whether or not someone is experiencing a vigilance decrement, even in an unseen task, and thus it is possible that a model could utilize a participant's baseline measure of performance to determine if that participant has dropped below a performance threshold; however, more work is necessary for proper implementation. Additionally, a regression model could be investigated to predict the measure of performance itself.

Lastly, EEG research into the vigilance decrement should overall move towards more multi-task experiments, task agnostic models, and dataset sharing. This research demonstrated that the neural correlates of the vigilance decrement span across different task types and can be utilized to detect the vigilance decrement across these task types; however, to further pinpoint which specific features span across all of the different types of vigilance tasks, additional experiments which utilize multiple tasks are needed. Dataset sharing through repositories such as Kaggle [70] or the UCI machine learning data repository [71] would also further enable future research into task agnostic models, as experiments with different tasks could be combined for model building and neural correlate analysis. Future experiments require time and funding; however, dataset sharing could quickly enable this research by utilizing existing datasets across many types of vigilance tasks (piloting of aircrafts, driving, air traffic control, etc.).

Author Contributions: A.K. is a PhD candidate performing this research as part of his dissertation research and is the investigator on the research as well as the primary author for the whole article. B.B. provided subject matter expertise on machine learning, and as A.K.'s as research advisor mentored this student research and assisted with editing the whole article. C.S.K. focused on the mathematical and theoretical mentorship and assisted with editing the whole article, with extra focus on Sections 4 and 5. M.M. provided human subject research expertise, with specific guidance around sustained attention type tasks and the vigilance decrement, and assisted with editing the whole article. All authors have read and agreed to the published version of the manuscript.

Funding: This research was sponsored and funded by James Lawton AFOSR/RTA, grant number F4FGA08305J006, "Information Acquisition Deficit Detection and Mitigation through Neurophysiological-sensed Operator Patterns".

Data Availability Statement: This work uses only de-identified existing data. A Human Research Protection Plan review determined this research to be not involving "human subjects" under US Common Rule (32 CFR 219) on 6 Jun 2020. The EEG datasets are owned by the United States Air Force Research Laboratory, 711th Human Performance Wing (HPW) and were collected through experiments performed by the 711th HPW in partnership with the University of Dayton. See [18,19] for details surrounding the experiment which produced the ATC and 3-stimulus Oddball Task dataset.

Acknowledgments: We would like to thank our sponsor, James Lawton AFOSR/RTA, for supporting this research. The views and conclusions contained in this document are those of the authors and should not be interpreted as representing the official policies, either expressed or implied, of the United States Air Force or the U.S. Government.

Conflicts of Interest: The authors declare no conflict of interest.

Abbreviations

The following abbreviations are used in this manuscript:

AE	Autoencoder
AI	Artificial Intelligence
ATC	Air Traffic Controller
BIS	Balanced Integration Score
CI	Confidence Interval
CNN	Convolutional Neural Network
CV	Cross-Validation
ECG	Electrocardiography
EEG	Electroencephalography
ELM	Extreme Learning Machine
EOG	Electrooculography
ERP	Event-Related Potential
FCN	Fully Connected Network
GRU	Gated Recurrent Unit
HEOG	Horizontal Electrooculography
HPW	Human Performance Wing
IES	Inverse Efficiency Score

L2PO-CV	Leave-Two-Participants-Out Cross-Validation
LSTM	Long Short-Term Memory
MLPNN	Multi-Layer Perceptron Neural Network
MSE	Mean Squared Error
PC	Proportion of Correct Responses
PSD	Power Spectral Density
PVT	Psychomotor Vigilance Test
RCS	Rate-Correct Score
ReLU	Rectified Linear Unit
RT	Response Time
TCN	Temporal Convolutional Network
TCN-AE	Temporal Convolutional Network Autoencoder
VEOG	Vertical Electrooculography

Appendix A

Preprocessing of EEG data was performed through script batch processing using EEGLAB [54] and consisted of a combination of best practice steps from both Makoto's preprocessing pipeline [55] and the PREP pipeline [56].

1. Modifed EEGLAB to use double precision, as single precision can destroy natural commutativity of the linear operations.
2. Imported data into EEGLAB and included reference channels based on the equipment used (e.g., Biosemi's 64 scalp electrode cap uses channels 65 and 66 as reference channels, which are electrodes placed on the mastoids specifically for the purpose of referencing).
3. Downsampled to 250 Hz for purpose of improving ICA decomposition by cutting off unnecessary high-frequency information and also to reduce data size.
4. High-pass filtered the data at 1 Hz to reduce baseline drift, to improve line-noise removal, and to improve ICA [72]. High-pass filter is done before line-noise removal and 1 Hz is used as we were not performing event-related potential (ERP) analysis, which could be affected by using a 1 Hz high-pass filter, and would require an alternate strategy.
5. Imported channel info using International 10–20 system to allow for re-referencing.
6. Removed line noise using CleanLine plugin (default 60Hz notch filter) [73].
7. Removed bad channels using EEGLAB clean_rawdata plugin patented by Christian Kothe [74], which utilizes Artifact Subspace Reconstruction.
8. Interpolated all removed channels to minimize a potential bias in the average referencing step.
9. Re-referenced data to the average. Mastoid referencing is not always sufficient [56], and re-referencing the data to the average helps suppress line noise that was not rejected by CleanLine [55].
10. Independent Component Analysis (ICA) "runica" "'infomax': (extended)" algorithm variant was executed with the vertical EOG (VEOG) electrode used as input for the function.
11. ICA results from Step 10 are used to remove artifact ICA components.

References

1. Lal, S.K.; Craig, A. A critical review of the psychophysiology of driver fatigue. *Biol. Psychol.* **2001**, *55*, 173–194. [CrossRef]
2. Ackerman, P.L.; Kanfer, R. Test length and cognitive fatigue: An empirical examination of effects on performance and test-taker reactions. *J. Exp. Psychol. Appl.* **2009**, *15*, 163. [CrossRef]
3. Parasuraman, R.; Warm, J.S.; Dember, W.N. Vigilance: Taxonomy and utility. In *Ergonomics and Human Factors*; Springer: New York, NY, USA, 1987; pp. 11–32.
4. Parasuraman, R.; Mouloua, M. *Automation and Human Performance: Theory and Applications*; Routledge: London, UK, 2018.
5. Warm, J.S.; Parasuraman, R.; Matthews, G. Vigilance requires hard mental work and is stressful. *Hum. Factors* **2008**, *50*, 433–441. [CrossRef] [PubMed]
6. Mackworth, N.H. The breakdown of vigilance during prolonged visual search. *Q. J. Exp. Psychol.* **1948**, *1*, 6–21. [CrossRef]

7. Sasahara, I.; Fujimura, N.; Nozawa, Y.; Furuhata, Y.; Sato, H. The effect of histidine on mental fatigue and cognitive performance in subjects with high fatigue and sleep disruption scores. *Physiol. Behav.* **2015**, *147*, 238–244. [CrossRef] [PubMed]
8. Smolders, K.C.; de Kort, Y.A. Bright light and mental fatigue: Effects on alertness, vitality, performance and physiological arousal. *J. Environ. Psychol.* **2014**, *39*, 77–91. [CrossRef]
9. Shigihara, Y.; Tanaka, M.; Ishii, A.; Tajima, S.; Kanai, E.; Funakura, M.; Watanabe, Y. Two different types of mental fatigue produce different styles of task performance. *Neurol. Psychiatry Brain Res.* **2013**, *19*, 5–11. [CrossRef]
10. Hogan, L.C.; Bell, M.; Olson, R. A preliminary investigation of the reinforcement function of signal detections in simulated baggage screening: Further support for the vigilance reinforcement hypothesis. *J. Organ. Behav. Manag.* **2009**, *29*, 6–18. [CrossRef]
11. Fisk, A.D.; Schneider, W. Control and automatic processing during tasks requiring sustained attention: A new approach to vigilance. *Hum. Factors* **1981**, *23*, 737–750. [CrossRef]
12. Ariga, A.; Lleras, A. Brief and rare mental "breaks" keep you focused: Deactivation and reactivation of task goals preempt vigilance decrements. *Cognition* **2011**, *118*, 439–443. [CrossRef]
13. Craig, A.; Tran, Y.; Wijesuriya, N.; Nguyen, H. Regional brain wave activity changes associated with fatigue. *Psychophysiology* **2012**, *49*, 574–582. [CrossRef] [PubMed]
14. Zhao, C.; Zhao, M.; Liu, J.; Zheng, C. Electroencephalogram and electrocardiograph assessment of mental fatigue in a driving simulator. *Accid. Anal. Prev.* **2012**, *45*, 83–90. [CrossRef] [PubMed]
15. Gao, Z.; Wang, X.; Yang, Y.; Mu, C.; Cai, Q.; Dang, W.; Zuo, S. EEG-based spatio–temporal convolutional neural network for driver fatigue evaluation. *IEEE Trans. Neural Netw. Learn. Syst.* **2019**, *30*, 2755–2763. [CrossRef]
16. Hu, X.; Lodewijks, G. Detecting fatigue in car drivers and aircraft pilots by using non-invasive measures: The value of differentiation of sleepiness and mental fatigue. *J. Saf. Res.* **2020**, *72*, 173–187. [CrossRef]
17. Yin, Z.; Zhang, J. Task-generic mental fatigue recognition based on neurophysiological signals and dynamical deep extreme learning machine. *Neurocomputing* **2018**, *283*, 266–281. [CrossRef]
18. Walsh, M.M.; Gunzelmann, G.; Anderson, J.R. Relationship of P3b single-trial latencies and response times in one, two, and three-stimulus oddball tasks. *Biol. Psychol.* **2017**, *123*, 47–61. [CrossRef] [PubMed]
19. Haubert, A.; Walsh, M.; Boyd, R.; Morris, M.; Wiedbusch, M.; Krusmark, M.; Gunzelmann, G. Relationship of event-related potentials to the vigilance decrement. *Front. Psychol.* **2018**, *9*, 237. [CrossRef]
20. Kamrud, A.; Borghetti, B.; Schubert Kabban, C. The Effects of Individual Differences, Non-Stationarity, and the Importance of Data Partitioning Decisions for Training and Testing of EEG Cross-Participant Models. *Sensors* **2021**, *21*, 3225. [CrossRef]
21. Parasuraman, R.; Davies, D. A taxonomic analysis of vigilance performance. In *Vigilance*; Springer: Boston, MA, USA, 1977; pp. 559–574.
22. See, J.E.; Howe, S.R.; Warm, J.S.; Dember, W.N. Meta-analysis of the sensitivity decrement in vigilance. *Psychol. Bull.* **1995**, *117*, 230. [CrossRef]
23. Oken, B.S.; Salinsky, M.C.; Elsas, S. Vigilance, alertness, or sustained attention: Physiological basis and measurement. *Clin. Neurophysiol.* **2006**, *117*, 1885–1901. [CrossRef]
24. Charbonnier, S.; Roy, R.N.; Bonnet, S.; Campagne, A. EEG index for control operators' mental fatigue monitoring using interactions between brain regions. *Expert Syst. Appl.* **2016**, *52*, 91–98. [CrossRef]
25. Gartenberg, D.; Veksler, B.Z.; Gunzelmann, G.; Trafton, J.G. An ACT-R process model of the signal duration phenomenon of vigilance. In *Proceedings of the Human Factors and Ergonomics Society Annual Meeting*; SAGE Publications Sage CA: Los Angeles, CA, USA, 2014; Volume 58, pp. 909–913
26. Desmond, P.; Matthews, G.; Bush, J. Sustained visual attention during simultaneous and successive vigilance tasks. In *Proceedings of the Human Factors and Ergonomics Society Annual Meeting*; SAGE Publications Sage CA: Los Angeles, CA, USA, 2001; Volume 45, pp. 1386–1389
27. Baker, C. Signal duration as a factor in vigilance tasks. *Science* **1963**, *141*, 1196–1197. [CrossRef]
28. Teo, G.; Szalma, J.L. The effects of task type and source complexity on vigilance performance, workload, and stress. In *Proceedings of the Human Factors and Ergonomics Society Annual Meeting*; SAGE Publications Sage CA: Los Angeles, CA, USA, 2011; Volume 55, pp. 1180–1184
29. Parasuraman, R.; Mouloua, M. Interaction of signal discriminability and task type in vigilance decrement. *Percept. Psychophys.* **1987**, *41*, 17–22. [CrossRef]
30. Lanzetta, T.M.; Dember, W.N.; Warm, J.S.; Berch, D.B. Effects of task type and stimulus heterogeneity on the event rate function in sustained attention. *Hum. Factors* **1987**, *29*, 625–633. [CrossRef] [PubMed]
31. Teichner, W.H. The detection of a simple visual signal as a function of time of watch. *Hum. Factors* **1974**, *16*, 339–352. [CrossRef] [PubMed]
32. Pachella, R.G. *The Interpretation of Reaction Time in Information Processing Research*; Technical Report; Michigan University Ann Arbor Human Performance Center: Ann Arbor, MI, USA, 1973.
33. Wickelgren, W.A. Speed-accuracy tradeoff and information processing dynamics. *Acta Psychol.* **1977**, *41*, 67–85. [CrossRef]
34. Townsend, J.T.; Ashby, F.G. Methods of modeling capacity in simple processing systems. In *Cognitive Theory*; Psychology Press: London, UK, 2014; pp. 211–252.
35. Woltz, D.J.; Was, C.A. Availability of related long-term memory during and after attention focus in working memory. *Mem. Cogn.* **2006**, *34*, 668–684. [CrossRef]

36. Liesefeld, H.R.; Janczyk, M. Combining speed and accuracy to control for speed-accuracy trade-offs (?). *Behav. Res. Methods* **2019**, *51*, 40–60. [CrossRef]
37. Mueller, S.T.; Alam, L.; Funke, G.J.; Linja, A.; Ibne Mamun, T.; Smith, S.L. Examining Methods for Combining Speed and Accuracy in a Go/No-Go Vigilance Task. In *Proceedings of the Human Factors and Ergonomics Society Annual Meeting*; SAGE Publications Sage CA: Los Angeles, CA, USA, 2020; Volume 64, pp. 1202–1206.
38. Gunzelmann, G.; Gross, J.B.; Gluck, K.A.; Dinges, D.F. Sleep deprivation and sustained attention performance: Integrating mathematical and cognitive modeling. *Cogn. Sci.* **2009**, *33*, 880–910. [CrossRef] [PubMed]
39. Caldwell, J.A.; Hall, K.K.; Erickson, B.S. EEG data collected from helicopter pilots in flight are sufficiently sensitive to detect increased fatigue from sleep deprivation. *Int. J. Aviat. Psychol.* **2002**, *12*, 19–32. [CrossRef]
40. Di Stasi, L.L.; Diaz-Piedra, C.; Suárez, J.; McCamy, M.B.; Martinez-Conde, S.; Roca-Dorda, J.; Catena, A. Task complexity modulates pilot electroencephalographic activity during real flights. *Psychophysiology* **2015**, *52*, 951–956. [CrossRef] [PubMed]
41. Cao, L.; Li, J.; Sun, Y.; Zhu, H.; Yan, C. EEG-based vigilance analysis by using fisher score and PCA algorithm. In Proceedings of the 2010 IEEE International Conference on Progress in Informatics and Computing, Shanghai, China, 10–12 December 2010; Volume 1, pp. 175–179.
42. Lin, C.T.; Wu, R.C.; Liang, S.F.; Chao, W.H.; Chen, Y.J.; Jung, T.P. EEG-based drowsiness estimation for safety driving using independent component analysis. *IEEE Trans. Circuits Syst. I Regul. Pap.* **2005**, *52*, 2726–2738.
43. Ji, H.; Li, J.; Cao, L.; Wang, D. A EEG-Based brain computer interface system towards applicable vigilance monitoring. In *Foundations of Intelligent Systems*; Springer: Berlin/Heidelberg, Germany, 2011; pp. 743–749
44. Borghini, G.; Vecchiato, G.; Toppi, J.; Astolfi, L.; Maglione, A.; Isabella, R.; Caltagirone, C.; Kong, W.; Wei, D.; Zhou, Z.; et al. Assessment of mental fatigue during car driving by using high resolution EEG activity and neurophysiologic indices. In Proceedings of the 2012 Annual International Conference of the IEEE Engineering in Medicine and Biology Society, San Diego, CA, USA, 28 August–1 September 2012; pp. 6442–6445.
45. Zhang, C.; Zheng, C.X.; Yu, X.L. Automatic recognition of cognitive fatigue from physiological indices by using wavelet packet transform and kernel learning algorithms. *Expert Syst. Appl.* **2009**, *36*, 4664–4671. [CrossRef]
46. Rosipal, R.; Peters, B.; Kecklund, G.; Åkerstedt, T.; Gruber, G.; Woertz, M.; Anderer, P.; Dorffner, G. EEG-based drivers' drowsiness monitoring using a hierarchical Gaussian mixture model. In Proceedings of the International Conference on Foundations of Augmented Cognition, Beijing, China, 22–27 July 2007; pp. 294–303
47. Al-Shargie, F.; Tariq, U.; Hassanin, O.; Mir, H.; Babiloni, F.; Al-Nashash, H. Brain connectivity analysis under semantic vigilance and enhanced mental states. *Brain Sci.* **2019**, *9*, 363. [CrossRef]
48. Al-Shargie, F.M.; Hassanin, O.; Tariq, U.; Al-Nashash, H. EEG-based semantic vigilance level classification using directed connectivity patterns and graph theory analysis. *IEEE Access* **2020**, *8*, 115941–115956. [CrossRef]
49. Hitchcock, E.M.; Warm, J.S.; Matthews, G.; Dember, W.N.; Shear, P.K.; Tripp, L.D.; Mayleben, D.W.; Parasuraman, R. Automation cueing modulates cerebral blood flow and vigilance in a simulated air traffic control task. *Theor. Issues Ergon. Sci.* **2003**, *4*, 89–112. [CrossRef]
50. Dinges, D.F.; Powell, J.W. Microcomputer analyses of performance on a portable, simple visual RT task during sustained operations. *Behav. Res. Methods Instrum. Comput.* **1985**, *17*, 652–655. [CrossRef]
51. Comerchero, M.D.; Polich, J. P3a and P3b from typical auditory and visual stimuli. *Clin. Neurophysiol.* **1999**, *110*, 24–30. [CrossRef]
52. Brainard, D.H. The psychophysics toolbox. *Spat. Vis.* **1997**, *10*, 433–436. [CrossRef]
53. Hitchcock, E.M.; Dember, W.N.; Warm, J.S.; Moroney, B.W.; See, J.E. Effects of cueing and knowledge of results on workload and boredom in sustained attention. *Hum. Factors* **1999**, *41*, 365–372. [CrossRef]
54. Delorme, A.; Makeig, S. EEGLAB: An open source toolbox for analysis of single-trial EEG dynamics including independent component analysis. *J. Neurosci. Methods* **2004**, *134*, 9–21. [CrossRef] [PubMed]
55. Miyakoshi, M. Makoto's Preprocessing Pipeline. 2020 Available online: https://sccn.ucsd.edu/wiki/Makoto's_preprocessing_pipeline (accessed on 1 May 2020).
56. Bigdely-Shamlo, N.; Mullen, T.; Kothe, C.; Su, K.M.; Robbins, K.A. The PREP pipeline: Standardized preprocessing for large-scale EEG analysis. *Front. Neuroinform.* **2015**, *9*, 16. [CrossRef] [PubMed]
57. Li, X.; Zhao, Z.; Song, D.; Zhang, Y.; Pan, J.; Wu, L.; Huo, J.; Niu, C.; Wang, D. Latent Factor Decoding of Multi-Channel EEG for Emotion Recognition Through Autoencoder-Like Neural Networks. *Front. Neurosci.* **2020**, *14*, 87. [CrossRef] [PubMed]
58. Prabhudesai, K.S.; Collins, L.M.; Mainsah, B.O. Automated feature learning using deep convolutional auto-encoder neural network for clustering electroencephalograms into sleep stages. In Proceedings of the 2019 9th International IEEE/EMBS Conference on Neural Engineering (NER), San Francisco, CA, USA, 20–23 March 2019; pp. 937–940.
59. Ayata, D.; Yaslan, Y.; Kamasak, M. Multi channel brain EEG signals based emotional arousal classification with unsupervised feature learning using autoencoders. In Proceedings of the 2017 25th Signal Processing and Communications Applications Conference (SIU), Antalya, Turkey, 15–18 May 2017; pp. 1–4.
60. Bai, S.; Kolter, J.Z.; Koltun, V. An empirical evaluation of generic convolutional and recurrent networks for sequence modeling. *arXiv* **2018**, arXiv:1803.01271.
61. Oord, A.v.d.; Dieleman, S.; Zen, H.; Simonyan, K.; Vinyals, O.; Graves, A.; Kalchbrenner, N.; Senior, A.; Kavukcuoglu, K. Wavenet: A generative model for raw audio. *arXiv* **2016**, arXiv:1609.03499.

62. Remy, P. Temporal Convolutional Networks for Keras. 2020. Available online: https://github.com/philipperemy/keras-tcn (accessed on 1 April 2021)
63. Goodfellow, I.; Bengio, Y.; Courville, A. *Deep Learning*; MIT Press: Cambridge, MA, USA, 2016.
64. Géron, A. *Hands-On Machine Learning with Scikit-Learn, Keras, and TensorFlow: Concepts, Tools, and Techniques to Build Intelligent Systems*; O'Reilly Media: Sebastopol, CA, USA, 2019.
65. Dertat, A. Applied Deep Learning-Part 3: Autoencoders. 2017. Available online: https://towardsdatascience.com/applied-deep-learning-part-3-autoencoders-1c083af4d798 (accessed on 1 April 2021).
66. Cohen, M.X. *Analyzing Neural Time Series Data: Theory and Practice*; MIT Press: Cambridge, MA, USA, 2014.
67. Kingma, D.P.; Ba, J. Adam: A method for stochastic optimization. *arXiv* **2014**, arXiv:1412.6980.
68. Thill, M.; Konen, W.; Bäck, T. Time Series Encodings with Temporal Convolutional Networks. In Proceedings of the International Conference on Bioinspired Methods and Their Applications, Brussels, Belgium, 19–20 November 2020; Springer: Berlin/Heidelberg, Germany, 2020; pp. 161–173
69. Agresti, A.; Coull, B.A. Approximate is better than "exact" for interval estimation of binomial proportions. *Am. Stat.* **1998**, *52*, 119–126.
70. Alphabet Inc. Kaggle. 2017. Available online: https://www.kaggle.com/datasets (accessed on 1 February 2021)
71. Dua, D.; Graff, C. UCI Machine Learning Repository. 2017. Available online: https://archive.ics.uci.edu/ml (accessed on 1 February 2021).
72. Winkler, I.; Debener, S.; Müller, K.R.; Tangermann, M. On the influence of high-pass filtering on ICA-based artifact reduction in EEG-ERP. In Proceedings of the 2015 37th Annual International Conference of the IEEE Engineering in Medicine and Biology Society (EMBC), Milan, Italy, 25–29 August 2015; pp. 4101–4105.
73. Mullen, T. CleanLine EEGLAB plugin. 2012. Available online: https://www.nitrc.org/projects/cleanline/ (accessed on 1 May 2020).
74. Kothe, C.A.E.; Jung, T.P. Artifact Removal Techniques with Signal Reconstruction; U.S. Patent Application No. 14/895,440, 28 April 2016.

Article

Automatic Speech Discrimination Assessment Methods Based on Event-Related Potentials (ERP)

Pimwipa Charuthamrong [1], Pasin Israsena [2], Solaphat Hemrungrojn [3] and Setha Pan-ngum [4,*

[1] Interdisciplinary Program of Biomedical Engineering, Faculty of Engineering, Chulalongkorn University, Pathumwan, Bangkok 10330, Thailand; pimwipa.c@alumni.chula.ac.th

[2] National Electronics and Computer Technology Center, 112 Thailand Science Park, Klong Luang, Pathumthani 12120, Thailand; pasin.israsena@nectec.or.th

[3] Department of Psychiatry, Faculty of Medicine, Chulalongkorn University, Pathumwan, Bangkok 10330, Thailand; solaphat.h@chula.ac.th

[4] Department of Computer Engineering, Faculty of Engineering, Chulalongkorn University, Pathumwan, Bangkok 10330, Thailand

* Correspondence: setha.p@chula.ac.th

Abstract: Speech discrimination is used by audiologists in diagnosing and determining treatment for hearing loss patients. Usually, assessing speech discrimination requires subjective responses. Using electroencephalography (EEG), a method that is based on event-related potentials (ERPs), could provide objective speech discrimination. In this work we proposed a visual-ERP-based method to assess speech discrimination using pictures that represent word meaning. The proposed method was implemented with three strategies, each with different number of pictures and test sequences. Machine learning was adopted to classify between the task conditions based on features that were extracted from EEG signals. The results from the proposed method were compared to that of a similar visual-ERP-based method using letters and a method that is based on the auditory mismatch negativity (MMN) component. The P3 component and the late positive potential (LPP) component were observed in the two visual-ERP-based methods while MMN was observed during the MMN-based method. A total of two out of three strategies of the proposed method, along with the MMN-based method, achieved approximately 80% average classification accuracy by a combination of support vector machine (SVM) and common spatial pattern (CSP). Potentially, these methods could serve as a pre-screening tool to make speech discrimination assessment more accessible, particularly in areas with a shortage of audiologists.

Keywords: EEG; ERP; speech discrimination; classifier

Citation: Charuthamrong, P.; Israsena, P.; Hemrungrojn, S.; Pan-ngum, S. Automatic Speech Discrimination Assessment Methods Based on Event-Related Potentials (ERP). *Sensors* **2022**, *22*, 2702. https://doi.org/10.3390/s22072702

Academic Editor: Yvonne Tran

Received: 8 February 2022
Accepted: 22 March 2022
Published: 1 April 2022

Publisher's Note: MDPI stays neutral with regard to jurisdictional claims in published maps and institutional affiliations.

1. Introduction

Pure-tone audiometry (PTA) and speech audiometry are routinely used in a clinical setting to assess auditory function [1]. PTA measures the minimum threshold level that can be heard by the user at different frequencies. As PTA only evaluates the absolute hearing threshold but not the ability to recognize speech, speech audiometry is used as a complement to PTA in order to measure different aspects of a patient's auditory function altogether. Speech audiometry commonly includes three speech tests: speech-detection threshold (SDT), speech reception threshold (SRT), and speech discrimination. SDT measures the threshold at which a patient can detect the presence of speech 50% of the time. SRT represents the threshold at which a patient can repeat 50% of the speech. Both SDT and SRT can be determined in a similar way to PTA but use speech instead of pure-tone sounds. Speech discrimination is more complex to determine. A commonly used method to assess speech discrimination includes presenting monosyllabic words at 50 dB above SRT and measure the percentage of correctly repeated words [2]. Speech discrimination scores, along with results from other tests, are used in diagnosing and determining treatment for hearing loss patients [2]. However, the behavioral assessment to test speech discrimination requires subjective responses which makes the assessment more difficult in some cases such as difficult-to-test patients or children. An electrophysiological method to assess speech discrimination provides an objective assessment and would be suitable in these situations.

Electroencephalography (EEG) measures the electrical potentials at the scalp which reflect the brain's electrical activity [3]. An EEG test usually includes placing electrodes on the scalp. The placement of the electrodes usually follows the 10–20 system [4] which indicates the different positions of the scalp using a combination of letters and numbers. For higher resolution, the ten percent electrode system [5] or the five percent electrode system [6] which can accommodate a larger number of electrodes could also be used. EEG signals have high temporal resolution compared to other modalities (e.g., functional magnetic resonance imaging or functional near infrared spectroscopy) and directly reflects neural activity [7]. For these reasons, EEG is a well-suited technique to study cognition as it can capture the rapid dynamics of cognitive processes which happen in the order of milliseconds. To analyze EEG signals, raw signals are often grouped into bands including delta (2–4 Hz), theta (4–8 Hz), (8–12 Hz), beta (15–30 Hz), and gamma (30–150 Hz) [7]. EEG changes that are triggered by specific events or stimuli are termed event-related potentials (ERPs). ERPs can be related to visual, auditory, or somatosensory stimuli [8]. ERP waveforms are obtained from averaging across many trials to extract the response that is related to the stimuli. The observed ERP waveforms that are recorded from the scalp usually represent the sum of multiple ERP components [9].

A commonly studied ERP component that is related to speech discrimination is the mismatch negativity (MMN) [10–12]. MMN is an ERP component that is elicited by a deviant stimulus that violates the representation of the standard stimulus that is formed by repetition [11]. MMN is conventionally studied using an oddball paradigm which usually involves a sequence of repeated standard stimuli, which occurs in most of the trials, sporadically interrupted by a deviant stimulus. This component has been used in various work that is related to auditory discrimination and speech discrimination [13–18]. The presence of MMN in response to the deviant stimuli in an oddball task indicates that the user can distinguish between the particular deviant and standard sounds that are used in the task. As MMN can be elicited even when users are not paying attention to the sound, an MMN-based method to assess speech discrimination is well-suited for a variety of participants, particularly those who have difficulties following instructions. However, an oddball task that is used to elicit MMN typically uses only one pair of sounds. Word lists that are used to test speech discrimination in Thai usually consist of 20–50 words [19–21]. For example, the word list that was proposed by Yimtae et al. [21] contained 24 words, the word list that was used by Visessumon [20] had 21 words, and the word list that was proposed by Hemakom et al. [19] comprised of 45 words. To test speech discrimination using an MMN-based method, the participants would be required to undergo many rounds of oddball tasks with different standard and deviant stimuli which would be time-consuming and repetitive. These limitations have led to various improvements and alternatives being proposed such as variations to the oddball paradigm [15,22], different analysis approaches [23,24], or an alternative method or marker [25,26].

Morikawa et al. [26] proposed an alternative speech discrimination assessment method that utilized visual ERP that was induced by visual stimuli that was associated with an auditory stimuli instead of auditory ERP. In their method, the participants listened to a sound corresponding to a Japanese letter and then were shown a picture of a Japanese letter. In 50% of the trials, the picture matched the letter that was presented. In the remaining trials, the picture matched another letter that had similar pronunciation and was often confused with the former letter. The participants were required to answer whether the picture and the sound matched. The study only included participants with normal hearing but some of the auditory stimuli that were used were manipulated to imitate hearing loss. They found that when the participants answered that the sound and picture matched (match condition), an ERP component called the P3 [27,28] was elicited at 290–400 ms after the visual stimulus onset. P3 was usually observed during stimulus discrimination and was theorized to be related to the brain activities that updated a mental model when a new stimulus was detected. When the participants answered that the sound and picture did not match (mismatch condition), a late positive potential (LPP) [29] was elicited at 480–570 ms after visual stimulus onset. LPP was assumed to be similar to the P3b component [28,30] and was elicited when there was a mismatch between the expectation and feedback. To classify between the match and mismatch conditions, Morikawa et al. [26] calculated a feature value from the average amplitude difference between the intervals where P3 and LPP were observed. This feature value was compared to a predetermined threshold value to separate between cases with successful and unsuccessful discrimination. They reported achieving 70.5% accuracy when one trial was used and more than 80% accuracy when four or more trials were averaged together. The promising result suggested that the proposed method might be a viable alternative to an MMN-based method. However, this method was only applicable to participants who can identify the Japanese letters. Therefore, the method might not be accessible to young children or illiterate people.

Currently, there is no widely used method to automatically assess speech discrimination. An ERP-based method provides an objective assessment of speech discrimination and could make the test more accessible, particularly in areas where there is a shortage of audiologists. An MMN-based method that is employed in many studies would be well-suited for assessing speech discrimination in children or patients who have difficulties with the behavioral test. However, as mentioned, this method might be time-consuming as many rounds of oddball task are needed in order to get an assessment that covers all the meaningful contrasts in a language. A possible alternative to an MMN-based method is a visual-ERP-based method that was proposed by Morikawa et al. [26] which was reported to have high accuracy. However, the use of Japanese letters limited the use of this method to literate patients only. Furthermore, there had not been an accuracy level obtained from an MMN-based method for comparison. Our research proposed a modification to the visual-ERP-based method by using pictures representing word meaning to make the method more accessible. This modified method was compared to the original visual-ERP-based method and an MMN-based method. In each method, machine learning techniques were employed to separate between the two different conditions. We hoped to recommend a suitable method that is based on ERP components, discrimination accuracy, and time efficiency. An overall framework combining a suitable ERP method and a classification technique into an automatic speech discrimination assessment system was also proposed.

2. Materials and Methods

2.1. Participants

A total of 30 native Thai volunteers participated in the research (mean age = 31 years, age range = 19–43 years, 13 males, 17 females). All of them had self-reported normal hearing and normal or corrected-to-normal vision. All the participants provided informed consent before beginning the research. The Research Ethics Review Committee for Research Involving Human at Chulalongkorn University approved the research protocol (protocol no. 171.1/63).

2.2. Procedure and Stimuli

There were two Thai words that were used in the experiment -/kài/(meaning chicken) and /kʰài/(meaning egg). Both words are basic nouns that are used in everyday life. The two words have the same vowel and tone but have different consonants. As Thai is a tonal language, tones are essential in distinguishing between words [31]. In this work, we aimed to test the ability to differentiate between words with consonant contrasts. Thus, the vowel and tone were the same for both words. The sounds were recorded by a native Thai woman and were obtained from The Thai Alphabets multimedia exhibit [32]. Each word had a duration of 500 ms with average sound pressure levels of approximately 60 dBSPL and 40 dBSPL as measured using a sound level meter. The stimuli were presented using PsychoPy [33].

The experiment consisted of three methods. From here on, the modified visual-ERP-based method that uses pictures of word meaning will be referred to as Method 1. The original visual-ERP-based method that was proposed by Morikawa et al. (2012) will be referred to as Method 2. The MMN-based method that only utilizes auditory stimuli will be referred to as Method 3.

The results from a pilot experiment suggested that using multiple pictures led to higher accuracy. However, using multiple pictures would also increase the experiment time considerably. Thus, we decided to further separate the picture method (Method 1) into three strategies. We aimed to compare between these strategies to find the optimal strategy in terms of both accuracy and time-efficiency. The three strategies included single-picture, multiple-pictures, and single-picture-with-expectation. We will call these Methods 1a, 1b, and 1c, respectively. The participants were randomly separated into three groups, with ten participants per group. Each group undertook either Method 1a, 1b, or 1c along with Method 2 and 3. The presentation order of the methods was counterbalanced across the participants. A laptop monitor and speaker were used to present the visual and auditory stimuli. The participants were seated approximately 0.3 m from the laptop and instructed to sit still to minimize movement artifacts.

Figure 1 shows the sequence of each trial in Method 1a, 1b, and 1c. For Method 1a, each trial started by playing a word through the laptop speaker for 500 ms. Then, a picture was shown for 1000 ms. There were 80 trials in this method, with the picture matching the meaning of the word in 40 trials. In the remaining trials, the picture did not match the meaning of the word. The participants were required to press '1' on a standard keyboard if they think the picture matched the word they heard and press '2' otherwise.

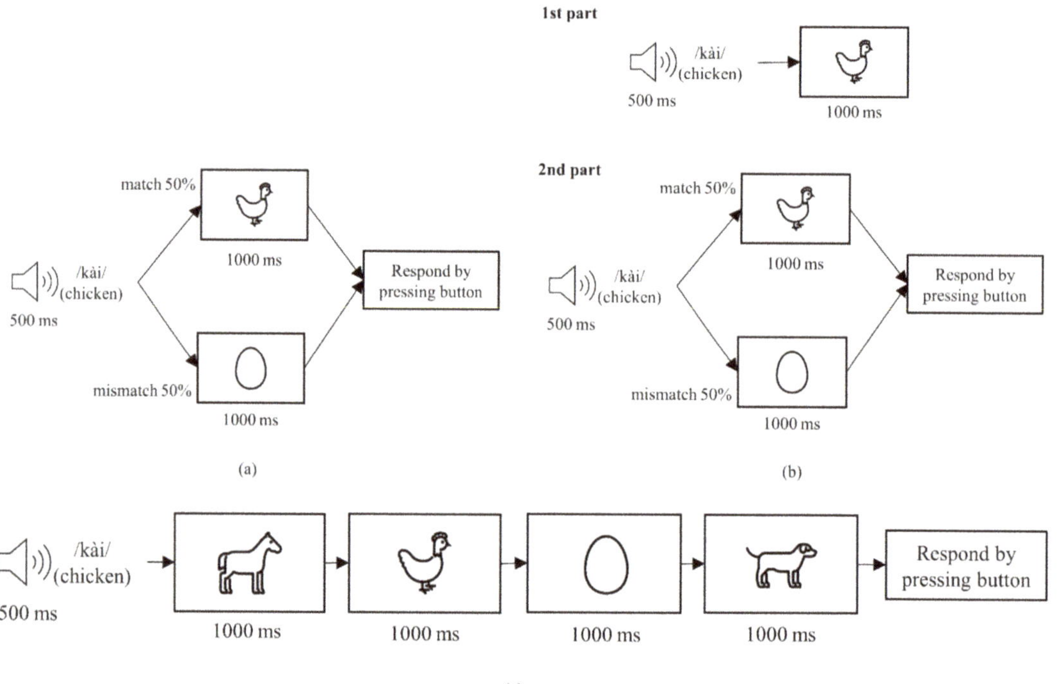

Figure 1. Event sequence for each trial in each strategy in Method 1 including (**a**) Method 1a; (**b**) Method 1c; (**c**) Method 1b.

In Method 1b, the participants listened to a 500 ms long word then were shown a sequence of four pictures. The pictures were shown one-by-one for 1000 ms each. In each trial, one picture matched the meaning of the word that was played. The order of the correct picture was randomized, appearing as the first (or second, third, or fourth) picture in one fourth of the trials (12 trials out of 48 total trials). After seeing all four pictures, the participants were required to press '1', '2', '3', or '4' on the keyboard according to the position of the correct picture.

Method 1c consisted of two parts. In the first part, a 500 ms long word was played before a picture corresponding to the meaning of the word was shown for 1000 ms. This sequence was repeated 10 times for each word in a random order. This first part was added to show the participants the correct picture to expect for each word. The second part was identical to Method 1a.

Figure 2 shows the sequence of each trial in Methods 2 and 3. Method 2 was slightly adjusted from the method that was proposed by Morikawa et al. [26]. Each trial started by playing a word through the laptop speaker for 500 ms then showing a picture for 1000 ms. In 40 out of 80 trials, the picture was of the word that was played, spelled out using a standard font. In the remaining trials, the picture was of another word. The participants were required to press '1' on the keyboard if they think the picture matched the word they heard and press '2' otherwise.

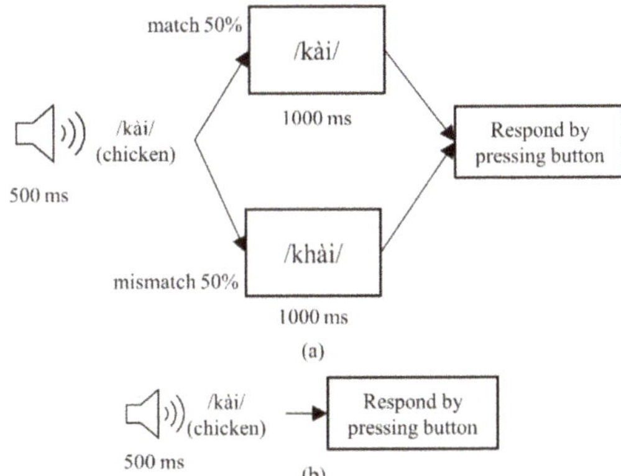

Figure 2. Event sequence for each trial in (**a**) Method 2; (**b**) Method 3.

Method 3 was an active oddball paradigm. In each trial, a word was played through the laptop speaker for 500 ms. On 80% of the trials (120 trials), the standard stimulus /kài/ was played. On the remaining 20% of the trials (30 trials), the deviant stimulus /k^{h}ài/ was played. The participants were required to press '1' on the keyboard when they heard the standard stimulus /kài/ and press '2' when they heard the deviant stimulus /k^{h}ài/.

2.3. EEG Recording and Analysis

Figure 3 shows the overall EEG recording system. EEG was recorded from eight positions (Fz, Cz, C3, C4, Pz, P3, P4, and Oz) according to the 10–20 system [4] using a g.SAHARA headset and a g.MOBIlab+ amplifier (g.tec medical engineering, Schiedlberg, Austria). The positions were chosen based on the scalp distribution of the ERP components that we expected to see. These included the MMN, P3, and LPP components. MMN had a frontocentral distribution and is prominent at Fz or Cz. P3 and LPP were reported to be most prominent at the parietal area. The signals were recorded with a sampling rate of 256 Hz.

The signal acquisition software, OpenViBE 3.0.0 [34], received the EEG signals via Bluetooth and combined the signals with the stimulus event data from the stimulus presentation software, PsychoPy 2020.1.2 [33], before outputting the EEG data with an event marker to be analyzed further. The EEG data was filtered with a bandpass filter. For stability and robustness, the filter used was an 847 point FIR filter with 1–40 Hz passband. It had transition frequencies of 1 Hz, −6 dB corner frequencies at 0.5 Hz and 40.5 Hz, and 50 dB attenuation at stopband. Then, it was passed through an EEGLAB [35] plugin function, clean_rawdata [36] to remove artifacts. The clean_rawdata function cleaned the EEG signals using the artifact subspace reconstruction (ASR) method [37]. The function rejected and reconstructed artifacts with variance more than 30 standard deviations (the cutoff parameter k = 30 was chosen according to [38]) away from the clean portions of the data. After that, the data were re-referenced to the common average reference and were extracted into epochs. For Method 1a, 1b, 1c, and 2, the interval from 0 to 900 ms after visual stimulus onset was extracted. For Method 3 the interval from 0 to 400 ms after stimulus onset was extracted. The −100 to 0 ms interval was used as a baseline. After removing the artifacts and extracting epochs, the data were manually inspected. Datasets that had less than half the original number of trials for each condition were excluded from further analysis.

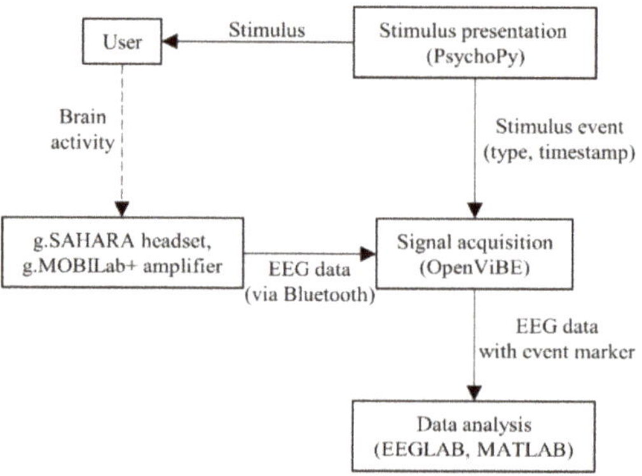

Figure 3. Overall EEG recording system [39].

Table 1 shows the amount of remaining data after preprocessing. For Method 1a, data from 9 subjects remained, including 328 epochs in the match condition and 320 epochs in the mismatch condition. For Method 1b, data from 10 subjects remained, including 389 epochs in the match condition and 1193 epochs in the mismatch condition. For Method 1c, data from 7 subjects remained, including 312 epochs in the match condition and 260 epochs in the mismatch condition. For Method 2, data from 25 subjects remained, including 881 epochs in the match condition and 879 epochs in the mismatch condition. For Method 3, data from 26 subjects remained, including 2795 epochs in the standard condition and 706 epochs in the deviant condition.

Table 1. Remaining data after preprocessing.

Method	Remaining Participants	Remaining Epochs
1a	9	Match: 328 Mismatch: 320
1b	10	Match: 389 Mismatch: 1193
1c	7	Match: 312 Mismatch: 260
2	25	Match: 881 Mismatch: 879
3	26	Standard: 2795 Deviant: 706

Figure 4 shows the overall EEG data processing. The EEG data were averaged according to match/mismatch or deviant/standard condition for each participant. The difference in waveforms between conditions were investigated by applying a paired *t*-test to the averaged EEG data for each participant in order to account for inter-subject variability. As beta and alpha waves were reported to be associated with attention [40], an average beta/alpha ratio was used to investigate the level of attention that was paid to each method. For each epoch, a beta/alpha ratio was calculated irrespective of the condition. Then, an average beta/alpha ratio was calculated for each participant and method. The Kruskal–Wallis test was applied to investigate the difference between beta/alpha ratios of each method. Post hoc analyses using Bonferroni correction for multiple comparisons were conducted when appropriate.

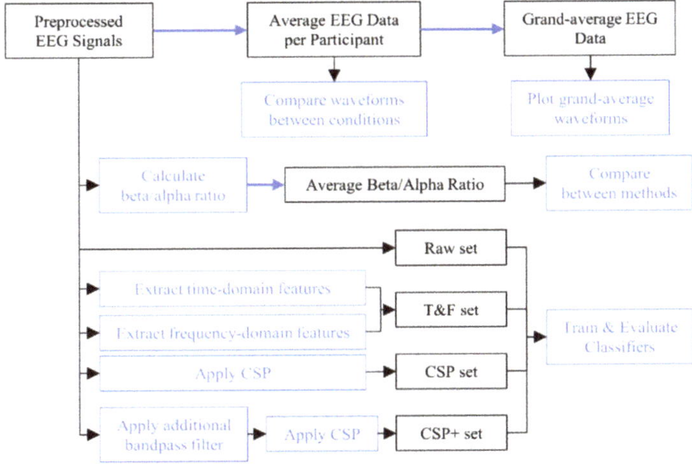

Figure 4. Overall EEG data processing. The black outlined boxes represent data while blue outlined boxes represent processes. Bold blue arrows represent averaging.

To classify between the conditions, several features were extracted from each epoch of EEG data. There were three types of features that were utilized including raw features, time-domain features, and frequency-domain features. The raw features were taken directly from the preprocessed data of each channel. The time-domain and frequency-domain features were extracted from two intervals in each trial. These intervals were 100–250 ms and 250–400 ms after stimulus onset for Method 3. For other methods, these intervals were 200–400 ms and 500–800 ms after visual stimulus onset. The intervals were chosen based on the result of the paired *t*-test and the expected ERP components for each method. The time-domain features included mean amplitude, variance, peak amplitude (maximum absolute amplitude), peak latency, maximal peak/amplitude ratio (MP ratio), positive area, and negative area. The frequency-domain features consisted of power in six frequency bands including Delta (1–4 Hz), Theta (4–8 Hz), Alpha (8–13 Hz), Beta (13–30 Hz), Gamma (30–40 Hz), and total band (1–40 Hz). A common spatial pattern (CSP) was applied to extract the CSP features. The extracted features were combined into four feature sets including raw features (Raw), time- and frequency-domain features (T&F), CSP features (CSP), and CSP features that were obtained after applying an additional bandpass filter (CSP+). The additional filter was a 16th order IIR filter with 1–30 Hz passband. It had 0.1 dB passband ripple and 60 dB attenuation at stopband. Each feature set was used to train and evaluate classifiers.

Figure 5 shows the process of training and evaluating classifiers in each fold of the 5-fold cross-validation framework. The dataset was separated into a training set (containing 80% of the data) and a test set (containing 20% of the data) with both sets having the same proportion of match/mismatch or standard/deviant conditions. Then, the features were extracted for both sets as described in the previous paragraph. For the CSP or CSP+ feature set, the training set was used to learn the CSP matrix, which was then applied to both the training set and test set. No data from the test set was used while learning the CSP matrix to prevent data leakage. After that, for the training set, the minority class (condition with lower number of epoch) was oversampled using random oversampling to achieve a 1:1 proportion between the conditions in the training data. Then, the training data were used to train either a linear discriminant analysis (LDA) or a support vector machine (SVM) classifier to classify whether the epoch was from a match or mismatch (or, for Method 3, standard or deviant) condition. The classifier was later evaluated using the test data and the results were recorded. The results from each fold were averaged and the classification accuracy was calculated from the number of correctly classified trials over the total number of trials.

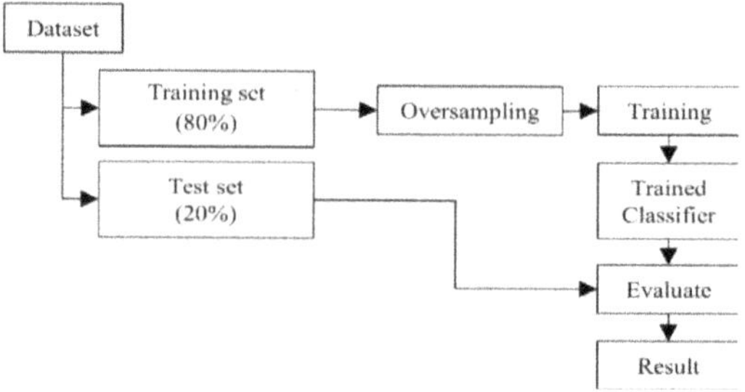

Figure 5. Process of training and evaluating the classifiers in each fold of the 5-fold cross-validation framework.

The average classification accuracies across the participants for each method and feature set were calculated and compared. The comparison was done using the Kruskal–Wallis test. An analysis of variance (ANOVA) was used to compare the average accuracy between each feature set. Additionally, the classification accuracy that was obtained from the SVM and LDA classifiers trained using the same feature set were compared using a paired *t*-test. Post hoc analyses using Bonferroni correction for multiple comparisons were conducted when appropriate.

3. Results

3.1. Behavioral Measures

Table 2 shows the average behavioral accuracy and response time for each method. The participants answered correctly most of the time in every method (98.47%, 98.54%, 99.29%, 98.55%, and 98.77% of the cases for Method 1a, 1b, 1c, 2, and 3 respectively). The participants rarely failed to answer correctly (less than 2% of the cases in every method). The average response time for each method was 0.672 s, 0.645 s, 0.473 s, 0.518 s, and 0.697 s, respectively. A one-way ANOVA showed that there was a significant difference in the average response time between at least two groups (F(4,72) = 3.68, p = 0.0088). Post hoc analyses using Bonferroni correction indicated a significant difference in the average response time between Method 2 and 3 (p = 0.0202).

Table 2. Mean (standard error) behavioral accuracy and response time for each method.

Method	Average Correct Answers (%)	Average Response Time (s)
1a	98.47 (0.62)	0.672 (0.12)
1b	98.54 (0.62)	0.645 (0.05)
1c	99.29 (0.37)	0.473 (0.03)
2	98.55 (0.44)	0.518 (0.03)
3	98.77 (0.48)	0.697 (0.04)

3.2. ERP Waveforms

Figure 6 shows the grand-average waveforms with standard error at Pz for each method. In Method 1a, a positive wave was observed at approximately 450–800 ms after stimulus onset in the mismatch condition. In Method 1b, no clear difference between the match and mismatch conditions were observed. We observed two intervals with difference between conditions in Method 1c. Positive waves were observed at approximately 200–400 ms and 500–750 ms after stimulus onset in the mismatch condition. In Method 2, a positive wave was elicited at approximately 500–900 ms after stimulus onset in the mismatch condition. In Method 3, a negative wave was observed at approximately 100–300 ms after stimulus onset in the deviant condition.

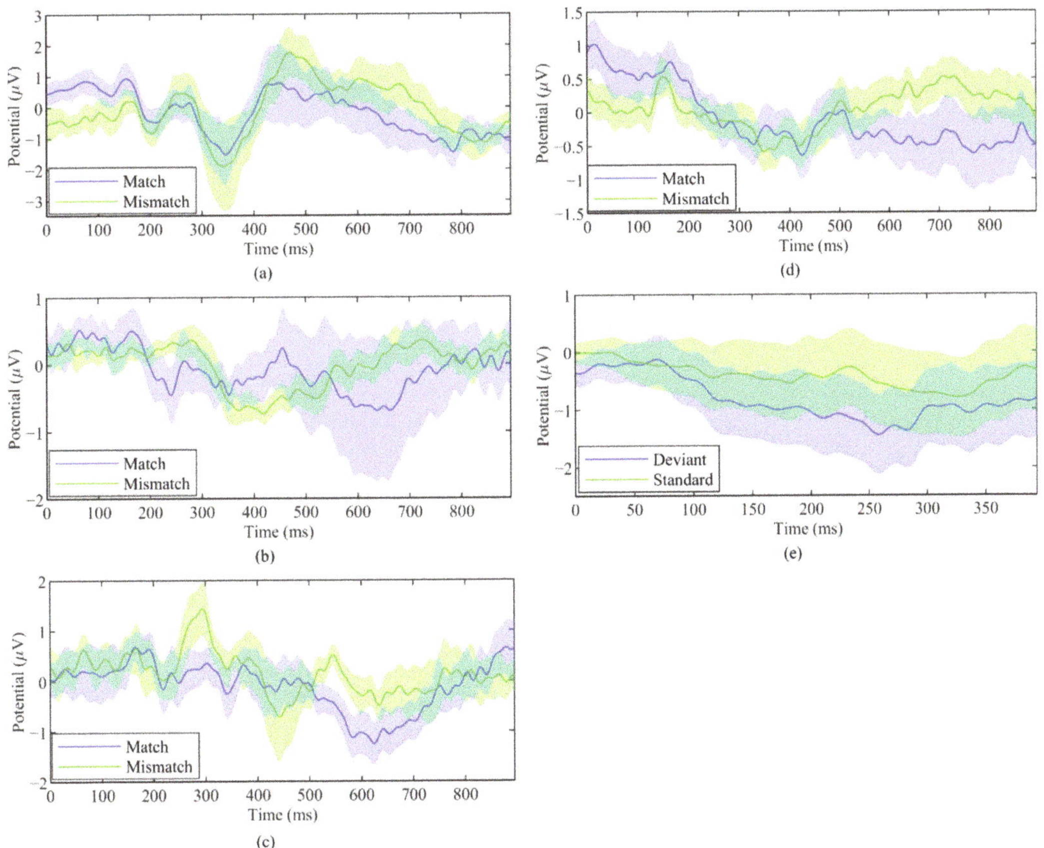

Figure 6. Grand-average waveforms with standard error at Pz (**a**) Method 1a; (**b**) Method 1b; (**c**) Method 1c; (**d**) Method 2; (**e**) Method 3.

Despite having no clear difference between waveforms of match and mismatch condition at Pz, Method 1b had a more prominent difference at Fz. As illustrated in Figure 7, a positive wave was observed at approximately 200–400 ms after stimulus onset in the match condition. In the mismatch condition, a positive wave was elicited at approximately 400–800 ms after stimulus onset.

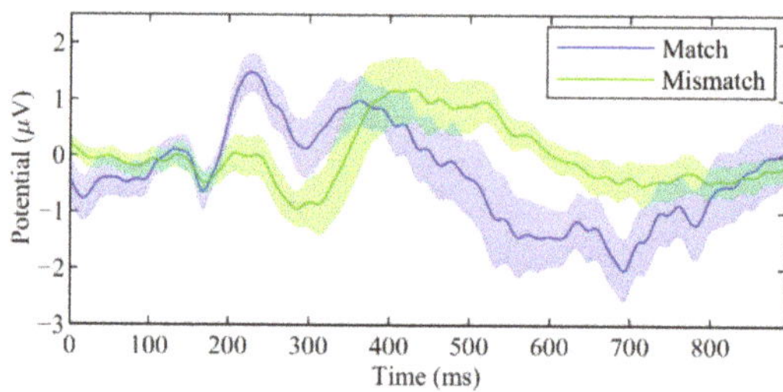

Figure 7. Grand-average waveforms with standard error at Fz in method 1b.

Table 3 shows the intervals with a significant difference, as indicated by a paired *t*-test, between the waveforms at Pz and Fz for each method. For Method 1a, a significant difference was found between the waveforms of the match and mismatch conditions during 23–78 ms, 141–145 ms, 160–168 ms, 488–512 ms, and 617–703 ms after stimulus onset at Pz. For Method 1b, a significant difference was found during 430–438 ms at Pz. However, at Fz, a significant difference was found during 195–281 ms, 309–336 ms, 492–609 ms, 617–625 ms, and 672–719 ms. For Method 1c, a significant difference was found during 535–555 ms at Pz and during 90–98 ms, 316–340 ms, 348–363 ms, 375–387 ms, 539–602 ms, 621–691 ms, and 699–711 ms at Fz. For Method 2, a significant difference was found during 4–39 ms, 59–90 ms, 191–203 ms, 633–762 ms, and 809–848 ms after stimulus onset at Pz. For Method 3, a significant difference was found between the waveforms of the standard and deviant conditions during 230–277 ms after stimulus onset at Pz.

Table 3. Intervals with a significant difference at Pz and Fz for each method.

Method	Interval with Significant Difference (at Pz)	Interval with Significant Difference (at Fz)
1a	23–78 ms	
	141–145 ms	
	160–168 ms	
	488–512 ms	
	617–703 ms	
1b	430–438 ms	
		195–281 ms
		309–336 ms
		492–609 ms
		617–625 ms
		672–719 ms
1c	535–555 ms	
		90–98 ms
		316–340 ms
		348–363 ms
		375–387 ms
		539–602 ms
		621–691 ms
		699–711 ms
2	4–39 ms	
	59–90 ms	
	191–203 ms	
	633–762 ms	
	809–848 ms	
3	230–277 ms	

The Kruskal–Wallis test indicated that no significant difference was found between the mean amplitude of the difference curve (mismatch condition–match condition) of

Method 1a, 1b, and 1c during the 200–400 ms interval (H(2,23) = 1.48, p = 0.4778) and 500–800 ms interval (H(2,23) = 5.00, p = 0.0820) at Pz. However, at Fz, the Kruskal–Wallis test indicated that there was a significant difference in the mean amplitude of the difference curve between at least two methods during the 500–800 ms interval (H(2,23) = 13.23, p = 0.0013). Post hoc analyses using the Bonferroni correction revealed a significant difference between Method 1a and 1b (p = 0.0049) and between Method 1b and 1c (p = 0.0075). No significant difference was found during the 200–400 ms interval at Fz (H(2,23) = 3.14, p = 0.2082).

3.3. Classification

Table 4 compares the average classification accuracy that was achieved by the SVM and LDA classifiers. Both classifiers were trained using the same set of features (time- and frequency-domain features). The LDA performed better in Method 1a and 2. Although, a significant difference was found only for Method 2 (t(24) = −2.49, p = 0.0203). The SVM performed better in the remaining methods with significant difference for Method 1b (t(9) = 5.55, p = 0.0004) and 3 (t(25) = 5.01, p = 0.00004). No significant difference was found between the classifier type for Method 1a (t(8) = −1.79, p = 0.1112) and 1c (t(6) = 1.42, p = 0.2061).

Table 4. The mean (standard error) classification accuracy from support vector machine (SVM) and linear discriminant analysis (LDA) classifiers for each method when using the time- and frequency-domain feature set.

Method	Classification Accuracy	
	SVM	**LDA**
1a	59.71 (4.85)	63.07 (4.03)
1b	72.73 (2.21)	62.01 (1.98)
1c	60.99 (2.78)	57.56 (3.16)
2	61.75 (2.03)	64.23 (2.11)
3	71.78 (1.94)	65.74 (1.95)

Table 5 and Figure 8 shows the average accuracy from different feature sets for each method when using SVM. Overall, using raw features produced the worst average accuracy. Using the time- and frequency-domain features resulted in slightly better accuracy. Applying CSP to transform the raw features caused the average accuracy to rise further. The highest average accuracy for each method was obtained when using the CSP+ feature set. A one-way ANOVA indicated that there was a significant difference in accuracy between at least two feature sets (F(3,304) = 83.18, p = 2.5765 × 10^{-39}). After post hoc comparison, significant difference was found between each pair of feature sets (p = 6.0125 × 10^{-7} between Raw and T&F sets, p = 3.4121 × 10^{-25} between Raw and CSP sets, p = 1.416 × 10^{-35} between Raw and CSP+ sets, p = 1.6078 × 10^{-8} between T&F and CSP sets, p = 1.7061 × 10^{-16} between T&F and CSP+ sets, and p = 0.028032 between CSP and CSP+ sets).

Table 5. The mean (standard error) classification accuracy from each feature set for each method when using SVM classifier.

Method	Classification Accuracy (%)			
	Raw	**T&F**	**CSP**	**CSP+**
1a	50.77 (2.23)	59.71 (4.85)	79.31 (4.11)	82.79 (3.90)
1b	64.77 (2.49)	72.73 (2.21)	80.24 (1.87)	83.31 (1.70)
1c	54.39 (2.60)	60.99 (2.78)	67.80 (3.36)	78.41 (3.37)
2	50.74 (1.54)	61.75 (2.03)	71.48 (2.16)	78.90 (2.10)
3	64.27 (0.99)	71.78 (1.94)	80.15 (1.52)	81.35 (1.43)

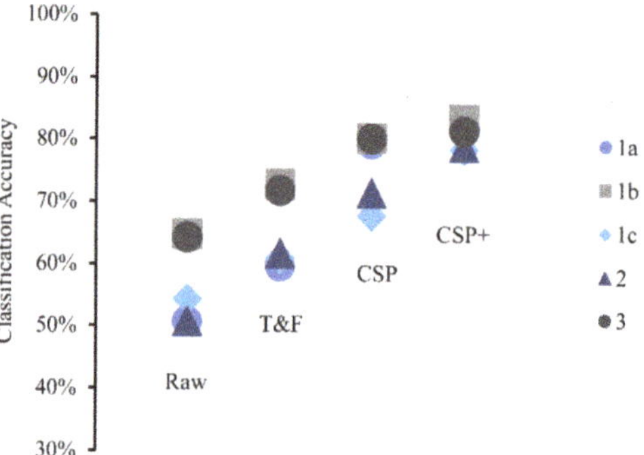

Figure 8. Comparison of the average accuracy between the different feature sets in each method.

The average accuracy was compared for each method, as shown in Figure 9a. When using the CSP set, Method 1b and 3 achieved the highest classification accuracy (80.24% and 80.15%, respectively). Method 1a performed slightly worse (79.31%), followed by Method 2 (71.48%). The lowest accuracy was obtained from Method 1c (67.80%). The Kruskal–Wallis test showed that there was a significant difference in the average accuracy between at least two methods (H(4,72) = 16.98, p = 0.002). Post hoc analyses using Bonferroni correction revealed a significant difference in average accuracy between Method 1c and 3 (p = 0.0378) and between method 2 and 3 (p = 0.0242). When using the CSP+ set, the average accuracy increased slightly; a similar trend was observed, although the differences were less pronounced. As shown in Figure 9b, the highest accuracy was obtained from Method 1b (83.31%), followed by Method 1a (82.79%) and Method 3 (81.35%). Method 2 and 1c produced slightly lower accuracy (78.90% and 78.41%, respectively). The result from the Kruskal–Wallis test indicated that no significant difference was found between the methods (H(4,72) = 2.63, p = 0.6218).

3.4. Attention

Figure 10 shows the average beta/alpha ratio at Cz for each method. Method 1b, 1c, and 2 have slightly higher beta/alpha ratio (0.9867, 1.0191, and 1.0092, respectively). Lower beta/alpha ratio was observed in Method 1a and 3 (0.8867 and 0.8002, respectively). The Kruskal–Wallis test revealed a significant difference between at least two methods (H(4,72) = 22.38, p = 0.0002). Post hoc analyses using Bonferroni correction showed a significant difference in average beta/alpha ratio between Method 1b and 3 (p = 0.0041), between Method 1c and 3 (p = 0.0366), and between Method 2 and 3 (p = 0.0012).

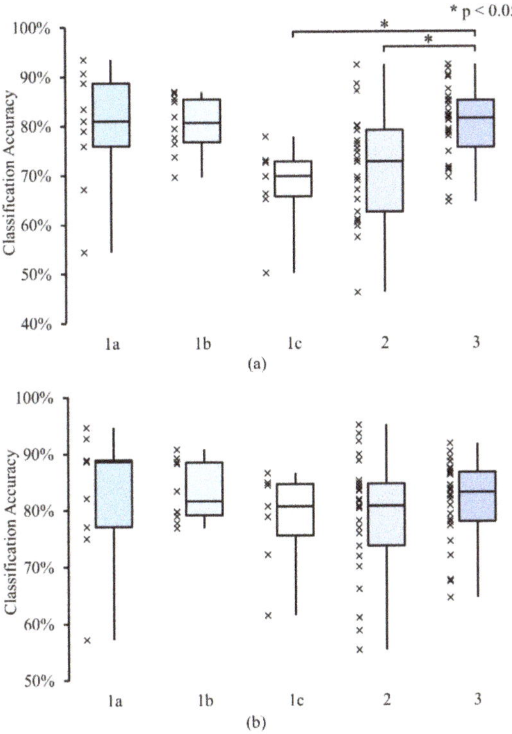

Figure 9. Box plot of the classification accuracy of each method using support vector machine (SVM) classifier (**a**) when using CSP feature set; (**b**) when using CSP+ feature set. The classification accuracies for individual participants are shown as x marks on the left of each box. An asterisk (*) denotes a significant difference in the classification accuracy between methods as indicated by the Kruskal–Wallis test with Bonferroni correction for multiple comparisons ($p < 0.05$).

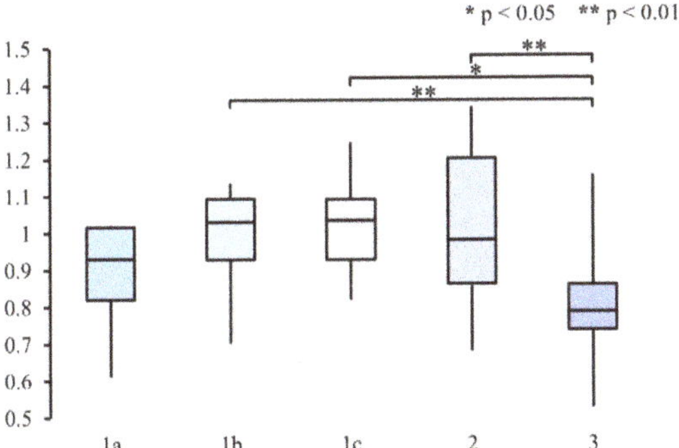

Figure 10. Box plot of the average beta/alpha ratio for each participant and method at Cz. Asterisks (*) denotes significant difference in beta/alpha ratio between the methods as indicated by the Kruskal–Wallis test with Bonferroni correction for multiple comparisons (* for $p < 0.05$ and ** for $p < 0.01$).

4. Discussion

Similar to Morikawa et al. [26], we found the LPP component during the mismatch condition in Method 1a, 1b, 1c, and 2. The component appeared as a positive waveform at approximately 500–800 ms after stimulus onset (Figures 6 and 7). LPP (also called P600 [41] or P3b [28]) is elicited when the presented stimulus is different from the expectation [29]. We also found a P3 component during the match condition, as seen from the positive waveform between 300–400 ms after stimulus onset (Figures 6 and 7). However, the difference in waveform was only significant in Method 1b and 1c. P3 is thought to be related to updating a mental representation of the incoming stimulus [28]. This component can be observed in various tasks such as oddball tasks [42,43], go/no-go or stop signal tasks [44–46], or identification tasks [47] during stimulus discrimination. P3 can be affected by stimulus probability and relevancy to the task. Method 1b included four pictures, from which the participants were required to select the correct one. This lower probability of the match stimulus resulted in a much clearer P3 component compared to all other methods. This result is in line with the findings that found increased P3 amplitude for rare stimuli compared to frequent stimuli [48,49]. We observed an MMN in response to deviant stimulus in Method 3 at approximately 100–300 ms after stimulus onset. This is consistent with the results from other works that utilizes MMN [13,15].

To classify between match/mismatch conditions, several feature sets were tested. The raw features produced the lowest classification accuracy compared to the other feature sets (Table 5). Combining the time-domain features and frequency-domain features into a feature set improved the accuracy slightly (Figure 8). However, classification accuracy from the time- and frequency-domain feature sets were below 80%. Thus, CSP was used to transform the raw features into a CSP feature set. An additional bandpass filter was applied to further improve the accuracy. CSP with the additional filter feature set produced the best accuracy compared to other feature sets. With this set, we obtained over 80% accuracy from Method 1a, 1b, and 3. This result is comparable to Morikawa et al. [26] which obtained over 80% accuracy when at least four trials were averaged.

Method 1b utilized four pictures to better differentiate between conditions. Thus, we expected this to result in much better classification accuracy compared to the other methods. When using the CSP+ feature set, we found that Method 1b performed best with 83.31% average classification accuracy. Method 1a and 3 also produced comparable average accuracies although with more variability of individual accuracies (Figure 9). Method 1c and 2 achieved slightly lower average accuracies (Figure 9). However, no significant difference between the average accuracy of each method was found when using this feature set. It is worth noting that this difference in accuracy between the methods was more pronounced when using Raw, T&F, or CSP feature sets (Figure 8). For example, the difference between the highest and lowest average accuracies (between Method 1b and 1c) increased from 4.9% with the CSP+ set to 12.44% with the CSP set. When using the CSP feature set, a significant difference was found between the average accuracy of Method 3 and 1c and between Method 3 and 2.

Despite being very similar to Method 1a, Method 1c produced the lowest accuracy compared to the rest. This went against our initial hypothesis that adding a prior section to set the expectation would increase the accuracy. This might be because of the participants getting confused as the two sections had different instructions. Method 2 was expected to achieve the same level of accuracy as experiment 1 in Morikawa et al. [26] as the design was similar. However, Method 2 achieved slightly lower accuracy at 78.90% compared to over 80% accuracy in Morikawa et al. [26]. This disparity might be because Morikawa et al. [26] used letters while our method used words. The participants had to evaluate the spelling of the words, which is relatively harder than identifying letters. This might cause more variation in ERP latency as each participant evaluates spelling using different techniques. Some people consider the spelling of the whole word while some people only look at the consonant. When averaged, this latency variation could cause ERP components to become less prominent.

To investigate the attention level in each method, we inspected the power in the beta and alpha bands. Higher attention is associated with an increase in beta power and a decrease in alpha power [40]. Thus, we used beta/alpha ratio to represent the attention level. Method 3, which included only auditory stimuli, had the lowest beta/alpha ratio

(Figure 10). The beta/alpha ratio of this method also had lower variability compared to the other methods, as can be seen from the shorter distance between the first and third quartile in the box plot (Figure 10). This lower ratio indicates that less attention was paid to the task. A significant difference was found between Method 3 and Method 1b, 1c, and 2 (Figure 10). The latter three methods utilized visual stimuli which resulted in higher attention being paid to the task. This is in line with how higher activation was observed in a visual attention task compared to an auditory attention task [50]. Compared to the other methods, Method 3 required less attention from the participants.

Method 1a and 1b used pictures matching the meaning of the words. This is similar to the method that audiologists use to assess speech discrimination in children. This should make it easier to compare the results from the ERP methods and conventional methods. Furthermore, existing materials [19,20,51] can be conveniently adapted for the ERP methods. Between these two methods, Method 1a used less time per trial while having comparable accuracy to Method 1b. Furthermore, Method 1b required the participants to remember the presentation order of the correct picture, which might make this method more sensitive to errors in some population groups such as elderly patients with cognitive decline. On the other hand, Method 3 was easier to perform. It required less attention than Method 1a and 1b, as seen from the lower beta/alpha ratio (Figure 10). Also, with some adjustment, this method could be performed using passive listening. Passive listening does not require the participants to pay attention to what they are listening to. Participants can even engage in other light activity such as reading or watching a video while listening. This means that it can be used with a wider group of people, including very young children or uncooperating patients. However, this method might get more time consuming as more word pairs are added to the test. All things considered, we recommend using either Method 1a or 3 depending on the target population. Method 1a might be a better choice in cases where existing materials can be adapted easily. However, Method 3 might be better for participants who have difficulties following instructions.

The participants answered correctly in most of the cases in all methods, in line with the fact that all the participants had normal hearing. The response time varied between methods with the highest average time in Method 3. This was surprising as Method 3 was the only method that utilized auditory stimulus only. Auditory stimulus usually had lower response time than visual stimulus [52,53]. It was possible that, because the words sounded very similar, they were difficult to differentiate, while the picture of word meaning did not resemble the other, making it easier to discern the correct answer in the visual methods. Thus, the average response time was higher in the auditory method and lower in the visual methods. The participants' opinion on each method varied. Some participants preferred the auditory-only method as they did not have to focus their attention. Others preferred having visual stimuli because they could differentiate between the conditions more easily with visual stimuli.

Using a cap with eight electrodes, our setup time was approximately 10–20 min. During this time, the participants were given instructions for each method. Some participants requested a very detailed explanation which could extend this setup time further. The two most promising methods, Method 1a and 3, took approximately five minutes to complete for one contrast repeated 40 times. During offline analysis, the result for these methods was obtained in under 10 s for each participant. In a real situation, more word pairs should be used but with lower repetition for each pair. Hypothetically, for 50 pairs, Method 1a would take under 10 min to complete if each pair was used twice. Method 3, using 20 trials per pair, could be completed in approximately 30 min. The experiment time for Method 3 could be reduced by adjusting the design. For example, using a double or multi-feature oddball paradigm [15,22]. The analysis time would vary depending on the hardware performance but should not take more than 30 min per patient. Overall, a speech discrimination assessment system using these methods could be done in under one hour.

The ERP-based methods of speech discrimination assessment that were investigated in this work could potentially be developed into an automated system that acts as a pre-screening tool. EEG is gradually becoming more accessible and user-friendly, as can be seen from the increased availability of consumer EEG devices such as Emotiv Insight (Emotiv, San Francisco, CA, USA), NeuroSky Mindwave (NeuroSky, San Jose, CA, USA), OpenBCI headset (OpenBCI, Brooklyn, NY, USA), etc. (for review of low-cost EEG headsets, see [54]). Such consumer EEG devices can be used by trained healthcare professionals (e.g., nurses, practical nurses, or health technicians) in local health centers to pre-screen patients using the automatic system. Then, if the result indicates speech discrimination problems, an in-person or telemedicine appointment with an audiologist may be scheduled. Employing such a system would make the assessment more accessible, especially in areas with a shortage of audiologists.

5. Limitations

This study has some limitations that warrant future research. Although the sample size used is not dissimilar to those in various BCI studies [55–57], it can still be considered rather small, especially for Method 1a, 1b, and 1c which involved 10 participants. Further studies with a larger sample size may help confirm the accuracy of the proposed methods. Furthermore, this experiment used only one pair of words with a consonant contrast to test the different methods. To cover all the meaningful contrast types in a language, a comprehensive word list [19–21,58] could be included. For example, in Thai, which is a tonal language, the word list should include all vowel and consonant groups with varied tones [19]. The method could also be further validated with evaluations that are made by audiologists in patients with hearing loss and other associated diseases. In addition, all the methods that were included in this work used active listening. The feasibility of a method utilizing passive listening should also be investigated. Näätänen et al. [11] suggested passive listening might produce clearer MMN waveform. However, there has not yet been any confirmation of any effect on classification accuracy. Passive listening allows the assessment to be applied to a much wider group of people. Thus, a passive listening method could be very useful if a comparable level of classification accuracy could be achieved.

6. Conclusions

This research compared several ERP-based methods for assessing speech discrimination. Of these, two methods are recommended. Both achieved 80% classification accuracy and required less time or effort than other methods. The first method used picture representing word meaning which allowed for easy adaptation of existing materials to be incorporated into the assessment. P3 and LPP were observed and used to classify whether the sound and picture matched. This method achieved 82.79% accuracy. The second recommended method used auditory stimuli only. MMN was elicited in response to the deviant stimuli and was used to classify between standard and deviant stimulus, achieving 81.35% accuracy. This method took longer to complete than the first but required less attention from participants. It is well-suited for use in cases where pictures are not available or where participants have difficulties following instructions.

Author Contributions: Conceptualization, P.C., P.I., S.H. and S.P.; methodology, P.C., P.I., S.H. and S.P.; software, P.C.; validation, P.C. and S.P.; formal analysis, P.C. and S.P.; investigation, P.C.; resources, P.I. and S.P.; data curation, P.C. and S.P.; writing—original draft preparation, P.C.; writing—review and editing, P.C., P.I., S.H. and S.P.; visualization, P.C.; supervision, P.I., S.H. and S.P.; project administration, P.C. and S.P.; funding acquisition, P.I. and S.P. All authors have read and agreed to the published version of the manuscript.

Funding: This research received no external funding.

Institutional Review Board Statement: The study was conducted in accordance with the Declaration of Helsinki and approved by the Ethics Committee of Chulalongkorn University (protocol code 171.1/63, 30 October 2020).

Informed Consent Statement: Informed consent was obtained from all subjects involved in the study.

Data Availability Statement: The data presented in this study are openly available in FigShare at https://doi.org/10.6084/m9.figshare.16786618.v2 (accessed on 24 January 2022).

Acknowledgments: The authors would like to thank Arpa Suwannarat and team for technical assistance and suggestions regarding EEG signal acquisition.

Conflicts of Interest: The authors declare no conflict of interest.

References

1. Chaix, B.; Lewis, R. Speech Audiometry. Available online: https://www.cochlea.eu/en/audiometry/subjective-measure/speech-audiometry (accessed on 18 January 2022).
2. Flint, P.W.; Haughey, B.H.; Robbins, K.T.; Lund, V.J.; Thomas, J.R.; Niparko, J.K.; Richardson, M.A.; Lesperance, M.M. *Cummings Otolaryngology-Head and Neck Surgery E-Book*; Head and Neck Surgery; Elsevier Health Sciences: Amsterdam, The Netherlands, 2010.
3. Siuly, S.; Li, Y.; Zhang, Y. *EEG Signal Analysis and Classification*; Health Information Science; Springer Nature: Berlin/Heidelberg, Germany, 2016. [CrossRef]
4. Jasper, H.H. The ten-twenty electrode system of the International Federation. *Electroencephalogr. Clin. Neurophysiol.* **1958**, *10*, 370–375. [CrossRef]
5. Chatrian, G.E.; Lettich, E.; Nelson, P.L. Ten percent electrode system for topographic studies of spontaneous and evoked EEG activities. *Am. J. EEG Technol.* **1985**, *25*, 83–92. [CrossRef]
6. Oostenveld, R.; Praamstra, P. The five percent electrode system for high-resolution EEG and ERP measurements. *Clin. Neurophysiol.* **2001**, *112*, 713–719. [CrossRef]
7. Cohen, M.X. *Analyzing Neural Time Series Data: Theory and Practice*; The MIT Press: Cambridge, MA, USA, 2014. [CrossRef]
8. Niedermeyer, E.; Schomer, D.L.; Da Silva, F.H.L. *Niedermeyer's Electroencephalography: Basic Principles, Clinical Applications, and Related Fields*; Wolters Kluwer Health/Lippincott Williams & Wilkins: Philadelphia, PA, USA, 2011.
9. Luck, S.J. *An Introduction to the Event-Related Potential Technique*, 2nd ed.; MIT Press: Cambridge, MA, USA, 2014.
10. Näätänen, R.; Gaillard, A.W.; Mäntysalo, S. Early selective-attention effect on evoked potential reinterpreted. *Acta Psychol.* **1978**, *42*, 313–329. [CrossRef]
11. Näätänen, R.; Paavilainen, P.; Rinne, T.; Alho, K. The mismatch negativity (MMN) in basic research of central auditory processing: A review. *Clin. Neurophysiol.* **2007**, *118*, 2544–2590. [CrossRef]
12. Näätänen, R.; Kreegipuu, K. The Mismatch Negativity (MMN). In *The Oxford Handbook of Event-Related Potential Components*; Oxford University Press: Oxford, UK, 2012. [CrossRef]
13. Koerner, T.K.; Zhang, Y.; Nelson, P.B.; Wang, B.; Zou, H. Neural indices of phonemic discrimination and sentence-level speech intelligibility in quiet and noise: A mismatch negativity study. *Hear. Res.* **2016**, *339*, 40–49. [CrossRef]
14. Kraus, N.; McGee, T.J.; Carrell, T.D.; Zecker, S.G.; Nicol, T.G.; Koch, D.B. Auditory neurophysiologic responses and discrimination deficits in children with learning problems. *Science* **1996**, *273*, 971–973. [CrossRef]
15. Pakarinen, S.; Takegata, R.; Rinne, T.; Huotilainen, M.; Näätänen, R. Measurement of extensive auditory discrimination profiles using the mismatch negativity (MMN) of the auditory event-related potential (ERP). *Clin. Neurophysiol.* **2007**, *118*, 177–185. [CrossRef]
16. Shestakova, A.; Brattico, E.; Huotilainen, M.; Galunov, V.; Soloviev, A.; Sams, M.; Ilmoniemi, R.J.; Näätänen, R. Abstract phoneme representations in the left temporal cortex: Magnetic mismatch negativity study. *NeuroReport* **2002**, *13*, 1813–1816. [CrossRef]
17. Sittiprapaporn, W.; Chindaduangratn, C.; Kotchabhakdi, N. Auditory preattentive processing of vowel change perception in single syllable. In Proceedings of the 42nd Kasetsart University Annual Conference, Bangkok, Thailand, 3–6 February 2004.
18. Virtala, P.; Partanen, E.; Tervaniemi, M.; Kujala, T. Neural discrimination of speech sound changes in a variable context occurs irrespective of attention and explicit awareness. *Biol. Psychol.* **2018**, *132*, 217–227. [CrossRef]
19. Hemakom, A.; Jitwiriyanont, S.; Rugchatjaroen, A.; Israsena, P. The development of Thai monosyllabic word and picture lists applicable to interactive speech audiometry in preschoolers. *Clin. Linguist. Phon.* **2021**, *35*, 809–828. [CrossRef] [PubMed]
20. Visessumon, P. Rama SD-III Words Picture Speech Discrimination Test for Children. Master's Thesis, Mahidol University, Library and Knowledge Center, Nakhon Pathom, Thailand, 1988.
21. Yimtae, K.; Israsena, P.; Thanawirattananit, P.; Seesutas, S.; Saibua, S.; Kasemsiri, P.; Noymai, A.; Soonrach, T. A Tablet-Based Mobile Hearing Screening System for Preschoolers: Design and Validation Study. *JMIR Mhealth Uhealth* **2018**, *6*, e186. [CrossRef] [PubMed]
22. Näätänen, R.; Pakarinen, S.; Rinne, T.; Takegata, R. The mismatch negativity (MMN): Towards the optimal paradigm. *Clin. Neurophysiol.* **2004**, *115*, 140–144. [CrossRef] [PubMed]
23. Ha, K.S.; Youn, T.; Kong, S.W.; Park, H.J.; Ha, T.H.; Kim, M.S.; Kwon, J.S. Optimized individual mismatch negativity source localization using a realistic head model and the Talairach coordinate system. *Brain Topogr.* **2003**, *15*, 233–238. [CrossRef]
24. Marco-Pallares, J.; Grau, C.; Ruffini, G. Combined ICA-LORETA analysis of mismatch negativity. *NeuroImage* **2005**, *25*, 471–477. [CrossRef]
25. Gilley, P.M.; Uhler, K.; Watson, K.; Yoshinaga-Itano, C. Spectral-temporal EEG dynamics of speech discrimination processing in infants during sleep. *BMC Neurosci.* **2017**, *18*, 34. [CrossRef]

26. Morikawa, K.; Kozuka, K.; Adachi, S. A Proposed Speech Discrimination Assessment Methodology Based on Event-Related Potentials to Visual Stimuli. *Int. J. E-Health Med. Commun.* **2012**, *3*, 19–35. [CrossRef]

27. Picton, T. The P300 Wave of the Human Event-Related Potential. *J. Clin. Neurophysiol. Off. Publ. Am. Electroencephalogr. Soc.* **1992**, *9*, 456–479. [CrossRef]

28. Polich, J. Updating P300: An integrative theory of P3a and P3b. *Clin. Neurophysiol. Off. J. Int. Fed. Clin. Neurophysiol.* **2007**, *118*, 2128–2148. [CrossRef]

29. Adachi, S.; Morikawa, K.; Nittono, H. Event-related potentials elicited by unexpected visual stimuli after voluntary actions. *Int. J. Psychophysiol.* **2007**, *66*, 238–243. [CrossRef]

30. Donchin, E.; Coles, M.G.H. Is the P300 component a manifestation of context updating? *Behav. Brain Sci.* **1988**, *11*, 357–374. [CrossRef]

31. Thubthong, N.; Kijsirikul, B. Tone Recognition Of Continuous Thai Speech Under Tonal Assimilation and Declination Effects Using Half-Tone Model. *Int. J. Uncertain. Fuzziness Knowl.-Based Syst.* **2002**, *9*, 815–825. [CrossRef]

32. Hoonchamlong, Y. The Thai Alphabets. Available online: http://thaiarc.tu.ac.th/thai/thindex.htm (accessed on 8 June 2020).

33. Peirce, J. Generating stimuli for neuroscience using PsychoPy. *Front. Neuroinform.* **2009**, *2*, 10. [CrossRef] [PubMed]

34. Renard, Y.; Lotte, F.; Gibert, G.; Congedo, M.; Maby, E.; Delannoy, V.; Bertrand, O.; Lécuyer, A. OpenViBE: An Open-Source Software Platform to Design, Test, and Use Brain–Computer Interfaces in Real and Virtual Environments. *Presence* **2010**, *19*, 35–53. [CrossRef]

35. Delorme, A.; Makeig, S. EEGLAB: An open source toolbox for analysis of single-trial EEG dynamics including independent component analysis. *J. Neurosci. Methods* **2004**, *134*, 9–21. [CrossRef] [PubMed]

36. Kothe, C.A.; Makeig, S. BCILAB: A platform for brain–computer interface development. *J. Neural Eng.* **2013**, *10*, 056014. [CrossRef]

37. KOTHE, C.A.E.; Jung, T.-P. Artifact Removal Techniques with Signal Reconstruction. U.S. Patent US20160113587A1, 3 June 2014.

38. Chang, C.Y.; Hsu, S.H.; Pion-Tonachini, L.; Jung, T.P. Evaluation of Artifact Subspace Reconstruction for Automatic EEG Artifact Removal. In Proceedings of the 2018 40th Annual International Conference of the IEEE Engineering in Medicine and Biology Society (EMBC), Honolulu, HI, USA, 18–21 July 2018; Volume 2018, pp. 1242–1245. [CrossRef]

39. Charuthamrong, P.; Israsena, P.; Hemrungrojn, S.; Pan-ngum, S. Active and Passive Oddball Paradigm for Automatic Speech Discrimination Assessment. In Proceedings of the 2021 4th International Conference on Bio-Engineering for Smart Technologies (BioSMART), Paris, France, 8–10 December 2021; pp. 1–3.

40. Israsena, P.; Hemrungrojn, S.; Sukwattanasinit, N.; Maes, M. Development and Evaluation of an Interactive Electro-Encephalogram-Based Neurofeedback System for Training Attention and Attention Defects in Children. *J. Med. Imaging Health Inform.* **2015**, *5*, 1045–1052. [CrossRef]

41. Friederici, A.D. Event-related brain potential studies in language. *Curr. Neurol. Neurosci. Rep.* **2004**, *4*, 466–470. [CrossRef]

42. Bennett, K.O.; Billings, C.J.; Molis, M.R.; Leek, M.R. Neural encoding and perception of speech signals in informational masking. *Ear Hear.* **2012**, *33*, 231–238. [CrossRef]

43. Koerner, T.K.; Zhang, Y.; Nelson, P.B.; Wang, B.; Zou, H. Neural indices of phonemic discrimination and sentence-level speech intelligibility in quiet and noise: A P3 study. *Hear. Res.* **2017**, *350*, 58–67. [CrossRef]

44. Chikara, R.K.; Komarov, O.; Ko, L.-W. Neural signature of event-related N200 and P300 modulation in parietal lobe during human response inhibition. *Int. J. Comput. Biol. Drug Des.* **2018**, *11*, 171–182. [CrossRef]

45. Enriquez-Geppert, S.; Konrad, C.; Pantev, C.; Huster, R.J. Conflict and inhibition differentially affect the N200/P300 complex in a combined go/nogo and stop-signal task. *NeuroImage* **2010**, *51*, 877–887. [CrossRef] [PubMed]

46. Keskin-Ergen, Y.; Tükel, R.; Aslantaş-Ertekin, B.; Ertekin, E.; Oflaz, S.; Devrim-Üçok, M. N2 and P3 potentials in early-onset and late-onset patients with obsessive-compulsive disorder. *Depress. Anxiety* **2014**, *31*, 997–1006. [CrossRef] [PubMed]

47. Kaplan-Neeman, R.; Kishon-Rabin, L.; Henkin, Y.; Muchnik, C. Identification of syllables in noise: Electrophysiological and behavioral correlates. *J. Acoust. Soc. Am.* **2006**, *120*, 926–933. [CrossRef]

48. Dimoska, A.; Johnstone, S.J. Effects of varying stop-signal probability on ERPs in the stop-signal task: Do they reflect variations in inhibitory processing or simply novelty effects? *Biol. Psychol.* **2008**, *77*, 324–336. [CrossRef] [PubMed]

49. Ramautar, J.R.; Kok, A.; Ridderinkhof, K.R. Effects of stop-signal probability in the stop-signal paradigm: The N2/P3 complex further validated. *Brain Cogn.* **2004**, *56*, 234–252. [CrossRef]

50. Rienäcker, F.; Van Gerven, P.W.M.; Jacobs, H.I.L.; Eck, J.; Van Heugten, C.M.; Guerreiro, M.J.S. The Neural Correlates of Visual and Auditory Cross-Modal Selective Attention in Aging. *Front. Aging Neurosci.* **2020**, *12*, 420. Available online: https://www.frontiersin.org/article/10.3389/fnagi.2020.498978 (accessed on 18 January 2022). [CrossRef]

51. Pronovost, W.; Dumbleton, C. A picture-type speech sound discrimination test. *J. Speech Hear. Disord.* **1953**, *18*, 258–266. [CrossRef]

52. Jain, A.; Bansal, R.; Kumar, A.; Singh, K.D. A comparative study of visual and auditory reaction times on the basis of gender and physical activity levels of medical first year students. *Int. J. Appl. Basic Med. Res.* **2015**, *5*, 124–127. [CrossRef]

53. Welford, W.; Brebner, J.M.; Kirby, N. *Reaction Times*; Stanford University: Stanford, CA, USA, 1980.

54. LaRocco, J.; Le, M.D.; Paeng, D.-G. A Systemic Review of Available Low-Cost EEG Headsets Used for Drowsiness Detection. *Front. Neuroinform.* **2020**, *14*, 553352. [CrossRef]

55. Bulea, T.C.; Prasad, S.; Kilicarslan, A.; Contreras-Vidal, J.L. Sitting and standing intention can be decoded from scalp EEG recorded prior to movement execution. *Front. Neurosci.* **2014**, *8*, 376. Available online: https://www.frontiersin.org/article/10.3389/fnins.2014.00376 (accessed on 18 January 2022). [CrossRef] [PubMed]

56. Min, B.-K.; Dähne, S.; Ahn, M.-H.; Noh, Y.-K.; Müller, K.-R. Decoding of top-down cognitive processing for SSVEP-controlled BMI. *Sci. Rep.* **2016**, *6*, 36267. [CrossRef] [PubMed]
57. Sburlea, A.I.; Montesano, L.; De la Cuerda, R.C.; Alguacil Diego, I.M.; Miangolarra-Page, J.C.; Minguez, J. Detecting intention to walk in stroke patients from pre-movement EEG correlates. *J. Neuroeng. Rehabil.* **2015**, *12*, 113. [CrossRef] [PubMed]
58. Runge, C.A.; Hosford-Dunn, H. Word Recognition Performance with Modified CID W-22 Word Lists. *J. Speech Lang. Hear. Res.* **1985**, *28*, 355–362. [CrossRef]

Article

EEG Evoked Potentials to Repetitive Transcranial Magnetic Stimulation in Normal Volunteers: Inhibitory TMS EEG Evoked Potentials

Jing Zhou, Adam Fogarty, Kristina Pfeifer, Jordan Seliger and Robert S. Fisher *

Department of Neurology and Neurological Sciences, Stanford University School of Medicine, Stanford, CA 94304, USA; zhou5@g.clemson.edu (J.Z.); afogarty@stanfordhealthcare.org (A.F.); kpfeifer@stanford.edu (K.P.); jseliger@stanford.edu (J.S.)
* Correspondence: robert.fisher@stanford.edu; Tel.: +1-721-5552; Fax: +1-650-721-4865

Abstract: The impact of repetitive magnetic stimulation (rTMS) on cortex varies with stimulation parameters, so it would be useful to develop a biomarker to rapidly judge effects on cortical activity, including regions other than motor cortex. This study evaluated rTMS-evoked EEG potentials (TEP) after 1 Hz of motor cortex stimulation. New features are controls for baseline amplitude and comparison to control groups of sham stimulation. We delivered 200 test pulses at 0.20 Hz before and after 1500 treatment pulses at 1 Hz. Sequences comprised AAA = active stimulation with the same coil for test–treat–test phases ($n = 22$); PPP = realistic placebo coil stimulation for all three phases ($n = 10$); and APA = active coil stimulation for tests and placebo coil stimulation for treatment ($n = 15$). Signal processing displayed the evoked EEG waveforms, and peaks were measured by software. ANCOVA was used to measure differences in TEP peak amplitudes in post-rTMS trials while controlling for pre-rTMS TEP peak amplitude. Post hoc analysis showed reduced P60 amplitude in the active (AAA) rTMS group versus the placebo (APA) group. The N100 peak showed a treatment effect compared to the placebo groups, but no pairwise post hoc differences. N40 showed a trend toward increase. Changes were seen in widespread EEG leads, mostly ipsilaterally. TMS-evoked EEG potentials showed reduction of the P60 peak and increase of the N100 peak, both possibly reflecting increased slow inhibition after 1 Hz of rTMS. TMS-EEG may be a useful biomarker to assay brain excitability at a seizure focus and elsewhere, but individual responses are highly variable, and the difficulty of distinguishing merged peaks complicates interpretation.

Keywords: transcranial magnetic stimulation; epilepsy; cerebral cortex stimulation; electromagnetic influence; neurostimulation

Citation: Zhou, J.; Fogarty, A.; Pfeifer, K.; Seliger, J.; Fisher, R.S. EEG Evoked Potentials to Repetitive Transcranial Magnetic Stimulation in Normal Volunteers: Inhibitory TMS EEG Evoked Potentials. *Sensors* **2022**, *22*, 1762. https://doi.org/10.3390/s22051762

Academic Editor: Yvonne Tran

Received: 27 December 2021
Accepted: 22 February 2022
Published: 24 February 2022

Publisher's Note: MDPI stays neutral with regard to jurisdictional claims in published maps and institutional affiliations.

1. Introduction

Transcranial magnetic stimulation (TMS) [1,2], repetitive transcranial magnetic stimulation (rTMS) [3,4], and intermittent or continuous theta burst stimulation [5] have been evaluated for therapeutic effects in numerous clinical conditions. Results vary and the optimal parameters of stimulation remain uncertain. For example, of seven controlled studies of rTMS as a treatment for epilepsy, two have been favorable for seizure improvement [6,7] and five documented transient, little, or no benefit against seizures [8–12]. Systematic testing of various stimulation protocols against different clinical outcomes is a lengthy and difficult process. Therefore, a biomarker able to efficiently assay the biological effect of rTMS would likely accelerate development of useful therapies.

Small variations of stimulation parameters or locations can lead to widely varying–sometimes opposite–clinical responses [13]. For example, low-frequency stimulation at 0.5–1 Hz depresses motor cortex excitability [14], whereas 5 Hz of stimulation increases excitability [15].

The most commonly used biomarker for effects of rTMS is the electromyogram (EMG)-evoked response in the hand, while stimulating contralateral motor cortex [16]. A sufficiently strong TMS stimulation delivered to motor cortex elicits a thumb or finger twitch and an EMG response recorded by a surface electrode on the hand. Cortical stimulation produces local excitation, followed by a silent period reflecting cortical inhibition [17]. A second TMS pulse delivered during the period of inhibition will produce a smaller EMG response in the hand, thereby allowing the ratio of EMG amplitudes of the second versus the first response to serve as a marker of induced cortical inhibition [16]. This method of estimating cortical inhibition only applies to motor cortex. However, the desired inhibitory effect of rTMS often is on another region of the brain, for example, the dorsolateral frontal cortex for treating depression [18] or a cortical seizure focus for treating epilepsy [19,20].

TMS evokes an electrical response in cortex that can be recorded by electroencephalography (EEG) electrodes [21–31]. Because the EEG signals are low amplitude and distorted by magnetic pulse artifact, signal averaging of multiple stimuli and digital signal processing methods are required to characterize the EEG response to TMS [32–41]. Nevertheless, TMS-evoked EEG potentials (TEPs) can be assessed in any region of cortex before and after a putative therapeutic maneuver. This potentially affords an opportunity to use TEPs as a biomarker of rTMS or TBS efficacy. Changes in TEPs in response to rTMS treatment were demonstrated by the present authors in a single case study of a patient with epilepsy, also correlating with improvement in seizures [42].

The usual TEP has negative (N) and positive (P) peaks at N20, P30, N40 [43] (sometimes labeled as N45), P60 (sometimes labeled P70), N100, P180, and N280 ms [44]. The N40 peak corresponds to the $GABA_A$ receptor-based fast IPSP [45–47]. Later peaks, including the P60 and N100 peaks, may correspond to the $GABA_B$ receptor-based slow IPSP [30,48,49]. Casula and colleagues [50] demonstrated that 1 Hz of rTMS in normal subjects increased the P60 and N100 peaks. Their study was carefully done, but only on 15 subjects and without a placebo-stimulation control group. In this study, we explored whether rTMS at 1 Hz can alter the N40, P60, and N100 peaks, with 47 subjects and two control groups, and we additionally examined cortical sources of the evoked potentials. The goal was to further develop TEPs as a biomarker for assaying regional cortical excitability alterations produced by rTMS in cortical regions that cannot be assayed by peripheral stimulation. This could be useful, for example, for testing excitability in cortical seizure foci or areas of cortical injury.

2. Materials and Methods

2.1. Subjects

Forty-seven healthy adult participants (age 21–67 years; mean = 32.8 ± 9.6 years) were recruited. Two participants reported left-handedness. All participants signed an informed consent form before participation in present study approved by the Stanford University Institutional Review Board (IRB). Exclusion criteria were extracted from the Rossi et al. (2009) article [51].

2.2. TMS Device and Coils

TMS was delivered using an EB Neuro ATES STM9000 magnetic stimulator (EB Neuro S.p.A., Florence, Italy) with the coil held tangentially to the skull, with its handle oriented 45 degrees from midline. TMS was delivered with the electrode cap on, as closely as possible without touching the electrodes. Active TMS was delivered with a 70-mm air-cooled figure-of-eight coil (B9621086004). (EB Neuro S.p.A., Florence, Italy) Pseudo-placebo stimulation was delivered with a visually identical 70-mm air-cooled figure-of-eight coil (B9621086009) (EB Neuro S.p.A., Florence, Italy). The placebo coil stimulated in a tangential plane to cortex, which reduced the effect, but still permitted some stimulation to provide a scalp sensation and mask the treatment group. Therefore, the treatment groups could be considered high versus low stimulation treatment arms, rather than active versus true placebo. A second control group consisting of exclusive use of the placebo coil for 0.20 Hz of TMS and 1 Hz of rTMS blocks was studied to account for possible group differences that might be derived from use of two different coils (Table 1). The group with the active coil for all stimulation was denoted as the full active AAA group ($n = 22$); full placebo as the PPP group ($n = 10$) and the group with the active coil for both sets of 0.20-Hz test TMS and placebo coil for the 1 Hz of rTMS was called the active–placebo–active (APA) group ($n = 15$).

Table 1. Test stages.

Group	n	Test Pre 0.20 Hz	rTMS 1 Hz	Test Post 0.20 Hz
Full active (AAA)	22	active	active	active
Full placebo (PPP)	10	placebo	placebo	placebo
Mixed (APA)	15	active	placebo	active

TMS sound artifact was masked via use of white noise played with sound canceling headphones. Volume was incrementally increased until participants reported that the TMS "click" was obscured. Continuous visualization of stimulation site in relation to individual cortical anatomy was ensured using an ATES Medica NetBrain Neuronavigation system. All three test stages followed the same procedure without revealing the coil type to the subjects. Figure 1 shows the experimental arrangement.

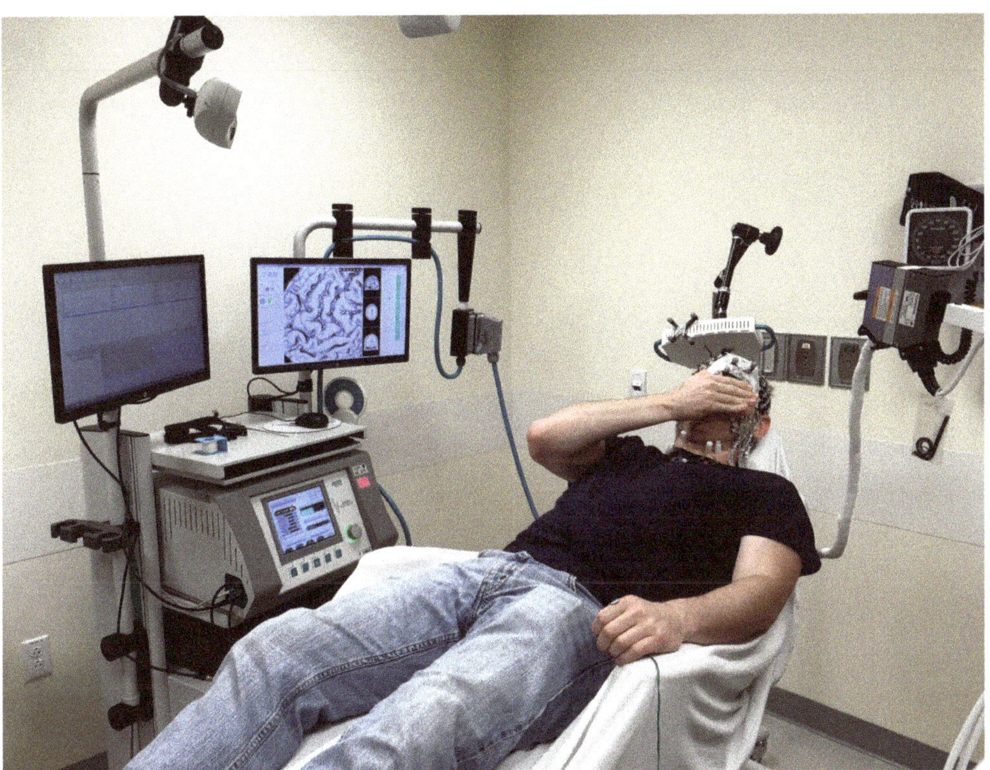

Figure 1. rTMS setup showing the subject in relation to the stimulator generator and coil, neuronavigation system, wire to the left thumb to stimulate and record EMG and (to the right) cable to the EEG machine.

2.3. EEG System

EEG was recorded using an Electrical Geodesic, Inc. 256-channel MicroCel sensor net. Elefix conductive paste was used on 77 electrodes. Gelled electrodes included the standard 10–20 montage electrodes as well as a denser cluster of electrodes in regions of interest near C3 and C4 (Figure 2). EEG data were recorded referenced to Cz and impedances were kept below 10 kΩ. EEG was sampled at 1 kHz and the amplifier was set to fast recovery. Electrodes were connected to scalp by conductive paste, comprising the 10–20 system and a dense array around C3 and C4, which was the stimulation site.

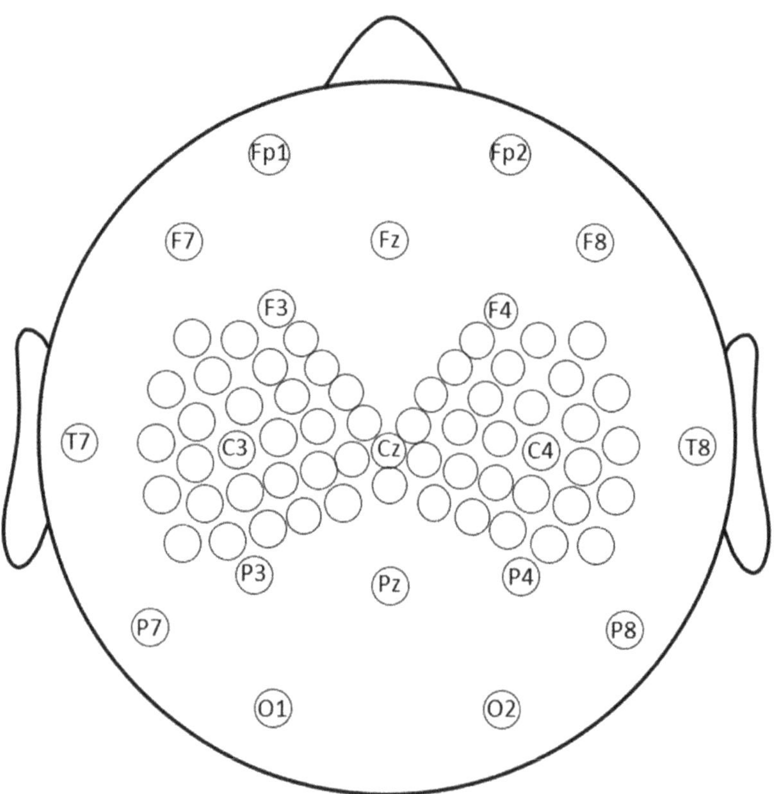

Figure 2. Topography of the recording electrodes. EEG was recorded using an Electrical Geodesic, Inc. 256-channel MicroCel sensor net. All electrodes on the sensor net were spaced 0.5–1.5 cm from each other at the center. Conductive paste was used on 76 electrodes including the standard 10–20 montage electrodes and a denser cluster of 27 electrodes near the TMS region of interest, collectively falling within 6 cm of C3 and C4, respectively.

2.4. Resting Motor Threshold (rMT)

Participants were seated in an adjustable chair with a headrest to keep the head stable for the duration of the study. The resting motor threshold (rMT) was determined with the EEG cap in place. The stimulation site was determined by finding location of the stimulation that evoked the largest movement in a participant's non-dominant hand. rMT was defined as the minimal stimulation intensity used to evoke a visible muscle twitch with time-locked EMG correlate in at least 5 out of 10 trials. When rMT could not be determined ($n = 6$), we set rMT as 65% of maximum stimulator output.

2.5. Repetitive Transcranial Magnetic Stimulation (rTMS)

Participants underwent the rTMS procedure in the late morning or early afternoon hours. Participants were asked to keep eyes closed throughout stimulation, but were kept awake for the duration of study, confirmed by EEG and behavioral monitoring. The experiment was limited to a single session that delivered rTMS to non-dominant hand region motor cortex (near the C4 electrode). Previous studies [52,53] provided evidence that non-dominant hand motor cortex and dominant hand motor cortex have similar resting motor thresholds; however, the non-dominant hand motor cortex may be more susceptible to inhibitory stimulation than is the dominant hand motor cortex.

Stimulations were divided into three separate blocks, all delivered with the electrode cap in place. The initial block of rTMS consisted of 200 single pulse rTMS (SpTMS) delivered at 0.20 Hz and 110% of rMT. The second block consisted of 1500 rTMS pulses at 1 Hz and 90% rMT. The rTMS pulses were divided into three sub-blocks of 500 pulses, separated by rest periods of 90–120 s to allow for coil cooling. The final stimulation block consisted of a second round of 200 SpTMS delivered at 0.20 Hz at 110% rMT.

2.6. Processing of EEG Data

EEG analysis was performed in MATLAB using EEG-LAB and the TMS-EEG signal analyzer, TESA [54], an open-source extension of EEGLAB. Order of operations for TMS-EEG analysis was the following: (1) EEG data were segmented from −600 ms before to +600 ms after the rTMS pulse. (2) Data were baseline corrected, based on EEG data occurring from −100 to −6 ms. (3) EEG data from −5 to +15 ms were removed and replaced with constant data extrapolated from the pre-artifact baseline, to eliminate the majority of the rTMS pulse artifact. (4) Data were visually inspected for profound artifacts (e.g., flat-lining or noise unrelated to the rTMS). Bad channels and trials were manually removed. (5) TESA performed a first pass of fast independent component analysis (ICA) to correct for rTMS-ringing artifact. (6) EEG data were band-pass filtered from 1–100 Hz and band-stop filtered from 59–61 Hz. (7) A second round of fast ICA was performed to remove remaining artifacts ICA components were grouped by TESA software into one of six categories, including electrode noise, eye-blink, muscle artifact linked to TMS, muscle artifact not linked to TMS, sensory artifact, and other. These were reviewed manually and accepted or rejected based upon topography, being in an isolated topographic island, localization only at sites of muscles or eye movement artifact, frequency spectrum, and waveform shape. When in doubt, potentials were included in the reconstruction. This was not done blinded as to treatment, but treatment was not actively considered during decisions about artifact. (8) Data were re-referenced to an average reference, and data were baseline corrected from −100 to −6 ms. (9) TEPs were averaged across all trials and the mean TEPs were then visualized.

2.7. Source Localization

To localize the cortical areas with EEG responses to left motor cortex magnetic stimulation, we reconstructed source activity in a manner comparable with methods used clinically for surgical evaluation of epilepsy patients [55] albeit with a standardized MRI to build the head model. Using trial group averaged pre-treatment TEP EEG signals, the distribution of current density in cortex over time was estimated using low-resolution brain electromagnetic tomography (LORETA) with loose constraints within the Brainstorm plugin for MATLAB. Results from LORETA were similar to those from LAURA (low-resolution electromagnetic tomography) so only results from LORETA were reported. This approximates the optimal current density at the cortical sources needed to produce a distribution of observed EEG potentials over the scalp.

2.8. Statistical Analysis

An analysis of covariance (ANCOVA) at $\alpha = 0.05$ was used to measure differences in TEP peak amplitudes in post-rTMS trials while controlling for pre-rTMS TEP peak amplitude differences. This allowed for adjustment of TEP amplitudes while accounting for any pre-existing differences between groups. If a significant effect was detected by the ANCOVA, Bonferroni-corrected post hoc analyses were conducted to decompose the effect. No ANCOVA assumptions were violated for the reported analyses.

3. Results

No clinical or electrographic seizures were induced by the stimulation. Except for occasional mild scalp discomfort, all were able to tolerate the procedure.

3.1. Motor-Evoked Potential (MEP)

The 47 subjects demonstrated a resting motor threshold of 67.47 ± 7.05 % of maximum machine output. There was no significant change in motor evoked potential (MEP) after 1 Hz of rTMS ($p = 0.26$, Cohen's d = 0.17). The MEP amplitude was not correlated significantly with the P30, N40 (Pearson r = -0.04, $p = 0.43$) or N100 (Pearson r = 0.32, $p = 0.07$) peaks or changes in peak amplitudes after rTMS (Pearson r = -0.01, $p = 0.49$ for N40 and Pearson r = 0.10, $p = 0.39$ for N100).

3.2. TEP Latencies and Amplitudes

Within individual subjects, TEP peak latencies were reliable with variations between pre- and post-rTMS treatments of no more than ± 4.18 ms. Between subjects, the peak latencies however showed substantial variability, with various individuals showing increases, decreases, or no change from pre -rTMS to post- rTMS.

Mean amplitudes (see Figure 3) for the test block TEPs in the fully active (AAA) group before and after 1 Hz of rTMS show that P60 and N100 amplitudes both became more negative (P60 amplitude decreased and N100 increased). The mixed (APA) group demonstrated increased (more negative) N40 amplitudes after 1 Hz of rTMS. The full placebo stimulation (PPP) group did not consistently have well-formed TEP components.

Full placebo stimulation (PPP) produced poorly formed low-amplitude (see Table 2) waveforms. Table 2 indicates that the amplitudes of the TEPs to the 0.2-Hz test pulses were generally similar before and after 1-Hz treatments, but the absolute amplitude of the all-placebo response was about 40% of those evoked by active stimulation. To determine the effect of active 1-Hz rTMS, independent of effects of the baseline amplitude of TEPs and of placebo stimulation, we performed ANCOVAs among the experimental conditions on post-rTMS amplitudes, while controlling for pre-rTMS TEP peak amplitudes, with post hoc testing conducted to decompose significant ANCOVAs (Figure 4).

Table 2. Baseline-adjusted amplitudes. All units are μV. SUM of AV represents the sum of the absolute values of the peak amplitudes and their standard error.

	AAA	**PPP**	**APA**
N20	-0.89 ± 0.38	-0.68 ± 0.59	-1.42 ± 0.48
P30	0.99 ± 0.39	0.62 ± 0.64	1.35 ± 0.46
N40	-1.17 ± 0.43	0.04 ± 0.68	-0.49 ± 0.50
P60 *	1.29 ± 0.24	0.86 ± 0.86	2.40 ± 0.31
N100 *	-3.26 ± 0.57	-1.04 ± 0.92	-1.37 ± 0.67
P180	1.32 ± 0.57	1.25 ± 0.84	1.65 ± 0.75
SUM of AV	8.91	3.48	8.68

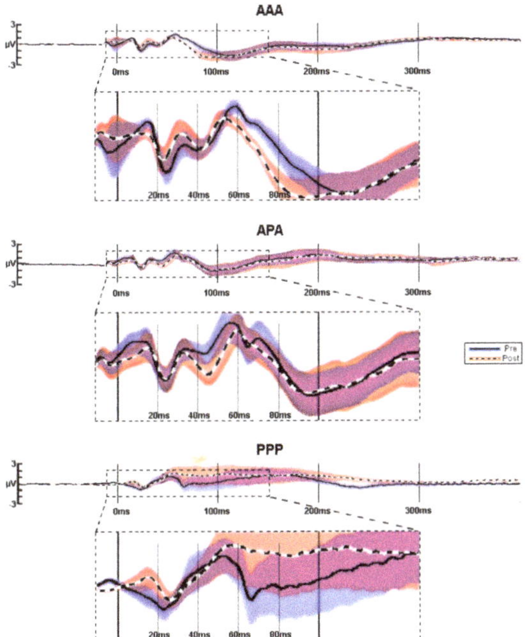

Figure 3. TEPs averaged at C4. Average pre (blue) vs. post (red) TEPs in response to 1-Hz rTMS for AAA ($n = 22$), APA ($n = 15$), and PPP ($n = 10$) groups. Since there was jitter in peak latencies between participants, the amplitudes of the grand averaged traces are lower than they would have been, had each peak been adjusted individual for peak latency (see Figure 3). Significance was not calculated for these raw averaged amplitudes, but on baseline-corrected peak amplitudes.

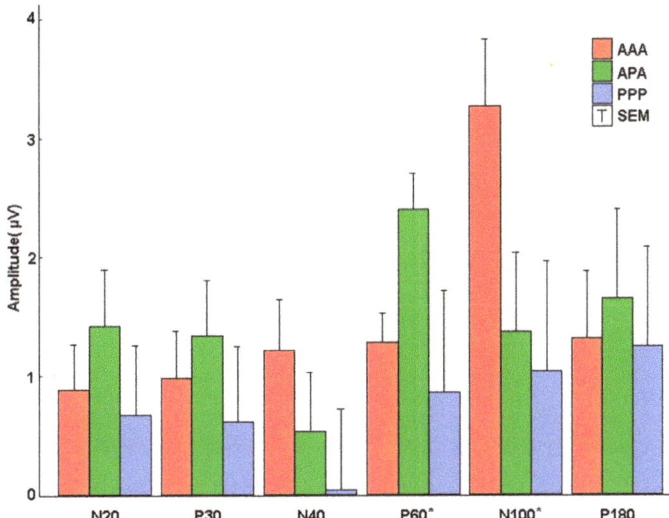

Figure 4. Absolute ANCOVA adjusted peak amplitudes. The amplitude of post-corrected for pre-rTMS TEP peaks for each experimental group. No significant changes before and after active 1-Hz rTMS were noted for N20 and P30. AAA showed a reduced P60 amplitude post rTMS compared to the APA stimulation group. AAA showed a more negative N100 amplitude after rTMS when compared to APA and PPP groups. (* $p < 0.05$). T-bars represent half the standard error.

3.3. N40

The baseline-corrected N40 peak amplitude increased in the AAA group after rTMS, but the change did not achieve significance according to the one-way analysis of covariance (ANCOVA), $F(2, 34) = 1.284$, $\eta p2 = 0.070$, $1-\beta = 0.259$, and $p = 0.290$. The N40 peak did not show a significant difference among groups, although the AAA group was descriptively larger after active stimulation (mean adjusted = -1.174, standard error (SE) = 0.426) compared to PPP (mean adjusted = 0.042, SE = 0.681), d = 0.676, and APA (mean adjusted = -0.493, SE = 0.500) d = 0.378 groups. The amplitude of APA was slightly more negative than PPP, d = 0.297.

3.4. P60

The amplitude-corrected P60 peak became less positive after active rTMS for the AAA group compared to the APA group. A one-way analysis of covariance (ANCOVA) of pre- versus post-TEP P60 amplitudes across the three experimental conditions, while controlling for pre-rTMS TEP P60 amplitudes revealed a significant effect of experimental group, $F(2, 39) = 5.494$, $\eta p2 = 0.220$, $1-\beta = 0.822$, and $p = 0.008$. Bonferroni-adjusted post hoc analyses showed a significant difference between the AAA condition (mean adjusted = 1.287, SE = 0.244), $t(39) = 2.889$, $p = 0.024$, d = 0.996 and the APA condition (mean adjusted = 2.398, SE = 0.308), There was also a significant difference between the PPP condition and the APA condition (mean adjusted = 0.860, SE = 0.403), $t(39) = 3.113$, $p = 0.016$, d = 0.1380. There was not a significant difference between the AAA and PPP conditions, $t(39) = 0.922$, $p = 1.000$, d = 0.383. Although the PPP group demonstrated a significantly more positive P60 peak amplitude after 1 Hz of rTMS in the averaged raw waveforms with the placebo coil, the placebo coil had a high degree of variation, obscuring potentially significant differences.

3.5. N100

The N100 peak increased with 1 Hz of rTMS in the AAA group. ANCOVA was conducted across the three experimental conditions on post-rTMS TEP N100 amplitudes controlling for pre-rTMS TEP N100 amplitudes. The analysis revealed a significant effect of experimental group, $F(2, 40) = 3.295$, $\eta p2 = 0.141$, $1-\beta = 0.593$, and $p = 0.047$, with N100 descriptively more negative amplitudes in the AAA group compared to the other groups after rTMS. However, Bonferroni-adjusted post hoc analyses failed to show significant differences between the APA condition (mean adjusted = -1.373, SE = 0.665), the AAA condition (mean adjusted = -3.262, SE = 0.565), and the PPP condition (mean adjusted = -1.041, SE = 0.921), all t-values ≤ 2.169, all p-values ≥ 0.109.

3.6. Topography of TEPs

rTMS caused topographically widespread TEP components. The early TEP waveforms between 10 and 30 ms had larger amplitudes at electrode locations near C4, but later components, while being visible contralaterally from the stimulation site, had their highest amplitude peaks in electrodes covering the stimulated cortical areas. Later waveforms such as the N100 and P180 were characterized by more profound bilateral distribution compared to earlier waveforms. When comparing TEP amplitudes pre-post changes at electrode sites distant from the stimulation site we found no significant changes in any groups (Figure 5).

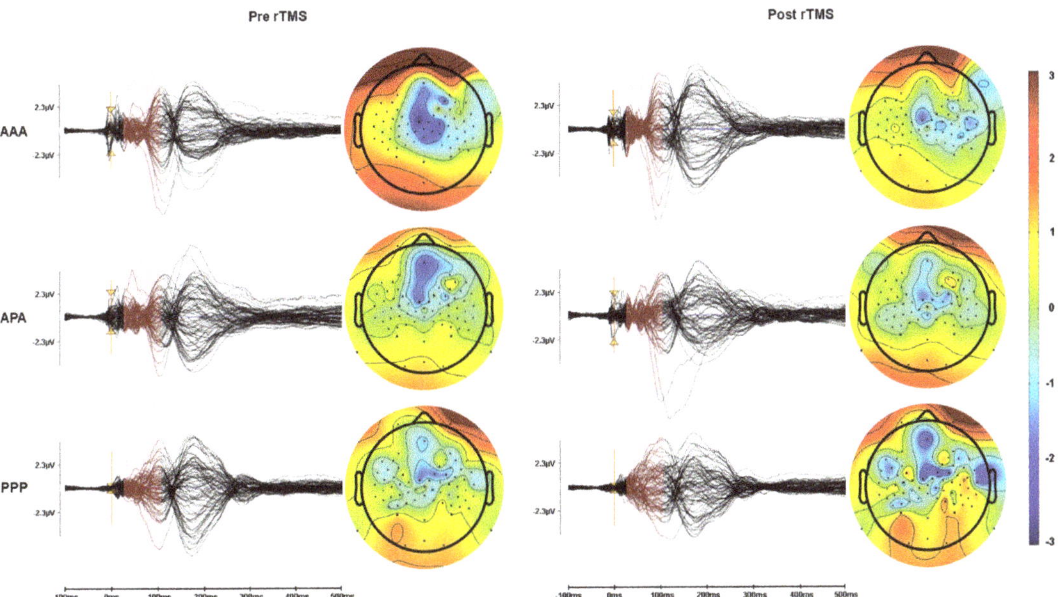

Figure 5. Group average TEP topography. Butterfly plot of EEG traces and topographic portrayal of TEP amplitude averaged from 20–100 ms in response to the stimulation of left motor cortex region. Left is left in the figure and right is right.

3.7. Source Analysis

The localized source activity for TEPs were shown to be likely evoked from focal areas near or under the site of stimulation. However, later waveforms (N100 and P180) affected a larger cortical area. At the initial stimulation, source activity shows a wide area effected by rTMS near the site of stimulation around the C4 electrode, including the ipsilateral precentral gyrus, superior frontal gyrus, and middle frontal gyrus, and to a lesser extent the ipsilateral postcentral gyrus and contralateral superior precentral and postcentral gyri (Figure 6). Source analysis of the P30 and N40 waveforms show activity generators predominantly anterior to the site of rTMS including the ipsilateral middle frontal and superior frontal gyri. The P60 waveform shows activity generators in the ipsilateral precentral, postcentral, supramarginal, and superior frontal gyri. The N100 waveform shows widespread activation, including generators in the temporal poles, frontal poles, superior frontal gyri, superior parietal lobes, ipsilateral pre- and post-central gyri, superior portions of the contralateral pre- and post-central gyri, and the ipsilateral middle frontal gyrus. The P180 waveform generators arise predominantly from the superior parietal lobes, ipsilateral superior frontal, middle frontal gyri, and the contralateral superior frontal lobe and, to a lesser extent, the contralateral pre- and post-central gyri, and the superior and middle temporal gyri.

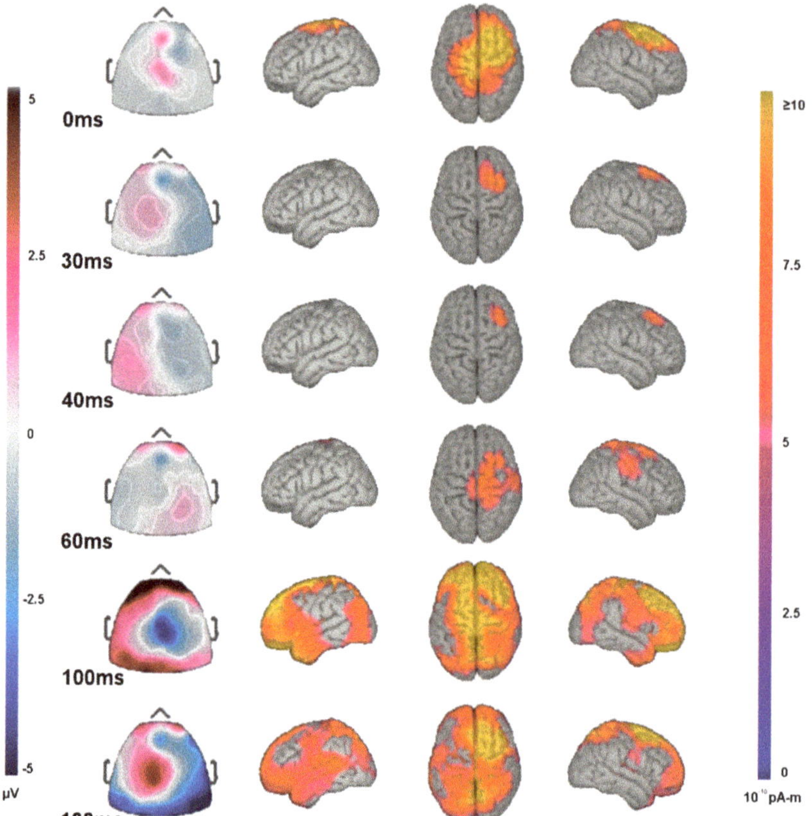

Figure 6. Source localization of TEP activity. Butterfly plot of EEG traces and topographic portrayal of TEP amplitude averaged from 20–100 ms in response to the stimulation of left motor cortex region. Right is on the left (MRI convention) and left is on the right.

4. Discussion

This study in normal volunteers confirms that 1 Hz of repetitive transcranial magnetic stimulation alters rTMS-evoked EEG potential (TEP) waveforms. These waveforms can then be rendered visible with signal averaging and processing [24,26,28,30,44,56]. Numerous studies [57,58] have evaluated short-interval intracortical inhibition (SICI, typically 1–5 ms) and long-interval intracortical inhibition (LICI, typically 50–200 ms) by measuring the EEG response to paired TMS pulses. Our study evaluated the effect of 1-Hz repetitive pulse trains on TEP waveforms, which is a much less commonly employed experimental paradigm than is paired-pulse stimulation, but one that might have greater potential for evaluation of different regions of cortex.

Previous related work includes a study by Casula and associates who found that rTMS increased the P60 and N100, but not the P30 or N40 [50]. That study did not control for pre-existing baseline amplitude differences and did not use a placebo comparison group. We observed a less positive (smaller) P60 and more negative (bigger) N100 after 1 Hz of rTMS. This partially confirmed the results found by Casula, while controlling for the confounding factor of highly variable TEP amplitudes among different subjects. The divergent findings of our study versus those of Casula regarding the P60 may be in part due to differences in our procedures. Casula and colleagues used 120% RMT and 50 single pulses to measure TEP amplitudes. We used 110% RMT and 200 single pulses to measure

TEP amplitudes, which might have improved the signal-to-noise ratio. Additionally, the present study evaluated a larger study population, with a different method of marking peak amplitudes, and addition of placebo comparison groups.

A control group with placebo stimulation is important to rule out nonspecific effects, because rTMS produces auditory and somatosensory components of the TEP that are not directly related to magnetic stimulation-induced cortical activity [59]. We were able to show group difference for several peaks generated by test stimulations at 0.20 Hz for placebo or active coils before and after 1 Hz of rTMS. Our placebo coil stimulation was able to evoke variable and low amplitude cortically-generated waveforms; therefore, our placebo stimulations might better be considered a low-dose comparator to active stimulation, meaning that comparisons of active to placebo stimulation might have underestimated the effects of active stimulation.

Only limited information is available about the physiological relevance of TEP peaks. Pharmacological studies suggest that the N40 is enhanced in humans with administration of diazepam, reflecting increased fast $GABA_A$ mediated inhibition [45–47,59,60]. The N100 peak increases with the $GABA_B$ agonist, baclofen [45,48] and decreases with presynaptic inhibitors of excitatory transmitter release [61] implying potential for serving as a marker of cortical inhibition that might be useful for reducing seizures [30]. The amplitude of the N100 peak might depend more on the ratio of GABA to glutamate than upon GABA alone [60]. However, increase in $GABA_B$-mediated synaptic inhibition might be expected to produce variable effects in different types of epilepsy. Absence seizures, for example, show spike-wave EEG discharges, with the wave component reflecting a significant component of slow $GABA_B$-mediated inhibitory potentials [62,63].

The P60 peak occurs at a time of long-interval intracortical inhibition, also mediated by $GABA_B$ receptors [45]. Excitatory transmission may play a role in generating the P60 peak, since the glutamate AMPA receptor antagonist, perampanel, suppresses P60 amplitude [61,64]. Rogasch [30] suggested that the P60 TEP could reflect a component of somatosensory feedback, but Cash [49] has argued that there is a significant cortical component of P60.

Identifying cortical sources of TEPs could be important for rTMS use in a clinical setting. Our study examined TEPs averaged across participants and modeled sources coming from a normalized atlas brain. The TEP activity localized near the stimulation site, but also with some distant activity, reflecting network spread. We again confirm the findings of Casula [50] and Bikmullina [65] that motor cortex rTMS influences TEPs over a wide region of ipsilateral cortex. In our study, the range of effect was from anterior temporal to occipital regions. Studies using motor assays have identified transcallosal inhibition provoked by contralateral motor cortex rTMS [66,67] and we can confirm a significant contralateral component of the TEPs. The widespread effects of rTMS to increase the N40 marker on fast inhibition might argue against the need for exact targeting of stimulation. However, enhancement of inhibition was maximal at the motor cortex stimulation site. TEP changes in this study did not correlate with amplitude of the motor evoked responses, as has been noted by others [28]. We did not systematically look for changes in motor threshold or paired-pulse inhibition at various inter-stimulus intervals. Studies using individual MRI brain modeling could take advantage of rTMS with cortical source modeling to create detailed individual connectivity maps by plotting propagation patterns of TEPs evoked from systematically chosen cortical areas.

Variability among participant's TEP responses is high [68], limiting the significance of post hoc pair-wise and group comparisons. However, individuals have relatively stable TEPs at 20, 30, 40, 60, and 100 ms [69], suggesting that TEPs might provide a useful biomarker for regional cortical excitability in specific patients. Of the TEPs we found the N100 to be the most viable biomarker for rTMS induced cortical inhibition. If TEPs are confirmed as a reliable surrogate marker for cortical inhibition in epilepsy patients, then TEP recordings could become significantly more efficient than seizure counts in screening the effects of anti-seizure therapies. Of course, any such findings would require validating the effect of a possible treatment on seizure counts.

Our study is subject to several interpretive limitations. Distinguishing continuous waveforms is problematic, because changes in adjacent peaks can be additive or subtractive, rather than independent. For this reason, we employed unbiased software measurements of amplitudes at selected peaks, rather than trough-to-peak values, but this may have been at the expense of evaluating effects on individual peaks. TMS-TEPs are subject to a wide variety of artifacts [70]. Late TEP components can possibly be influenced by improper masking of the loud rTMS "click" that occurs during stimulation [71]. We attempted to account for this by use of white noise played via sound cancelling earbuds, but we cannot guarantee that the rTMS "click" was completely obscured. We have no evidence that the changes in TEP waveforms correlate with any clinical or even biological effect. Examining grand averaged source space activity likely misses the nuances that would be critical to clinical application of rTMS source reconstruction. rTMS-evoked EEG potentials might differ from normal volunteers versus those with epilepsy or other neurological diseases when stimulating the areas of neurological abnormality. Our experiments do not document the durability and replicability of TEP changes. Future clarification of these issues will further the use of rTMS-evoked EEG potentials as biomarkers for cortical excitability in non-motor cortex.

5. Conclusions

As described by several prior studies, rTMS evokes measurable EEG potentials, discernible after suitable processing. A positive peak at 60 ms and a negative peak at 100 ms are each altered by 1 Hz of repetitive TMS at motor cortex, with P60 decreased and N100 increased. N40 showed a non-significant trend towards an increase. Effects on other evoked peaks are variable. While several of our findings are confirmatory of previous work, new features include reliable persistence of evoked EEG potential changes in response to rTMS when controlling for highly variable initial amplitude and as compared to sham-stimulation controls. Comparing changes in the EEG, not only to before-after rTMS, but to sham stimulation controls confirm the important increase of the N100 potential, but are less confirmatory of P60 peak changes. Our topographic dipole analysis documents that the largest effects of 1-Hz rTMS on TEPs are manifest early and close to the stimulation site. These changes in rTMS-evoked EEGs may reflect increased $GABA_B$-mediated inhibition in specific brain regions.

Limitations of this technology are several, including difficulties in isolating individual EEG peaks and intra-subject variability, often requiring population averages and statistics to visualize changes. However, measurements of rTMS-evoked EEG changes are not restricted to motor cortex, so they may serve as useful biomarkers for cortical excitability at a seizure focus.

Author Contributions: Conceptualization, R.S.F. and J.Z.; methodology, J.Z. and R.S.F.; software, J.Z.; validation, all authors; formal analysis, J.Z. and R.S.F.; investigation, R.S.F. and J.Z.; resources, R.S.F.; data curation, J.Z. and A.F.; writing—original draft preparation, R.S.F. and J.Z.; writing—review and editing, all authors; visualization, J.Z., A.F. and R.S.F.; supervision, R.S.F.; project administration, R.S.F.; funding acquisition, R.S.F. All authors have read and agreed to the published version of the manuscript.

Funding: This research was funded from the Steve Chen Philanthropic Fund for Epilepsy Research; James and Carrie Anderson Research Fund; Maslah Saul MD Chair.

Institutional Review Board Statement: The Stanford IRB approved the research project "Transcranial Magnetic Stimulation and Brain Excitability" protocol #37619 to continue to 15 June 2022, Assurance FWA00000935 (SU).

Informed Consent Statement: Informed consent was obtained from all subjects involved in the study.

Data Availability Statement: Data may be provided upon request to Mr. Adam Fogarty at afogarty@stanfordhealthcare.org, depending upon circumstances of the request.

Acknowledgments: R.S.F. was supported by the Maslah Saul MD Chair, James and Carrie Anderson Epilepsy Research Fund, and the Susan Horngren Fund. J.Z. and the overall research project was supported by the Steve Chen Fund for Epilepsy Research.

Conflicts of Interest: None of the authors report any conflict of interest relevant to this study. Non-relevant declarations are: Dr. Fisher was lead investigator of a Medtronic grant to Stanford to study deep brain stimulation and he consults for Medtronic. Dr. Fisher holds stock or options in Avails Medical, Cerebral Therapeutics, Eysz, Irody, SmartWatch, Zeto.

References

1. Barker, A.T.; Jalinous, R.; Freeston, I.L. Non-invasive magnetic stimulation of human motor cortex. *Lancet* **1985**, *1*, 1106–1107. [CrossRef]
2. Tremblay, S.; Rogasch, N.C.; Premoli, I.; Blumberger, D.M.; Casarotto, S.; Chen, R.; Di Lazzaro, V.; Farzan, F.; Ferrarelli, F.; Fitzgerald, P.B.; et al. Clinical utility and prospective of TMS-EEG. *Clin. Neurophysiol.* **2019**, *130*, 802–844. [CrossRef] [PubMed]
3. Hemond, C.C.; Fregni, F. Transcranial magnetic stimulation in neurology: What we have learned from randomized controlled studies. *Neuromodul. Technol. Neural Interface* **2007**, *10*, 333–344. [CrossRef]
4. Kobayashi, M.; Pascual-Leone, A. Transcranial magnetic stimulation in neurology. *Lancet Neurol.* **2003**, *2*, 145–156. [CrossRef]
5. Wischnewski, M.; Schutter, D.J. Efficacy and Time Course of Theta Burst Stimulation in Healthy Humans. *Brain Stimul.* **2015**, *8*, 685–692. [CrossRef]
6. Fregni, F.; Otachi, P.T.; Do Valle, A.; Boggio, P.S.; Thut, G.; Rigonatti, S.P.; Pascual-Leone, A.; Valente, K.D. A randomized clinical trial of repetitive transcranial magnetic stimulation in patients with refractory epilepsy. *Ann. Neurol.* **2006**, *60*, 447–455. [CrossRef]
7. Sun, W.; Mao, W.; Meng, X.; Wang, D.; Qiao, L.; Tao, W.; Li, L.; Jia, X.; Han, C.; Fu, M.; et al. Low-frequency repetitive transcranial magnetic stimulation for the treatment of refractory partial epilepsy: A controlled clinical study. *Epilepsia* **2012**, *53*, 1782–1789. [CrossRef]
8. Cantello, R.; Rossi, S.; Varrasi, C.; Ulivelli, M.; Civardi, C.; Bartalini, S.; Vatti, G.; Cincotta, M.; Borgheresi, A.; Zaccara, G.; et al. Slow repetitive TMS for drug-resistant epilepsy: Clinical and EEG findings of a placebo-controlled trial. *Epilepsia* **2007**, *48*, 366–374. [CrossRef]
9. Theodore, W.H.; Hunter, K.; Chen, R.; Vega-Bermudez, F.; Boroojerdi, B.; Reeves-Tyer, P.; Werhahn, K.; Kelley, K.R.; Cohen, L. Transcranial magnetic stimulation for the treatment of seizures: A controlled study. *Neurology* **2002**, *59*, 560–562. [CrossRef]
10. Joo, E.Y.; Han, S.J.; Chung, S.H.; Cho, J.W.; Seo, D.W.; Hong, S.B. Antiepileptic effects of low-frequency repetitive transcranial magnetic stimulation by different stimulation durations and locations. *Clin. Neurophysiol.* **2007**, *118*, 702–708. [CrossRef]
11. Tergau, F.; Naumann, U.; Paulus, W.; Steinhoff, B.J. Low-frequency repetitive transcranial magnetic stimulation improves intractable epilepsy. *Lancet* **1999**, *353*, 2209. [CrossRef]
12. Seynaeve, L.; Devroye, A.; Dupont, P.; Van Paesschen, W. Randomized crossover sham-controlled clinical trial of targeted low-frequency transcranial magnetic stimulation comparing a figure-8 and a round coil to treat refractory neocortical epilepsy. *Epilepsia* **2016**, *57*, 141–150. [CrossRef] [PubMed]
13. Thickbroom, G.W. Transcranial magnetic stimulation and synaptic plasticity: Experimental framework and human models. *Exp. Brain Res.* **2007**, *180*, 583–593. [CrossRef]
14. Chen, R.; Classen, J.; Gerloff, C.; Celnik, P.; Wassermann, E.M.; Hallett, M.; Cohen, L.G. Depression of motor cortex excitability by low-frequency transcranial magnetic stimulation. *Neurology* **1997**, *48*, 1398–1403. [CrossRef] [PubMed]
15. Berardelli, A.; Inghilleri, M.; Rothwell, J.C.; Romeo, S.; Currà, A.; Gilio, F.; Modugno, N.; Manfredi, M. Facilitation of muscle evoked responses after repetitive cortical stimulation in man. *Exp. Brain Res.* **1998**, *122*, 79–84. [CrossRef] [PubMed]
16. Kujirai, T.; Caramia, M.D.; Rothwell, J.C.; Day, B.L.; Thompson, P.D.; Ferbert, A.; Wroe, S.; Asselman, P.; Marsden, C.D. Corticocortical inhibition in human motor cortex. *J. Physiol.* **1993**, *471*, 501–519. [CrossRef]
17. Ertaş, N.K.; Gül, G.; Altunhalka, A.; Kirbas, D. Cortical silent period following transcranial magnetic stimulation in epileptic patients. *Epileptic Disord.* **2000**, *2*, 137–140.
18. Bakker, N.; Shahab, S.; Giacobbe, P.; Blumberger, D.M.; Daskalakis, Z.J.; Kennedy, S.H.; Downar, J. rTMS of the dorsomedial prefrontal cortex for major depression: Safety, tolerability, effectiveness, and outcome predictors for 10 Hz versus intermittent theta-burst stimulation. *Brain Stimul.* **2015**, *8*, 208–215. [CrossRef]
19. Sun, W.; Fu, W.; Mao, W.; Wang, D.; Wang, Y. Low-frequency repetitive transcranial magnetic stimulation for the treatment of refractory partial epilepsy. *Clin. EEG Neurosci.* **2011**, *42*, 40–44. [CrossRef]
20. Badawy, R.A.; Strigaro, G.; Cantello, R. TMS, cortical excitability and epilepsy: The clinical impact. *Epilepsy Res.* **2014**, *108*, 153–161. [CrossRef]
21. Hill, A.T.; Rogasch, N.C.; Fitzgerald, P.B.; Hoy, K.E. TMS-EEG: A window into the neurophysiological effects of transcranial electrical stimulation in non-motor brain regions. *Neurosci. Biobehav. Rev.* **2016**, *64*, 175–184. [CrossRef] [PubMed]
22. Ilmoniemi, R.J.; Kicić, D. Methodology for combined TMS and EEG. *Brain Topogr.* **2010**, *22*, 233–248. [CrossRef]

23. Kimiskidis, V.K. Transcranial magnetic stimulation (TMS) coupled with electroencephalography (EEG): Biomarker of the future. *Rev. Neurol.* **2016**, *172*, 123–126. [CrossRef] [PubMed]

24. Cracco, R.Q.; Amassian, V.E.; Maccabee, P.J.; Cracco, J.B. Comparison of human transcallosal responses evoked by magnetic coil and electrical stimulation. *Electroencephalogr. Clin. Neurophysiol./Evoked Potentials Sect.* **1989**, *74*, 417–424. [CrossRef]

25. Farzan, F.; Barr, M.S.; Levinson, A.J.; Chen, R.; Wong, W.; Fitzgerald, P.B.; Daskalakis, Z.J. Reliability of long-interval cortical inhibition in healthy human subjects: A TMS-EEG study. *J. Neurophysiol.* **2010**, *104*, 1339–1346. [CrossRef] [PubMed]

26. Ferreri, F.; Rossini, P.M. TMS and TMS-EEG techniques in the study of the excitability, connectivity, and plasticity of the human motor cortex. *Rev. Neurosci.* **2013**, *24*, 431–442. [CrossRef]

27. Fitzgerald, P.B. TMS-EEG: A technique that has come of age? *Clin. Neurophysiol.* **2010**, *121*, 265–267. [CrossRef]

28. Fitzgerald, P.B.; Daskalakis, Z.J.; Hoy, K.; Farzan, F.; Upton, D.J.; Cooper, N.R.; Maller, J.J. Cortical inhibition in motor and non-motor regions: A combined TMS-EEG study. *Clin. EEG Neurosci.* **2008**, *39*, 112–117. [CrossRef]

29. Schulze-Bonhage, A.; Feldwisch-Drentrup, H.; Ihle, M. The role of high-quality EEG databases in the improvement and assessment of seizure prediction methods. *Epilepsy Behav.* **2011**, *22*, S88–S93. [CrossRef] [PubMed]

30. Rogasch, N.C.; Daskalakis, Z.J.; Fitzgerald, P.B. Mechanisms underlying long-interval cortical inhibition in the human motor cortex: A TMS-EEG study. *J. Neurophysiol.* **2013**, *109*, 89–98. [CrossRef]

31. Thut, G.; Miniussi, C. New insights into rhythmic brain activity from TMS-EEG studies. *Trends Cogn. Sci.* **2009**, *13*, 182–189. [CrossRef] [PubMed]

32. Ilmoniemi, R.J.; Hernandez-Pavon, J.C.; Makela, N.N.; Metsomaa, J.; Mutanen, T.P.; Stenroos, M.; Sarvas, J. Dealing with artifacts in TMS-evoked EEG. In Proceedings of the 2015 37th Annual International Conference of the IEEE Engineering in Medicine and Biology Society (EMBC), Milano, Italy, 25–29 August 2015; pp. 230–233. [CrossRef]

33. Mäki, H.; Ilmoniemi, R.J. Projecting out muscle artifacts from TMS-evoked EEG. *Neuroimage* **2011**, *54*, 2706–2710. [CrossRef] [PubMed]

34. Morbidi, F.; Garulli, A.; Prattichizzo, D.; Rizzo, C.; Manganotti, P.; Rossi, S. Off-line removal of TMS-induced artifacts on human electroencephalography by Kalman filter. *J. Neurosci. Methods* **2007**, *162*, 293–302. [CrossRef] [PubMed]

35. Mutanen, T.; Mäki, H.; Ilmoniemi, R.J. The effect of stimulus parameters on TMS-EEG muscle artifacts. *Brain Stimul.* **2013**, *6*, 371–376. [CrossRef] [PubMed]

36. Mutanen, T.P.; Kukkonen, M.; Nieminen, J.O.; Stenroos, M.; Sarvas, J.; Ilmoniemi, R.J. Recovering TMS-evoked EEG responses masked by muscle artifacts. *Neuroimage* **2016**, *139*, 157–166. [CrossRef] [PubMed]

37. Rogasch, N.C.; Thomson, R.H.; Daskalakis, Z.J.; Fitzgerald, P.B. Short-latency artifacts associated with concurrent TMS-EEG. *Brain Stimul.* **2013**, *6*, 868–876. [CrossRef]

38. Rogasch, N.C.; Thomson, R.H.; Farzan, F.; Fitzgibbon, B.M.; Bailey, N.W.; Hernandez-Pavon, J.C.; Daskalakis, Z.J.; Fitzgerald, P.B. Removing artefacts from TMS-EEG recordings using independent component analysis: Importance for assessing prefrontal and motor cortex network properties. *Neuroimage* **2014**, *101*, 425–439. [CrossRef]

39. Sekiguchi, H.; Takeuchi, S.; Kadota, H.; Kohno, Y.; Nakajima, Y. TMS-induced artifacts on EEG can be reduced by rearrangement of the electrode's lead wire before recording. *Clin. Neurophysiol.* **2011**, *122*, 984–990. [CrossRef]

40. Ter Braack, E.M.; de Jonge, B.; van Putten, M.J. Reduction of TMS induced artifacts in EEG using principal component analysis. *IEEE Trans. Neural Syst. Rehabil. Eng.* **2013**, *21*, 376–382. [CrossRef]

41. Van Doren, J.; Langguth, B.; Schecklmann, M. TMS-related potentials and artifacts in combined TMS-EEG measurements: Comparison of three different TMS devices. *Neurophysiol. Clin.* **2015**, *45*, 159–166. [CrossRef]

42. Fisher, R.; Zhou, J.; Fogarty, A.; Joshi, A.; Markert, M.; Deutsch, G.K.; Velez, M. Repetitive transcranial magnetic stimulation directed to a seizure focus localized by high-density EEG: A case report. *Epilepsy Behav. Case Rep.* **2018**, *10*, 47–53. [CrossRef] [PubMed]

43. Bonato, C.; Miniussi, C.; Rossini, P.M. Transcranial magnetic stimulation and cortical evoked potentials: A TMS/EEG co-registration study. *Clin. Neurophysiol.* **2006**, *117*, 1699–1707. [CrossRef] [PubMed]

44. Farzan, F.; Vernet, M.; Shafi, M.M.; Rotenberg, A.; Daskalakis, Z.J.; Pascual-Leone, A. Characterizing and Modulating Brain Circuitry through Transcranial Magnetic Stimulation Combined with Electroencephalography. *Front. Neural Circuits* **2016**, *10*, 73. [CrossRef]

45. Premoli, I.; Castellanos, N.; Rivolta, D.; Belardinelli, P.; Bajo, R.; Zipser, C.; Espenhahn, S.; Heidegger, T.; Müller-Dahlhaus, F.; Ziemann, U. TMS-EEG signatures of GABAergic neurotransmission in the human cortex. *J. Neurosci.* **2014**, *34*, 5603–5612. [CrossRef]

46. Fitzgerald, P.B.; Maller, J.J.; Hoy, K.; Farzan, F.; Daskalakis, Z.J. GABA and cortical inhibition in motor and non-motor regions using combined TMS-EEG: A time analysis. *Clin. Neurophysiol.* **2009**, *120*, 1706–1710. [CrossRef] [PubMed]

47. Darmani, G.; Zipser, C.M.; Böhmer, G.M.; Deschet, K.; Müller-Dahlhaus, F.; Belardinelli, P.; Schwab, M.; Ziemann, U. Effects of the Selective α5-GABAAR Antagonist S44819 on Excitability in the Human Brain: A TMS-EMG and TMS-EEG Phase I Study. *J. Neurosci.* **2016**, *36*, 12312–12320. [CrossRef]

48. Premoli, I.; Rivolta, D.; Espenhahn, S.; Castellanos, N.; Belardinelli, P.; Ziemann, U.; Müller-Dahlhaus, F. Characterization of GABAB-receptor mediated neurotransmission in the human cortex by paired-pulse TMS-EEG. *Neuroimage* **2014**, *103*, 152–162. [CrossRef] [PubMed]

49. Cash, R.F.; Noda, Y.; Zomorrodi, R.; Radhu, N.; Farzan, F.; Rajji, T.K.; Fitzgerald, P.B.; Chen, R.; Daskalakis, Z.J.; Blumberger, D.M. Characterization of Glutamatergic and GABA. *Neuropsychopharmacology* **2017**, *42*, 502–511. [CrossRef]
50. Casula, E.P.; Tarantino, V.; Basso, D.; Arcara, G.; Marino, G.; Toffolo, G.M.; Rothwell, J.C.; Bisiacchi, P.S. Low-frequency rTMS inhibitory effects in the primary motor cortex: Insights from TMS-evoked potentials. *Neuroimage* **2014**, *98*, 225–232. [CrossRef]
51. Rossi, S.; Hallett, M.; Rossini, P.M.; Pascual-Leone, A.; Safety of TMS Consensus Group. Safety, ethical considerations, and application guidelines for the use of transcranial magnetic stimulation in clinical practice and research. *Clin. Neurophysiol.* **2009**, *120*, 2008–2039. [CrossRef]
52. Ridding, M.C.; Flavel, S.C. Induction of plasticity in the dominant and non-dominant motor cortices of humans. *Exp. Brain Res.* **2006**, *171*, 551–557. [CrossRef] [PubMed]
53. Poole, B.J.; Mather, M.; Livesey, E.J.; Harris, I.M.; Harris, J.A. Motor-evoked potentials reveal functional differences between dominant and non-dominant motor cortices during response preparation. *Cortex* **2018**, *103*, 1–12. [CrossRef] [PubMed]
54. Rogasch, N.C.; Sullivan, C.; Thomson, R.H.; Rose, N.S.; Bailey, N.W.; Fitzgerald, P.B.; Farzan, F.; Hernandez-Pavon, J.C. Analysing concurrent transcranial magnetic stimulation and electroencephalographic data: A review and introduction to the open-source TESA software. *NeuroImage* **2017**, *147*, 934–951. [CrossRef] [PubMed]
55. Kuo, C.C.; Tucker, D.M.; Luu, P.; Jenson, K.; Tsai, J.J.; Ojemann, J.G.; Holmes, M.D. EEG source imaging of epileptic activity at seizure onset. *Epilepsy Res.* **2018**, *146*, 160–171. [CrossRef]
56. Daskalakis, Z.J.; Farzan, F.; Barr, M.S.; Maller, J.J.; Chen, R.; Fitzgerald, P.B. Long-interval cortical inhibition from the dorsolateral prefrontal cortex: A TMS-EEG study. *Neuropsychopharmacology* **2008**, *33*, 2860–2869. [CrossRef]
57. Casarotto, S.; Romero Lauro, L.J.; Bellina, V.; Casali, A.G.; Rosanova, M.; Pigorini, A.; Defendi, S.; Mariotti, M.; Massimini, M. EEG responses to TMS are sensitive to changes in the perturbation parameters and repeatable over time. *PLoS ONE* **2010**, *5*, e10281. [CrossRef]
58. Rogasch, N.C.; Fitzgerald, P.B. Assessing cortical network properties using TMS-EEG. *Hum. Brain Mapp.* **2013**, *34*, 1652–1669. [CrossRef]
59. Conde, V.; Tomasevic, L.; Akopian, I.; Stanek, K.; Saturnino, G.B.; Thielscher, A.; Bergmann, T.O.; Siebner, H.R. The non-transcranial TMS-evoked potential is an inherent source of ambiguity in TMS-EEG studies. *Neuroimage* **2019**, *185*, 300–312. [CrossRef]
60. Du, X.; Rowland, L.M.; Summerfelt, A.; Wijtenburg, A.; Chiappelli, J.; Wisner, K.; Kochunov, P.; Choa, F.S.; Hong, L.E. TMS evoked N100 reflects local GABA and glutamate balance. *Brain Stimul.* **2018**, *11*, 1071–1079. [CrossRef]
61. Darmani, G.; Ziemann, U. Pharmacophysiology of TMS-evoked EEG potentials: A mini-review. *Brain Stimul.* **2019**, *12*, 829–831. [CrossRef]
62. Staak, R.; Pape, H.C. Contribution of GABA(A) and GABA(B) receptors to thalamic neuronal activity during spontaneous absence seizures in rats. *J. Neurosci.* **2001**, *21*, 1378–1384. [CrossRef] [PubMed]
63. Snead, O.C.; Hechler, V.; Vergnes, M.; Marescaux, C.; Maitre, M. Increased gamma-hydroxybutyric acid receptors in thalamus of a genetic animal model of petit mal epilepsy. *Epilepsy Res.* **1990**, *7*, 121–128. [CrossRef]
64. Belardinelli, P.; König, F.; Liang, C.; Premoli, I.; Desideri, D.; Müller-Dahlhaus, F.; Gordon, P.C.; Zipser, C.; Zrenner, C.; Ziemann, U. TMS-EEG signatures of glutamatergic neurotransmission in human cortex. *Sci. Rep.* **2021**, *11*, 8159. [CrossRef] [PubMed]
65. Bikmullina, R.; Kicić, D.; Carlson, S.; Nikulin, V.V. Electrophysiological correlates of short-latency afferent inhibition: A combined EEG and TMS study. *Exp. Brain Res.* **2009**, *194*, 517–526. [CrossRef] [PubMed]
66. Takeuchi, N.; Ikoma, K.; Chuma, T.; Matsuo, Y. Measurement of transcallosal inhibition in traumatic brain injury by transcranial magnetic stimulation. *Brain Inj.* **2006**, *20*, 991–996. [CrossRef]
67. Trompetto, C.; Bove, M.; Marinelli, L.; Avanzino, L.; Buccolieri, A.; Abbruzzese, G. Suppression of the transcallosal motor output: A transcranial magnetic stimulation study in healthy subjects. *Exp. Brain Res.* **2004**, *158*, 133–140. [CrossRef]
68. Terranova, C.; Rizzo, V.; Cacciola, A.; Chillemi, G.; Calamuneri, A.; Milardi, D.; Quartarone, A. Is There a Future for Non-invasive Brain Stimulation as a Therapeutic Tool? *Front. Neurol.* **2018**, *9*, 1146. [CrossRef]
69. Kerwin, L.J.; Keller, C.J.; Wu, W.; Narayan, M.; Etkin, A. Test-retest reliability of transcranial magnetic stimulation EEG evoked potentials. *Brain Stimul.* **2018**, *11*, 536–544. [CrossRef]
70. Varone, G.; Hussain, Z.; Sheikh, Z.; Howard, A.; Boulila, W.; Mahmud, M.; Howard, N.; Morabito, F.C.; Hussain, A. Real-Time Artifacts Reduction during TMS-EEG Co-Registration: A Comprehensive Review on Technologies and Procedures. *Sensors* **2021**, *21*, 637. [CrossRef]
71. Rocchi, L.; Di Santo, A.; Brown, K.; Ibáñez, J.; Casula, E.; Rawji, V.; Di Lazzaro, V.; Koch, G.; Rothwell, J. Disentangling EEG responses to TMS due to cortical and peripheral activations. *Brain Stimul.* **2021**, *14*, 4–18. [CrossRef]

MDPI

St. Alban-Anlage 66

4052 Basel

Switzerland

Tel. +41 61 683 77 34

Fax +41 61 302 89 18

www.mdpi.com

Sensors Editorial Office

E-mail: sensors@mdpi.com

www.mdpi.com/journal/sensors